☐ Procedure for solving a linear programming problem geometrically, p. 279

☐ The simplex method, p. 299

☐ Procedure for determining a clique in a digraph, p. 342

☐ Procedure for determining if a digraph is strongly connected, p. 345

☐ Procedure for computing the line of best fit, p. 369

☐ First procedure for computing the steady-state vector of a regular transition matrix, p. 390

☐ Second procedure for computing the steady-state vector of a regular transition matrix, p. 390

☐ Procedure for obtaining the general solution to $\mathbf{X}' = \mathbf{AX}$, where $\mathbf{A}$ is diagonalizable, p. 403

☐ Gaussian elimination method, p. 419

☐ Jacobi iteration method for solving a linear system, p. 424

☐ Gauss–Seidel iteration method for solving a linear system, p. 425

☐ Procedure for obtaining the $\mathbf{LU}$-factorization of a matrix, p. 433

☐ The power method for diagonalizing a matrix, p. 438

☐ Jacobi's method for diagonalizing a symmetric matrix, pp. 441–442

Introductory
Linear
Algebra
with
Applications

FOURTH EDITION

Bernard Kolman

Drexel University

INTRODUCTORY LINEAR ALGEBRA

WITH

APPLICATIONS

Macmillan Publishing Company
New York
Collier Macmillan Publishers
London

Copyright © 1988, KTI

Printed in the United States of America

All rights reserved. No part of this book may be reproduced or transmitted in any form or by any means, electronic or mechanical, including photocopying, recording, or any information storage and retrieval system, without permission in writing from the publisher.

Earlier editions copyright © 1976, 1980 by Bernard Kolman. Copyright © 1984 by KTI.

Macmillan Publishing Company
866 Third Avenue, New York, New York 10022

Collier Macmillan Canada, Inc.

Library of Congress Cataloging-in-Publication Data

Kolman, Bernard, 1932–
   Introductory linear algebra with applications.

   Includes index.
   1. Algebras, Linear.   I. Title.
QA184.K67   1988   512.5   87-20256
ISBN 0-02-366091-0

Printing: 6 7 8   Year: 1 2 3 4 5 6 7

Cover art, *Berc* by Vasarely, Courtesy of Vasarely Center, New York.

ISBN 0-02-366091-0

To the memory of Lillie
and to
Lisa and Stephen

Preface

Linear algebra has now become a standard part of the undergraduate mathematical training of a diverse number of students for at least two reasons. First, few subjects can claim to have such widespread applications in other areas of mathematics—multivariable calculus, differential equations, and probability theory, for example—as well as in physics, biology, chemistry, economics, psychology, sociology, and all fields of engineering. Second, the subject provides the student with his or her first introduction to postulational or axiomatic mathematics.

Both the author and the publisher have been very pleased by the widespread acceptance of the first three editions of this book. Although a great many changes have been made in this edition, my objective has remained the same as in the first two editions: *to develop a textbook that will help the instructor to teach and the student to learn the basic ideas of linear algebra and to see some of its applications*. To achieve this objective, author and publisher have made extensive use of faculty and student suggestions to develop the following helpful features.

FEATURES

Presentation

In a semester or quarter course in linear algebra the instructor and student are frequently faced with several problems. The important subject of eigenvalues is often studied in too much haste to do the material justice. The student seldom sees any meaningful applications of linear algebra, and he or she rarely gets any feel for the widespread use, in conjunction with computers, of the new area of numerical linear algebra. In this book we have tried to deal with these problems.

This book presents a brief introduction to linear algebra and to some of its significant applications and is designed for a course at the freshman or sophomore level. If most of the applications are covered, there is enough material for a year course. This book can also be used for a second course in linear algebra covering applications of the subject. Calculus is not a prerequisite although examples and exercises using very basic calculus are included (these are labeled **"Calculus Required"**).

The emphasis is on the computational and geometrical aspects of the subject, keeping the abstraction down to a minimum. Thus we sometimes omit proofs of difficult or less rewarding theorems, while amply illustrating them with examples. The proofs that are included are presented at a level appropriate for the student. We have also concentrated our attention on the essential areas of linear algebra; the book does not attempt to cover the subject exhaustively.

Material Covered

Part I covers the basic linear algebra material. **Chapter 1** deals with matrices and their properties; **Chapter 2** covers determinants in a brief manner. In **Chapter 3,** on vectors and vector spaces, and **Chapter 4,** on linear transformations and matrices, we explore some of the geometric ideas of the subject. We first cover vectors in the plane and then n-vectors before getting to the general notion of a vector space. Thus R^n is viewed as a generalization of R^2 and R^3, the familiar 2- and 3-spaces. Section 4.4, which is new, presents some elementary material in computer graphics. **Chapter 5,** on eigenvalues and eigenvectors, settles the diagonalization problem for symmetric matrices. This chapter can be covered without first covering Chapter 4, which can be omitted. Short applications to a wide variety of areas have been included in the first five chapters. More extensive applications comprise Part II of the book. A suggested pace for covering Part I appears on page x.

Part II consists of Chapters 6 and 7. **Chapter 6** contains an introduction to linear programming, an extremely important application of linear algebra. **Chapter 7** covers several other diverse applications of linear algebra: lines and planes, quadratic forms, graph theory, the theory of games, least squares, linear economic models, Markov processes, the Fibonacci sequence, and differential equations (Calculus Required). The applications in Chapters 1 through 7 deal with problems in mathematics, physics, biology, the social sciences, management, business, and economics.

In Part III, which consists of **Chapter 8** on numerical linear algebra, we deal rather briefly with numerical methods commonly used in linear algebra to solve linear systems of equations and to find eigenvalues and eigenvectors of matrices. Section 8.3, which is new, presents an introduction to **LU**-factorization. These methods are widely used in conjunction with computers. However, the examples and exercises presented here can all be worked with microcomputers and hand-held calculators.

Chapters 6, 7, and 8 are almost entirely independent the one exception being Section 7.4, "The Theory of Games," which requires Chapter 6 as preliminary study. The first five chapters, which constitute the basic linear algebra material, can be comfortably covered in a semester or quarter (see a suggested pace on page x), and there should be time left over to do some of the applications. These can either be covered after the first five chapters have been studied or they can be taken up after the required material has been developed. The chart on page x gives the prerequisites for each of the applications. Moreover, some of the topics, in particular those in Chapters 7 and 8, can also be used as independent student projects. The author's classroom experience with the latter approach has met with favorable student reaction. Thus the instructor can be quite selective both in the choice of material and in the method of study of the applications.

Exercises

The exercises in this book are grouped into two classes. The first class, "Exercises," contains routine exercises. The second class, "Theoretical Exercises," contains exercises that fill in gaps in some of the proofs and amplify material in the text. These exercises can be used to raise the level of the course and to challenge the more capable and interested student. Answers to the odd-numbered numerical exercises appear at the end of the book. New to this edition is a comprehensive review of Part I, consisting of 50 true–false questions (all answers appear in the back of the book). An Answer Manual, containing answers to the even-numbered exercises and solutions to the theoretical exercises, is available (to instructors only) at no cost from the publisher. The Student Solutions Manual, prepared by David R. Hill, Temple University, contains answers to all odd-numbered numerical and theoretical exercises.

End of Chapter Material

Every chapter contains a summary of **Key Ideas for Review,** a set of Supplementary Exercises (odd answers appear in the back of the book), and a chapter test (all answers appear in the back of the book).

The Computer in Linear Algebra

It would be difficult, at this time, to ignore the use of the computer in linear algebra, since most computations involving linear algebra in real problems are carried out on computers. This relationship is briefly explored in the appendix. First, we provide a list of 66 *computer projects* grouped by sections corresponding to the text. These projects are of widely differing degrees of difficulty, and the student is expected to have adequate programming skills to handle these tasks. Most of these projects can be programmed on microcomputers and hand-held programmable calculators. Next, we discuss the availability of *canned software* to implement linear algebra techniques. The use of some of these programs requires some programming skills, although many others merely require that the user know how to use a terminal to enter the data. Finally, we outline some of the features of APL, a programming language that is especially suitable for matrix manipulations, and is easy to learn. We also make several remarks about BASIC, a general-purpose computer language that is being widely used with microcomputers and time-sharing systems.

PREREQUISITE

APPLICATION

Chapter 6	Section 1.4
Section 7.1	Chapters 2 and 3
Section 7.2	Section 5.2
Section 7.3	Section 1.3
Section 7.4	Chapter 6
Section 7.5	Sections 1.4 and 1.5
Section 7.6	Chapter 1
Section 7.7	Section 1.4
Section 7.8	Section 5.1
Section 7.9	Section 5.1, Calculus
Section 8.1	None
Section 8.2	Section 1.4
Section 8.3	Sections 1.5 and 8.2
Section 8.4	Chapter 5

SUGGESTED PACE FOR BASIC MATERIAL, PART I

Chapter 1	7 lectures
Chapter 2	4 lectures
Chapter 3	10 lectures (omit Section 3.3)
Chapter 4	5 lectures (omit Section 4.4)
Chapter 5	5 lectures
	31 lectures

Acknowledgments

I am pleased to express my thanks to the following people who thoroughly reviewed the entire manuscript: in the first edition—William Arendt, of the University of Missouri; and David A. Shedler, of Virginia Commonwealth University. In the second edition—Gerald E. Bergum, of South Dakota State University; James O. Brooks, of Villanova University; Frank R. DeMeyer, of Colorado State University; Joseph Malkevitch, of York College of The City University of New York; Harry W. McLaughlin, of Rensselaer Polytechnic Institute; and Lynn Arthur Steen, of St. Olaf's College. In the third edition—Jerry Goldman, of DePaul University; David R. Hill, of Temple University; Allan Krall, of The Pennsylvania State University at University Park; Stanley Lukawecki, of Clemson University; David Royster, of The University of North Carolina; Sandra Welch, of Stephen F. Austin State University; and Paul Zweir, of Calvin College. In the fourth edition—William G. Vick, Broome Community College; Carrol G. Wells, Western Kentucky University; Andre L. Yandl, Seattle University; and Lance L. Littlejohn, Utah State University.

I also wish to thank the following for their help with selected portions of the manuscript: Thomas L. Bartlow, of Villanova University; Robert E. Beck, of Villanova University; Robin Clark, Drexel University; the late Charles S. Duris, of Drexel University; Herman E. Gollwitzer, of Drexel University; Oded Kariv, of Technion, Israel Institute of Technology; Michael L. Levitan, of Villanova University; and the late John H. Staib, of Drexel University.

The numerous suggestions, comments, and criticisms of these people greatly improved the manuscript. To all of them goes a sincere expression of gratitude.

I am deeply grateful to David R. Hill, Temple University, for developing the Supplementary Exercises, Appendix A, Section 4.4 (computer graphics), and Section 8.3 (**LU**-factorization). He also prepared the Answer Manual and the Student Solutions Manual, and critically read galley and page proofs. It was a pleasure working with him.

I thank William F. Trench, Trinity University, and Robert C. Busby, Drexel University, for suggesting the procedure described in Section 3.6 for finding a basis for a vector space spanned by a given set of vectors; the students at Drexel University who used the earlier editions; and the instructors from a number of institutions who communicated to me their classroom experiences with the earlier editions. I also thank John Edenhofner, Jr., for providing numerical solutions to many of the exercises.

Finally, I thank Robert Clark, Mathematics Editor; Elaine W. Wetterau, Production Supervisor; and the entire staff of the Macmillan Publishing Company for their interest and unfailing cooperation during the conception, design, production, and marketing phases of this book.

B. K.

New Features in Fourth Edition

- ☐ An Appendix, *Complex Numbers,* presents in a brief but thorough manner the use of complex numbers in linear algebra.
- ☐ Section 8.3 (new) on **LU**-decomposition.
- ☐ Section 4.4 (new) on computer graphics.
- ☐ Additional material on computational linear algebra.
- ☐ New material on linear algebra software.
- ☐ More exercises at all levels.
- ☐ More figures and illustrative examples.
- ☐ A new and extremely simple procedure for finding a basis for a vector space that consists of a subset of a given set of vectors spanning the space has been presented in Section 3.6.
- ☐ A comprehensive review of Part I.
- ☐ Additional computer projects.
- ☐ Improved clarity of writing in many places.

Supplementary Materials

The **Student Solutions Manual,** prepared by David R. Hill, Temple University, contains solutions to all odd-numbered numerical and theoretical exercises. It is available from the publisher.

The Answer Manual (available gratis to instructors only) contains answers to all even-numbered exercises and solutions to all theoretical exercises.

Contents

▌▌ APPLICATIONS

III NUMERICAL LINEAR ALGEBRA

Introductory
Linear
Algebra
with
Applications

INTRODUCTORY LINEAR ALGEBRA

1

Linear Equations and Matrices

1.1

LINEAR SYSTEMS

A good many problems in the natural and social sciences as well as in engineering and science deal with equations relating two sets of variables. An equation of the type

$$y = ax,$$

expressing the variable y in terms of the variable x and the constant a, is called a **linear equation**. The word *linear* is used here because the graph of the equation above is a straight line. Similarly, the equation

$$b = a_1x_1 + a_2x_2 + \cdots + a_nx_n, \tag{1}$$

expressing b in terms of the variables $x_1, x_2, \ldots, x_n$ and the known constants $a_1, a_2, \ldots, a_n$, is called a **linear equation**. In many applications we are given b and the constants $a_1, a_2, \ldots, a_n$ and must find numbers $x_1, x_2, \ldots, x_n$ satisfying (1).

A **solution** to a linear equation (1) is a sequence of n numbers $s_1, s_2, \ldots, s_n$, which has the property that (1) is satisfied when $x_1 = s_1, x_2 = s_2, \ldots, x_n = s_n$ are substituted in (1).

Thus $x_1 = 2$, $x_2 = 3$, and $x_3 = -4$ is a solution to the linear equation

$$6x_1 - 3x_2 + 4x_3 = -13,$$

because

$$6(2) - 3(3) + 4(-4) = -13.$$

This is not the only solution to the given linear equation, since $x_1 = 3$, $x_2 = 1$, and $x_3 = -7$ is another solution.

More generally, a **system of m linear equations in n unknowns**, or simply a **linear system**, is a set of m linear equations each in n unknowns. A linear system can be conveniently denoted by

$$
\begin{aligned}
a_{11}x_1 + a_{12}x_2 + \cdots + a_{1n}x_n &= b_1 \\
a_{21}x_1 + a_{22}x_2 + \cdots + a_{2n}x_n &= b_2 \\
\vdots \qquad \vdots \qquad\qquad \vdots \qquad\quad \vdots \\
a_{m1}x_1 + a_{m2}x_2 + \cdots + a_{mn}x_n &= b_m.
\end{aligned}
\tag{2}
$$

The two subscripts i and j are used as follows. The first subscript i indicates that we are dealing with the ith equation, while the second subscript j is associated with the jth variable x_j. Thus the ith equation is

$$a_{i1}x_1 + a_{i2}x_2 + \cdots + a_{in}x_n = b_i.$$

In (2) the a_{ij} are known constants. Given values of $b_1, b_2, \ldots, b_m$, we want to find values of $x_1, x_2, \ldots, x_n$ that will satisfy each equation in (2).

A **solution** to a linear system (2) is a sequence of n numbers $s_1, s_2, \ldots, s_n$, which has the property that each equation in (2) is satisfied when $x_1 = s_1$, $x_2 = s_2, \ldots, x_n = s_n$ are substituted in (2).

To find solutions to a linear system, we shall use a technique called the **method of elimination**. That is, we eliminate some of the unknowns by adding a multiple of one equation to another equation. Most readers have had some experience with this technique in high school algebra courses. Most likely, the reader has confined his or her earlier work with this method to linear systems in which $m = n$, that is, linear systems having as many equations as unknowns. In this course we shall broaden our outlook by dealing with systems in which we have $m = n$, $m < n$, and $m > n$. Indeed, there are numerous applications in which $m \neq n$. If we deal with two, three, or four unknowns, we shall often write them as x, y, z, and w. In this section we use the method of elimination as it was studied in high school. In Section 1.4 we shall look at this method in a much more systematic manner.

Example 1 Consider the linear system

$$
\begin{aligned}
x - 3y &= -3 \\
2x + y &= 8.
\end{aligned}
\tag{3}
$$

To eliminate x, we subtract twice the first equation from the second, obtaining

$$7y = 14,$$

an equation having no x term. We have eliminated the unknown x. Then solving for y, we have

$$y = 2,$$

and substituting into the first equation of (3), we obtain

$$x = 3.$$

To check that $x = 3$, $y = 2$ is a solution to (3), we verify that these values of x and y satisfy *each* of the equations in the given linear system. Thus $x = 3$, $y = 2$ is the only solution to (3).

Example 2　Consider the linear system

$$\begin{aligned} x - 3y &= -7 \\ 2x - 6y &= 7. \end{aligned} \tag{4}$$

Again, we decide to eliminate x. We subtract twice the first equation from the second one, obtaining

$$0 = 21,$$

which makes no sense. This means that (4) has no solution. We might have come to the same conclusion from observing that in (4) the left side of the second equation is twice the left side of the first equation, but the right side of the second equation is not twice the right side of the first equation.

Example 3　Consider the linear system

$$\begin{aligned} x + 2y + 3z &= 6 \\ 2x - 3y + 2z &= 14 \\ 3x + y - z &= -2. \end{aligned} \tag{5}$$

To eliminate x, we subtract twice the first equation from the second one and three times the first equation from the third one, obtaining

$$\begin{aligned} -7y - 4z &= 2 \\ -5y - 10z &= -20. \end{aligned} \tag{6}$$

This is a system of two equations in the unknowns y and z. We multiply the second equation of (6) by $-\frac{1}{5}$, obtaining

$$\begin{aligned} -7y - 4z &= 2 \\ y + 2z &= 4, \end{aligned}$$

which we write, by interchanging equations, as

$$\begin{aligned} y + 2z &= 4 \\ -7y - 4z &= 2. \end{aligned} \tag{7}$$

We now eliminate y in (7) by adding 7 times the first equation to the second one, to obtain

$$10z = 30,$$

or

$$z = 3. \tag{8}$$

Substituting this value of z into the first equation of (7), we find $y = -2$. Substituting these values of y and z into the first equation of (5), we find $x = 1$. To check that $x = 1$, $y = -2$, $z = 3$ is a solution to (5), we verify that these values of x, y, and z satisfy *each* of the equations in (5). Thus $x = 1$, $y = -2$, $z = 3$ is the unique solution to the linear system (5). We might further observe that our elimination procedure has effectively produced the following linear system:

$$\begin{aligned} x + 2y + 3z &= 6 \\ y + 2z &= 4 \\ z &= 3, \end{aligned} \tag{9}$$

obtained by using the first equations of (5) and (7) as well as (8). The importance of the procedure lies in the fact that the linear systems (5) and (9) have exactly the same solutions. System (9) has the advantage that it can be solved quite easily, giving the foregoing values for x, y, and z.

Example 4 Consider the linear system

$$\begin{aligned} x + 2y - 3z &= -4 \\ 2x + y - 3z &= 4. \end{aligned} \tag{10}$$

Eliminating x, we subtract twice the first equation from the second one, to obtain

$$-3y + 3z = 12. \tag{11}$$

We must now solve (11). A solution is

$$y = z - 4,$$

where z can be any real number. Then from the first equation of (10),

$$\begin{aligned} x &= -4 - 2y + 3z \\ &= -4 - 2(z - 4) + 3z \\ &= z + 4. \end{aligned}$$

Thus a solution to the linear system (10) is

$$\begin{aligned} x &= z + 4 \\ y &= z - 4 \\ z &= \text{any real number.} \end{aligned}$$

This means that the linear system (10) has infinitely many solutions. Every time we assign a value to z, we obtain another solution to (10). Thus, if $z = 1$, then

$$x = 5, \qquad y = -3, \qquad \text{and} \qquad z = 1$$

is a solution, while if $z = -2$, then

$$x = 2, \qquad y = -6, \qquad \text{and} \qquad z = -2$$

is another solution.

Example 5 Consider the linear system

$$
\begin{array}{rl}
x + 2y = & 10 \\
2x - 2y = & -4 \\
3x + 5y = & 26.
\end{array}
\tag{12}
$$

Eliminating x, we subtract twice the first equation from the second one to obtain $-6y = -24$, or

$$y = 4.$$

Subtracting three times the first equation from the third one, we obtain $-y = -4$, or

$$y = 4.$$

Thus we have been led to the system

$$
\begin{array}{rl}
x + 2y = & 10 \\
y = & 4 \\
y = & 4,
\end{array}
\tag{13}
$$

which has the same solutions as (12). Substituting $y = 4$ in the first equation of (13), we obtain $x = 2$. Hence $x = 2$, $y = 4$ is a solution to (12).

Example 6 Consider the linear system

$$
\begin{array}{rl}
x + 2y = & 10 \\
2x - 2y = & -4 \\
3x + 5y = & 20.
\end{array}
\tag{14}
$$

To eliminate x, we subtract twice the first equation from the second one to obtain $-6y = -24$, or

$$y = 4.$$

Subtracting three times the first equation from the third one, we obtain $-y = -10$, or

$$y = 10.$$

Thus we have been led to the system

$$
\begin{array}{rl}
x + 2y = & 10 \\
y = & 4 \\
y = & 10,
\end{array}
\tag{15}
$$

which has the same solutions as (14). Since (15) has no solutions, we conclude that (14) has no solutions.

These examples suggest that a linear system may have a unique solution, no solution, or infinitely many solutions.

We have seen that the method of elimination consists of repeatedly performing the following operations:

1. Interchange two equations.
2. Multiply an equation by a nonzero constant.
3. Add a multiple of one equation to another equation.

It is not difficult to show (Exercises T.1 through T.3) that the method of elimination yields another linear system having exactly the same solutions as the given system. The new linear system can then be solved quite readily.

Consider now a linear system of two equations in the unknowns x and y:

$$a_1x + a_2y = c_1$$
$$b_1x + b_2y = c_2. \tag{16}$$

The graph of each of these equations is a straight line, which we denote by l_1 and l_2, respectively. If $x = s_1$, $y = s_2$ is a solution to the linear system (16), then the point (s_1, s_2) lies on both lines l_1 and l_2. Conversely, if the point (s_1, s_2) lies on both lines l_1 and l_2, then $x = s_1$, $y = s_2$ is a solution to the linear system (16). Thus we are led geometrically to the same three possibilities mentioned above (Figure 1.1).

1. The system has a unique solution; that is, the lines l_1 and l_2 intersect at exactly one point.
2. The system has no solution; that is, the lines l_1 and l_2 do not intersect.
3. The system has infinitely many solutions; that is, the lines l_1 and l_2 coincide.

Example 7
(Application to
Production
Planning)
A manufacturer makes three different types of chemical products: A, B, and C. Each product must go through two processing machines: X and Y. The products require the following times in machines X and Y:

1. One ton of product A requires 2 hours in machine X and 2 hours in machine Y.
2. One ton of product B requires 3 hours in machine X and 2 hours in machine Y.
3. One ton of product C requires 4 hours in machine X and 3 hours in machine Y.

(a) A unique solution. (b) No solution. (c) Infinitely many solutions.

Figure 1.1

Machine X is available 80 hours per week and machine Y is available 60 hours per week. Since management does not want to keep the expensive machines X and Y idle, it would like to know how many tons of each product to make so that the machines are fully utilized. It is assumed that the manufacturer can sell as much of the products as is made.

To solve this problem, we let x_1, x_2, and x_3 denote the number of tons of products A, B, and C, respectively, to be made. The number of hours that machine X will be used is

$$2x_1 + 3x_2 + 4x_3,$$

which must equal 80. Thus we have

$$2x_1 + 3x_2 + 4x_3 = 80.$$

Similarly, the number of hours that machine Y will be used is 60, so we have

$$2x_1 + 2x_2 + 3x_3 = 60.$$

Mathematically, our problem is to find nonnegative values of x_1, x_2, and x_3 so that

$$2x_1 + 3x_2 + 4x_3 = 80$$
$$2x_1 + 2x_2 + 3x_3 = 60.$$

This linear system has infinitely many solutions. Following the method of Example 4, we see that all solutions are given by

$$x_1 = \frac{20 - x_3}{2}$$

$$x_2 = 20 - x_3$$
$$x_3 = \text{any real number such that } 0 \leqslant x_3 \leqslant 20,$$

since we must have $x_1 \geqslant 0$, $x_2 \geqslant 0$, and $x_3 \geqslant 0$. When $x_3 = 10$, we have

$$x_1 = 5, \qquad x_2 = 10, \qquad x_3 = 10$$

while

$$x_1 = \tfrac{13}{2}, \qquad x_2 = 13, \qquad x_3 = 7$$

when $x_3 = 7$.

The reader should observe that one solution is just as good as the other. There is no best solution unless additional information or restrictions are given.

1.1 EXERCISES

In Exercises 1 through 14, solve the given linear system by the method of elimination.

1. $x + 2y = 8$
    $3x - 4y = 4.$

2. $2x - 3y + 4z = -12$
    $x - 2y + z = -5$
    $3x + y + 2z = 1.$

3. $3x + 2y + z = 2$
$4x + 2y + 2z = 8$
$x - y + z = 4.$

4. $x + y = 5$
$3x + 3y = 10.$

5. $2x + 4y + 6z = -12$
$2x - 3y - 4z = 15$
$3x + 4y + 5z = -8.$

6. $x + y - 2z = 5$
$2x + 3y + 4z = 2.$

7. $x + 4y - z = 12$
$3x + 8y - 2z = 4.$

8. $3x + 4y - z = 8$
$6x + 8y - 2z = 3.$

9. $x + y + 3z = 12$
$2x + 2y + 6z = 6.$

10. $x + y = 1$
$2x - y = 5$
$3x + 4y = 2.$

11. $2x + 3y = 13$
$x - 2y = 3$
$5x + 2y = 27.$

12. $x - 5y = 6$
$3x + 2y = 1$
$5x + 2y = 1.$

13. $x + 3y = -4$
$2x + 5y = -8$
$x + 3y = -5.$

14. $2x + 3y - z = 6$
$2x - y + 2z = -8$
$3x - y + z = -7.$

15. An oil refinery produces low-sulfur and high-sulfur fuel. Each ton of low-sulfur fuel requires 5 minutes in the blending plant and 4 minutes in the refining plant; each ton of high-sulfur fuel requires 4 minutes in the blending plant and 2 minutes in the refining plant. If the blending plant is available for 3 hours and the refining plant is available for 2 hours, how many tons of each type of fuel should be manufactured so that the plants are fully utilized?

16. A plastics manufacturer makes two types of plastic: regular and special. Each ton of regular plastic requires 2 hours in plant A and 5 hours in plant B; each ton of special plastic requires 2 hours in plant A and 3 hours in plant B. If plant A is available 8 hours per day and plant B is available 15 hours per day, how many tons of each type of plastic can be made daily so that the plants are fully utilized?

17. A dietician is preparing a meal consisting of foods A, B, and C. Each ounce of food A contains 2 units of protein, 3 units of fat, and 4 units of carbohydrate. Each ounce of food B contains 3 units of protein, 2 units of fat, and 1 unit of carbohydrate. Each unit of food C contains 3 units of protein, 3 units of fat, and 2 units of carbohydrate. If the meal must provide exactly 25 units of protein, 24 units of fat, and 21 units of carbohydrate, how many ounces of each type of food should be used?

18. A manufacturer makes 2-minute, 6-minute, and 9-minute film developers. Each ton of 2-minute developer requires 6 minutes in plant A and 24 minutes in plant B. Each ton of 6-minute developer requires 12 minutes in plant A and 12 minutes in plant B. Each ton of 9-minute developer requires 12 minutes in plant A and 12 minutes in plant B. If plant A is available 10 hours per day and plant B is available 16 hours per day, how many tons of each type of developer can be produced so that the plants are fully utilized?

THEORETICAL EXERCISES

T.1. Show that the linear system obtained by interchanging two equations in (2) has exactly the same solutions as (2).

T.2. Show that the linear system obtained by replacing an equation in (2) by a nonzero constant multiple of the equation has exactly the same solutions as (2).

T.3. Show that the linear system obtained by replacing an equation in (2) by itself plus a multiple of another equation in (2) has exactly the same solutions as (2).

1.2

MATRICES

If we examine the method of elimination described in Section 1.1, we make the following observation. Only the numbers in front of the unknowns x_1, x_2, . . . , x_n are being changed as we perform the steps in the method of elimination. Thus we might think of looking for a way of writing a linear system without having to carry along the unknowns. Matrices enable us to do this—that is, to write linear systems in a compact form that makes it easier to automate the elimination method on an electronic computer in order to obtain a fast and efficient procedure for finding solutions. Their use is not, however, merely that of a convenient notation. We now develop operations on matrices and will work with matrices according to the rules they obey; this will enable us to solve systems of linear equations and to do other computational problems in a fast and efficient manner. Of course, as any good definition should do, the notion of a matrix provides not only a new way of looking at old problems, but also gives rise to a great many new questions, some of which we study in this book.

Definition An $m \times n$ **matrix A** is a rectangular array of mn real (or complex) numbers arranged in m horizontal **rows** and n vertical **columns**:

$$\mathbf{A} = \begin{bmatrix} a_{11} & a_{12} & \cdots & a_{1n} \\ a_{21} & a_{22} & \cdots & a_{2n} \\ \vdots & \vdots & & \vdots \\ a_{m1} & a_{m2} & \cdots & a_{mn} \end{bmatrix}. \tag{1}$$

The i**th row** of **A** is

$$[a_{i1} \quad a_{i2} \quad \cdots \quad a_{in}] \qquad (1 \le i \le m);$$

the j**th column** of **A** is

$$\begin{bmatrix} a_{1j} \\ a_{2j} \\ \vdots \\ a_{mj} \end{bmatrix} \qquad (1 \le j \le n).$$

We shall say that $\mathbf{A}$ is $\boldsymbol{m}$ by $\boldsymbol{n}$ ($\boldsymbol{m} \times \boldsymbol{n}$). If $m = n$, we say that $\mathbf{A}$ is a **square matrix of order n**, and that the numbers $a_{11}, a_{22}, \ldots, a_{nn}$ form the **main diagonal** of $\mathbf{A}$. We refer to the number a_{ij}, which is in the ith row and jth column of $\mathbf{A}$, as the $\boldsymbol{i}, \boldsymbol{j}$**th element of** $\mathbf{A}$, or the $(\boldsymbol{i}, \boldsymbol{j})$ **entry of** $\mathbf{A}$, and we often write (1) as

$$\mathbf{A} = [a_{ij}].$$

For the sake of simplicity, we restrict our attention in this book to matrices all of whose elements are real numbers. However, matrices with complex entries are studied and are important in applications.

Example 1 Let

$$\mathbf{A} = \begin{bmatrix} 1 & 2 & 3 \\ -1 & 0 & 1 \end{bmatrix}, \quad \mathbf{B} = \begin{bmatrix} 1 & 4 \\ 2 & -3 \end{bmatrix}, \quad \mathbf{C} = \begin{bmatrix} 1 \\ -1 \\ 2 \end{bmatrix},$$

$$\mathbf{D} = \begin{bmatrix} 1 & 1 & 0 \\ 2 & 0 & 1 \\ 3 & -1 & 2 \end{bmatrix}, \quad \mathbf{E} = [3], \quad \text{and} \quad \mathbf{F} = [-1 \quad 0 \quad 2].$$

Then $\mathbf{A}$ is a 2×3 matrix with $a_{12} = 2$, $a_{13} = 3$, $a_{22} = 0$, and $a_{23} = 1$; $\mathbf{B}$ is a 2×2 matrix with $b_{11} = 1$, $b_{12} = 4$, $b_{21} = 2$, and $b_{22} = -3$; $\mathbf{C}$ is a 3×1 matrix with $c_{11} = 1$, $c_{21} = -1$, and $c_{31} = 2$; $\mathbf{D}$ is a 3×3 matrix; $\mathbf{E}$ is a 1×1 matrix; and $\mathbf{F}$ is a 1×3 matrix. In $\mathbf{D}$, the elements $d_{11} = 1$, $d_{22} = 0$, and $d_{33} = 2$ form the main diagonal.

Definition A square matrix $\mathbf{A} = [a_{ij}]$ for which every term off the main diagonal is zero, that is, $a_{ij} = 0$ for $i \neq j$, is called a **diagonal matrix**.

Example 2

$$\mathbf{G} = \begin{bmatrix} 4 & 0 \\ 0 & -2 \end{bmatrix} \quad \text{and} \quad \mathbf{H} = \begin{bmatrix} -3 & 0 & 0 \\ 0 & -2 & 0 \\ 0 & 0 & 4 \end{bmatrix}$$

are diagonal matrices.

Definition A diagonal matrix $\mathbf{A} = [a_{ij}]$, for which all terms on the main diagonal are equal, that is, $a_{ij} = c$ for $i = j$ and $a_{ij} = 0$ for $i \neq j$, is called a **scalar matrix**.

Example 3 The following are scalar matrices:

$$\mathbf{I}_3 = \begin{bmatrix} 1 & 0 & 0 \\ 0 & 1 & 0 \\ 0 & 0 & 1 \end{bmatrix}, \quad \mathbf{J} = \begin{bmatrix} -2 & 0 \\ 0 & -2 \end{bmatrix}.$$

Example 4
(Tabular
Display of
Data)
The following matrix gives the airline distances between the indicated cities (in statute miles).

	London	Madrid	New York	Tokyo
London	0	785	3469	5959
Madrid	785	0	3593	6706
New York	3469	3593	0	6757
Tokyo	5959	6706	6757	0

Example 5
(Production)
Suppose that a manufacturer has four plants each of which makes three products. If we let a_{ij} denote the amount of product i made by plant j in 1 week, then the 3×4 matrix

Plant 1	Plant 2	Plant 3	Plant 4	
560	360	380	0	Product 1
340	450	420	80	Product 2
280	270	210	380	Product 3

gives the manufacturer's production for the week.

Whenever a new object is introduced in mathematics, we must determine when two such objects are equal. For example, in the set of all rational numbers, the numbers $\frac{2}{3}$ and $\frac{4}{6}$ are called equal although they are not represented in the same manner. What we have in mind is the definition that a/b equals c/d when $ad = bc$. Accordingly, we now have the following definition.

Definition Two $m \times n$ matrices $\mathbf{A} = [a_{ij}]$ and $\mathbf{B} = [b_{ij}]$ are said to be **equal** if $a_{ij} = b_{ij}$, $1 \leqslant i \leqslant m$, $1 \leqslant j \leqslant n$, that is, if corresponding elements agree.

Example 6 The matrices

$$\mathbf{A} = \begin{bmatrix} 1 & 2 & -1 \\ 2 & -3 & 4 \\ 0 & -4 & 5 \end{bmatrix} \quad \text{and} \quad \mathbf{B} = \begin{bmatrix} 1 & 2 & w \\ 2 & x & 4 \\ y & -4 & z \end{bmatrix}$$

are equal if $w = -1$, $x = -3$, $y = 0$, and $z = 5$.

We shall now define a number of operations that will produce new matrices out of given matrices. These operations are useful in the applications of matrices.

Matrix Addition

Definition If $\mathbf{A} = [a_{ij}]$ and $\mathbf{B} = [b_{ij}]$ are $m \times n$ matrices, then the **sum** of $\mathbf{A}$ and $\mathbf{B}$ is the $m \times n$ matrix $\mathbf{C} = [c_{ij}]$, defined by

$$c_{ij} = a_{ij} + b_{ij} \qquad (1 \leq i \leq m, 1 \leq j \leq n).$$

That is, **C** is obtained by adding corresponding elements of **A** and **B**.

Example 7 Let

$$\mathbf{A} = \begin{bmatrix} 1 & -2 & 4 \\ 2 & -1 & 3 \end{bmatrix} \quad \text{and} \quad \mathbf{B} = \begin{bmatrix} 0 & 2 & -4 \\ 1 & 3 & 1 \end{bmatrix}.$$

Then

$$\mathbf{A} + \mathbf{B} = \begin{bmatrix} 1+0 & -2+2 & 4+(-4) \\ 2+1 & -1+3 & 3+1 \end{bmatrix} = \begin{bmatrix} 1 & 0 & 0 \\ 3 & 2 & 4 \end{bmatrix}.$$

It should be noted that the sum of the matrices **A** and **B** is defined only when **A** and **B** have the same number of rows and the same number of columns, that is, only when **A** and **B** are of the same size. We shall now establish the convention that when **A** + **B** is formed, both **A** and **B** are of the same size. The basic properties of matrix addition are considered in the following section and are similar to those satisfied by the real numbers.

Example 8
(Production) A manufacturer of a certain product makes three models, A, B, and C. Each model is partially made in factory F_1 in Taiwan and then finished in factory F_2 in the United States. The total cost of each product consists of the manufacturing cost and the shipping cost. Then the costs at each factory (in dollars) can be described by the 3×2 matrices $\mathbf{F}_1$ and $\mathbf{F}_2$:

$$\mathbf{F}_1 = \begin{bmatrix} 32 & 40 \\ 50 & 80 \\ 70 & 20 \end{bmatrix} \begin{matrix} \text{Model } A \\ \text{Model } B \\ \text{Model } C \end{matrix}$$

Manufacturing cost, Shipping cost

$$\mathbf{F}_2 = \begin{bmatrix} 40 & 60 \\ 50 & 50 \\ 130 & 20 \end{bmatrix} \begin{matrix} \text{Model } A \\ \text{Model } B \\ \text{Model } C \end{matrix}$$

Manufacturing cost, Shipping cost

The matrix $\mathbf{F}_1 + \mathbf{F}_2$ gives the total manufacturing and shipping costs for each product. Thus the total manufacturing and shipping costs of a model C product are \$200 and \$40, respectively.

Matrix Multiplication

Definition If $\mathbf{A} = [a_{ij}]$ is an $m \times p$ matrix and $\mathbf{B} = [b_{ij}]$ is a $p \times n$ matrix, then the **product** of $\mathbf{A}$ and $\mathbf{B}$ is the $m \times n$ matrix $\mathbf{C} = [c_{ij}]$, defined by

$$c_{ij} = a_{i1}b_{1j} + a_{i2}b_{2j} + \cdots + a_{ip}b_{pj} \qquad (1 \leqslant i \leqslant m, \, 1 \leqslant j \leqslant n). \qquad (2)$$

Equation (2) says that the i, jth element in the product matrix is computed by adding all products obtained by multiplying each element in the ith row of $\mathbf{A}$ by the corresponding element in the jth column of $\mathbf{B}$; this is shown in Figure 1.2. Thus the product of $\mathbf{A}$ and $\mathbf{B}$ is defined only when the number of rows of $\mathbf{B}$ is exactly the same as the number of columns of $\mathbf{A}$.

Figure 1.2

One might ask why matrix equality and matrix addition are defined in such a natural way while matrix multiplication appears to be much more complicated. Only a thorough understanding of the composition of functions and the relationship that exists between matrices and what shall be called linear transformations would show that the foregoing definition of multiplication is the natural one.

Example 9 Let

$$\mathbf{A} = \begin{bmatrix} 1 & 2 & -1 \\ 3 & 1 & 4 \end{bmatrix} \quad \text{and} \quad \mathbf{B} = \begin{bmatrix} -2 & 5 \\ 4 & -3 \\ 2 & 1 \end{bmatrix}.$$

Then

$$\mathbf{AB} = \begin{bmatrix} (1)(-2) + (2)(4) + (-1)(2) & (1)(5) + (2)(-3) + (-1)(1) \\ (3)(-2) + (1)(4) + (4)(2) & (3)(5) + (1)(-3) + (4)(1) \end{bmatrix}$$

$$= \begin{bmatrix} 4 & -2 \\ 6 & 16 \end{bmatrix}.$$

The basic properties of matrix multiplication will be considered in the following section. However, multiplication of matrices requires much more care than their addition, since the algebraic properties of matrix multiplication differ from those satisfied by the real numbers. Part of the problem is due to the fact that AB is defined only when the number of columns of A is the same as the number of rows of B. Thus, if A is an $m \times p$ matrix and B is a $p \times n$ matrix, then AB is an $m \times n$ matrix. What about BA? Four different situations may occur:

1. BA may not be defined; this will take place if $n \neq m$.
2. If BA is defined, which means that $m = n$, then BA is $p \times p$ while AB is $m \times m$; thus, if $m \neq p$, AB and BA are of different sizes.
3. If AB and BA are both of the same size, they may be equal.
4. If AB and BA are both of the same size, they may be unequal.

Example 10 If A is a 2×3 matrix and B is a 3×4 matrix, then AB is a 2×4 matrix while BA is undefined.

Example 11 Let A be 2×3 and let B be 3×2. Then AB is 2×2 while BA is 3×3.

Example 12 Let

$$A = \begin{bmatrix} 1 & 2 \\ -1 & 3 \end{bmatrix} \quad \text{and} \quad B = \begin{bmatrix} 2 & 1 \\ 0 & 1 \end{bmatrix}.$$

Then

$$AB = \begin{bmatrix} 2 & 3 \\ -2 & 2 \end{bmatrix} \quad \text{while} \quad BA = \begin{bmatrix} 1 & 7 \\ -1 & 3 \end{bmatrix}.$$

Thus $AB \neq BA$.

It is sometimes useful to be able to find a column in the matrix product AB without having to multiply the two matrices. It is not difficult to show (Exercise T.5) that the jth column of the matrix product AB is equal to the matrix product AB_j, where B_j is the jth column of B.

Example 13 Let

$$A = \begin{bmatrix} 1 & 2 \\ 3 & 4 \\ -1 & 5 \end{bmatrix} \quad \text{and} \quad B = \begin{bmatrix} -2 & 3 & 4 \\ 3 & 2 & 1 \end{bmatrix}.$$

Then the second column of AB is

$$AB_2 = \begin{bmatrix} 1 & 2 \\ 3 & 4 \\ -1 & 5 \end{bmatrix} \begin{bmatrix} 3 \\ 2 \end{bmatrix} = \begin{bmatrix} 7 \\ 17 \\ 7 \end{bmatrix}.$$

Example 14 Pesticides are sprayed on plants to eliminate harmful insects. However, some of the
(Ecology) pesticide is absorbed by the plant. The pesticides are absorbed by herbivores when they
eat the plants that have been sprayed. To determine the amount of pesticide absorbed by a
herbivore, we proceed as follows. Suppose that we have three pesticides and four plants.
Let a_{ij} denote the amount of pesticide i (in milligrams) that has been absorbed by plant j.
This information can be represented by the matrix

$$\mathbf{A} = \begin{array}{cccc} \text{Plant 1} & \text{Plant 2} & \text{Plant 3} & \text{Plant 4} \\ \begin{bmatrix} 2 & 3 & 4 & 3 \\ 3 & 2 & 2 & 5 \\ 4 & 1 & 6 & 4 \end{bmatrix} & & & \begin{array}{l} \text{Pesticide 1} \\ \text{Pesticide 2} \\ \text{Pesticide 3} \end{array} \end{array}$$

Now suppose that we have three herbivores, and let b_{ij} denote the number of plants of type
i that a herbivore of type j eats per month. This information can be represented by the
matrix

$$\mathbf{B} = \begin{array}{ccc} \text{Herbivore 1} & \text{Herbivore 2} & \text{Herbivore 3} \\ \begin{bmatrix} 20 & 12 & 8 \\ 28 & 15 & 15 \\ 30 & 12 & 10 \\ 40 & 16 & 20 \end{bmatrix} & & \begin{array}{l} \text{Plant 1} \\ \text{Plant 2} \\ \text{Plant 3} \\ \text{Plant 4} \end{array} \end{array}$$

The (i, j) entry in $\mathbf{AB}$ gives the amount of pesticide of type i that animal j has absorbed.
Thus, if $i = 2$ and $j = 3$, the $(2, 3)$ entry in $\mathbf{AB}$ is

$$3(8) + 2(15) + 2(10) + 5(20) = 174 \text{ mg of pesticide 2 absorbed by herbivore 3.}$$

If we now have p carnivores (such as man) who eat the herbivores, we can repeat the
analysis to find out how much of each pesticide has been absorbed by each carnivore.

Linear Systems

Consider the linear system of m equations in n unknowns

$$\begin{aligned} a_{11}x_1 + a_{12}x_2 + \cdots + a_{1n}x_n &= b_1 \\ a_{21}x_1 + a_{22}x_2 + \cdots + a_{2n}x_n &= b_2 \\ &\vdots \\ a_{m1}x_1 + a_{m2}x_2 + \cdots + a_{mn}x_n &= b_m. \end{aligned} \tag{3}$$

Now define the following matrices:

$$\mathbf{A} = \begin{bmatrix} a_{11} & a_{12} & \cdots & a_{1n} \\ a_{21} & a_{22} & \cdots & a_{2n} \\ \vdots & \vdots & & \vdots \\ a_{m1} & a_{m2} & \cdots & a_{mn} \end{bmatrix}, \quad \mathbf{X} = \begin{bmatrix} x_1 \\ x_2 \\ \vdots \\ x_n \end{bmatrix}, \quad \mathbf{B} = \begin{bmatrix} b_1 \\ b_2 \\ \vdots \\ b_m \end{bmatrix}.$$

Then the linear system (3) can be written in matrix form as

$$\mathbf{AX} = \mathbf{B}.$$

The matrix $\mathbf{A}$ is called the **coefficient matrix** of the linear system (3) and the matrix

$$\begin{bmatrix} a_{11} & a_{12} & \cdots & a_{1n} & b_1 \\ a_{21} & a_{22} & \cdots & a_{2n} & b_2 \\ \vdots & \vdots & & \vdots & \vdots \\ a_{m1} & a_{m2} & \cdots & a_{mn} & b_m \end{bmatrix},$$

obtained by adjoining $\mathbf{B}$ to $\mathbf{A}$, is called the **augmented matrix** of the linear system (3). The augmented matrix of (3) will be written as $[\mathbf{A} \mid \mathbf{B}]$. Conversely, any matrix with more than one column can be thought of as the augmented matrix of a linear system. The coefficient and augmented matrices will play key roles in our method for solving linear systems.

Example 15 Consider the linear system

$$\begin{aligned} 2x + 3y - 4z &= 5 \\ -2x \qquad\;\; + z &= 7 \\ 3x + 2y + 2z &= 3. \end{aligned}$$

Letting

$$\mathbf{A} = \begin{bmatrix} 2 & 3 & -4 \\ -2 & 0 & 1 \\ 3 & 2 & 2 \end{bmatrix}, \qquad \mathbf{X} = \begin{bmatrix} x \\ y \\ z \end{bmatrix}, \qquad \text{and} \qquad \mathbf{B} = \begin{bmatrix} 5 \\ 7 \\ 3 \end{bmatrix},$$

we can write the given linear system in matrix form as

$$\mathbf{AX} = \mathbf{B}.$$

The coefficient matrix is $\mathbf{A}$ and the augmented matrix is

$$\begin{bmatrix} 2 & 3 & -4 & 5 \\ -2 & 0 & 1 & 7 \\ 3 & 2 & 2 & 3 \end{bmatrix}.$$

Example 16 The matrix

$$\begin{bmatrix} 2 & -1 & 3 & 4 \\ 3 & 0 & 2 & 5 \end{bmatrix}$$

is the augmented matrix of the linear system

$$\begin{aligned} 2x - y + 3z &= 4 \\ 3x \qquad\;\; + 2z &= 5. \end{aligned}$$

Scalar Multiplication

Definition If $\mathbf{A} = [a_{ij}]$ is an $m \times n$ matrix and r is a real number, then the **scalar multiple** of $\mathbf{A}$ by r, $r\mathbf{A}$, is the $m \times n$ matrix $\mathbf{B} = [b_{ij}]$, where

$$b_{ij} = ra_{ij} \qquad (1 \leqslant i \leqslant m, \ 1 \leqslant j \leqslant n).$$

That is, $\mathbf{B}$ is obtained by multiplying each element of $\mathbf{A}$ by r.

Example 17 If $r = -3$ and

$$\mathbf{A} = \begin{bmatrix} 4 & -2 & 3 \\ 2 & -5 & 0 \\ 3 & 6 & -2 \end{bmatrix},$$

then

$$r\mathbf{A} = -3\begin{bmatrix} 4 & -2 & 3 \\ 2 & -5 & 0 \\ 3 & 6 & -2 \end{bmatrix} = \begin{bmatrix} (-3)(4) & (-3)(-2) & (-3)(3) \\ (-3)(2) & (-3)(-5) & (-3)(0) \\ (-3)(3) & (-3)(6) & (-3)(-2) \end{bmatrix} = \begin{bmatrix} -12 & 6 & -9 \\ -6 & 15 & 0 \\ -9 & -18 & 6 \end{bmatrix}.$$

The Transpose of a Matrix

Definition If $\mathbf{A} = [a_{ij}]$ is an $m \times n$ matrix, then the $n \times m$ matrix $\mathbf{A}^T = [a_{ij}^T]$, where

$$a_{ij}^T = a_{ji} \qquad (1 \leqslant i \leqslant m, \ 1 \leqslant j \leqslant n)$$

is called the **transpose** of $\mathbf{A}$. Thus the transpose of $\mathbf{A}$ is obtained by interchanging the rows and columns of $\mathbf{A}$.

Example 18 Let

$$\mathbf{A} = \begin{bmatrix} 4 & -2 & 3 \\ 0 & 5 & -2 \end{bmatrix}, \qquad \mathbf{B} = \begin{bmatrix} 6 & 2 & -4 \\ 3 & -1 & 2 \\ 0 & 4 & 3 \end{bmatrix}, \qquad \mathbf{C} = \begin{bmatrix} 5 & 4 \\ -3 & 2 \\ 2 & -3 \end{bmatrix},$$

$$\mathbf{D} = [3 \quad -5 \quad 1], \quad \text{and} \quad \mathbf{E} = \begin{bmatrix} 2 \\ -1 \\ 3 \end{bmatrix}.$$

Then

$$\mathbf{A}^T = \begin{bmatrix} 4 & 0 \\ -2 & 5 \\ 3 & -2 \end{bmatrix}, \qquad \mathbf{B}^T = \begin{bmatrix} 6 & 3 & 0 \\ 2 & -1 & 4 \\ -4 & 2 & 3 \end{bmatrix},$$

$$\mathbf{C}^T = \begin{bmatrix} 5 & -3 & 2 \\ 4 & 2 & -3 \end{bmatrix}, \qquad \mathbf{D}^T = \begin{bmatrix} 3 \\ -5 \\ 1 \end{bmatrix}, \qquad \text{and} \qquad \mathbf{E}^T = [2 \quad -1 \quad 3].$$

Summation Notation

We shall occasionally use the **summation notation** and we now review this useful and compact notation, which is widely used in mathematics.

By $\displaystyle\sum_{i=1}^{n} a_i$ we mean

$$a_1 + a_2 + \cdots + a_n.$$

The letter i is called the **index of summation**; it is a dummy variable that can be replaced by another letter. Hence we can write

$$\sum_{i=1}^{n} a_i = \sum_{j=1}^{n} a_j = \sum_{k=1}^{n} a_k.$$

Example 19 If

$$a_1 = 3, \qquad a_2 = 4, \qquad a_3 = 5, \qquad \text{and} \qquad a_4 = 8,$$

then

$$\sum_{i=1}^{4} a_i = 3 + 4 + 5 + 8 = 20.$$

Example 20

By $\displaystyle\sum_{i=1}^{n} r_i a_i$ we mean

$$r_1 a_1 + r_2 a_2 + \cdots + r_n a_n.$$

It is not difficult to show (Exercise T.7) that the summation notation satisfies the following properties:

$$\text{(i)} \ \sum_{i=1}^{n} (r_i + s_i) a_i = \sum_{i=1}^{n} r_i a_i + \sum_{i=1}^{n} s_i a_i.$$

$$\text{(ii)} \ \sum_{i=1}^{n} c(r_i a_i) = c \left(\sum_{i=1}^{n} r_i a_i \right).$$

Example 21 We can write Equation (2), for the i, jth element in the product of the matrices $\mathbf{A}$ and $\mathbf{B}$, in terms of the summation notation as

$$c_{ij} = \sum_{k=1}^{p} a_{ik} b_{kj} \qquad (1 \leqslant i \leqslant m, \ 1 \leqslant j \leqslant n).$$

It is also possible to form double sums. Thus by $\sum\limits_{j=1}^{m} \sum\limits_{i=1}^{n} a_{ij}$ we mean that we first sum on i and then sum the resulting expression on j.

Example 22 If $n = 2$ and $m = 3$, we have

$$\sum_{j=1}^{3} \sum_{i=1}^{2} a_{ij} = \sum_{j=1}^{3} (a_{1j} + a_{2j})$$

$$= (a_{11} + a_{21}) + (a_{12} + a_{22}) + (a_{13} + a_{23}) \qquad (4)$$

$$\sum_{i=1}^{2} \sum_{j=1}^{3} a_{ij} = \sum_{i=1}^{2} (a_{i1} + a_{i2} + a_{i3})$$

$$= (a_{11} + a_{12} + a_{13}) + (a_{21} + a_{22} + a_{23})$$
$$= \text{right side of (4).}$$

It is not difficult to show, in general (Exercise T.8), that

$$\sum_{i=1}^{n} \sum_{j=1}^{m} a_{ij} = \sum_{j=1}^{m} \sum_{i=1}^{n} a_{ij}. \qquad (5)$$

Equation (5) can be interpreted as follows. If we add up the entries in each row of a matrix and then add the resulting numbers, we obtain the same result as when we add up the entries in each column of the matrix and then add the resulting numbers.

1.2 EXERCISES

1. Let

$$\mathbf{A} = \begin{bmatrix} 2 & -3 & 5 \\ 6 & -5 & 4 \end{bmatrix}, \qquad \mathbf{B} = \begin{bmatrix} 4 \\ -3 \\ 5 \end{bmatrix},$$

and

$$\mathbf{C} = \begin{bmatrix} 7 & 3 & 2 \\ -4 & 3 & 5 \\ 6 & 1 & -1 \end{bmatrix}.$$

(a) What is a_{12}, a_{22}, a_{23}?
(b) What is b_{11}, b_{31}?
(c) What is c_{13}, c_{31}, c_{33}?

In Exercises 2 through 7, let

$$\mathbf{A} = \begin{bmatrix} 1 & 2 & 3 \\ 2 & 1 & 4 \end{bmatrix}, \qquad \mathbf{B} = \begin{bmatrix} 1 & 0 \\ 2 & 1 \\ 3 & 2 \end{bmatrix},$$

$$\mathbf{C} = \begin{bmatrix} 3 & -1 & 3 \\ 4 & 1 & 5 \\ 2 & 1 & 3 \end{bmatrix}, \qquad \mathbf{D} = \begin{bmatrix} 3 & -2 \\ 2 & 4 \end{bmatrix},$$

$$\mathbf{E} = \begin{bmatrix} 2 & -4 & 5 \\ 0 & 1 & 4 \\ 3 & 2 & 1 \end{bmatrix},$$

and

$$\mathbf{F} = \begin{bmatrix} -4 & 5 \\ 2 & 3 \end{bmatrix}.$$

2. If possible, compute
(a) $\mathbf{C} + \mathbf{E}$.
(b) $\mathbf{AB}$ and $\mathbf{BA}$.
(c) $2\mathbf{D} - 3\mathbf{F}$.
(d) $\mathbf{CB} + \mathbf{D}$.
(e) $\mathbf{AB} + \mathbf{DF}$.
(f) $(3)(2\mathbf{A})$ and $6\mathbf{A}$.

3. If possible, compute
(a) $\mathbf{A(BD)}$ and $\mathbf{(AB)D}$.
(b) $\mathbf{A(C+E)}$ and $\mathbf{AC + AE}$.
(c) $\mathbf{3A + 2A}$ and $\mathbf{5A}$.
(d) $\mathbf{DF + AB}$.
(e) $\mathbf{EF + 2A}$.
(f) $\mathbf{(-4)(3C)}$ and $\mathbf{(-12)C}$.

4. If possible, compute
(a) $\mathbf{2B - 3F}$.
(b) $\mathbf{EB + FA}$.
(c) $\mathbf{A(B + D)}$.
(d) $\mathbf{(D + F)A}$.
(e) $\mathbf{3F + 4F}$ and $\mathbf{7F}$.
(f) $\mathbf{AC + DE}$.

5. If possible, compute
(a) $\mathbf{A}^T$ and $\mathbf{(A}^T)^T$.
(b) $\mathbf{(C + E)}^T$ and $\mathbf{C}^T + \mathbf{E}^T$.
(c) $\mathbf{(AB)}^T$ and $\mathbf{B}^T\mathbf{A}^T$.
(d) $\mathbf{(2D + 3F)}^T$.
(e) $\mathbf{(2BC + F)}^T$.
(f) $\mathbf{B}^T\mathbf{C} + \mathbf{A}$.

6. If possible, compute
(a) $\mathbf{(3C - 2E)}^T\mathbf{B}$.
(b) $\mathbf{A}^T\mathbf{(D + F)}$.
(c) $\mathbf{(2E)A}^T$.
(d) $\mathbf{(BC)}^T$ and $\mathbf{C}^T\mathbf{B}^T$.
(e) $\mathbf{(B}^T + \mathbf{A})C$.
(f) $\mathbf{(D}^T + \mathbf{E})F$.

7. If $\mathbf{I} = \begin{bmatrix} 1 & 0 \\ 0 & 1 \end{bmatrix}$, compute $\mathbf{DI}$ and $\mathbf{ID}$.

8. If $\begin{bmatrix} a + b & c + d \\ c - d & a - b \end{bmatrix} = \begin{bmatrix} 4 & 6 \\ 10 & 2 \end{bmatrix}$, find a, b, c, and d.

9. If $\begin{bmatrix} a + 2b & 2a - b \\ 2c + d & c - 2d \end{bmatrix} = \begin{bmatrix} 4 & -2 \\ 4 & -3 \end{bmatrix}$, find a, b, c, and d.

10. Let $\mathbf{A} = \begin{bmatrix} 1 & 2 \\ 3 & 2 \end{bmatrix}$ and $\mathbf{B} = \begin{bmatrix} 2 & -1 \\ -3 & 4 \end{bmatrix}$. Show that $\mathbf{AB} \neq \mathbf{BA}$.

11. If $\mathbf{A}$ is the matrix in Exercise 1 and $\mathbf{0}$ is the 3×2 matrix every one of whose entries is zero, compute $\mathbf{A0}$.

12. Consider the following linear system:

$$\begin{aligned} 2x \quad\quad\quad + w &= 7 \\ 3x + 2y + 3z \quad\quad &= -2 \\ 2x + 3y - 4z \quad\quad &= 3 \\ x + \quad\quad 3z \quad\quad &= 5. \end{aligned}$$

(a) Find the coefficient matrix.
(b) Write the linear system in matrix form.
(c) Find the augmented matrix.

13. Write the linear system with augmented matrix

$$\begin{bmatrix} -2 & -1 & 0 & 4 & \vdots & 5 \\ -3 & 2 & 7 & 8 & \vdots & 3 \\ 1 & 0 & 0 & 2 & \vdots & 4 \\ 3 & 0 & 1 & 3 & \vdots & 6 \end{bmatrix}.$$

14. Write the linear system with augmented matrix

$$\begin{bmatrix} 2 & 0 & -4 & \vdots & 3 \\ 0 & 1 & 2 & \vdots & 5 \\ 1 & 3 & 4 & \vdots & -1 \end{bmatrix}.$$

15. Consider the following linear system:

$$\begin{aligned} 3x - y + 2z &= 4 \\ 2x + y \quad\quad &= 2 \\ y + 3z &= 7 \\ 4x \quad\quad - z &= 4. \end{aligned}$$

(a) Find the coefficient matrix.
(b) Write the linear system in matrix form.
(c) Find the augmented matrix.

In Exercises 16 and 17, let

$$\mathbf{A} = \begin{bmatrix} 1 & -1 & 2 \\ 3 & 2 & 4 \\ 4 & -2 & 3 \\ 2 & 1 & 5 \end{bmatrix}$$

and

$$\mathbf{B} = \begin{bmatrix} 1 & 0 & -1 & 2 \\ 3 & 3 & -3 & 4 \\ 4 & 2 & 5 & 1 \end{bmatrix}.$$

16. Using the method in Example 13, compute the following columns of **AB**:
(a) The first column.
(b) The third column.

17. Using the method of Example 13, compute the following columns of **AB**:
(a) The second column.
(b) The fourth column.

18. (Manufacturing Costs). A furniture manufacturer makes chairs and tables, each of which must go through an assembly process and a finishing process. The times required for these processes are given (in hours) by the matrix

	Assembly process	Finishing process	
A =	2	2	Chair
	3	4	Table

The manufacturer has a plant in Salt Lake City and another in Chicago. The hourly rates for each of the processes are given (in dollars) by the matrix

	Salt Lake City	Chicago	
B =	9	10	Assembly process
	10	12	Finishing process

What do the entries in the matrix product **AB** tell the manufacturer?

19. (Ecology—Pollution). A manufacturer makes two kinds of products, *P* and *Q*, at each of two plants, *X* and *Y*. In making these products, the pollutants sulfur dioxide, nitric oxide, and particulate matter are produced. The amounts of pollutants produced are given (in kilograms) by the matrix

	Sulfur dioxide	Nitric oxide	Particulate matter	
A =	300	100	150	Product *P*
	200	250	400	Product *Q*

State and federal ordinances require that these pollutants be removed. The daily cost of removing each kilogram of pollutant is given (in dollars) by the matrix

	Plant *X*	Plant *Y*	
B =	8	12	Sulfur dioxide
	7	9	Nitric oxide
	15	10	Particulate matter

What do the entries in the matrix product **AB** tell the manufacturer?

20. (Medicine). A diet research project consists of adults and children of both sexes. The composition of the participants in the project is given by the matrix

	Adults	Children	
A =	80	120	Male
	100	200	Female

The number of daily grams of protein, fat, and carbohydrate consumed by each child and adult is given by the matrix

	Protein	Fat	Carbohydrate	
B =	20	20	20	Adult
	10	20	30	Child

(a) How many grams of protein are consumed daily by the males in the project?
(b) How many grams of fat are consumed daily by the females in the project?

21. (Business). A photography business has a store in each of the following cities: New York, Denver, and Los Angeles. A particular make of camera is available in automatic, semiautomatic, and non-automatic models. Moreover, each camera has a matched flash unit and a camera is usually sold together with the corresponding flash unit. The selling prices of the cameras and flash units are given (in dollars) by the matrix

	Automatic	Semiautomatic	Non-automatic	
A =	200	150	120	Camera
	50	40	25	Flash unit

The number of sets (camera and flash unit) available at each store is given by the matrix

	New York	Denver	Los Angeles	
B =	220	180	100	Automatic
	300	250	120	Semiautomatic
	120	320	250	Nonautomatic

(a) What is the total value of the cameras in New York?

(b) What is the total value of the flash units in Los Angeles?

THEORETICAL EXERCISES

T.1. (a) Prove that if **A** has a row of zeros, then **AB** has a row of zeros.

(b) Prove that if **B** has a column of zeros, then **AB** has a column of zeros.

T.2. Prove that the sum and product of diagonal matrices is diagonal.

T.3. Prove that the sum and product of scalar matrices is scalar.

T.4. A matrix $\mathbf{A} = [a_{ij}]$ is called **upper triangular** if $a_{ij} = 0$ for $i > j$. It is called **lower triangular** if $a_{ij} = 0$ for $i < j$.

(a) Prove that the sum and product of upper triangular matrices is upper triangular.

(b) Prove that the sum and product of lower triangular matrices is lower triangular.

T.5. Show that the jth column of the matrix product **AB** is equal to the matrix product $\mathbf{AB}_j$, where $\mathbf{B}_j$ is the jth column of **B**.

T.6. Let **A** be an $m \times n$ matrix whose entries are real numbers. Show that if $\mathbf{AA}^T = \mathbf{0}$ (the $m \times n$ matrix all of whose entries are zero), then $\mathbf{A} = \mathbf{0}$.

T.7. Show that the summation notation satisfies the following properties:

(i) $\displaystyle\sum_{i=1}^{n} (r_i + s_i)a_i = \sum_{i=1}^{n} r_i a_i + \sum_{i=1}^{n} s_i a_i.$

(ii) $\displaystyle\sum_{i=1}^{n} c(r_i a_i) = c\left(\sum_{i=1}^{n} r_i a_i \right).$

T.8. Show that

$$\sum_{i=1}^{n} \sum_{j=1}^{m} a_{ij} = \sum_{j=1}^{m} \sum_{i=1}^{n} a_{ij}.$$

1.3

PROPERTIES OF MATRIX OPERATIONS

In this section we consider the algebraic properties of the matrix operations just defined. Many of these properties are similar to familiar properties of the real numbers. However, there will be striking differences between the set of real numbers and the set of matrices in their algebraic behavior under certain operations, for example, under multiplication (as seen in Section 1.2). Most of the properties will be stated as theorems, whose proofs will be left as exercises.

Theorem 1.1
(Properties of Matrix Addition)

(a) $\mathbf{A} + \mathbf{B} = \mathbf{B} + \mathbf{A}$.

(b) $\mathbf{A} + (\mathbf{B} + \mathbf{C}) = (\mathbf{A} + \mathbf{B}) + \mathbf{C}$.

(c) *There is a unique $m \times n$ matrix $\mathbf{0}$ such that*

$$\mathbf{A} + \mathbf{0} = \mathbf{A} \tag{1}$$

*for any $m \times n$ matrix $\mathbf{A}$. The matrix $\mathbf{0}$ is called the $m \times n$ **zero matrix***

(d) *For each m × n matrix* **A**, *there is a unique m × n matrix* **D** *such that*

$$\mathbf{A} + \mathbf{D} = \mathbf{0}. \tag{2}$$

We shall write **D** *as* −**A** *so that* (2) *can be written as*

$$\mathbf{A} + (-\mathbf{A}) = \mathbf{0}.$$

The matrix −**A** *is called the* **negative** *of* **A**.

Proof

(a) To establish (a), we must prove that the i, jth element of **A** + **B** equals the i, jth element of **B** + **A**. The i, jth element of **A** + **B** is $a_{ij} + b_{ij}$; the i, jth element of **B** + **A** is $b_{ij} + a_{ij}$. Since

$$a_{ij} + b_{ij} = b_{ij} + a_{ij} \qquad (1 \le i \le m, \, 1 \le j \le n),$$

the result follows.

(b) Exercise T.1.

(c) Let **U** = $[u_{ij}]$. Then

$$\mathbf{A} + \mathbf{U} = \mathbf{A}$$

if and only if†

$$a_{ij} + u_{ij} = a_{ij},$$

which holds if and only if $u_{ij} = 0$. Thus **U** is the $m \times n$ matrix all of whose entries are zero; **U** is denoted by **0**.

(d) Exercise T.1.

Example 1 To illustrate (c) of Theorem 1.1, we note that the 2×2 zero matrix is

$$\begin{bmatrix} 0 & 0 \\ 0 & 0 \end{bmatrix}.$$

If

$$\mathbf{A} = \begin{bmatrix} 4 & -1 \\ 2 & 3 \end{bmatrix},$$

we have

$$\begin{bmatrix} 4 & -1 \\ 2 & 3 \end{bmatrix} + \begin{bmatrix} 0 & 0 \\ 0 & 0 \end{bmatrix} = \begin{bmatrix} 4 + 0 & -1 + 0 \\ 2 + 0 & 3 + 0 \end{bmatrix} = \begin{bmatrix} 4 & -1 \\ 2 & 3 \end{bmatrix}.$$

The 2×3 zero matrix is

$$\begin{bmatrix} 0 & 0 & 0 \\ 0 & 0 & 0 \end{bmatrix}.$$

---

† The connector "if and only if" means that both statements are true or both statements are false. Thus (1) if **A** + **U** = **A**, then $a_{ij} + u_{ij} = a_{ij}$ and (2) if $a_{ij} + u_{ij} = a_{ij}$, then **A** + **U** = **A**.

Example 2 To illustrate (d) of Theorem 1.1, let

$$\mathbf{A} = \begin{bmatrix} 2 & 3 & 4 \\ -4 & 5 & -2 \end{bmatrix}.$$

Then

$$-\mathbf{A} = \begin{bmatrix} -2 & -3 & -4 \\ 4 & -5 & 2 \end{bmatrix}.$$

We shall write $\mathbf{A} + (-\mathbf{B})$ as $\mathbf{A} - \mathbf{B}$, and refer to this as the **difference** between $\mathbf{A}$ and $\mathbf{B}$.

Example 3 Let

$$\mathbf{A} = \begin{bmatrix} 3 & -2 & 5 \\ -1 & 2 & 3 \end{bmatrix} \quad \text{and} \quad \mathbf{B} = \begin{bmatrix} 2 & 3 & 2 \\ -3 & 4 & 6 \end{bmatrix}.$$

Then

$$\mathbf{A} - \mathbf{B} = \begin{bmatrix} 3-2 & -2-3 & 5-2 \\ -1+3 & 2-4 & 3-6 \end{bmatrix} = \begin{bmatrix} 1 & -5 & 3 \\ 2 & -2 & -3 \end{bmatrix}.$$

Theorem 1.2 (Properties of Matrix Multiplication)

(a) *If* $\mathbf{A}$, $\mathbf{B}$, *and* $\mathbf{C}$ *are of the appropriate sizes, then*

$$\mathbf{A(BC)} = \mathbf{(AB)C}.$$

(b) *If* $\mathbf{A}$, $\mathbf{B}$, *and* $\mathbf{C}$ *are of the appropriate sizes, then*

$$\mathbf{A(B + C)} = \mathbf{AB} + \mathbf{AC}.$$

(c) *If* $\mathbf{A}$, $\mathbf{B}$, *and* $\mathbf{C}$ *are of the appropriate sizes, then*

$$\mathbf{(A + B)C} = \mathbf{AC} + \mathbf{BC}.$$

Proof
(a) We omit a general proof. Exercise T.2 asks the reader to prove the result for a specific case.
(b) Exercise T.3.
(c) Exercise T.3.

Example 4 Let

$$\mathbf{A} = \begin{bmatrix} 5 & 2 & 3 \\ 2 & -3 & 4 \end{bmatrix}, \quad \mathbf{B} = \begin{bmatrix} 2 & -1 & 1 & 0 \\ 0 & 2 & 2 & 2 \\ 3 & 0 & -1 & 3 \end{bmatrix},$$

and

$$C = \begin{bmatrix} 1 & 0 & 2 \\ 2 & -3 & 0 \\ 0 & 0 & 3 \\ 2 & 1 & 0 \end{bmatrix}.$$

Then

$$A(BC) = \begin{bmatrix} 5 & 2 & 3 \\ 2 & -3 & 4 \end{bmatrix} \begin{bmatrix} 0 & 3 & 7 \\ 8 & -4 & 6 \\ 9 & 3 & 3 \end{bmatrix} = \begin{bmatrix} 43 & 16 & 56 \\ 12 & 30 & 8 \end{bmatrix}$$

and

$$(AB)C = \begin{bmatrix} 19 & -1 & 6 & 13 \\ 16 & -8 & -8 & 6 \end{bmatrix} \begin{bmatrix} 1 & 0 & 2 \\ 2 & -3 & 0 \\ 0 & 0 & 3 \\ 2 & 1 & 0 \end{bmatrix} = \begin{bmatrix} 43 & 16 & 56 \\ 12 & 30 & 8 \end{bmatrix}.$$

Example 5 Let

$$A = \begin{bmatrix} 2 & 2 & 3 \\ 3 & -1 & 2 \end{bmatrix}, \quad B = \begin{bmatrix} 1 & 0 \\ 2 & 2 \\ 3 & -1 \end{bmatrix}, \quad \text{and} \quad C = \begin{bmatrix} -1 & 2 \\ 1 & 0 \\ 2 & -2 \end{bmatrix}.$$

Then

$$A(B + C) = \begin{bmatrix} 2 & 2 & 3 \\ 3 & -1 & 2 \end{bmatrix} \begin{bmatrix} 0 & 2 \\ 3 & 2 \\ 5 & -3 \end{bmatrix} = \begin{bmatrix} 21 & -1 \\ 7 & -2 \end{bmatrix}$$

and

$$AB + AC = \begin{bmatrix} 15 & 1 \\ 7 & -4 \end{bmatrix} + \begin{bmatrix} 6 & -2 \\ 0 & 2 \end{bmatrix} = \begin{bmatrix} 21 & -1 \\ 7 & -2 \end{bmatrix}.$$

The $n \times n$ scalar matrix

$$I_n = \begin{bmatrix} 1 & 0 & \cdots & 0 \\ 0 & 1 & \cdots & 0 \\ \vdots & \vdots & & \vdots \\ 0 & 0 & \cdots & 1 \end{bmatrix},$$

all of whose diagonal elements are 1, is called the **identity matrix of order n.** If A is an $m \times n$ matrix, then it is easy to verify (Exercise T.4) that

$$I_m A = A I_n = A.$$

It is easy to see that every $n \times n$ scalar matrix can be written as rI_n for some r.

Example 6 The identity matrix I_2 of order 2 is

$$I_2 = \begin{bmatrix} 1 & 0 \\ 0 & 1 \end{bmatrix}.$$

If

$$A = \begin{bmatrix} 4 & -2 & 3 \\ 5 & 0 & 2 \end{bmatrix},$$

then

$$I_2A = A.$$

The identity matrix I_3 of order 3 is

$$I_3 = \begin{bmatrix} 1 & 0 & 0 \\ 0 & 1 & 0 \\ 0 & 0 & 1 \end{bmatrix}.$$

Hence

$$AI_3 = A.$$

Suppose that **A** is a square matrix. If p is a positive integer, then we define

$$A^p = \underbrace{A \cdot A \cdots A}_{p \text{ factors}}.$$

If **A** is $n \times n$, we also define

$$A^0 = I_n.$$

For nonnegative integers p and q, the familiar laws of exponents for the real numbers can also be proved for matrix multiplication (Exercise T.5) of a square matrix **A**:

$$A^pA^q = A^{p+q}$$

and

$$(A^p)^q = A^{pq}.$$

It should be noted that the rule

$$(AB)^p = A^pB^p$$

does not hold for square matrices in general. However, if **AB** = **BA**, then this rule does hold. (Exercise T.6).

We now note two other peculiarities of matrix multiplication. If a and b are real numbers, then $ab = 0$ can hold only if a or b is zero. However, this is not true for matrices.

Example 7 If

$$A = \begin{bmatrix} 1 & 2 \\ 2 & 4 \end{bmatrix} \quad \text{and} \quad B = \begin{bmatrix} 4 & -6 \\ -2 & 3 \end{bmatrix},$$

then neither **A** nor **B** is the zero matrix, but

$$\mathbf{AB} = \begin{bmatrix} 0 & 0 \\ 0 & 0 \end{bmatrix}.$$

If a, b and c are real numbers for which $ab = ac$ and $a \neq 0$, it then follows that $b = c$. That is, we can cancel a out. However, the cancellation law does not hold for matrices, as the following example shows.

Example 8 If

$$\mathbf{A} = \begin{bmatrix} 1 & 2 \\ 2 & 4 \end{bmatrix}, \quad \mathbf{B} = \begin{bmatrix} 2 & 1 \\ 3 & 2 \end{bmatrix}, \quad \text{and} \quad \mathbf{C} = \begin{bmatrix} -2 & 7 \\ 5 & -1 \end{bmatrix},$$

then

$$\mathbf{AB} = \mathbf{AC} = \begin{bmatrix} 8 & 5 \\ 16 & 10 \end{bmatrix},$$

but **B** ≠ **C**.

Example 9
(Business) Suppose that only two rival companies, R and S, manufacture a certain product. Each year, company R keeps $\frac{1}{4}$ of its customers while $\frac{3}{4}$ switch to S. Each year, S keeps $\frac{2}{3}$ of its customers while $\frac{1}{3}$ switch to R. This information can be displayed in matrix form as

$$\mathbf{A} = \begin{matrix} & R & S \\ & \begin{bmatrix} \frac{1}{4} & \frac{1}{3} \\ \frac{3}{4} & \frac{2}{3} \end{bmatrix} & \begin{matrix} R \\ S \end{matrix} \end{matrix}.$$

When manufacture of the product first starts, R has $\frac{3}{5}$ of the market (the market is the total number of customers) while S has $\frac{2}{5}$ of the market. We denote the initial distribution of the market by

$$\mathbf{X}_0 = \begin{bmatrix} \frac{3}{5} \\ \frac{2}{5} \end{bmatrix}.$$

One year later, the distribution of the market is

$$\mathbf{X}_1 = \mathbf{AX}_0 = \begin{bmatrix} \frac{1}{4} & \frac{1}{3} \\ \frac{3}{4} & \frac{2}{3} \end{bmatrix} \begin{bmatrix} \frac{3}{5} \\ \frac{2}{5} \end{bmatrix}$$

$$= \begin{bmatrix} \frac{1}{4}(\frac{3}{5}) + \frac{1}{3}(\frac{2}{5}) \\ \frac{3}{4}(\frac{3}{5}) + \frac{2}{3}(\frac{2}{5}) \end{bmatrix} = \begin{bmatrix} \frac{17}{60} \\ \frac{43}{60} \end{bmatrix}.$$

This can be readily seen as follows. Suppose that the initial market consists of k people, say $k = 12,000$, and no change in this number occurs with time. Then, initially, R has $\frac{3}{5}k$ customers, and S has $\frac{2}{5}k$ customers. At the end of the first year, R keeps $\frac{1}{4}$ of its customers and gains $\frac{1}{3}$ of S's customers. Thus R has

$$\tfrac{1}{4}(\tfrac{3}{5}k) + \tfrac{1}{3}(\tfrac{2}{5}k) = [\tfrac{1}{4}(\tfrac{3}{5}) + \tfrac{1}{3}(\tfrac{2}{5})]k = \tfrac{17}{60}k \text{ customers.}$$

When $k = 12,000$, R has $\frac{17}{60}$ (12,000) = 3400 customers. Similarly, at the end of the first year, S keeps $\frac{2}{3}$ of its customers and gains $\frac{3}{4}$ of R's customers. Thus S has

$$\tfrac{3}{4}(\tfrac{3}{5}k) + \tfrac{2}{3}(\tfrac{2}{5}k) = [\tfrac{3}{4}(\tfrac{3}{5}) + \tfrac{2}{3}(\tfrac{2}{5})]k = \tfrac{43}{60}k \text{ customers.}$$

When $k = 12,000$, S has $\frac{43}{60}$ (12,000) = 8600 customers. Similarly, at the end of 2 years, the distribution of the market will be given by $\mathbf{X}_2 = \mathbf{AX}_1 = \mathbf{A}(\mathbf{AX}_0) = \mathbf{A}^2\mathbf{X}_0$.

If $\mathbf{X}_0 = \begin{bmatrix} a \\ b \end{bmatrix}$, can we determine a and b so that the distribution will be the same from year to year? When this happens, the distribution of the market is said to be **stable**. We proceed as follows. Since R and S control the entire market, we must have

$$a + b = 1. \tag{3}$$

We also want the distribution after 1 year to be unchanged. Hence

$$\mathbf{AX}_0 = \mathbf{X}_0$$

or

$$\begin{bmatrix} \tfrac{1}{4} & \tfrac{1}{3} \\ \tfrac{3}{4} & \tfrac{2}{3} \end{bmatrix} \begin{bmatrix} a \\ b \end{bmatrix} = \begin{bmatrix} a \\ b \end{bmatrix}.$$

Then

$$\tfrac{1}{4}a + \tfrac{1}{3}b = a$$
$$\tfrac{3}{4}a + \tfrac{2}{3}b = b$$

or

$$-\tfrac{3}{4}a + \tfrac{1}{3}b = 0$$
$$\tfrac{3}{4}a - \tfrac{1}{3}b = 0. \tag{4}$$

From Equations (3) and (4), we find (verify) that

$$a = \tfrac{4}{13} \quad \text{and} \quad b = \tfrac{9}{13}.$$

The problem just described is an example of a **Markov chain**. We shall return to this topic in Section 7.7, which the reader may pursue at this time.

Theorem 1.3
(Properties of Scalar Multiplication)
If r and s are real numbers and **A** *and* **B** *are matrices, then*
(a) $r(s\mathbf{A}) = (rs)\mathbf{A}$.
(b) $(r + s)\mathbf{A} = r\mathbf{A} + s\mathbf{A}$.
(c) $r(\mathbf{A} + \mathbf{B}) = r\mathbf{A} + r\mathbf{B}$.
(d) $\mathbf{A}(r\mathbf{B}) = r(\mathbf{AB}) = (r\mathbf{A})\mathbf{B}$.

Proof
Exercise T.8.

Example 10 Let $r = -2$,

$$\mathbf{A} = \begin{bmatrix} 1 & 2 & 3 \\ -2 & 0 & 1 \end{bmatrix} \quad \text{and} \quad \mathbf{B} = \begin{bmatrix} 2 & -1 \\ 1 & 4 \\ 0 & -2 \end{bmatrix}.$$

Then

$$\mathbf{A}(r\mathbf{B}) = \begin{bmatrix} 1 & 2 & 3 \\ -2 & 0 & 1 \end{bmatrix} \begin{bmatrix} -4 & 2 \\ -2 & -8 \\ 0 & 4 \end{bmatrix} = \begin{bmatrix} -8 & -2 \\ 8 & 0 \end{bmatrix}$$

and

$$r(\mathbf{AB}) = (-2) \begin{bmatrix} 4 & 1 \\ -4 & 0 \end{bmatrix} = \begin{bmatrix} -8 & -2 \\ 8 & 0 \end{bmatrix},$$

which illustrates (d) of Theorem 1.3.

It is easy to show that $(-1)\mathbf{A} = -\mathbf{A}$ (Exercise T.9).

Theorem 1.4
(Properties
of Transpose)
If r is a scalar and $\mathbf{A}$ *and* $\mathbf{B}$ *are matrices, then*
(a) $(\mathbf{A}^T)^T = \mathbf{A}$.
(b) $(\mathbf{A} + \mathbf{B})^T = \mathbf{A}^T + \mathbf{B}^T$.
(c) $(\mathbf{AB})^T = \mathbf{B}^T\mathbf{A}^T$.
(d) $(r\mathbf{A})^T = r\mathbf{A}^T$.

Proof
We leave the proofs of (a), (b), and (d) as an exercise (Exercise T.10) and only prove (c) here. Thus let $\mathbf{A} = [a_{ij}]$ be $m \times p$ and let $\mathbf{B} = [b_{ij}]$ be $p \times n$. The i, jth element of $(\mathbf{AB})^T$ is c_{ij}^T. Now

$$c_{ij}^T = c_{ji} = a_{j1}b_{1i} + a_{j2}b_{2i} + \cdots + a_{jp}b_{pi}$$
$$= a_{1j}^T b_{i1}^T + a_{2j}^T b_{i2}^T + \cdots + a_{pj}^T b_{ip}^T$$
$$= b_{i1}^T a_{1j}^T + b_{i2}^T a_{2j}^T + \cdots + b_{ip}^T a_{pj}^T,$$

which is the i, jth element of $\mathbf{B}^T\mathbf{A}^T$.

Example 11 Let

$$\mathbf{A} = \begin{bmatrix} 1 & 3 & 2 \\ 2 & -1 & 3 \end{bmatrix} \quad \text{and} \quad \mathbf{B} = \begin{bmatrix} 0 & 1 \\ 2 & 2 \\ 3 & -1 \end{bmatrix}.$$

Then

$$(\mathbf{AB})^T = \begin{bmatrix} 12 & 7 \\ 5 & -3 \end{bmatrix}$$

and

$$\mathbf{B}^T\mathbf{A}^T = \begin{bmatrix} 0 & 2 & 3 \\ 1 & 2 & -1 \end{bmatrix} \begin{bmatrix} 1 & 2 \\ 3 & -1 \\ 2 & 3 \end{bmatrix} = \begin{bmatrix} 12 & 7 \\ 5 & -3 \end{bmatrix}.$$

Definition   A matrix $\mathbf{A} = [a_{ij}]$ is called **symmetric** if

$$\mathbf{A}^T = \mathbf{A}.$$

That is, $\mathbf{A}$ is symmetric if it is a square matrix for which

$$a_{ij} = a_{ji} \qquad \text{(Exercise T.11)}.$$

If $\mathbf{A}$ is symmetric, then the elements of $\mathbf{A}$ are symmetric with respect to the main diagonal of $\mathbf{A}$.

Example 12   The matrices

$$\mathbf{A} = \begin{bmatrix} 1 & 2 & 3 \\ 2 & 4 & 5 \\ 3 & 5 & 6 \end{bmatrix} \quad \text{and} \quad \mathbf{I}_3 = \begin{bmatrix} 1 & 0 & 0 \\ 0 & 1 & 0 \\ 0 & 0 & 1 \end{bmatrix}$$

are symmetric.

The matrix in Example 4 of Section 1.2 is symmetric.

OTHER APPLICATIONS.   At this point, the reader may also turn to a more extensive application: Section 7.3, Graph Theory. This new area of applied mathematics is being widely used in the solution of many problems in business, the social sciences, and the physical sciences.

1.3 EXERCISES

1. Verify Theorem 1.1 for

$$\mathbf{A} = \begin{bmatrix} 1 & 2 & -2 \\ 3 & 4 & 5 \end{bmatrix}, \qquad \mathbf{B} = \begin{bmatrix} 2 & 0 & 1 \\ 3 & -2 & 5 \end{bmatrix},$$

and

$$\mathbf{C} = \begin{bmatrix} -4 & -6 & 1 \\ 2 & 3 & 0 \end{bmatrix}.$$

2. Verify (a) of Theorem 1.2 for

$$\mathbf{A} = \begin{bmatrix} 1 & 3 \\ 2 & -1 \end{bmatrix}, \qquad \mathbf{B} = \begin{bmatrix} -1 & 3 & 2 \\ 1 & -3 & 4 \end{bmatrix},$$

and

$$\mathbf{C} = \begin{bmatrix} 1 & 0 \\ 3 & -1 \\ 1 & 2 \end{bmatrix}.$$

3. Verify (b) of Theorem 1.2 for

$$\mathbf{A} = \begin{bmatrix} 1 & -3 \\ -3 & 4 \end{bmatrix}, \qquad \mathbf{B} = \begin{bmatrix} 2 & -3 & 2 \\ 3 & -1 & -2 \end{bmatrix},$$

and

$$\mathbf{C} = \begin{bmatrix} 0 & 1 & 2 \\ 1 & 3 & -2 \end{bmatrix}.$$

4. Verify (a), (b), and (c) of Theorem 1.3 for $r = 6$, $s = -2$, and

$$\mathbf{A} = \begin{bmatrix} 4 & 2 \\ 1 & -3 \end{bmatrix}, \qquad \mathbf{B} = \begin{bmatrix} 0 & 2 \\ -4 & 3 \end{bmatrix}.$$

5. Verify (d) of Theorem 1.3 for $r = -3$ and

$$\mathbf{A} = \begin{bmatrix} 1 & 3 \\ 2 & -1 \end{bmatrix}, \qquad \mathbf{B} = \begin{bmatrix} -1 & 3 & 2 \\ 1 & -3 & 4 \end{bmatrix}.$$

6. Verify (b) and (d) of Theorem 1.4 for $r = -4$ and

$$\mathbf{A} = \begin{bmatrix} 1 & 3 & 2 \\ 2 & 1 & -3 \end{bmatrix}, \qquad \mathbf{B} = \begin{bmatrix} 4 & 2 & -1 \\ -2 & 1 & 5 \end{bmatrix}.$$

7. Verify (c) of Theorem 1.4 for

$$\mathbf{A} = \begin{bmatrix} 1 & 3 & 2 \\ 2 & 1 & -3 \end{bmatrix}, \qquad \mathbf{B} = \begin{bmatrix} 3 & -1 \\ 2 & 4 \\ 1 & 2 \end{bmatrix}.$$

8. If $\mathbf{A} = \begin{bmatrix} -2 & 3 \\ 2 & -3 \end{bmatrix}$ and $\mathbf{B} = \begin{bmatrix} 3 & 6 \\ 2 & 4 \end{bmatrix}$, show that $\mathbf{AB} = \mathbf{0}$.

9. If $\mathbf{A} = \begin{bmatrix} -2 & 3 \\ 2 & -3 \end{bmatrix}$, $\mathbf{B} = \begin{bmatrix} -1 & 3 \\ 2 & 0 \end{bmatrix}$, and $\mathbf{C} = \begin{bmatrix} -4 & -3 \\ 0 & -4 \end{bmatrix}$, show that $\mathbf{AB} = \mathbf{AC}$.

10. If $\mathbf{A} = \begin{bmatrix} 0 & 1 \\ 1 & 0 \end{bmatrix}$, show that $\mathbf{A}^2 = \mathbf{I}_2$.

11. Let $\mathbf{A} = \begin{bmatrix} 4 & 2 \\ 1 & 3 \end{bmatrix}$. Find (a) $\mathbf{A}^2 + 3\mathbf{A}$, (b) $2\mathbf{A}^3 + 3\mathbf{A}^2 + 4\mathbf{A} + 5\mathbf{I}_2$.

12. Let $\mathbf{A} = \begin{bmatrix} 1 & -1 \\ 2 & 3 \end{bmatrix}$. Find

(a) $\mathbf{A}^2 - 2\mathbf{A}$. (b) $3\mathbf{A}^3 - 2\mathbf{A}^2 + 5\mathbf{A} - 4\mathbf{I}_2$.

13. Suppose that the matrix $\mathbf{A}$ in Example 9 is

$$\mathbf{A} = \begin{bmatrix} \frac{1}{3} & \frac{2}{5} \\ \frac{2}{3} & \frac{3}{5} \end{bmatrix} \qquad \text{and} \qquad \mathbf{X}_0 = \begin{bmatrix} \frac{2}{3} \\ \frac{1}{3} \end{bmatrix}.$$

(a) Find the distribution of the market after 1 year.
(b) Find the stable distribution of the market.

14. Consider two quick food companies, M and N. Each year, company M keeps $\frac{1}{3}$ of its customers, while $\frac{2}{3}$ switch to N. Each year, N keeps $\frac{1}{2}$ of its customers, while $\frac{1}{2}$ switch to M. Suppose that the initial distribution of the market is given by

$$\mathbf{X}_0 = \begin{bmatrix} \frac{1}{3} \\ \frac{2}{3} \end{bmatrix}.$$

(a) Find the distribution of the market after 1 year.
(b) Find the stable distribution of the market.

THEORETICAL EXERCISES

T.1. Prove properties (b) and (d) of Theorem 1.1.

T.2. If $\mathbf{A} = [a_{ij}]$ is a 2×3 matrix, $\mathbf{B} = [b_{ij}]$ is a 3×4 matrix, and $\mathbf{C} = [c_{ij}]$ is a 4×3 matrix, show that

$$\mathbf{A}(\mathbf{BC}) = (\mathbf{AB})\mathbf{C}.$$

T.3. Prove properties (b) and (c) of Theorem 1.2.

T.4. If $\mathbf{A}$ is an $m \times n$ matrix, show that

$$\mathbf{AI}_n = \mathbf{I}_m\mathbf{A} = \mathbf{A}.$$

T.5. Let p and q be nonnegative integers and let $\mathbf{A}$ be a square matrix. Show that

$$\mathbf{A}^p\mathbf{A}^q = \mathbf{A}^{p+q} \qquad \text{and} \qquad (\mathbf{A}^p)^q = \mathbf{A}^{pq}.$$

T.6. If $\mathbf{AB} = \mathbf{BA}$, and p is a nonnegative integer, show that $(\mathbf{AB})^p = \mathbf{A}^p\mathbf{B}^p$.

T.7. If p is a nonnegative integer and c is a scalar, show that

$$(c\mathbf{A})^p = c^p\mathbf{A}^p.$$

T.8. Prove Theorem 1.3.

T.9. Show that $(-1)\mathbf{A} = -\mathbf{A}$.

T.10. Complete the proof of Theorem 1.4.

T.11. Show that a square matrix $\mathbf{A}$ is symmetric if and only if $a_{ij} = a_{ji}$ for all i, j.

T.12. Prove that if $\mathbf{A}$ is symmetric, then $\mathbf{A}^T$ is symmetric.

T.13. Let $\mathbf{A}$ and $\mathbf{B}$ be symmetric matrices.
(a) Show that $\mathbf{A} + \mathbf{B}$ is symmetric.
(b) Show that $\mathbf{AB}$ is symmetric if and only if $\mathbf{AB} = \mathbf{BA}$.

T.14. A matrix $\mathbf{A} = [a_{ij}]$ is called **skew symmetric** if $\mathbf{A}^T = -\mathbf{A}$. Show that $\mathbf{A}$ is skew symmetric if and only if $a_{ij} = -a_{ji}$ for all i, j.

T.15. If $\mathbf{A}$ is an $n \times n$ matrix, prove that
(a) $\mathbf{AA}^T$ and $\mathbf{A}^T\mathbf{A}$ are symmetric.
(b) $\mathbf{A} + \mathbf{A}^T$ is symmetric.
(c) $\mathbf{A} - \mathbf{A}^T$ is skew symmetric.

T.16. Prove that if $\mathbf{A}$ is an $n \times n$ matrix, then $\mathbf{A}$ can be written uniquely as $\mathbf{A} = \mathbf{S} + \mathbf{K}$, where $\mathbf{S}$ is symmetric and $\mathbf{K}$ is skew symmetric.

T.17. Prove that if $\mathbf{A}$ is an $n \times n$ scalar matrix, then $\mathbf{A} = r\mathbf{I}_n$ for some real number r.

T.18. Show that
$$\mathbf{I}_n^T = \mathbf{I}_n.$$

T.19. Let A be an $m \times n$ matrix. Show that if $r\mathbf{A} = \mathbf{0}$, then $r = 0$ or $\mathbf{A} = \mathbf{0}$.

T.20. Show that if $\mathbf{AX} = \mathbf{B}$ is a linear system that has more than one solution, then it has infinitely many solutions.
(*Hint:* If $\mathbf{X}_1$ and $\mathbf{X}_2$ are solutions, consider $\mathbf{X}_3 = r\mathbf{X}_1 + s\mathbf{X}_2$, where $r + s = 1$.)

T.21. Determine all 2×2 matrices $\mathbf{A}$ such that $\mathbf{AB} = \mathbf{BA}$ for any 2×2 matrix $\mathbf{B}$.

1.4

SOLUTIONS OF EQUATIONS

In this section we shall systematize the familiar method of elimination of unknowns (discussed in Section 1.1) for the solution of linear systems and thus obtain a useful method for solving such systems. This method starts with the augmented matrix of the given linear system and obtains a matrix of a certain special form. This new matrix represents a linear system that has exactly the same solutions as the given system. For example, if

$$\begin{bmatrix} 1 & 0 & 0 & 2 & \vdots & 4 \\ 0 & 1 & 0 & -1 & \vdots & -5 \\ 0 & 0 & 1 & 3 & \vdots & 6 \end{bmatrix}$$

represents the augmented matrix of a linear system, then the solution is easily found from the corresponding equations

$$\begin{aligned} x_1 \quad\quad + 2x_4 &= 4 \\ x_2 \quad - x_4 &= -5 \\ x_3 + 3x_4 &= 6. \end{aligned}$$

The task of this section is to manipulate the augmented matrix representing a given linear system into a form from which the solution can easily be found.

Definition   An $m \times n$ matrix is said to be in **reduced row echelon form** when it satisfies the following properties:

(a) All rows consisting entirely of zeros, if any, are at the bottom of the matrix.

(b) The first nonzero entry in each row that does not consist entirely of zeros is a 1, called the **leading entry** of its row.

(c) If rows i and $i + 1$ are two successive rows that do not consist entirely of zeros, then the leading entry of row $i + 1$ is to the right of the leading entry of row i.

(d) If a column contains a leading entry of some row, then all other entries in that column are zero.

Note that a matrix in reduced row echelon form might not have any rows that consist entirely of zeros.

Example 1 The matrices

$$A = \begin{bmatrix} 1 & 0 & 0 & 4 \\ 0 & 1 & 0 & 5 \\ 0 & 0 & 1 & 2 \end{bmatrix}, \quad B = \begin{bmatrix} 1 & 2 & 0 & 0 & 2 \\ 0 & 0 & 1 & 0 & 1 \\ 0 & 0 & 0 & 1 & 0 \end{bmatrix},$$

and

$$C = \begin{bmatrix} 1 & 0 & 0 & 3 & 0 \\ 0 & 0 & 1 & 0 & 0 \\ 0 & 0 & 0 & 0 & 1 \\ 0 & 0 & 0 & 0 & 0 \\ 0 & 0 & 0 & 0 & 0 \end{bmatrix}$$

are in reduced row echelon form.

Example 2 The matrices

$$A = \begin{bmatrix} 1 & 2 & 0 & 4 \\ 0 & 0 & 0 & 0 \\ 0 & 0 & 1 & -3 \end{bmatrix}, \quad B = \begin{bmatrix} 1 & 0 & 3 & 4 \\ 0 & 2 & -2 & 5 \\ 0 & 0 & 1 & 2 \end{bmatrix},$$

$$C = \begin{bmatrix} 1 & 0 & 3 & 4 \\ 0 & 1 & -2 & 5 \\ 0 & 1 & 2 & 2 \\ 0 & 0 & 0 & 0 \end{bmatrix}, \quad \text{and} \quad D = \begin{bmatrix} 1 & 2 & 3 & 4 \\ 0 & 1 & -2 & 5 \\ 0 & 0 & 1 & 2 \\ 0 & 0 & 0 & 0 \end{bmatrix}$$

are not in reduced row echelon form, since they fail to satisfy (a), (b), (c), and (d), respectively.

Observe that when a matrix is in reduced row echelon form, the leading entries of its nonzero rows form a staircase (''echelon'') pattern.

We shall now turn to the discussion of how to transform a given matrix to a matrix in reduced row echelon form.

Definition   An **elementary row operation** on an $m \times n$ matrix $\mathbf{A} = [a_{ij}]$ is any one of the following operations:

(a) Interchange rows r and s of $\mathbf{A}$. That is, replace $a_{r1}, a_{r2}, \ldots, a_{rn}$ by $a_{s1}, a_{s2}, \ldots, a_{sn}$ and $a_{s1}, a_{s2}, \ldots, a_{sn}$ by $a_{r1}, a_{r2}, \ldots, a_{rn}$.

(b) Multiply row r of $\mathbf{A}$ by $c \neq 0$. That is, replace $a_{r1}, a_{r2}, \ldots, a_{rn}$ by $ca_{r1}, ca_{r2}, \ldots, ca_{rn}$.

(c) Add d times row r of $\mathbf{A}$ to row s of $\mathbf{A}$, $r \neq s$. That is, replace $a_{s1}, a_{s2}, \ldots, a_{sn}$ by $a_{s1} + da_{r1}, a_{s2} + da_{r2}, \ldots, a_{sn} + da_{rn}$.

Observe that when a matrix is viewed as the augmented matrix of a linear system, the elementary row operations are equivalent, respectively, to interchanging two equations, multiplying an equation by a nonzero constant, and adding a multiple of one equation to another equation.

Example 3   Let

$$\mathbf{A} = \begin{bmatrix} 0 & 0 & 1 & 2 \\ 2 & 3 & 0 & -2 \\ 3 & 3 & 6 & -9 \end{bmatrix}.$$

Interchanging rows 1 and 3 of $\mathbf{A}$, we obtain

$$\mathbf{B} = \begin{bmatrix} 3 & 3 & 6 & -9 \\ 2 & 3 & 0 & -2 \\ 0 & 0 & 1 & 2 \end{bmatrix}.$$

Multiplying the third row of $\mathbf{A}$ by $\frac{1}{3}$, we obtain

$$\mathbf{C} = \begin{bmatrix} 0 & 0 & 1 & 2 \\ 2 & 3 & 0 & -2 \\ 1 & 1 & 2 & -3 \end{bmatrix}.$$

Adding -2 times row 2 of $\mathbf{A}$ to row 3 of $\mathbf{A}$, we obtain

$$\mathbf{D} = \begin{bmatrix} 0 & 0 & 1 & 2 \\ 2 & 3 & 0 & -2 \\ -1 & -3 & 6 & -5 \end{bmatrix}.$$

Definition   An $m \times n$ matrix $\mathbf{A}$ is said to be **row equivalent** to an $m \times n$ matrix $\mathbf{B}$ if $\mathbf{B}$ can be obtained by applying a finite sequence of elementary row operations to $\mathbf{A}$.

Example 4   The matrix

$$\mathbf{A} = \begin{bmatrix} 1 & 2 & 4 & 3 \\ 2 & 1 & 3 & 2 \\ 1 & -1 & 2 & 3 \end{bmatrix} \text{ is row equivalent to } \mathbf{D} = \begin{bmatrix} 2 & 4 & 8 & 6 \\ 1 & -1 & 2 & 3 \\ 4 & -1 & 7 & 8 \end{bmatrix},$$

because if we add twice row 3 of **A** to its second row, we obtain

$$\mathbf{B} = \begin{bmatrix} 1 & 2 & 4 & 3 \\ 4 & -1 & 7 & 8 \\ 1 & -1 & 2 & 3 \end{bmatrix}.$$

Interchanging rows 2 and 3 of **B**, we obtain

$$\mathbf{C} = \begin{bmatrix} 1 & 2 & 4 & 3 \\ 1 & -1 & 2 & 3 \\ 4 & -1 & 7 & 8 \end{bmatrix}.$$

Multiplying row 1 of **C** by 2, we obtain **D**.

It is easy to show (Exercise T.2) that (1) every matrix is row equivalent to itself; (2) if **A** is row equivalent to **B**, then **B** is row equivalent to **A**; and (3) if **A** is row equivalent to **B** and **B** is row equivalent to **C**, then **A** is row equivalent to **C**. In view of (2), both statements, "**A** is row equivalent to **B**" and "**B** is row equivalent to **A**," can be replaced by "**A** and **B** are row equivalent."

Theorem 1.5 *Every nonzero m × n matrix is row equivalent to a unique matrix in reduced row echelon form.*

We shall illustrate the proof of the theorem by giving the steps that must be carried out on a specific matrix **A** to obtain a matrix in reduced row echelon form that is row equivalent to **A**. We omit the proof that the matrix thus obtained is unique. We use the following example to illustrate the steps involved.

Example 5

$$\mathbf{A} = \begin{bmatrix} 0 & 2 & 3 & -4 & 1 \\ 0 & 0 & 2 & 3 & 4 \\ 2 & 2 & -5 & 2 & 4 \\ 2 & 0 & -6 & 9 & 7 \end{bmatrix}.$$

The procedure for transforming a matrix to reduced row echelon form is as follows.

PROCEDURE	EXAMPLE
STEP 1. Find the first (counting from left to right) column in **A** not all of whose entries are zero. This column is called the **pivotal column**.	$$\mathbf{A} = \begin{bmatrix} 0 & 2 & 3 & -4 & 1 \\ 0 & 0 & 2 & 3 & 4 \\ 2 & 2 & -5 & 2 & 4 \\ 2 & 0 & -6 & 9 & 7 \end{bmatrix}.$$ Pivotal ⎯ column of **A**

STEP 2. Identify the first (counting from top to bottom) nonzero entry in the pivotal column. This element is called the **pivot**, which we circle in $\mathbf{A}$.

$$\mathbf{A} \underset{\mathbf{Pivot}}{\longrightarrow} \begin{bmatrix} 0 & 2 & 3 & -4 & 1 \\ 0 & 0 & 2 & 3 & 4 \\ ② & 2 & -5 & 2 & 4 \\ 2 & 0 & -6 & 9 & 7 \end{bmatrix}.$$

STEP 3. Interchange, if necessary, the first row with the row where the pivot occurs so that the pivot is now in the first row. Call the new matrix $\mathbf{A}_1$.

$$\mathbf{A}_1 = \begin{bmatrix} ② & 2 & -5 & 2 & 4 \\ 0 & 0 & 2 & 3 & 4 \\ 0 & 2 & 3 & -4 & 1 \\ 2 & 0 & -6 & 9 & 7 \end{bmatrix}.$$

The first and third rows of $\mathbf{A}$ were interchanged.

STEP 4. Divide the first row of $\mathbf{A}_1$ by the pivot. Thus the entry in the first row and pivotal column (where the pivot was located) is now a 1. Call the new matrix $\mathbf{A}_2$.

$$\mathbf{A}_2 = \begin{bmatrix} 1 & 1 & -\frac{5}{2} & 1 & 2 \\ 0 & 0 & 2 & 3 & 4 \\ 0 & 2 & 3 & -4 & 1 \\ 2 & 0 & -6 & 9 & 7 \end{bmatrix}.$$

The first row of $\mathbf{A}_1$ was divided by 2.

STEP 5. Add multiples of the first row of $\mathbf{A}_2$ to all other rows to make all entries in the pivotal column, except the entry where the pivot was located, equal to zero. Thus all entries in the pivotal column and rows 2, 3, . . . , m, are zero. Call the new matrix $\mathbf{A}_3$.

$$\mathbf{A}_3 = \begin{bmatrix} 1 & 1 & -\frac{5}{2} & 1 & 2 \\ 0 & 0 & 2 & 3 & 4 \\ 0 & 2 & 3 & -4 & 1 \\ 0 & -2 & -1 & 7 & 3 \end{bmatrix}.$$

-2 times the first row of $\mathbf{A}_2$ was added to its fourth row.

STEP 6. Identify $\mathbf{B}$ as the $(m - 1) \times n$ submatrix of $\mathbf{A}_3$ obtained by deleting the first row of $\mathbf{A}_3$; do not erase the first row of $\mathbf{A}_3$. Repeat steps 1 through 5 on $\mathbf{B}$.

$$\mathbf{B} = \begin{bmatrix} 1 & 1 & -\frac{5}{2} & 1 & 2 \\ 0 & 0 & 2 & 3 & 4 \\ 0 & ② & 3 & -4 & 1 \\ 0 & -2 & -1 & 7 & 3 \end{bmatrix}.$$

Pivotal column of $\mathbf{B}$
Pivot

$$\mathbf{B}_1 = \begin{bmatrix} 1 & 1 & -\frac{5}{2} & 1 & 2 \\ 0 & ② & 3 & -4 & 1 \\ 0 & 0 & 2 & 3 & 4 \\ 0 & -2 & -1 & 7 & 3 \end{bmatrix}.$$

The first and second rows of $\mathbf{B}$ were interchanged.

$$\mathbf{B}_2 = \begin{bmatrix} 1 & 1 & -\frac{5}{2} & 1 & 2 \\ 0 & 1 & \frac{3}{2} & -2 & \frac{1}{2} \\ 0 & 0 & 2 & 3 & 4 \\ 0 & -2 & -1 & 7 & 3 \end{bmatrix}.$$

The first row of $\mathbf{B}_1$ was divided by 2.

$$\mathbf{B}_3 = \begin{bmatrix} 1 & 1 & -\frac{5}{2} & 1 & 2 \\ 0 & 1 & \frac{3}{2} & -2 & \frac{1}{2} \\ 0 & 0 & 2 & 3 & 4 \\ 0 & 0 & 2 & 3 & 4 \end{bmatrix}.$$

2 times the first row of $\mathbf{B}_2$ was added to its third row.

STEP 7. Add multiples of the first row of $\mathbf{B}_3$ to all the rows of $\mathbf{A}_3$ above the first row of $\mathbf{B}_3$ so that all entries in the pivotal column, except for the pivot, become zero.

$$\mathbf{B}_3 = \begin{bmatrix} 1 & 0 & -4 & 3 & \frac{3}{2} \\ 0 & 1 & \frac{3}{2} & -2 & \frac{1}{2} \\ 0 & 0 & 2 & 3 & 4 \\ 0 & 0 & 2 & 3 & 4 \end{bmatrix}.$$

(-1) times the first row of $\mathbf{B}_3$ was added to the shaded row.

STEP 8. Identify $\mathbf{C}$ as the $(m - 2) \times n$ submatrix of $\mathbf{B}_3$ obtained by deleting the first row of $\mathbf{B}_3$; do not erase the first row of $\mathbf{B}_3$. Repeat Steps 1–7 on $\mathbf{C}$.

$$\mathbf{C} = \begin{bmatrix} 1 & 0 & -4 & 3 & \frac{3}{2} \\ 0 & 1 & \frac{3}{2} & -2 & \frac{1}{2} \\ 0 & 0 & ② & 3 & 4 \\ 0 & 0 & 2 & 3 & 4 \end{bmatrix}.$$

Pivotal column of $\mathbf{C}$

Pivot

$$\mathbf{C_1} = \mathbf{C_2} = \begin{bmatrix} 1 & 0 & -4 & 3 & \frac{3}{2} \\ 0 & 1 & \frac{3}{2} & -2 & \frac{1}{2} \\ 0 & 0 & 1 & \frac{3}{2} & 2 \\ 0 & 0 & 2 & 3 & 4 \end{bmatrix}.$$

No rows of $\mathbf{C}$ had to be interchanged. The first row of $\mathbf{C}$ was divided by 2.

$$\mathbf{C_3} = \begin{bmatrix} 1 & 0 & -4 & 3 & \frac{3}{2} \\ 0 & 1 & \frac{3}{2} & -2 & \frac{1}{2} \\ 0 & 0 & 1 & \frac{3}{2} & 2 \\ 0 & 0 & 0 & 0 & 0 \end{bmatrix}.$$

-2 times the first row of $\mathbf{C_2}$ was added to its second row.

$$\mathbf{C_3} = \begin{bmatrix} 1 & 0 & 0 & 9 & \frac{19}{2} \\ 0 & 1 & 0 & -\frac{17}{4} & -\frac{5}{2} \\ 0 & 0 & 1 & \frac{3}{2} & 2 \\ 0 & 0 & 0 & 0 & 0 \end{bmatrix}.$$

4 times the first row of $\mathbf{C_3}$ was added to the first shaded row; $-\frac{3}{2}$ times the first row of $\mathbf{C_3}$ was added to the second shaded row.

The final matrix,

$$\begin{bmatrix} 1 & 0 & 0 & 9 & \frac{19}{2} \\ 0 & 1 & 0 & -\frac{17}{4} & -\frac{5}{2} \\ 0 & 0 & 1 & \frac{3}{2} & 2 \\ 0 & 0 & 0 & 0 & 0 \end{bmatrix},$$

is in reduced row echelon form.

The process just illustrated is used as many times as necessary until the matrix in reduced row echelon form is obtained.

We now apply these results to the solution of linear systems.

Theorem 1.6 *Let* **AX** = **B** *and* **CX** = **D** *be two linear systems each of m equations in n unknowns. If the augmented matrices* [**A** ⋮ **B**] *and* [**C** ⋮ **D**] *of these systems are row equivalent, then both linear systems have exactly the same solutions.*

Proof

This follows from the definition of row equivalence and from the fact that the three elementary row operations on the augmented matrix turn out to be the three manipulations on a linear system, discussed in Section 1.1, yielding a linear system having the same solutions as the given system. We also note that if one system has no solution, then the other system has no solution.

Corollary 1.1 *If* **A** *and* **C** *are row equivalent m* × *n matrices, then the linear systems* **AX** = **0** *and* **CX** = **0** *have exactly the same solutions.*

Proof
Exercise T.3.

The **Gauss–Jordan reduction** procedure for solving the linear system **AX** = **B** is as follows.

STEP 1. Form the augmented matrix [**A** ⋮ **B**].

STEP 2. Transform the augmented matrix to reduced row echelon form by using elementary row operations.

STEP 3. The linear system that corresponds to the matrix in reduced row echelon form that has been obtained in Step 2 has exactly the same solutions as the given linear system. For each nonzero row of the matrix in reduced row echelon form, solve the corresponding equation for the unknown that corresponds to the leading entry of the row. The rows consisting entirely of zeros can be ignored, since the corresponding equation will be satisfied for any values of the unknowns.

The following examples illustrate the Gauss–Jordan reduction procedure.

Example 6 Solve the linear system

$$\begin{align}
x + 2y + 3z &= 9 \\
2x - y + z &= 8 \\
3x \qquad - z &= 3.
\end{align} \tag{1}$$

by Gauss–Jordan reduction.

Solution

STEP 1. The augmented matrix of this linear system is

$$\begin{bmatrix} 1 & 2 & 3 & | & 9 \\ 2 & -1 & 1 & | & 8 \\ 3 & 0 & -1 & | & 3 \end{bmatrix}.$$

STEP 2. The augmented matrix is row equivalent to the matrix (verify)

$$\begin{bmatrix} 1 & 0 & 0 & | & 2 \\ 0 & 1 & 0 & | & -1 \\ 0 & 0 & 1 & | & 3 \end{bmatrix} \tag{2}$$

in reduced row echelon form.

STEP 3. The linear system represented by (2) is

$$\begin{aligned} x & & & = 2 \\ & y & & = -1 \\ & & z & = 3 \end{aligned}$$

so that the unique solution to the given linear system (1) is

$$\begin{aligned} x &= 2 \\ y &= -1 \\ z &= 3. \end{aligned}$$

Example 7 Solve the linear system

$$\begin{aligned} x + y + 2z - 5w &= 3 \\ 2x + 5y - z - 9w &= -3 \\ 2x + y - z + 3w &= -11 \\ x - 3y + 2z + 7w &= -5. \end{aligned} \tag{3}$$

by Gauss†–Jordan‡ reduction.

† Carl Friedrich Gauss (1777–1855) was born into a poor working class family in Brunswick, Germany, and died in Göttingen, Germany, the most famous mathematician in the world. He was a child prodigy with a genius that did not impress his father, who called him a "star-gazer." However, his teachers were impressed enough to arrange for the Duke of Brunswick to provide a scholarship for Gauss at the local secondary school. As a teenager there, he made original discoveries in number theory and began to speculate about non-Euclidean geometry. His scientific publications include important contributions in number theory, mathematical astronomy, mathematical geography, statistics, differential geometry, and magnetism. His diaries and private notes contain many other discoveries that he never published.

An austere, conservative man who had few friends and whose private life was generally unhappy, he was very concerned that proper credit be given for scientific discoveries. When he relied on the results of others, he was careful to acknowledge them, and when others independently discovered results in his private notes he was quick to claim priority.

In his research he used a method of calculation that later generations generalized to row reduction of matrices and named in his honor although the method was used in China almost 2000 years earlier.

‡ Wilhelm Jordan (1842–1899) was born in southern Germany. He attended college in Stuttgart and in 1868 became full professor of geodesy at the technical college in Karlsruhe, Germany. He participated in surveying several regions of Germany. Jordan was a prolific writer whose major work, *Handbuch der Vermessungskunde*

Solution

STEP 1. The augmented matrix of this linear system is

$$\left[\begin{array}{cccc:c} 1 & 1 & 2 & -5 & 3 \\ 2 & 5 & -1 & -9 & -3 \\ 2 & 1 & -1 & 3 & -11 \\ 1 & -3 & 2 & 7 & -5 \end{array}\right].$$

STEP 2. The augmented matrix is row equivalent to the matrix (verify)

$$\left[\begin{array}{cccc:c} 1 & 0 & 0 & 2 & -5 \\ 0 & 1 & 0 & -3 & 2 \\ 0 & 0 & 1 & -2 & 3 \\ 0 & 0 & 0 & 0 & 0 \end{array}\right] \tag{4}$$

in reduced row echelon form.

STEP 3. The linear system represented by (4) is

$$\begin{aligned} x \quad\quad\quad + 2w &= -5 \\ y \quad\quad - 3w &= 2 \\ z - 2w &= 3. \end{aligned}$$

The row in (4) consisting entirely of zeros has been ignored.

Solving each equation for the unknown that corresponds to the leading entry in each row of (4), we obtain

$$\begin{aligned} x &= -5 - 2r \\ y &= 2 + 3r \\ z &= 3 + 2r \\ w &= r, \end{aligned} \tag{5}$$

where r is any real number. Thus (5) is the solution to the given linear system. Since r can be assigned any real number, the given linear system (3) has infinitely many solutions.

Example 8 Solve the linear system

$$\begin{aligned} x_1 + 2x_2 \quad\quad - 3x_4 + x_5 \quad\quad &= 2 \\ x_1 + 2x_2 + x_3 - 3x_4 + x_5 + 2x_6 &= 3 \\ x_1 + 2x_2 \quad\quad - 3x_4 + 2x_5 + x_6 &= 4 \\ 3x_1 + 6x_2 + x_3 - 9x_4 + 4x_5 + 3x_6 &= 9. \end{aligned} \tag{6}$$

by Gauss–Jordan reduction.

(*Handbook of Geodesy*), was translated into French, Italian, and Russian. He was considered a superb writer and an excellent teacher. Unfortunately, the Gauss–Jordan reduction method has been widely attributed to Camille Jordan (1838–1922), a well-known French mathematician. Moreover, it seems that the method was also discovered independently at the same time by B. I. Clasen, a priest who lived in Luxembourg. This biographical sketch is based on an excellent article: S. C. Althoen and R. McLaughlin, ''Gauss–Jordan reduction: a brief history,'' *MAA Monthly*, 94 (1987), 130–142.

Solution

STEP 1. The augmented matrix of this linear system is

$$\left[\begin{array}{cccccc|c} 1 & 2 & 0 & -3 & 1 & 0 & 2 \\ 1 & 2 & 1 & -3 & 1 & 2 & 3 \\ 1 & 2 & 0 & -3 & 2 & 1 & 4 \\ 3 & 6 & 1 & -9 & 4 & 3 & 9 \end{array}\right].$$

STEP 2. The augmented matrix is now equivalent to the matrix (verify)

$$\left[\begin{array}{cccccc|c} 1 & 2 & 0 & -3 & 0 & -1 & 0 \\ 0 & 0 & 1 & 0 & 0 & 2 & 1 \\ 0 & 0 & 0 & 0 & 1 & 1 & 2 \\ 0 & 0 & 0 & 0 & 0 & 0 & 0 \end{array}\right]. \tag{7}$$

STEP 3. The linear system represented by (7) is

$$\begin{aligned} x_1 + 2x_2 \quad - 3x_4 \quad - \quad x_6 &= 0 \\ x_3 \qquad\qquad + 2x_6 &= 1 \\ x_5 + \quad x_6 &= 2. \end{aligned}$$

Solving each equation for the unknown that corresponds to the leading entry in each row of (7), we obtain

$$\begin{aligned} x_1 &= r + 3s - 2t \\ x_2 &= t \\ x_3 &= 1 - 2r \\ x_4 &= s \\ x_5 &= 2 - r \\ x_6 &= r, \end{aligned} \tag{8}$$

where r, s, and t are any real numbers. Since r, s, and t can be assigned any real numbers, the given linear system (6) has infinitely many solutions. Thus (8) is the solution to the given linear system (6).

Example 9 Solve the linear system

$$\begin{aligned} x + 2y + 3z + 4w &= 5 \\ x + 3y + 5z + 7w &= 11 \\ x \qquad - z - 2w &= -6 \end{aligned} \tag{9}$$

by Gauss–Jordan reduction.

Solution

STEP 1. The augmented matrix of this linear system is

$$\left[\begin{array}{cccc|c} 1 & 2 & 3 & 4 & 5 \\ 1 & 3 & 5 & 7 & 11 \\ 1 & 0 & -1 & -2 & -6 \end{array}\right].$$

STEP 2. The augmented matrix is row equivalent to the matrix (verify)

$$\begin{bmatrix} 1 & 0 & -1 & -2 & \vdots & 0 \\ 0 & 1 & 2 & 3 & \vdots & 0 \\ 0 & 0 & 0 & 0 & \vdots & 1 \end{bmatrix}. \tag{10}$$

STEP 3. The last equation of the linear system represented by (10) is

$$0x + 0y + 0z + 0w = 1,$$

which has no solution for any x, y, z, and w. Consequently, the given linear system (9) has no solution.

The last example is characteristic of the way in which a linear system has no solution. That is, a linear system $\mathbf{AX} = \mathbf{B}$ has no solution if and only if its augmented matrix is row equivalent to a matrix in reduced row echelon form, which has a row whose first n elements are zero and whose $(n + 1)$st element is 1 (Exercise T.4). The Gauss–Jordan reduction method is frequently used in actual practice, and computer codes implementing it are widely available.

If a matrix satisfies the first three properties in the definition of a matrix in reduced row echelon form, it is said to be in **row echelon form**. A method called **Gaussian elimination**, which is used to solve the linear system $\mathbf{AX} = \mathbf{B}$, transforms the augmented matrix $[\mathbf{A} \mid \mathbf{B}]$ to row echelon form and then uses **back substitution** to obtain the solution. This procedure, somewhat more efficient than Gauss–Jordan reduction, is discussed in Section 8.2.

The linear systems of Examples 6, 7, and 8 each had at least one solution, while the system in Example 9 had no solution. Linear systems with at least one solution are called **consistent** and linear systems with no solutions are called **inconsistent**. Every inconsistent linear system results in the situation illustrated in Example 9.

Homogeneous Systems

A linear system of the form

$$\begin{matrix} a_{11}x_1 + a_{12}x_2 + \cdots + a_{1n}x_n = 0 \\ a_{21}x_1 + a_{22}x_2 + \cdots + a_{2n}x_n = 0 \\ \vdots \qquad \vdots \qquad \qquad \vdots \qquad \vdots \\ a_{m1}x_1 + a_{m2}x_2 + \cdots + a_{mn}x_n = 0 \end{matrix} \tag{11}$$

is called a **homogeneous system**. We can also write (11) in matrix form as

$$\mathbf{AX} = \mathbf{0}. \tag{12}$$

The solution,

$$x_1 = x_2 = \cdots = x_n = 0,$$

to the homogeneous system (12) is called the **trivial solution**. A solution $x_1, x_2, \ldots, x_n$ to a homogeneous system in which not all the x_i are zero is called a **nontrivial solution**.

We see that a homogeneous system is always consistent, since it always has the trivial solution.

Example 10 Consider the homogeneous system

$$
\begin{aligned}
x + 2y + 3z &= 0 \\
-x + 3y + 2z &= 0 \\
2x + y - 2z &= 0.
\end{aligned}
\tag{13}
$$

The augmented matrix of this system,

$$
\begin{bmatrix}
1 & 2 & 3 & \vdots & 0 \\
-1 & 3 & 2 & \vdots & 0 \\
2 & 1 & -2 & \vdots & 0
\end{bmatrix},
$$

is row equivalent to (verify)

$$
\begin{bmatrix}
1 & 0 & 0 & \vdots & 0 \\
0 & 1 & 0 & \vdots & 0 \\
0 & 0 & 1 & \vdots & 0
\end{bmatrix},
$$

which is in reduced row echelon form. Hence the solution to (13) is

$$
x = y = z = 0,
$$

which means that the given homogeneous system (13) has only the trivial solution.

Example 11 Consider the homogeneous system

$$
\begin{aligned}
x + y + z + w &= 0 \\
x \qquad\quad + w &= 0 \\
x + 2y + z \qquad &= 0.
\end{aligned}
\tag{14}
$$

The augmented matrix of this system,

$$
\begin{bmatrix}
1 & 1 & 1 & 1 & \vdots & 0 \\
1 & 0 & 0 & 1 & \vdots & 0 \\
1 & 2 & 1 & 0 & \vdots & 0
\end{bmatrix},
$$

is row equivalent to (verify)

$$
\begin{bmatrix}
1 & 0 & 0 & 1 & \vdots & 0 \\
0 & 1 & 0 & -1 & \vdots & 0 \\
0 & 0 & 1 & 1 & \vdots & 0
\end{bmatrix},
$$

which is in reduced row echelon form. Hence the solution to (14) is

$$
\begin{aligned}
x &= -r \\
y &= r \\
z &= -r \\
w &= r,
\end{aligned}
$$

where r is any real number.

Example 11 shows that a homogeneous system may have a nontrivial solution. The following theorem tells of one case when this occurs.

Theorem 1.7 *A homogeneous system of m equations in n unknowns always has a nontrivial solution if $m < n$, that is, if the number of unknowns exceeds the number of equations.*

Proof

Let **C** be the matrix in reduced row echelon form which is row equivalent to **A**. Then the homogeneous systems **AX** = **0** and **CX** = **0** are equivalent. If we let r be the number of nonzero rows of **C**, then $r \leq m$. If $m < n$, we conclude that $r < n$. We are then solving r equations in n unknowns and can solve for r unknowns in terms of the remaining $n - r$ unknowns, the latter being free to take on any values we please. Thus, by letting one of these $n - r$ unknowns be nonzero, we obtain a nontrivial solution of **CX** = **0** and thus of **AX** = **0**.

APPLICATIONS.   At this point the reader may turn to Chapter 6, Linear Programming; Section 7.5, Least Squares (first part); Section 7.7, Markov Chains, Section 8.2, Linear Systems (from a computational point of view) and Section 8.3, **LU**-Decomposition.

1.4 EXERCISES

1. Which of the following matrices are in reduced row echelon form?

$$
A = \begin{bmatrix} 1 & 0 & 0 & 0 & -3 \\ 0 & 0 & 1 & 0 & 4 \\ 0 & 0 & 0 & 1 & 2 \end{bmatrix},
$$

$$
B = \begin{bmatrix} 0 & 1 & 0 & 0 & 5 \\ 0 & 0 & 1 & 0 & -4 \\ 0 & 0 & 0 & -1 & 3 \end{bmatrix},
$$

$$
C = \begin{bmatrix} 0 & 1 & 0 & 0 & 5 \\ 0 & 0 & 1 & 0 & 4 \\ 0 & 1 & 0 & -2 & 3 \end{bmatrix},
$$

$$
D = \begin{bmatrix} 0 & 1 & 0 & 0 & 2 \\ 0 & 0 & 0 & 0 & -1 \\ 0 & 0 & 0 & 1 & 4 \\ 0 & 0 & 0 & 0 & 0 \\ 0 & 0 & 0 & 0 & 1 \end{bmatrix},
$$

$$
E = \begin{bmatrix} 1 & 0 & 0 & 0 & 2 \\ 0 & 0 & 1 & 0 & 0 \\ 0 & 0 & 0 & 1 & 3 \\ 0 & 0 & 0 & 0 & 0 \end{bmatrix},
$$

$$
F = \begin{bmatrix} 0 & 0 & 0 & 0 & 0 \\ 0 & 0 & 1 & 2 & -3 \\ 0 & 0 & 0 & 1 & 0 \\ 0 & 0 & 0 & 0 & 0 \end{bmatrix},
$$

$$G = \begin{bmatrix} 1 & 0 & 0 & 0 & 1 \\ 0 & 1 & 0 & 0 & 2 \\ 0 & 0 & 0 & 1 & -1 \\ 0 & 0 & 0 & 0 & 0 \end{bmatrix},$$

$$H = \begin{bmatrix} 1 & 0 & 0 & 1 \\ 0 & 1 & 0 & 2 \\ 0 & 0 & 0 & -1 \\ 0 & 0 & 0 & 0 \end{bmatrix}.$$

2. Let

$$A = \begin{bmatrix} 1 & 0 & 3 \\ -3 & 1 & 4 \\ 4 & 2 & 2 \\ 5 & -1 & 5 \end{bmatrix}.$$

Find the matrices obtained by performing the following elementary row operations on **A**.
(a) Interchanging the second and fourth rows.
(b) Multiplying the third row by 3.
(c) Adding -3 times the first row to the fourth row.

3. Let

$$A = \begin{bmatrix} 2 & 0 & 4 & 2 \\ 3 & -2 & 5 & 6 \\ -1 & 3 & 1 & 1 \end{bmatrix}.$$

Find the matrices obtained by performing the following elementary row operations on **A**.
(a) Interchanging the second and third rows.
(b) Multiplying the second row by -4.
(c) Adding 2 times the third row to the first row.

4. Find three matrices that are row equivalent to

$$A = \begin{bmatrix} 2 & -1 & 3 & 4 \\ 0 & 1 & 2 & -1 \\ 5 & 2 & -3 & 4 \end{bmatrix}.$$

5. Find three matrices that are row equivalent to

$$\begin{bmatrix} 4 & 3 & 7 & 5 \\ -1 & 2 & -1 & 3 \\ 2 & 0 & 1 & 4 \end{bmatrix}.$$

6. If

$$A = \begin{bmatrix} 0 & 0 & -1 & 2 & 3 \\ 0 & 2 & 3 & 4 & 5 \\ 0 & 1 & 3 & -1 & 2 \\ 0 & 3 & 2 & 4 & 1 \end{bmatrix},$$

find a matrix **C** in reduced row echelon form that is row equivalent to **A**.

7. If

$$A = \begin{bmatrix} 1 & -2 & 0 & 2 \\ 2 & -3 & -1 & 5 \\ 1 & 3 & 2 & 5 \\ 1 & 1 & 0 & 2 \end{bmatrix},$$

find a matrix **C** in reduced row echelon form that is row equivalent to **A**.

In Exercises 8 through 10, find all solutions to the given linear systems.

8. (a) $\begin{aligned} x + y + 2z &= -1 \\ x - 2y + z &= -5 \\ 3x + y + z &= 3. \end{aligned}$

(b) $\begin{aligned} x + y + 3z + 2w &= 7 \\ 2x - y + 4w &= 8 \\ 3y + 6z &= 8. \end{aligned}$

(c) $\begin{aligned} x + 2y - 4z &= 3 \\ x - 2y + 3z &= -1 \\ 2x + 3y - z &= 5 \\ 4x + 3y - 2z &= 7 \\ 5x + 2y - 6z &= 7. \end{aligned}$

9. (a) $\begin{aligned} x + y + 2z + 3w &= 13 \\ x - 2y + z + w &= 8 \\ 3x + y + z - w &= 1. \end{aligned}$

(b) $\begin{aligned} x + y + z &= 1 \\ x + y - 2z &= 3 \\ 2x + y + z &= 2. \end{aligned}$

(c) $\begin{aligned} 2x + y + z - 2w &= 1 \\ 3x - 2y + z - 6w &= -2 \\ x + y - z - w &= -1 \\ 6x + z - 9w &= -2 \\ 5x - y + 2z - 8w &= 3. \end{aligned}$

10. (a) $2x - y + z = 3$
         $x - 3y + z = 4$
         $-5x \quad\quad - 2z = -5.$

(b) $x + y + z + w = 6$
     $2x + y - z \quad\quad = 3$
     $3x + y \quad\quad + 2w = 6.$

(c) $2x - y + z = 3$
     $3x + y - 2z = -2$
     $x - y + z = 7$
     $x + 5y + 7z = 13$
     $x - 7y - 5z = 12.$

In Exercises 11 through 14, find all values of a for which the resulting linear system has (a) no solution, (b) a unique solution, and (c) infinitely many solutions.

11. $x + y - \quad\quad z = 2$
    $x + 2y + \quad\quad z = 3$
    $x + y + (a^2 - 5) z = a.$

12. $x + y + \quad\quad z = 2$
    $2x + 3y + \quad\quad 2z = 5$
    $2x + 3y + (a^2 - 1) z = a + 1.$

13. $x + y + \quad\quad z = 2$
    $x + 2y + \quad\quad z = 3$
    $x + y + (a^2 - 5) z = a.$

14. $x + \quad\quad y = 3$
    $x + (a^2 - 8) y = a.$

In Exercises 15 through 18, solve the linear system with the given augmented matrix.

15. (a) $\begin{bmatrix} 1 & 1 & 1 & \vdots & 0 \\ 1 & 1 & 0 & \vdots & 3 \\ 0 & 1 & 1 & \vdots & 1 \end{bmatrix}.$

(b) $\begin{bmatrix} 1 & 2 & 3 & \vdots & 0 \\ 1 & 1 & 1 & \vdots & 0 \\ 1 & 1 & 2 & \vdots & 0 \\ 1 & 3 & 3 & \vdots & 0 \end{bmatrix}.$

16. (a) $\begin{bmatrix} 1 & 2 & 3 & \vdots & 0 \\ 1 & 1 & 1 & \vdots & 0 \\ 5 & 7 & 9 & \vdots & 0 \end{bmatrix}.$

(b) $\begin{bmatrix} 1 & 2 & 1 & \vdots & 7 \\ 2 & 0 & 1 & \vdots & 4 \\ 1 & 0 & 2 & \vdots & 5 \\ 1 & 2 & 3 & \vdots & 11 \\ 2 & 1 & 4 & \vdots & 12 \end{bmatrix}.$

17. (a) $\begin{bmatrix} 1 & 2 & 3 & 1 & \vdots & 8 \\ 1 & 3 & 0 & 1 & \vdots & 7 \\ 1 & 0 & 2 & 1 & \vdots & 3 \end{bmatrix}.$

(b) $\begin{bmatrix} 1 & -2 & 3 & \vdots & 4 \\ 2 & -1 & -3 & \vdots & 5 \\ 3 & 0 & 1 & \vdots & 2 \\ 3 & -3 & 0 & \vdots & 7 \end{bmatrix}.$

18. (a) $\begin{bmatrix} 4 & 2 & -1 & \vdots & 5 \\ 3 & 3 & 6 & \vdots & 1 \\ 5 & 1 & -8 & \vdots & 8 \end{bmatrix}.$

(b) $\begin{bmatrix} 1 & 1 & 3 & -3 & \vdots & 0 \\ 0 & 2 & 1 & -3 & \vdots & 3 \\ 1 & 0 & 2 & -1 & \vdots & -1 \end{bmatrix}.$

In Exercises 19 and 20, let

$$A = \begin{bmatrix} 1 & 0 & 5 \\ 1 & 1 & 1 \\ 0 & 1 & -4 \end{bmatrix}.$$

19. Find a nontrivial solution to the homogeneous system $(-4I_3 - A)X = 0$.

20. Find a nontrivial solution to the homogeneous system $(2I_3 - A)X = 0$.

21. A furniture manufacturer makes chairs, coffee tables, and dining-room tables. Each chair requires 10 minutes of sanding, 6 minutes of staining, and 12 minutes of varnishing. Each coffee table requires 12 minutes of sanding, 8 minutes of staining, and 12 minutes of varnishing. Each dining-room table requires 15 minutes of sanding, 12 minutes of staining, and 18 minutes of varnishing. The sanding

bench is available 16 hours per week, the staining bench 11 hours per week, and the varnishing bench 18 hours per week. How many (per week) of each type of furniture should be made so that the benches are fully utilized?

22. A book publisher publishes a potential best seller in three different bindings: paperback, book club, and deluxe. Each paperback book requires 1 minute for sewing and 2 minutes for gluing. Each book club book requires 2 minutes for sewing and 4 minutes for gluing. Each deluxe book requires 3 minutes for sewing and 5 minutes for gluing. If the sewing plant is available 6 hours per day and the gluing plant is available 11 hours per day, how many books of each type can be produced per day so that the plants are fully utilized?

THEORETICAL EXERCISES

T.1. Show that properties (a), (b), and (c) alone [excluding (d)] of the definition of the reduced row echelon form of a matrix **A** imply that if a column of **A** contains a leading entry of some row, then all other entries in that column *below the leading entry* are zero.

T.2. Prove:
   (a) Every matrix is row equivalent to itself.
   (b) If **A** is row equivalent to **B**, then **B** is row equivalent to **A**.
   (c) If **A** is row equivalent to **B** and **B** is row equivalent to **C**, then **A** is row equivalent to **C**.

T.3. Prove Corollary 1.1.

T.4. Prove that the linear system $\mathbf{AX} = \mathbf{B}$ has no solution if and only if its augmented matrix is row equivalent to a matrix in reduced row echelon form that has a row whose first n elements are zero and whose $(n + 1)$st element is 1.

T.5. Let

$$\mathbf{A} = \begin{bmatrix} a & b \\ c & d \end{bmatrix}.$$

Show that **A** is row equivalent to $\mathbf{I}_2$ if and only if $ad - bc \neq 0$.

T.6. Let

$$\mathbf{A} = \begin{bmatrix} a & b \\ c & d \end{bmatrix}.$$

Show that the homogeneous system $\mathbf{AX} = \mathbf{0}$ has only the trivial solution if and only if $ad - bc \neq 0$.

T.7. Let **A** be an $n \times n$ matrix in reduced row echelon form. Show that if $\mathbf{A} \neq \mathbf{I}_n$, then **A** has a row consisting entirely of zeros.

T.8. Show that the values of λ for which the homogeneous system

$$(a - \lambda)\, x + \qquad by = 0$$
$$cx + (d - \lambda)y = 0$$

has a nontrivial solution satisfy the equation $(a - \lambda)(d - \lambda) - bc = 0$. (*Hint:* See Exercise T.6.)

T.9. Let $\mathbf{AX} = \mathbf{B}, \mathbf{B} \neq \mathbf{0}$ be a consistent linear system.
   (a) Show that if $\mathbf{X}_1$ is a solution to the linear system $\mathbf{AX} = \mathbf{B}$ and $\mathbf{Y}_1$ is a solution to the associated homogeneous system $\mathbf{AX} = \mathbf{0}$, then $\mathbf{X}_1 + \mathbf{Y}_1$ is a solution to the system $\mathbf{AX} = \mathbf{B}$.
   (b) Show that every solution **X** to $\mathbf{AX} = \mathbf{B}$ can be written as $\mathbf{X}_1 + \mathbf{Y}_1$, where $\mathbf{X}_1$ is a particular solution to $\mathbf{AX} = \mathbf{B}$ and $\mathbf{Y}_1$ is a solution to $\mathbf{AX} = \mathbf{0}$. [*Hint:* Let $\mathbf{X} = \mathbf{X}_1 + (\mathbf{X} - \mathbf{X}_1)$.]

1.5

THE INVERSE OF A MATRIX

In this section we restrict our attention to square matrices and formulate the notion corresponding to the reciprocal of a nonzero real number.

Definition   An $n \times n$ matrix $\mathbf{A}$ is called **nonsingular** (or **invertible**) if there exists an $n \times n$ matrix $\mathbf{B}$ such that

$$\mathbf{AB} = \mathbf{BA} = \mathbf{I}_n.$$

The matrix $\mathbf{B}$ is called an **inverse** of $\mathbf{A}$. If there exists no such matrix $\mathbf{B}$, then $\mathbf{A}$ is called **singular** (or **noninvertible**).

Example 1   Let

$$\mathbf{A} = \begin{bmatrix} 2 & 3 \\ 2 & 2 \end{bmatrix} \quad \text{and} \quad \mathbf{B} = \begin{bmatrix} -1 & \frac{3}{2} \\ 1 & -1 \end{bmatrix}.$$

Since

$$\mathbf{AB} = \mathbf{BA} = \mathbf{I}_2,$$

we conclude that $\mathbf{B}$ is an inverse of $\mathbf{A}$ and that $\mathbf{A}$ is nonsingular.

Theorem 1.8   *If a matrix has an inverse, then the inverse is unique.*

Proof
Let $\mathbf{B}$ and $\mathbf{C}$ be inverses of $\mathbf{A}$. Then

$$\mathbf{BA} = \mathbf{AC} = \mathbf{I}_n.$$

Therefore,

$$\mathbf{B} = \mathbf{BI}_n = \mathbf{B(AC)} = \mathbf{(BA)C} = \mathbf{I}_n\mathbf{C} = \mathbf{C},$$

which completes the proof.

We shall now write the inverse of $\mathbf{A}$, if it exists, as $\mathbf{A}^{-1}$. Thus

$$\mathbf{AA}^{-1} = \mathbf{A}^{-1}\mathbf{A} = \mathbf{I}_n.$$

Example 2   Let

$$\mathbf{A} = \begin{bmatrix} 1 & 2 \\ 3 & 4 \end{bmatrix}.$$

To find $\mathbf{A}^{-1}$, we let

$$\mathbf{A}^{-1} = \begin{bmatrix} a & b \\ c & d \end{bmatrix}.$$

Then we must have

$$\mathbf{A}\mathbf{A}^{-1} = \begin{bmatrix} 1 & 2 \\ 3 & 4 \end{bmatrix}\begin{bmatrix} a & b \\ c & d \end{bmatrix} = \mathbf{I}_2 = \begin{bmatrix} 1 & 0 \\ 0 & 1 \end{bmatrix}$$

so that

$$\begin{bmatrix} a + 2c & b + 2d \\ 3a + 4c & 3b + 4d \end{bmatrix} = \begin{bmatrix} 1 & 0 \\ 0 & 1 \end{bmatrix}.$$

Equating corresponding entries of these two matrices, we obtain the linear systems

$$\begin{array}{ccc} a + 2c = 1 & & b + 2d = 0 \\ 3a + 4c = 0 & \text{and} & 3b + 4d = 1. \end{array}$$

The solutions are (verify) $a = -2$, $c = \frac{3}{2}$, $b = 1$, and $d = -\frac{1}{2}$. Moreover, since the matrix

$$\begin{bmatrix} a & b \\ c & d \end{bmatrix} = \begin{bmatrix} -2 & 1 \\ \frac{3}{2} & -\frac{1}{2} \end{bmatrix}$$

also satisfies the property that

$$\begin{bmatrix} -2 & 1 \\ \frac{3}{2} & -\frac{1}{2} \end{bmatrix}\begin{bmatrix} 1 & 2 \\ 3 & 4 \end{bmatrix} = \begin{bmatrix} 1 & 0 \\ 0 & 1 \end{bmatrix},$$

we conclude that $\mathbf{A}$ is nonsingular and that

$$\mathbf{A}^{-1} = \begin{bmatrix} -2 & 1 \\ \frac{3}{2} & -\frac{1}{2} \end{bmatrix}.$$

Not every matrix has an inverse. For instance, consider the following example.

Example 3 Let

$$\mathbf{A} = \begin{bmatrix} 1 & 2 \\ 2 & 4 \end{bmatrix}.$$

To find $\mathbf{A}^{-1}$, we let

$$\mathbf{A}^{-1} = \begin{bmatrix} a & b \\ c & d \end{bmatrix}.$$

Then we must have

$$\mathbf{A}\mathbf{A}^{-1} = \begin{bmatrix} 1 & 2 \\ 2 & 4 \end{bmatrix}\begin{bmatrix} a & b \\ c & d \end{bmatrix} = \mathbf{I}_2 = \begin{bmatrix} 1 & 0 \\ 0 & 1 \end{bmatrix}$$

so that

$$\begin{bmatrix} a + 2c & b + 2d \\ 2a + 4c & 2b + 4d \end{bmatrix} = \begin{bmatrix} 1 & 0 \\ 0 & 1 \end{bmatrix}.$$

Equating corresponding entries of these two matrices, we obtain the linear systems

$$\begin{array}{ccc} a + 2c = 1 & & b + 2d = 0 \\ & \text{and} & \\ 2a + 4c = 0 & & 2b + 4d = 1. \end{array}$$

These linear systems have no solutions, so **A** has no inverse. Hence **A** is a singular matrix.

The method used in Example 2 to find the inverse of a matrix is not a very efficient one. We shall soon modify it and thereby obtain a much faster method. We first establish several properties of matrices.

Theorem 1.9
(Properties of
the Inverse)

(a) *If* **A** *is a nonsingular matrix, then* $\mathbf{A}^{-1}$ *is nonsingular and*

$$(\mathbf{A}^{-1})^{-1} = \mathbf{A}.$$

(b) *If* **A** *and* **B** *are nonsingular matrices, then* **AB** *is nonsingular and*

$$(\mathbf{AB})^{-1} = \mathbf{B}^{-1}\mathbf{A}^{-1}.$$

(c) *If* **A** *is a nonsingular matrix, then*

$$(\mathbf{A}^T)^{-1} = (\mathbf{A}^{-1})^T.$$

Proof

(a) $\mathbf{A}^{-1}$ is nonsingular if we can find a matrix **B** such that

$$\mathbf{A}^{-1}\mathbf{B} = \mathbf{B}\mathbf{A}^{-1} = \mathbf{I}_n.$$

Now

$$\mathbf{A}^{-1}\mathbf{A} = \mathbf{A}\mathbf{A}^{-1} = \mathbf{I}_n.$$

Thus **B** = **A** is the inverse of $\mathbf{A}^{-1}$, and hence

$$(\mathbf{A}^{-1})^{-1} = \mathbf{A}.$$

(b) We have

$$(\mathbf{AB})(\mathbf{B}^{-1}\mathbf{A}^{-1}) = \mathbf{A}(\mathbf{BB}^{-1})\mathbf{A}^{-1} = \mathbf{AI}_n\mathbf{A}^{-1} = \mathbf{AA}^{-1} = \mathbf{I}_n$$

and

$$(\mathbf{B}^{-1}\mathbf{A}^{-1})(\mathbf{AB}) = \mathbf{B}^{-1}(\mathbf{A}^{-1}\mathbf{A})\mathbf{B} = \mathbf{B}^{-1}\mathbf{I}_n\mathbf{B} = \mathbf{B}^{-1}\mathbf{B} = \mathbf{I}_n.$$

Therefore, **AB** is nonsingular. Since the inverse of a matrix is unique, we conclude that

$$(\mathbf{AB})^{-1} = \mathbf{B}^{-1}\mathbf{A}^{-1}.$$

(c) We have

$$AA^{-1} = A^{-1}A = I_n.$$

Taking transposes, we obtain

$$(AA^{-1})^T = (A^{-1}A)^T = I_n^T = I_n.$$

Then

$$(A^{-1})^TA^T = A^T(A^{-1})^T = I_n.$$

This equation implies that

$$(A^T)^{-1} = (A^{-1})^T.$$

Example 4 If

$$A = \begin{bmatrix} 1 & 2 \\ 3 & 4 \end{bmatrix},$$

then from Example 2

$$A^{-1} = \begin{bmatrix} -2 & 1 \\ \frac{3}{2} & -\frac{1}{2} \end{bmatrix} \quad \text{and} \quad (A^{-1})^T = \begin{bmatrix} -2 & \frac{3}{2} \\ 1 & -\frac{1}{2} \end{bmatrix}.$$

Also (verify)

$$A^T = \begin{bmatrix} 1 & 3 \\ 2 & 4 \end{bmatrix} \quad \text{and} \quad (A^T)^{-1} = \begin{bmatrix} -2 & \frac{3}{2} \\ 1 & -\frac{1}{2} \end{bmatrix}.$$

Corollary 1.2 *If $A_1, A_2, \ldots, A_r$ are $n \times n$ nonsingular matrices, then $A_1A_2 \cdots A_r$ is nonsingular and*

$$(A_1A_2 \cdots A_r)^{-1} = A_r^{-1}A_{r-1}^{-1} \cdots A_1^{-1}.$$

Proof
Exercise T.2.

Earlier, we defined a matrix **B** to be the inverse of **A** if $AB = BA = I_n$.
    The following theorem, whose proof we omit, shows that one of these equations follows from the other.

Theorem 1.10 *Suppose that A and B are $n \times n$ matrices.*
    (a) *If*

$$AB = I_n,$$

   then $BA = I_n$.
    (b) *If*

$$BA = I_n,$$

   then $AB = I_n$.

A Practical Method for Finding $\mathbf{A}^{-1}$

We shall now develop a practical method for finding $\mathbf{A}^{-1}$. If $\mathbf{A}$ is a given $n \times n$ matrix, we are looking for an $n \times n$ matrix $\mathbf{B} = [b_{ij}]$ such that

$$\mathbf{AB} = \mathbf{BA} = \mathbf{I}_n.$$

Let the columns of $\mathbf{B}$ be denoted by the $n \times 1$ matrices $\mathbf{X}_1, \mathbf{X}_2, \ldots, \mathbf{X}_n$, where

$$\mathbf{X}_j = \begin{bmatrix} b_{1j} \\ b_{2j} \\ \vdots \\ b_{ij} \\ \vdots \\ b_{nj} \end{bmatrix} \qquad (1 \leqslant j \leqslant n).$$

Let the columns of $\mathbf{I}_n$ be denoted by the $n \times 1$ matrices $\mathbf{E}_1, \mathbf{E}_2, \ldots, \mathbf{E}_n$. Thus

$$\mathbf{E}_j = \begin{bmatrix} 0 \\ 0 \\ \vdots \\ 1 \\ 0 \\ \vdots \\ 0 \end{bmatrix} \longleftarrow j\text{th row.}$$

By Exercise T.5 of Section 1.2 the jth column of $\mathbf{AB}$ is the $n \times 1$ matrix $\mathbf{AX}_j$. Since equal matrices must agree column by column, it follows that the problem of finding an $n \times n$ matrix $\mathbf{B} = \mathbf{A}^{-1}$ such that

$$\mathbf{AB} = \mathbf{I}_n \tag{1}$$

is equivalent to the problem of finding n matrices (each $n \times 1$) $\mathbf{X}_1, \mathbf{X}_2, \ldots, \mathbf{X}_n$ such that

$$\mathbf{AX}_j = \mathbf{E}_j \qquad (1 \leqslant j \leqslant n). \tag{2}$$

Thus finding $\mathbf{B}$ is equivalent to solving n linear systems (each is n equations in n unknowns). This is precisely what we did in Example 2. Each of these systems can be solved by the Gauss–Jordan reduction method. To solve the first linear system, we form the augmented matrix $[\mathbf{A} \mid \mathbf{E}_1]$ and put it into reduced row echelon form. We do the same with

$$[\mathbf{A} \mid \mathbf{E}_2], \ldots, [\mathbf{A} \mid \mathbf{E}_n].$$

However, if we observe that the coefficient matrix of each of these n linear systems is always $\mathbf{A}$, we can solve all these systems simultaneously. We form the $n \times 2n$ matrix

$$[\mathbf{A} \mid \mathbf{E}_1\ \mathbf{E}_2 \cdots \mathbf{E}_n] = [\mathbf{A} \mid \mathbf{I}_n]$$

and transform it into reduced row echelon form $[\mathbf{C} \mid \mathbf{D}]$. The $n \times n$ matrix $\mathbf{C}$ is the reduced row echelon form matrix that is row equivalent to $\mathbf{A}$. Let $\mathbf{D}_1, \mathbf{D}_2, \ldots, \mathbf{D}_n$ be the n columns of $\mathbf{D}$. Then the matrix $[\mathbf{C} \mid \mathbf{D}]$ gives rise to the n linear systems

$$\mathbf{C}\mathbf{X}_j = \mathbf{D}_j \qquad (1 \leqslant j \leqslant n) \tag{3}$$

or to the matrix equation

$$\mathbf{C}\mathbf{B} = \mathbf{D}. \tag{4}$$

There are now two possible cases.

CASE 1. $\mathbf{C} = \mathbf{I}_n$. Then Equation (3) becomes

$$\mathbf{I}_n\mathbf{X}_j = \mathbf{X}_j = \mathbf{D}_j,$$

and $\mathbf{B} = \mathbf{D}$, so we have obtained $\mathbf{A}^{-1}$.

CASE 2. $\mathbf{C} \neq \mathbf{I}_n$. It then follows from Exercise T.7 in Section 1.4 that $\mathbf{C}$ has a row consisting entirely of zeros. From Exercise T.1 in Section 1.2, we observe that the product $\mathbf{C}\mathbf{B}$ in Equation (4) has a row of zeros. The matrix $\mathbf{D}$ in (4) arose from $\mathbf{I}_n$ by a sequence of elementary row operations, and it is intuitively clear that $\mathbf{D}$ cannot have a row of zeros. The statement that $\mathbf{D}$ cannot have a row of zeros can be rigorously established at this point, but we shall ask the reader to accept the argument now. In Section 2.2, an argument using determinants will show the validity of the result. Thus one of the equations $\mathbf{C}\mathbf{X}_j = \mathbf{D}_j$ has no solution, so $\mathbf{A}\mathbf{X}_j = \mathbf{E}_j$ has no solution and $\mathbf{A}$ is singular in this case.

The practical procedure for computing the inverse of matrix $\mathbf{A}$ is as follows.

STEP 1. Form the $n \times 2n$ matrix $[\mathbf{A} \mid \mathbf{I}_n]$ obtained by adjoining the identity matrix $\mathbf{I}_n$ to the given matrix $\mathbf{A}$.

STEP 2. Transform the matrix obtained in Step 1 to reduced row echelon form by using elementary row operations. Remember that whatever we do to a row of $\mathbf{A}$ we also do to the corresponding row of $\mathbf{I}_n$.

STEP 3. Suppose that Step 2 has produced the matrix $[\mathbf{C} \mid \mathbf{D}]$ in reduced row echelon form.
(a) If $\mathbf{C} = \mathbf{I}_n$, then $\mathbf{D} = \mathbf{A}^{-1}$.
(b) If $\mathbf{C} \neq \mathbf{I}_n$, then $\mathbf{C}$ has a row of zeros. In this case $\mathbf{A}$ is singular and $\mathbf{A}^{-1}$ does not exist.

Example 5 Find the inverse of the matrix

$$\mathbf{A} = \begin{bmatrix} 1 & 1 & 1 \\ 0 & 2 & 3 \\ 5 & 5 & 1 \end{bmatrix}.$$

Solution

STEP 1. The 3×6 matrix $[\mathbf{A} \mathrel{\vdots} \mathbf{I}_3]$ is

$$[\mathbf{A} \mathrel{\vdots} \mathbf{I}_3] = \begin{bmatrix} \overset{\mathbf{A}}{} & & & \overset{\mathbf{I}_3}{} & & \\ 1 & 1 & 1 & 1 & 0 & 0 \\ 0 & 2 & 3 & 0 & 1 & 0 \\ 5 & 5 & 1 & 0 & 0 & 1 \end{bmatrix}.$$

STEP 2. We transform the matrix obtained in Step 1 to reduced row echelon form:

A			**I₃**			
1	1	1	1	0	0	Subtract 5 times the first row
0	2	3	0	1	0	from the third row to obtain
5	5	1	0	0	1	

1	1	1	1	0	0	Divide the second row by 2
0	2	3	0	1	0	to obtain
0	0	-4	-5	0	1	

1	1	1	1	0	0	Subtract the second row from
0	1	$\frac{3}{2}$	0	$\frac{1}{2}$	0	the first row to obtain
0	0	-4	-5	0	1	

1	0	$-\frac{1}{2}$	1	$-\frac{1}{2}$	0	Multiply the third row by $-\frac{1}{4}$
0	1	$\frac{3}{2}$	0	$\frac{1}{2}$	0	to obtain
0	0	-4	-5	0	1	

1	0	$-\frac{1}{2}$	1	$-\frac{1}{2}$	0	Add $-\frac{3}{2}$ times the third row to
0	1	$\frac{3}{2}$	0	$\frac{1}{2}$	0	the second row to obtain
0	0	1	$\frac{5}{4}$	0	$-\frac{1}{4}$	

1	0	$-\frac{1}{2}$	1	$-\frac{1}{2}$	0	Add $\frac{1}{2}$ times the third row to
0	1	0	$-\frac{15}{8}$	$\frac{1}{2}$	$\frac{3}{8}$	the first row to obtain
0	0	1	$\frac{5}{4}$	0	$-\frac{1}{4}$	

1	0	0	$\frac{13}{8}$	$-\frac{1}{2}$	$-\frac{1}{8}$	
0	1	0	$-\frac{15}{8}$	$\frac{1}{2}$	$\frac{3}{8}$	
0	0	1	$\frac{5}{4}$	0	$-\frac{1}{4}$	

STEP 3. Since $\mathbf{C} = \mathbf{I}_3$, we conclude that $\mathbf{D} = \mathbf{A}^{-1}$.

Hence

$$\mathbf{A}^{-1} = \begin{bmatrix} \frac{13}{8} & -\frac{1}{2} & -\frac{1}{8} \\ -\frac{15}{8} & \frac{1}{2} & \frac{3}{8} \\ \frac{5}{4} & 0 & -\frac{1}{4} \end{bmatrix}.$$

It is easy to verify that $\mathbf{A}\mathbf{A}^{-1} = \mathbf{A}^{-1}\mathbf{A} = \mathbf{I}_3$.

If the reduced row echelon matrix under **A** has a row of zeros, then **A** is singular. Since each matrix under **A** is row equivalent to **A**, once a matrix under **A** has a row of zeros, every subsequent matrix that is row equivalent to **A** will have a row of zeros. Thus we can stop the procedure as soon as we find a matrix **F** that is row equivalent to **A** and has a row of zeros. In this case $\mathbf{A}^{-1}$ does not exist.

Example 6 Find the inverse of the matrix

$$\mathbf{A} = \begin{bmatrix} 1 & 2 & -3 \\ 1 & -2 & 1 \\ 5 & -2 & -3 \end{bmatrix}.$$

Solution

STEP 1. The 3×6 matrix $[\mathbf{A} \mid \mathbf{I}_3]$ is

$$[\mathbf{A} \mid \mathbf{I}_3] = \left[\begin{array}{ccc|ccc} 1 & 2 & -3 & 1 & 0 & 0 \\ 1 & -2 & 1 & 0 & 1 & 0 \\ 5 & -2 & -3 & 0 & 0 & 1 \end{array}\right].$$

STEP 2. We transform the matrix obtained in Step 1 to reduced row echelon form. To find $\mathbf{A}^{-1}$, we proceed as follows:

$$\left[\begin{array}{ccc|ccc} 1 & 2 & -3 & 1 & 0 & 0 \\ 1 & -2 & 1 & 0 & 1 & 0 \\ 5 & -2 & -3 & 0 & 0 & 1 \end{array}\right]$$

Subtract the first row from the second row to obtain

$$\left[\begin{array}{ccc|ccc} 1 & 2 & -3 & 1 & 0 & 0 \\ 0 & -4 & 4 & -1 & 1 & 0 \\ 5 & -2 & -3 & 0 & 0 & 1 \end{array}\right]$$

Subtract 5 times the first row from the third row to obtain

$$\left[\begin{array}{ccc|ccc} 1 & 2 & -3 & 1 & 0 & 0 \\ 0 & -4 & 4 & -1 & 1 & 0 \\ 0 & -12 & 12 & -5 & 0 & 1 \end{array}\right]$$

Subtract 3 times the second row from the third row to obtain

$$\left[\begin{array}{ccc|ccc} 1 & 2 & -3 & 1 & 0 & 0 \\ 0 & -4 & 4 & -1 & 1 & 0 \\ 0 & 0 & 0 & -2 & -3 & 1 \end{array}\right]$$

At this point **A** is row equivalent to

$$\mathbf{F} = \begin{bmatrix} 1 & 2 & -3 \\ 0 & -4 & 4 \\ 0 & 0 & 0 \end{bmatrix}.$$

Since **F** has a row of zeros, we stop and conclude that **A** is a singular matrix.

Observe that to find $\mathbf{A}^{-1}$ we do not have to determine, in advance, whether or not it exists. We merely start the procedure given above and either obtain $\mathbf{A}^{-1}$ or find out that **A** is singular.

The foregoing discussion for the practical method of obtaining $\mathbf{A}^{-1}$ has actually established the following theorem.

Theorem 1.11 *An $n \times n$ matrix is nonsingular if and only if it is row equivalent to $\mathbf{I}_n$.*

Linear Systems and Inverses

If **A** is an $n \times n$ matrix, then the linear system $\mathbf{AX} = \mathbf{B}$ is a system of n equations in n unknowns. Suppose that **A** is nonsingular. Then $\mathbf{A}^{-1}$ exists and we can multiply $\mathbf{AX} = \mathbf{B}$ by $\mathbf{A}^{-1}$ on both sides, obtaining

$$\mathbf{A}^{-1}(\mathbf{AX}) = \mathbf{A}^{-1}\mathbf{B},$$

or

$$\mathbf{I}_n\mathbf{X} = \mathbf{X} = \mathbf{A}^{-1}\mathbf{B}.$$

Moreover, $\mathbf{X} = \mathbf{A}^{-1}\mathbf{B}$ is clearly a solution to the given linear system. Thus, if **A** is nonsingular, we have a unique solution.

APPLICATIONS.   This method is useful in industrial problems. Many physical models are described by linear systems. This means that if n values are used as inputs (which can be arranged as the $n \times 1$ matrix **X**), then m values are obtained as outputs (which can be arranged as the $m \times 1$ matrix **B**) by the rule $\mathbf{AX} = \mathbf{B}$. The matrix **A** is inherently tied to the process. Thus suppose that a chemical process has a certain matrix **A** associated with it. Any change in the process may result in a new matrix. In fact, we speak of a **black box,** meaning that the internal structure of the process does not interest us. The problem frequently encountered in systems analysis is that of determining the input to be used to obtain a desired output. That is, we want to solve the linear system $\mathbf{AX} = \mathbf{B}$ for **X** as we vary **B**. If **A** is a nonsingular square matrix, an efficient way of handling this is as follows. Compute $\mathbf{A}^{-1}$ once; then whenever we change **B**, we find the corresponding solution **X** by forming $\mathbf{A}^{-1}\mathbf{B}$.

Example 7   Consider an industrial process whose matrix is the matrix **A** of Example 5. If **B** is the output matrix

$$\begin{bmatrix} 8 \\ 24 \\ 8 \end{bmatrix},$$

then the input matrix **X** is the solution to the linear system $\mathbf{AX} = \mathbf{B}$. Then

$$\mathbf{X} = \mathbf{A}^{-1}\mathbf{B} = \begin{bmatrix} \frac{13}{8} & -\frac{1}{2} & -\frac{1}{8} \\ -\frac{15}{8} & \frac{1}{2} & \frac{3}{8} \\ \frac{5}{4} & 0 & -\frac{1}{4} \end{bmatrix}\begin{bmatrix} 8 \\ 24 \\ 8 \end{bmatrix} = \begin{bmatrix} 0 \\ 0 \\ 8 \end{bmatrix}.$$

On the other hand, if **B** is the output matrix

$$\begin{bmatrix} 16 \\ 8 \\ 8 \end{bmatrix},$$

then

$$\mathbf{X} = \mathbf{A}^{-1}\begin{bmatrix} 16 \\ 8 \\ 8 \end{bmatrix} = \begin{bmatrix} 21 \\ -23 \\ 18 \end{bmatrix}.$$

Theorem 1.12   *If **A** is an $n \times n$ matrix, the homogeneous system*

$$\mathbf{AX} = \mathbf{0} \tag{5}$$

*has a nontrivial solution if and only if **A** is singular.*

Proof

Suppose that **A** is nonsingular. Then $\mathbf{A}^{-1}$ exists, and multiplying both sides of (5) by $\mathbf{A}^{-1}$, we have

$$\mathbf{A}^{-1}(\mathbf{AX}) = \mathbf{A}^{-1}\mathbf{0}$$
$$(\mathbf{A}^{-1}\mathbf{A})\mathbf{X} = \mathbf{0}$$
$$\mathbf{I}_n\mathbf{X} = \mathbf{0}$$
$$\mathbf{X} = \mathbf{0}.$$

Hence the only solution to (5) is $\mathbf{X} = \mathbf{0}$.

We leave the proof of the converse—if **A** is singular, then (5) has a nontrivial solution—as an exercise (Exercise T.3).

Example 8   Consider the homogeneous system $\mathbf{AX} = \mathbf{0}$, where **A** is the matrix of Example 5. Since **A** is nonsingular,

$$\mathbf{X} = \mathbf{A}^{-1}\mathbf{0} = \mathbf{0}.$$

We could also solve the given system by Gauss–Jordan reduction. In this case we find that the matrix in reduced row echelon form that is row equivalent to the augmented matrix of the given system,

$$\left[\begin{array}{ccc:c} 1 & 1 & 1 & 0 \\ 0 & 2 & 3 & 0 \\ 5 & 5 & 1 & 0 \end{array}\right],$$

is

$$\left[\begin{array}{ccc:c} 1 & 0 & 0 & 0 \\ 0 & 1 & 0 & 0 \\ 0 & 0 & 1 & 0 \end{array}\right],$$

which again shows that the solution is

$$\mathbf{X} = \mathbf{0}.$$

Example 9   Consider the homogeneous system $\mathbf{AX} = \mathbf{0}$, where $\mathbf{A}$ is the singular matrix of Example 6. In this case the matrix in reduced row echelon form that is row equivalent to the augmented matrix of the given system,

$$\left[\begin{array}{ccc:c} 1 & 2 & -3 & 0 \\ 1 & -2 & 1 & 0 \\ 5 & -2 & -3 & 0 \end{array}\right],$$

is

$$\left[\begin{array}{ccc:c} 1 & 0 & -1 & 0 \\ 0 & 1 & -1 & 0 \\ 0 & 0 & 0 & 0 \end{array}\right],$$

which implies that

$$x = r$$
$$y = r$$
$$z = r,$$

where r is any real number. Thus the given system has a nontrivial solution.

The proof of the following theorem is left as an exercise (Supplementary Exercise T.13).

Theorem 1.13   *If $\mathbf{A}$ is an $n \times n$ matrix, then $\mathbf{A}$ is nonsingular if and only if the linear system $\mathbf{AX} = \mathbf{B}$ has a unique solution for every $n \times 1$ matrix $\mathbf{B}$.*

We may summarize our results on homogeneous systems and nonsingular matrices by noting that the following statements are equivalent:

1. $\mathbf{A}$ is nonsingular.
2. $\mathbf{AX} = \mathbf{0}$ has only the trivial solution.
3. $\mathbf{A}$ is row equivalent to $\mathbf{I}_n$.
4. The linear system $\mathbf{AX} = \mathbf{B}$ has a unique solution for every $n \times 1$ matrix $\mathbf{B}$.

OTHER APPLICATIONS.   At this point the reader may turn to Section 7.5, Least Squares (second part); and Section 7.6, Linear Economic Models.

1.5 EXERCISES

In Exercises 1 through 4, use the method of Examples 2 and 3.

1. Show that

$$\begin{bmatrix} 2 & 1 \\ -2 & 3 \end{bmatrix}$$

is nonsingular.

2. Show that

$$\begin{bmatrix} 2 & 1 \\ -4 & -2 \end{bmatrix}$$

is singular.

3. Is the matrix

$$\begin{bmatrix} 1 & 1 \\ 3 & 4 \end{bmatrix}$$

singular or nonsingular? If it is nonsingular, find its inverse.

4. Is the matrix

$$\begin{bmatrix} 1 & 2 & -1 \\ 3 & 2 & 3 \\ 2 & 2 & 1 \end{bmatrix}$$

singular or nonsingular? If it is nonsingular, find its inverse.

In Exercises 5 through 10, find the inverses of the given matrices, if possible:

5. (a) $\begin{bmatrix} 1 & 3 \\ -2 & 6 \end{bmatrix}$.   (b) $\begin{bmatrix} 1 & 2 & 3 \\ 1 & 1 & 2 \\ 0 & 1 & 2 \end{bmatrix}$.

(c) $\begin{bmatrix} 1 & 1 & 1 & 1 \\ 1 & 2 & -1 & 2 \\ 1 & -1 & 2 & 1 \\ 1 & 3 & 3 & 2 \end{bmatrix}$.

6. (a) $\begin{bmatrix} 1 & 3 \\ 2 & 6 \end{bmatrix}$.   (b) $\begin{bmatrix} 1 & 2 & 3 \\ 0 & 2 & 3 \\ 1 & 2 & 4 \end{bmatrix}$.

(c) $\begin{bmatrix} 1 & 1 & 2 & 1 \\ 0 & -2 & 0 & 0 \\ 1 & 2 & 1 & -2 \\ 0 & 3 & 2 & 1 \end{bmatrix}$.

7. (a) $\begin{bmatrix} 1 & 3 \\ 2 & 4 \end{bmatrix}$.   (b) $\begin{bmatrix} 1 & 1 & 1 & 1 \\ 1 & 3 & 1 & 2 \\ 1 & 2 & -1 & 1 \\ 5 & 9 & 1 & 6 \end{bmatrix}$.

(c) $\begin{bmatrix} 1 & 2 & 1 \\ 1 & 3 & 2 \\ 1 & 0 & 1 \end{bmatrix}$.

8. (a) $\begin{bmatrix} 1 & 1 & 1 \\ 1 & 2 & 3 \\ 0 & 1 & 1 \end{bmatrix}$.   (b) $\begin{bmatrix} 1 & 2 & 2 \\ 1 & 3 & 1 \\ 1 & 3 & 2 \end{bmatrix}$.

(c) $\begin{bmatrix} 1 & 2 & 3 \\ 1 & 1 & 2 \\ 0 & 1 & 1 \end{bmatrix}$.

9. (a) $\begin{bmatrix} 1 & 2 & -3 & 1 \\ -1 & 3 & -3 & -2 \\ 2 & 0 & 1 & 5 \\ 3 & 1 & -2 & 5 \end{bmatrix}$.   (b) $\begin{bmatrix} 3 & 1 & 2 \\ 2 & 1 & 2 \\ 1 & 2 & 2 \end{bmatrix}$.

(c) $\begin{bmatrix} 1 & 2 & 3 \\ 1 & 1 & 2 \\ 1 & 1 & 0 \end{bmatrix}$.

10. (a) $\begin{bmatrix} 2 & 1 & 3 \\ 0 & 1 & 2 \\ 1 & 0 & 3 \end{bmatrix}$.   (b) $\begin{bmatrix} 1 & -1 & 2 & 3 \\ 4 & 1 & 2 & 0 \\ 2 & -1 & 3 & 1 \\ 4 & 2 & 1 & -5 \end{bmatrix}$.

(c) $\begin{bmatrix} 2 & 1 & -2 \\ 3 & 4 & 6 \\ 7 & 6 & 2 \end{bmatrix}$.

11. Which of the following linear systems have a nontrivial solution?

(a) $x + 2y + 3z = 0$ (b) $2x + y - z = 0$
     $2y + 2z = 0$ $x - 2y - 3z = 0$
     $x + 2y + 3z = 0.$ $-3x - y + 2z = 0.$

12. Which of the following linear systems have a nontrivial solution?

(a) $x + y + 2z = 0$ (b) $x - y + z = 0$
     $2x + y + z = 0$ $2x + y = 0$
     $3x - y + z = 0.$ $2x - 2y + 2z = 0.$

(c) $2x - y + 5z = 0$
     $3x + 2y - 3z = 0$
     $x - y + 4z = 0.$

13. If

$$A^{-1} = \begin{bmatrix} 2 & 3 \\ 1 & 4 \end{bmatrix},$$

find A.

14. If

$$A^{-1} = \begin{bmatrix} 3 & 4 \\ -1 & -1 \end{bmatrix},$$

find A.

15. Show that a matrix which has a row or a column consisting entirely of zeros must be singular.

16. Find all values of a for which the inverse of

$$A = \begin{bmatrix} 1 & 1 & 0 \\ 1 & 0 & 0 \\ 1 & 2 & a \end{bmatrix}$$

exists. What is A^{-1}?

17. Consider an industrial process whose matrix is

$$A = \begin{bmatrix} 2 & 1 & 3 \\ 3 & 2 & -1 \\ 2 & 1 & 1 \end{bmatrix}.$$

Find the input matrix for each of the following output matrices:

(a) $\begin{bmatrix} 30 \\ 20 \\ 10 \end{bmatrix}$. (b) $\begin{bmatrix} 12 \\ 8 \\ 14 \end{bmatrix}$.

18. Suppose that $A = \begin{bmatrix} 1 & 3 \\ 2 & 7 \end{bmatrix}$.

(a) Find A^{-1}.
(b) Find $(A^T)^{-1}$. How do $(A^T)^{-1}$ and A^{-1} compare?

19. Is the inverse of a nonsingular symmetric matrix always symmetric? Explain.

20. (a) Is $(A + B)^{-1} = A^{-1} + B^{-1}$?

(b) Is $(cA)^{-1} = \dfrac{1}{c} A^{-1}$?

21. For what values of λ does the homogeneous system

$$(\lambda - 1) x + \qquad\quad 2y = 0$$
$$2x + (\lambda - 1) y = 0$$

have a nontrivial solution?

22. If A and B are nonsingular, are $A + B$, $A - B$, and $-A$ nonsingular? Explain.

23. If

$$D = \begin{bmatrix} 4 & 0 & 0 \\ 0 & -2 & 0 \\ 0 & 0 & 3 \end{bmatrix},$$

find D^{-1}.

24. If

$$A^{-1} = \begin{bmatrix} 3 & 2 \\ 1 & 3 \end{bmatrix} \quad \text{and} \quad B^{-1} = \begin{bmatrix} 2 & 5 \\ 3 & -2 \end{bmatrix},$$

find $(AB)^{-1}$.

25. Solve $AX = B$ for X if

$$A^{-1} = \begin{bmatrix} 2 & 3 \\ 4 & 1 \end{bmatrix} \quad \text{and} \quad B = \begin{bmatrix} 5 \\ 3 \end{bmatrix}.$$

THEORETICAL EXERCISES

T.1. Suppose that **A** and **B** are square matrices and **AB** = **0**. If **B** is nonsingular, find **A**.

T.2. Prove Corollary 1.2.

T.3. Let **A** be an $n \times n$ matrix. Prove that if **A** is singular, then the homogeneous system **AX** = **0** has a nontrivial solution.
(*Hint*: Use Theorem 1.11.)

T.4. Prove that the matrix

$$\mathbf{A} = \begin{bmatrix} a & b \\ c & d \end{bmatrix}$$

is nonsingular if and only if $ad - bc \neq 0$. If this condition holds, show that

$$\mathbf{A}^{-1} = \begin{bmatrix} \dfrac{d}{ad - bc} & \dfrac{-b}{ad - bc} \\ \dfrac{-c}{ad - bc} & \dfrac{a}{ad - bc} \end{bmatrix}.$$

T.5. Describe all skew symmetric (see Exercise T.14 in Section 1.3) scalar (see Section 1.2) matrices.

T.6. Describe all matrices that are both upper and lower triangular (see Exercise T.4 in Section 1.2).

T.7. Show that if **A** is singular and **AX** = **B**, **B** ≠ **0** has one solution, then it has infinitely many.
(*Hint*: Use Exercise T.9 in Section 1.4.)

KEY IDEAS FOR REVIEW

☐ **Method of elimination.** To solve a linear system, repeatedly perform the following operations.
1. Interchange two equations.
2. Multiply an equation by a nonzero constant.
3. Add a multiple of one equation to another equation.

☐ **Matrix operations.** **Addition** (see page 13); **multiplication** (see page 15); **scalar multiplication** (see page 19); **transpose** (see page 19).

☐ **Theorem 1.1.** Properties of matrix addition. See page 24.

☐ **Theorem 1.2.** Properties of matrix multiplication. See page 26.

☐ **Theorem 1.3.** Properties of scalar multiplication. See page 30.

☐ **Theorem 1.4.** Properties of transpose. See page 31.

☐ **Reduced row echelon form.** See page 34.

☐ Procedure for transforming a matrix to reduced row echelon form. See pages 37–39.

☐ **Gauss–Jordan reduction procedure** (for solving the linear system **AX** = **B**). See page 41.

☐ **Theorem 1.7.** A homogeneous system of m equations in n unknowns always has a nontrivial solution if $m < n$.

☐ **Theorem 1.9.** Properties of the inverse. See page 53.

☐ Practical method for finding $\mathbf{A}^{-1}$. See page 56.

☐ **Theorem 1.11.** An $n \times n$ matrix is nonsingular if and only if it is row equivalent to $\mathbf{I}_n$.

☐ **Theorem 1.12.** If **A** is an $n \times n$ matrix, the homogeneous system **AX** = **0** has a nontrivial solution if and only if **A** is singular.

☐ The following statements are equivalent:
1. **A** is nonsingular.
2. **AX** = **0** has only the trivial solution.
3. **A** is row equivalent to $\mathbf{I}_n$.
4. The linear system **AX** = **B** has a unique solution for every $n \times 1$ matrix **B**.

SUPPLEMENTARY EXERCISES

In Exercises 1 through 3, let

$$\mathbf{A} = \begin{bmatrix} 1 & 2 \\ 3 & -2 \end{bmatrix}, \quad \mathbf{B} = \begin{bmatrix} 3 & -5 \\ 2 & 4 \end{bmatrix}, \quad \text{and}$$

$$\mathbf{C} = \begin{bmatrix} 4 & 1 \\ 3 & 2 \end{bmatrix}.$$

1. Compute $2\mathbf{A} + \mathbf{BC}$ if possible.

2. Compute $\mathbf{A}^2 - 2\mathbf{A} + 3\mathbf{I}_2$ if possible.

3. Compute $\mathbf{A}^T + \mathbf{B}^T\mathbf{C}$ if possible.

4. (a) If $\mathbf{A}$ and $\mathbf{B}$ are $n \times n$ matrices, when is

$$(\mathbf{A} + \mathbf{B})(\mathbf{A} - \mathbf{B}) = \mathbf{A}^2 - \mathbf{B}^2?$$

(b) Let $\mathbf{A}$, $\mathbf{B}$, and $\mathbf{C}$ be $n \times n$ matrices such that $\mathbf{AC} = \mathbf{CA}$ and $\mathbf{BC} = \mathbf{CB}$. Verify that $(\mathbf{AB})\mathbf{C} = \mathbf{C}(\mathbf{AB})$.

5. (a) Write the augmented matrix of the linear system

$$\begin{aligned} x_1 + 2x_2 - x_3 + x_4 &= 7 \\ 2x_1 - x_2 \quad\quad + 2x_4 &= -8. \end{aligned}$$

(b) Write the linear system whose augmented matrix is

$$\begin{bmatrix} 3 & 2 & \vdots & -4 \\ 5 & 1 & \vdots & 2 \\ 3 & 2 & \vdots & 6 \end{bmatrix}.$$

6. If

$$\mathbf{A} = \begin{bmatrix} 1 & 3 & -2 & 5 \\ 2 & 1 & 3 & 2 \\ 4 & 7 & -1 & -8 \\ -3 & 1 & -8 & 1 \end{bmatrix},$$

find a matrix $\mathbf{C}$ in reduced row echelon form that is row equivalent to $\mathbf{A}$.

7. Find all solutions to the linear system

$$\begin{aligned} x + y - z &= 5 \\ 2x + y + z &= 2 \\ x - y - 2z &= 3. \end{aligned}$$

8. Find all solutions to the linear system

$$\begin{aligned} x + y - z + w &= 3 \\ 2x - y - z + 2w &= 4 \\ - 3y + z &= -2 \\ -3x + 3y + z - 3w &= -5. \end{aligned}$$

9. Find all values of a for which the resulting system has (a) no solution, (b) a unique solution, and (c) infinitely many solutions.

$$\begin{aligned} x + y - \quad\quad z &= 3 \\ x - y + \quad\quad 3z &= 4 \\ x + y + (a^2 - 10)z &= a. \end{aligned}$$

10. Find an equation relating b_1, b_2, and b_3 so that the linear system with augmented matrix

$$\begin{bmatrix} 1 & 1 & -2 & \vdots & b_1 \\ 2 & -1 & -1 & \vdots & b_2 \\ 4 & 1 & -5 & \vdots & b_3 \end{bmatrix}$$

has a solution.

11. If

$$\mathbf{A} = \begin{bmatrix} 5 & 3 & 1 \\ 0 & 4 & 2 \\ 0 & 0 & 4 \end{bmatrix}$$

and $\lambda = 4$, find all solutions to the homogeneous system $(\lambda \mathbf{I}_3 - \mathbf{A})\mathbf{X} = \mathbf{0}$.

12. For what values of a is the linear system

$$\begin{aligned} x_1 + \quad\quad x_3 &= a^2 \\ 2x_1 + x_2 + 3x_3 &= -3a \\ 3x_1 + x_2 + 4x_3 &= -2 \end{aligned}$$

consistent?

13. If possible, find the inverse of the matrix

$$\begin{bmatrix} 1 & 2 & 3 \\ 2 & 5 & 3 \\ 1 & 0 & 8 \end{bmatrix}.$$

14. If possible, find the inverse of the matrix

$$\begin{bmatrix} -1 & 2 & 1 \\ 1 & 0 & 1 \\ -3 & 2 & 3 \end{bmatrix}.$$

15. Does the homogeneous system

$$\begin{bmatrix} 1 & -1 & 3 \\ 1 & 2 & -3 \\ 2 & 1 & 0 \end{bmatrix}$$

have a nontrivial solution?

16. If

$$\mathbf{A}^{-1} = \begin{bmatrix} 1 & 2 & -1 \\ 3 & 4 & 2 \\ 0 & 1 & -2 \end{bmatrix}$$

and

$$\mathbf{B}^{-1} = \begin{bmatrix} 0 & 1 & 1 \\ 1 & 0 & 1 \\ -2 & 3 & 2 \end{bmatrix},$$

Compute $(\mathbf{AB})^{-1}$.

17. Solve $\mathbf{AX} = \mathbf{B}$ for $\mathbf{X}$ if

$$\mathbf{A}^{-1} = \begin{bmatrix} 1 & 2 & 0 \\ 0 & 1 & 0 \\ 3 & 1 & -1 \end{bmatrix} \quad \text{and} \quad \mathbf{B} = \begin{bmatrix} 2 \\ 1 \\ 3 \end{bmatrix}.$$

18. For what value(s) of λ does the homogeneous system

$$\begin{aligned} (\lambda - 2)x + \quad 2y &= 0 \\ 2x + (\lambda - 2)y &= 0 \end{aligned}$$

have a nontrivial solution?

19. If $\mathbf{A}$ is an $n \times n$ matrix and $\mathbf{A}^4 = \mathbf{0}$, verify that $(\mathbf{I}_n - \mathbf{A})^{-1} = \mathbf{I}_n + \mathbf{A} + \mathbf{A}^2 + \mathbf{A}^3$.

20. If $\mathbf{A}$ is a nonsingular $n \times n$ matrix and c is a nonzero scalar, what is $(c\mathbf{A})^{-1}$?

21. For what values of a does the linear system

$$\begin{aligned} x + \quad y &= 3 \\ 5x + 5y &= a \end{aligned}$$

have (a) no solution; (b) exactly one solution; (c) infinitely many solutions?

22. Find all values of a for which the following linear systems have solutions.

(a) $\begin{aligned} x + 2y + \quad z &= a^2 \\ x + \quad y + 3z &= a \\ 3x + 4y + 7z &= 8. \end{aligned}$ (b) $\begin{aligned} x + 2y + \quad z &= a^2 \\ x + \quad y + 3z &= a \\ 3x + 4y + 8z &= 8. \end{aligned}$

23. Find all values of a for which the following homogeneous system has nontrivial solutions.

$$\begin{aligned} (1 - a)x + \quad z &= 0 \\ -ay + \quad z &= 0 \\ y - az &= 0. \end{aligned}$$

24. Determine the number of entries on or above the main diagonal of a $k \times k$ matrix when (a) $k = 2$; (b) $k = 3$; (c) $k = 4$; (d) $k = n$.

25. Let

$$\mathbf{A} = \begin{bmatrix} 0 & 2 \\ 0 & 5 \end{bmatrix}.$$

(a) Find a $2 \times k$ matrix $\mathbf{B} \neq \mathbf{0}$ such that $\mathbf{AB} = \mathbf{0}$ for $k = 1, 2, 3, 4$.

(b) Are your answers to part (a) unique? Explain.

26. Find all 2×2 matrices with real entries of the form

$$\mathbf{A} = \begin{bmatrix} a & b \\ 0 & c \end{bmatrix}$$

such that $\mathbf{A}^2 = \mathbf{I}_2$.

27. An $n \times n$ matrix $\mathbf{A}$ (with real entries) is called a **square root** of the $n \times n$ matrix $\mathbf{B}$ (with real entries) if $\mathbf{A}^2 = \mathbf{B}$.

(a) Find a square root of $\mathbf{B} = \begin{bmatrix} 1 & 1 \\ 0 & 1 \end{bmatrix}$.

(b) Find a square root of $\mathbf{B} = \begin{bmatrix} 1 & 0 & 0 \\ 0 & 0 & 0 \\ 0 & 0 & 0 \end{bmatrix}$.

(c) Find a square root of $\mathbf{B} = \mathbf{I}_4$.

(d) Show that there is no square root of

$$\mathbf{B} = \begin{bmatrix} 0 & 1 \\ 0 & 0 \end{bmatrix}.$$

28. Compute the trace (see Supplementary Exercise T.1) of each of the following matrices.

(a) $\begin{bmatrix} 1 & 0 \\ 2 & 3 \end{bmatrix}$.   (b) $\begin{bmatrix} 2 & 2 & 3 \\ 2 & 4 & 4 \\ 3 & -2 & -5 \end{bmatrix}$.   (c) $\begin{bmatrix} 1 & 0 & 0 \\ 0 & 1 & 0 \\ 0 & 0 & 1 \end{bmatrix}$.

THEORETICAL EXERCISES

T.1. If $A = [a_{ij}]$ is an $n \times n$ matrix, then the **trace** of A, $\text{Tr}(A)$, is defined as the sum of all elements on the main diagonal of A, $\text{Tr}(A) = a_{11} + a_{22} + \cdots + a_{nn}$. Prove
   (a) $\text{Tr}(cA) = c\,\text{Tr}(A)$, where c is a real number.
   (b) $\text{Tr}(A + B) = \text{Tr}(A) + \text{Tr}(B)$.
   (c) $\text{Tr}(AB) = \text{Tr}(BA)$.
   (d) $\text{Tr}(A^T) = \text{Tr}(A)$

T.2. Show that there are no 2×2 matrices A and B such that $AB - BA = \begin{bmatrix} 1 & 0 \\ 0 & 1 \end{bmatrix}$.

T.3. If A is an $n \times n$ matrix, show that $(A^k)^T = (A^T)^k$ for any positive integer k.

T.4. Prove that every symmetric upper (or lower) triangular matrix (see Exercise T.4 in Section 1.2) is diagonal.

T.5. Let A be an upper triangular matrix. Show that A is nonsingular if and only if all the entries on the main diagonal of A are nonzero.

T.6. Let A be an $m \times n$ matrix. Show that A is row equivalent to 0 if and only if $A = 0$.

T.7. Let A and B be row equivalent $n \times n$ matrices. Prove that A is nonsingular if and only if B is nonsingular.

T.8. Let A and B be $n \times n$ matrices. Show that if AB is nonsingular, then A and B must be nonsingular. (*Hint:* Use Theorem 1.12.)

T.9. Show that if A is skew symmetric (see Exercise T.14 in Section 1.3), then the elements on the main diagonal of A are all zero.

T.10. Prove that if A is skew symmetric and nonsingular, then A^{-1} is skew symmetric.

T.11. If A is an $n \times n$ matrix, then A is called **idempotent** if $A^2 = A$.
   (a) Verify that I_n and 0 are idempotent.
   (b) Find an idempotent matrix that is not I_n or 0.
   (c) Prove that the only $n \times n$ nonsingular idempotent matrix is I_n.

T.12. If A is an $n \times n$ matrix, then A is called **nilpotent** if $A^k = 0$ for some positive integer k.
   (a) Prove that every nilpotent matrix is singular.
   (b) Verify that $A = \begin{bmatrix} 0 & 1 & 1 \\ 0 & 0 & 1 \\ 0 & 0 & 0 \end{bmatrix}$ is nilpotent.
   (c) If A is nilpotent, prove that $I_n - A$ is nonsingular. [*Hint:* Find $(I_n - A)^{-1}$ in the cases $A^k = 0$, $k = 1, 2, \ldots$, and look for a pattern.]

T.13. Prove Theorem 1.13.

CHAPTER TEST

1. Find all solutions to the linear system
$$\begin{aligned} x_1 + x_2 + x_3 - 2x_4 &= 3 \\ 2x_1 + x_2 + 3x_3 + 2x_4 &= 5 \\ -x_2 + x_3 + 6x_4 &= 3. \end{aligned}$$

2. Find all values of a for which the resulting linear system has (a) no solution, (b) a unique solution, and (c) infinitely many solutions.
$$\begin{aligned} x + \qquad z &= 4 \\ 2x + y + 3z &= 5 \\ -3x - 3y + (a^2 - 5a)z &= a - 8. \end{aligned}$$

3. If possible, find the inverse of the following matrix.
$$\begin{bmatrix} 1 & 2 & -1 \\ 0 & 1 & 1 \\ 1 & 0 & -1 \end{bmatrix}.$$

4. If
$$A = \begin{bmatrix} -1 & -2 \\ -2 & 2 \end{bmatrix},$$
find all values of λ for which the homogeneous system $(\lambda I_2 - A)X = 0$ has a nontrivial solution.

5. (a) If

$$A^{-1} = \begin{bmatrix} 1 & 3 & 0 \\ 0 & 1 & 1 \\ 1 & -1 & 4 \end{bmatrix}$$

and

$$B^{-1} = \begin{bmatrix} 2 & 1 & 1 \\ 0 & 0 & -2 \\ 1 & 1 & -1 \end{bmatrix},$$

compute $(AB)^{-1}$.

(b) Solve $AX = B$ for X if

$$A^{-1} = \begin{bmatrix} 1 & 0 & -2 \\ 2 & 1 & 3 \\ 4 & 2 & 5 \end{bmatrix} \quad \text{and} \quad B = \begin{bmatrix} 2 \\ 1 \\ 3 \end{bmatrix}.$$

6. Answer each of the following as true or false.

(a) If A and B are $n \times n$ matrices, then $(A + B)(A + B) = A^2 + 2AB + B^2$.

✱(b) If X_1 and X_2 are solutions to the linear system $AX = B$, then $X_3 = \frac{1}{4}X_1 + \frac{3}{4}X_2$ is also a solution to $AX = B$.

(c) If A is a nonsingular matrix, then the homogeneous system $AX = 0$ has a nontrivial solution.

(d) A homogeneous system of three equations in four unknowns has a nontrivial solution.

(e) If A, B, and C are $n \times n$ nonsingular matrices, then $(ABC)^{-1} = C^{-1}A^{-1}B^{-1}$.

2

Determinants

DEFINITION AND PROPERTIES

In this section we define the notion of a determinant and study some of its properties. Determinants arose first in the solution of linear systems. Although the method given in Chapter 1 for solving such systems is much more efficient than those involving determinants, determinants are useful in other aspects of linear algebra; some of these areas will be considered in Chapter 5. First, we deal rather briefly with permutations, which are used in our definition of determinant. In this chapter, by matrix we mean a square matrix, and every matrix considered in the chapter is a square matrix.

Definition Let $S = \{1, 2, \ldots, n\}$ be the set of integers from 1 to n, arranged in ascending order. A rearrangement $j_1 j_2 \cdots j_n$ of the elements of **S** is called a **permutation** of S. Thus 4231 is a permutation of $S = \{1, 2, 3, 4\}$.

We can put any one of the n elements of S in first position, any one of the remaining $n - 1$ elements in second position, any one of the remaining $n - 2$ elements in third position, and so on, until the nth position can only be filled by the last remaining element. Thus there are

$$n(n - 1)(n - 2) \cdots 2 \cdot 1 \tag{1}$$

permutations of S; we denote the set of all permutations of S by S_n.

69

The expression in Equation (1) is denoted

$$n!, \qquad n \text{ factorial}.$$

We have

$$
\begin{aligned}
1! &= 1 \\
2! &= 2 \cdot 1 = 2 \\
3! &= 3 \cdot 2 \cdot 1 = 6 \\
4! &= 4 \cdot 3 \cdot 2 \cdot 1 = 24 \\
5! &= 5 \cdot 4 \cdot 3 \cdot 2 \cdot 1 = 120 \\
6! &= 6 \cdot 5 \cdot 4 \cdot 3 \cdot 2 \cdot 1 = 720 \\
7! &= 7 \cdot 6 \cdot 5 \cdot 4 \cdot 3 \cdot 2 \cdot 1 = 5040 \\
8! &= 8 \cdot 7 \cdot 6 \cdot 5 \cdot 4 \cdot 3 \cdot 2 \cdot 1 = 40{,}320 \\
9! &= 9 \cdot 8 \cdot 7 \cdot 6 \cdot 5 \cdot 4 \cdot 3 \cdot 2 \cdot 1 = 362{,}880.
\end{aligned}
$$

Example 1 S_1 consists of only $1! = 1$ permutation of the set $\{1\}$, namely, 1; S_2 consists of $2! = 2$ permutations of the set $\{1, 2\}$, namely, 12 and 21; and S_3 consists of $3! = 3 \cdot 2 \cdot 1 = 6$ permutations of the set $\{1, 2, 3\}$, namely, 123, 231, 312, 132, 213, and 321.

A permutation $j_1 j_2 \cdots j_n$ of $S = \{1, 2, \ldots, n\}$ is said to have an **inversion** if a larger integer j_r precedes a smaller one j_s. A permutation is called **even** or **odd** according to whether the total number of inversions in it is even or odd.

If $n \geqslant 2$, it can be shown that S_n has $n!/2$ even permutations and an equal number of odd permutations.

Example 2 In S_2, the permutation 12 is even, since it has no inversions; the permutation 21 is odd, since it has one inversion.

Example 3 The even permutations in S_3 are 123 (no inversions), 231 (two inversions: 21 and 31); and 312 (two inversions: 31 and 32). The odd permutations in S_3 are 132 (one inversion: 32); 213 (one inversion: 21); and 321 (three inversions: 32, 31, and 21).

Definition Let $\mathbf{A} = [a_{ij}]$ be an $n \times n$ matrix. We define the **determinant** of $\mathbf{A}$ (written $|\mathbf{A}|$) by

$$|\mathbf{A}| = \Sigma \, (\pm) a_{1j_1} a_{2j_2} \cdots a_{nj_n}, \tag{2}$$

where the summation ranges over all permutations $j_1 j_2 \cdots j_n$ of the set $S = \{1, 2, \ldots, n\}$. The sign if taken as $+$ or $-$ according to whether the permutation $j_1 j_2 \cdots j_n$ is even or odd.

In each term $(\pm) a_{1j_1} a_{2j_2} \cdots a_{nj_n}$ of $|\mathbf{A}|$, the row subscripts are in their natural order, whereas the column subscripts are in the order $j_1 j_2 \cdots j_n$. Since the permutation $j_1 j_2 \cdots j_n$ is merely a rearrangement of the numbers from 1 to n, it has no repeats. Thus each term in $|\mathbf{A}|$ is a product of n elements of $\mathbf{A}$ each with its appropriate sign, with exactly one element from each row and exactly one element from each column. Since we sum over all the permutations of the set, $S = \{1, 2, \ldots, n\}$, $|\mathbf{A}|$ has $n!$ terms in the sum.

Example 4 If $\mathbf{A} = [a_{11}]$ is a 1×1 matrix, then S_1 has only one permutation in it, the identity permutation 1, which is even. Thus $|\mathbf{A}| = a_{11}$.

Example 5 If

$$\mathbf{A} = \begin{bmatrix} a_{11} & a_{12} \\ a_{21} & a_{22} \end{bmatrix}$$

is a 2×2 matrix, then to obtain $|\mathbf{A}|$ we write down the terms

$$a_{1\_}a_{2\_} \quad \text{and} \quad a_{1\_}a_{2\_},$$

and fill in the blanks with all possible elements of S_2; these are 12 and 21. Since 12 is an even permutation, the term $a_{11}a_{22}$ has a $+$ sign associated with it; since 21 is an odd permutation, the term $a_{12}a_{21}$ has a $-$ sign associated with it. Hence

$$|\mathbf{A}| = a_{11}a_{22} - a_{12}a_{21}.$$

We can also obtain $|\mathbf{A}|$ by forming the product of the entries on the line from left to right in the following diagram and subtracting from this number the product of the entries on the line from right to left.

$$\begin{matrix} a_{11} & a_{12} \\ a_{21} & a_{22}. \end{matrix}$$

Thus, if

$$\mathbf{A} = \begin{bmatrix} 2 & -3 \\ 4 & 5 \end{bmatrix},$$

Then $|\mathbf{A}| = (2)(5) - (-3)(4) = 22$.

Example 6 If

$$\mathbf{A} = \begin{bmatrix} a_{11} & a_{12} & a_{13} \\ a_{21} & a_{22} & a_{23} \\ a_{31} & a_{32} & a_{33} \end{bmatrix},$$

then, to compute $|\mathbf{A}|$, we write down the six terms

$$a_{1\_}a_{2\_}a_{3\_}, \quad a_{1\_}a_{2\_}a_{3\_}, \quad a_{1\_}a_{2\_}a_{3\_}, \quad a_{1\_}a_{2\_}a_{3\_},$$
$$a_{1\_}a_{2\_}a_{3\_}, \quad \text{and} \quad a_{1\_}a_{2\_}a_{3\_}.$$

All the elements of S_3 are used to fill in the blanks, and if we prefix each term by $+$ or by $-$ according to whether the permutation used is even or odd, we find that

$$|\mathbf{A}| = a_{11}a_{22}a_{33} + a_{12}a_{23}a_{31} + a_{13}a_{21}a_{32} - a_{11}a_{23}a_{32} \\ - a_{12}a_{21}a_{33} - a_{13}a_{22}a_{31}. \tag{3}$$

We can also obtain $|\mathbf{A}|$ as follows. Repeat the first and second columns of $\mathbf{A}$ as shown below. Form the sum of the products of the entries on the lines from left to right, and subtract from this number the products of the entries on the lines from right to left (verify).

$$
\begin{array}{ccccc}
a_{11} & a_{12} & a_{13} & a_{11} & a_{12} \\
a_{21} & a_{22} & a_{23} & a_{24} & a_{22} \\
a_{31} & a_{32} & a_{33} & a_{31} & a_{32}.
\end{array}
$$

It should be emphasized that for $n \geq 4$, there is no "easy" method for evaluating $|\mathbf{A}|$ as in Examples 5 and 6.

Example 7 Let

$$
\mathbf{A} = \begin{bmatrix} 1 & 2 & 3 \\ 2 & 1 & 3 \\ 3 & 1 & 2 \end{bmatrix}.
$$

Evaluate $|\mathbf{A}|$.

Solution. Substituting in (3), we find that

$$
\begin{aligned}
|\mathbf{A}| &= (1)(1)(2) + (2)(3)(3) + (3)(2)(1) \\
&\quad - (1)(3)(1) - (2)(2)(2) - (3)(1)(3) = 6.
\end{aligned}
$$

We could obtain the same result by using the easy method above (verify).

It may already have struck the reader that this is an extremely tedious way of computing determinants for a sizable value of n. In fact, $10! = 3.6288 \times 10^6$ and $20! = 2.4329 \times 10^{18}$, enormous numbers. We shall soon develop a number of properties satisfied by determinants, which will greatly reduce the computational effort.

Permutations are studied to some depth in abstract algebra courses and in courses dealing with group theory. We shall not make use of permutations in our methods for computing determinants. We require the following property of permutations. If we interchange two numbers in the permutation $j_1 j_2 \cdots j_n$, then the number of inversions is either increased or decreased by an odd number (Exercise T.1).

Example 8 The number of inversions in the permutation 54132 is 8. The number of inversions in the permutation 52134 is 5. The permutation 52134 was obtained from 54132 by interchanging 2 and 4. The number of inversions differs by 3, an odd number.

Properties of Determinants

Theorem 2.1 *The determinants of a matrix and its transpose are equal.*

Proof
Let $\mathbf{A} = [a_{ij}]$ and let $\mathbf{A}^T = [b_{ij}]$, where $b_{ij} = a_{ji}$ ($1 \leq i \leq n$, $1 \leq j \leq n$). Then from (2) we have

$$|\mathbf{A}^T| = \Sigma \ (\pm)b_{1j_1}b_{2j_2} \cdots b_{nj_n} = \Sigma \ (\pm)a_{j_11}a_{j_22} \cdots a_{j_nn}. \tag{4}$$

We can now rearrange the factors in the term $a_{j_11}a_{j_22} \cdots a_{j_nn}$ so that the row indices are in their natural order. Thus

$$b_{1j_1}b_{2j_2} \cdots b_{nj_n} = a_{j_11}a_{j_22} \cdots a_{j_nn} = a_{1k_1}a_{2k_2} \cdots a_{nk_n}.$$

We shall state, without proof, that the permutation $k_1k_2 \cdots k_n$, which determines the sign associated with $a_{1k_1}a_{2k_2} \cdots a_{nk_n}$, and $j_1j_2 \cdots j_n$, which determines the sign associated with $a_{1j_1}a_{2j_2} \cdots a_{nj_n}$, are both even or both odd. As an example,

$$b_{13}b_{24}b_{35}b_{41}b_{52} = a_{31}a_{42}a_{53}a_{14}a_{25} = a_{14}a_{25}a_{31}a_{42}a_{53};$$

the number of inversions in the permutation 45123 is 6 and the number of inversions in the permutation 34512 is also 6. Since the terms and corresponding signs in (2) and (4) agree, we conclude that $|\mathbf{A}| = |\mathbf{A}^T|$.

Example 9 Let $\mathbf{A}$ be the matrix of Example 7. Then

$$\mathbf{A}^T = \begin{bmatrix} 1 & 2 & 3 \\ 2 & 1 & 1 \\ 3 & 3 & 2 \end{bmatrix}.$$

Substituting in (3), we find that

$$\begin{aligned} |\mathbf{A}^T| &= \ (1)(1)(2) + (2)(1)(3) + (3)(2)(3) - (1)(1)(3) \\ &\quad - (2)(2)(2) - (3)(1)(3) \\ &= 6 = |\mathbf{A}|. \end{aligned}$$

Theorem 2.1 will enable us to replace "row" by "column" in many of the additional properties of determinants; we see how to do this in the following theorem.

Theorem 2.2 *If matrix $\mathbf{B}$ results from matrix $\mathbf{A}$ by interchanging two rows (columns) of $\mathbf{A}$, then $|\mathbf{B}| = -|\mathbf{A}|$.*

Proof

Suppose that $\mathbf{B}$ arises from $\mathbf{A}$ by interchanging rows r and s of $\mathbf{A}$, say $r < s$. Then we have $b_{rj} = a_{sj}$, $b_{sj} = a_{rj}$, and $b_{ij} = a_{ij}$ for $i \neq r$, $i \neq s$. Now

$$\begin{aligned} |\mathbf{B}| &= \Sigma \ (\pm)b_{1j_1}b_{2j_2} \cdots b_{rj_r} \cdots b_{sj_s} \cdots b_{nj_n} \\ &= \Sigma \ (\pm)a_{1j_1}a_{2j_2} \cdots a_{sj_r} \cdots a_{rj_s} \cdots a_{nj_n} \\ &= \Sigma \ (\pm)a_{1j_1}a_{2j_2} \cdots a_{rj_s} \cdots a_{sj_r} \cdots a_{nj_n}. \end{aligned}$$

The permutation $j_1j_2 \cdots j_s \cdots j_r \cdots j_n$ results from the permutation $j_1j_2 \cdots j_r \cdots j_s \cdots j_n$ by an interchange of two numbers; the number of inversions in the former differs by an odd number from the number of inversions in the latter (see Exercise T.1). This means that the sign of each term in $|\mathbf{B}|$ is the negative of the sign of the corresponding term in $|\mathbf{A}|$. Hence $|\mathbf{B}| = -|\mathbf{A}|$.

Now suppose that $\mathbf{B}$ is obtained from $\mathbf{A}$ by interchanging two columns of $\mathbf{A}$. Then $\mathbf{B}^T$ is obtained from $\mathbf{A}^T$ by interchanging two rows of $\mathbf{A}^T$. So $|\mathbf{B}^T| = -|\mathbf{A}^T|$, but $|\mathbf{B}^T| = |\mathbf{B}|$ and $|\mathbf{A}^T| = |\mathbf{A}|$. Hence $|\mathbf{B}| = -|\mathbf{A}|$.

In the results to follow, proofs will be given only for the rows of $\mathbf{A}$; the proofs of the corresponding column case proceed as at the end of the proof of Theorem 2.2.

Example 10 We have

$$\begin{vmatrix} 2 & -1 \\ 3 & 2 \end{vmatrix} = 7 \quad \text{and} \quad \begin{vmatrix} 3 & 2 \\ 2 & -1 \end{vmatrix} = -7.$$

Theorem 2.3 *If two rows (columns) of* $\mathbf{A}$ *are equal, then* $|\mathbf{A}| = 0$.

Proof
Suppose that rows r and s of $\mathbf{A}$ are equal. Interchange rows r and s of $\mathbf{A}$ to obtain a matrix $\mathbf{B}$. Then $|\mathbf{B}| = -|\mathbf{A}|$. On the other hand, $\mathbf{B} = \mathbf{A}$, so $|\mathbf{B}| = |\mathbf{A}|$. Thus $|\mathbf{A}| = -|\mathbf{A}|$, so $|\mathbf{A}| = 0$.

Example 11 Using Theorem 2.3, it follows that

$$\begin{vmatrix} 1 & 2 & 3 \\ -1 & 0 & 7 \\ 1 & 2 & 3 \end{vmatrix} = 0.$$

Theorem 2.4 *If a row (column) of* $\mathbf{A}$ *consists entirely of zeros, then* $|\mathbf{A}| = 0$.

Proof
Let the rth row of $\mathbf{A}$ consist entirely of zeros. Since each term in the definition for the determinant of $\mathbf{A}$ contains a factor from the rth row, each term in $|\mathbf{A}|$ is zero. Hence $|\mathbf{A}| = 0$.

Example 12 Using Theorem 2.4, it follows that

$$\begin{vmatrix} 1 & 2 & 3 \\ 4 & 5 & 6 \\ 0 & 0 & 0 \end{vmatrix} = 0.$$

Theorem 2.5 *If* $\mathbf{B}$ *is obtained from* $\mathbf{A}$ *by multiplying a row (column) of* $\mathbf{A}$ *by a real number* c, *then* $|\mathbf{B}| = c|\mathbf{A}|$.

Proof
Suppose that the rth row of $\mathbf{A} = [a_{ij}]$ is multiplied by c to obtain $\mathbf{B} = [b_{ij}]$. Then $b_{ij} = a_{ij}$ if $i \neq r$ and $b_{rj} = ca_{rj}$. We obtain $|\mathbf{B}|$ from Equation (2) as

$$\begin{aligned} |\mathbf{B}| &= \Sigma \, (\pm) b_{1j_1} b_{2j_2} \cdots b_{rj_r} \cdots b_{nj_n} \\ &= \Sigma \, (\pm) a_{1j_1} a_{2j_2} \cdots (ca_{rj_r}) \cdots a_{nj_n} \\ &= c(\Sigma \, (\pm) a_{1j_1} a_{2j_2} \cdots a_{rj_r} \cdots a_{nj_n}) = c|\mathbf{A}|. \end{aligned}$$

We can use Theorem 2.5 to simplify the computation of $|\mathbf{A}|$ by factoring out common factors from rows and columns of $\mathbf{A}$.

Example 13   We have

$$\begin{vmatrix} 2 & 6 \\ 1 & 12 \end{vmatrix} = 2 \begin{vmatrix} 1 & 3 \\ 1 & 12 \end{vmatrix} = (2)(3) \begin{vmatrix} 1 & 1 \\ 1 & 4 \end{vmatrix} = 6(4 - 1) = 18.$$

Example 14   We have

$$\begin{vmatrix} 1 & 2 & 3 \\ 1 & 5 & 3 \\ 2 & 8 & 6 \end{vmatrix} = 2 \begin{vmatrix} 1 & 2 & 3 \\ 1 & 5 & 3 \\ 1 & 4 & 3 \end{vmatrix} = (2)(3) \begin{vmatrix} 1 & 2 & 1 \\ 1 & 5 & 1 \\ 1 & 4 & 1 \end{vmatrix} = (2)(3)(0) = 0.$$

Here we first factored out 2 from the third row, then 3 from the third column, and then used Theorem 2.3, since the first and third columns are equal.

Theorem 2.6   *If* $\mathbf{B} = [b_{ij}]$ *is obtained from* $\mathbf{A} = [a_{ij}]$ *by adding to each element of the* rth *row (column) of* $\mathbf{A}$ *a constant* c *times the corresponding element of the* sth *row (column)* $r \neq s$ *of* $\mathbf{A}$, *then* $|\mathbf{B}| = |\mathbf{A}|$.

Proof

We prove the theorem for rows. We have $b_{ij} = a_{ij}$ for $i \neq r$, and $b_{rj} = a_{rj} + ca_{sj}$, $r \neq s$, say $r < s$. Then

$$\begin{aligned}
|\mathbf{B}| &= \Sigma (\pm)b_{1j_1}b_{2j_2} \cdots b_{rj_r} \cdots b_{nj_n} \\
&= \Sigma (\pm)a_{1j_1}a_{2j_2} \cdots (a_{rj_r} + ca_{sj_r}) \cdots a_{sj_s} \cdots a_{nj_n} \\
&= \Sigma (\pm)a_{1j_1}a_{2j_2} \cdots a_{rj_r} \cdots a_{sj_s} \cdots a_{nj_n} \\
&\quad + \Sigma (\pm)a_{1j_1}a_{2j_2} \cdots (ca_{sj_r}) \cdots a_{sj_s} \cdots a_{nj_n}.
\end{aligned}$$

The first sum in this last expression is $|\mathbf{A}|$; the second sum can be written as $c[\Sigma (\pm)a_{1j_1}a_{2j_2} \cdots a_{sj_r} \cdots a_{sj_s} \cdots a_{nj_n}]$. Note that

$$\Sigma (\pm)a_{1j_1}a_{2j_2} \cdots a_{sj_r} \cdots a_{sj_s} \cdots a_{nj_n} = \begin{vmatrix} a_{11} & a_{12} & \cdots & a_{1n} \\ a_{21} & a_{22} & \cdots & a_{2n} \\ \vdots & \vdots & & \vdots \\ a_{s1} & a_{s2} & \cdots & a_{sn} \\ \vdots & \vdots & & \vdots \\ a_{s1} & a_{s2} & \cdots & a_{sn} \\ \vdots & \vdots & & \vdots \\ a_{n1} & a_{n2} & \cdots & a_{nn} \end{vmatrix} \begin{matrix} \\ \\ \\ \leftarrow r\text{th row} \\ \\ \leftarrow s\text{th row} \\ \\ \end{matrix}$$

$$= 0,$$

because there are two equal rows.

Hence $|\mathbf{B}| = |\mathbf{A}| + 0 = |\mathbf{A}|$.

Example 15 We have

$$\begin{vmatrix} 1 & 2 & 3 \\ 2 & -1 & 3 \\ 1 & 0 & 1 \end{vmatrix} = \begin{vmatrix} 5 & 0 & 9 \\ 2 & -1 & 3 \\ 1 & 0 & 1 \end{vmatrix},$$

obtained by adding twice the second row to its first row. By applying the definition of determinant to the second determinant, both are seen to have the value 4.

Theorem 2.7 *If a matrix* $\mathbf{A} = [a_{ij}]$ *is upper (lower) triangular (see Exercise T.4, Section 1.2), then*

$$|\mathbf{A}| = a_{11}a_{22} \cdots a_{nn};$$

that is, the determinant of a triangular matrix is the product of the elements on the main diagonal.

Proof

Let $\mathbf{A} = [a_{ij}]$ be upper triangular (that is, $a_{ij} = 0$ for $i > j$). Then a term $a_{1j_1}a_{2j_2} \cdots a_{nj_n}$ in the expression for $|\mathbf{A}|$ can be nonzero only for $1 \leq j_1, 2 \leq j_2, \cdots, n \leq j_n$. Now $j_1 j_2 \cdots j_n$ must be a permutation, or rearrangement, of $\{1, 2, \ldots, n\}$. Hence we must have $j_n = n, j_{n-1} = n - 1, \ldots, j_2 = 2, j_1 = 1$. Thus the only term of $|\mathbf{A}|$ that can be nonzero is the product of the elements on the main diagonal of $\mathbf{A}$. Since the permutation $12 \cdots n$ has no inversions, the sign associated with it is $+$. Hence $|\mathbf{A}| = a_{11}a_{22} \cdots a_{nn}$.

We leave the proof of the lower triangular case to the reader (Exercise T.2). ∎

Theorems 2.2, 2.5, 2.6, and 2.7 are very useful in the evaluation of $|\mathbf{A}|$. What we do is transform $\mathbf{A}$ by means of our elementary row operations to a triangular matrix. Of course, we must keep track of how the determinant of the resulting matrices changes as we perform the elementary row operations.

Example 16 We have

$$\begin{vmatrix} 4 & 3 & 2 \\ 3 & -2 & 5 \\ 2 & 4 & 6 \end{vmatrix}_{\frac{1}{2}r_3 \to r_3} = 2\begin{vmatrix} 4 & 3 & 2 \\ 3 & -2 & 5 \\ 1 & 2 & 3 \end{vmatrix}_{r_1 \leftrightarrow r_3} = -2\begin{vmatrix} 1 & 2 & 3 \\ 3 & -2 & 5 \\ 4 & 3 & 2 \end{vmatrix}_{r_2 - 3r_1 \to r_2}$$

$$= -2\begin{vmatrix} 1 & 2 & 3 \\ 0 & -8 & -4 \\ 4 & 3 & 2 \end{vmatrix}_{r_3 - 4r_1 \to r_3} = -2\begin{vmatrix} 1 & 2 & 3 \\ 0 & -8 & -4 \\ 0 & -5 & -10 \end{vmatrix}_{\frac{1}{4}r_2 \to r_2}$$

$$= (-2)(4)\begin{vmatrix} 1 & 2 & 3 \\ 0 & -2 & -1 \\ 0 & -5 & -10 \end{vmatrix}_{\frac{1}{5}r_3 \to r_3} = (-2)(4)(5)\begin{vmatrix} 1 & 2 & 3 \\ 0 & -2 & -1 \\ 0 & -1 & -2 \end{vmatrix}_{r_3 - \frac{1}{2}r_2 \to r_3}$$

$$= (-2)(4)(5)\begin{vmatrix} 1 & 2 & 3 \\ 0 & -2 & -1 \\ 0 & 0 & -\frac{3}{2} \end{vmatrix} = (-2)(4)(5)(1)(-2)(-\tfrac{3}{2}) = -120.$$

Here $\frac{1}{2}\mathbf{r}_3 \to \mathbf{r}_3$ means that $\frac{1}{2}$ times the third row $\mathbf{r}_3$ replaces the third row; $\mathbf{r}_1 \leftrightarrow \mathbf{r}_3$ means that the first row $\mathbf{r}_1$ and the third row are interchanged; $\mathbf{r}_2 - 3\mathbf{r}_1 \to \mathbf{r}_2$ means that the second row $\mathbf{r}_2$ minus three times the first row replaces the second row.

We shall omit the proof of the following theorem.

Theorem 2.8 *The determinant of a product of two matrices is the product of their determinants; that is,*

$$|\mathbf{AB}| = |\mathbf{A}||\mathbf{B}|.$$

Example 17 Let

$$\mathbf{A} = \begin{bmatrix} 1 & 2 \\ 3 & 4 \end{bmatrix} \quad \text{and} \quad \mathbf{B} = \begin{bmatrix} 2 & -1 \\ 1 & 2 \end{bmatrix}.$$

Then

$$|\mathbf{A}| = -2 \quad \text{and} \quad |\mathbf{B}| = 5.$$

Also,

$$\mathbf{AB} = \begin{bmatrix} 4 & 3 \\ 10 & 5 \end{bmatrix}$$

and

$$|\mathbf{AB}| = -10 = |\mathbf{A}||\mathbf{B}|.$$

Corollary 2.1 *If $\mathbf{A}$ is nonsingular, then $|\mathbf{A}| \neq 0$ and*

$$|\mathbf{A}^{-1}| = \frac{1}{|\mathbf{A}|}.$$

Proof
Exercise T.4.

Example 18 Let

$$\mathbf{A} = \begin{bmatrix} 1 & 2 \\ 3 & 4 \end{bmatrix}.$$

Then $|\mathbf{A}| = -2$ and

$$\mathbf{A}^{-1} = \begin{bmatrix} -2 & 1 \\ \frac{3}{2} & -\frac{1}{2} \end{bmatrix}.$$

Now

$$|\mathbf{A}^{-1}| = -\frac{1}{2} = \frac{1}{|\mathbf{A}|}.$$

2.1 EXERCISES

1. Find the number of inversions in each of the following permutations of $S = \{1, 2, 3, 4, 5\}$.
   (a) 52134.   (b) 45213.   (c) 42135.
   (d) 13542.   (e) 35241.   (f) 12345.

2. Determine whether each of the following permutations of $S = \{1, 2, 3, 4\}$ is even or odd.
   (a) 4213.   (b) 1243.   (c) 1234.
   (d) 3214.   (e) 1423.   (f) 2431.

3. Determine the sign associated with each of the following permutations of $S = \{1, 2, 3, 4, 5\}$.
   (a) 25431.   (b) 31245.   (c) 21345.
   (d) 52341.   (e) 34125.   (f) 41253.

4. In each of the following pairs of permutations of $S = \{1, 2, 3, 4, 5, 6\}$, verify that the number of inversions differs by an odd number.
   (a) 436215 and 416235.
   (b) 623415 and 523416.
   (c) 321564 and 341562.
   (d) 123564 and 423561.

In Exercises 5 and 6, evaluate the determinants using Equation (2).

5. (a) $\begin{vmatrix} 2 & -1 \\ 3 & 2 \end{vmatrix}$.   (b) $\begin{vmatrix} 0 & 3 & 0 \\ 2 & 0 & 0 \\ 0 & 0 & -5 \end{vmatrix}$.

   (c) $\begin{vmatrix} 4 & 2 & 0 \\ 0 & -2 & 5 \\ 0 & 0 & 3 \end{vmatrix}$.   (d) $\begin{vmatrix} 4 & 2 & 2 & 0 \\ 2 & 0 & 0 & 0 \\ 3 & 0 & 0 & 1 \\ 0 & 0 & 1 & 0 \end{vmatrix}$.

6. (a) $\begin{vmatrix} 2 & 1 \\ 4 & 3 \end{vmatrix}$.   (b) $\begin{vmatrix} 0 & 0 & -2 \\ 0 & 3 & 0 \\ 4 & 0 & 0 \end{vmatrix}$.

   (c) $\begin{vmatrix} 3 & 4 & 2 \\ 2 & 5 & 0 \\ 3 & 0 & 0 \end{vmatrix}$.   (d) $\begin{vmatrix} -4 & 2 & 0 & 0 \\ 2 & 3 & 1 & 0 \\ 3 & 1 & 0 & 2 \\ 1 & 3 & 0 & 3 \end{vmatrix}$.

7. Let $\mathbf{A} = [a_{ij}]$ be a 4×4 matrix. Write the general expression for $|\mathbf{A}|$ using Equation (2).

8. If

$$|\mathbf{A}| = \begin{vmatrix} a_1 & a_2 & a_3 \\ b_1 & b_2 & b_3 \\ c_1 & c_2 & c_3 \end{vmatrix} = -4,$$

find the determinants of the following matrices:

$$\mathbf{B} = \begin{bmatrix} a_3 & a_2 & a_1 \\ b_3 & b_2 & b_1 \\ c_3 & c_2 & c_1 \end{bmatrix},$$

$$\mathbf{C} = \begin{bmatrix} a_1 & a_2 & a_3 \\ b_1 & b_2 & b_3 \\ 2c_1 & 2c_2 & 2c_3 \end{bmatrix},$$

and

$$\mathbf{D} = \begin{bmatrix} a_1 & a_2 & a_3 \\ b_1 + 4c_1 & b_2 + 4c_2 & b_3 + 4c_3 \\ c_1 & c_2 & c_3 \end{bmatrix}.$$

9. If

$$|\mathbf{A}| = \begin{vmatrix} a_1 & a_2 & a_3 \\ b_1 & b_2 & b_3 \\ c_1 & c_2 & c_3 \end{vmatrix} = 3,$$

find the determinants of the following matrices:

$$\mathbf{B} = \begin{bmatrix} a_1 + 2b_1 - 3c_1 & a_2 + 2b_2 - 3c_2 & a_3 + 2b_3 - 3c_3 \\ b_1 & b_2 & b_3 \\ c_1 & c_2 & c_3 \end{bmatrix},$$

$$\mathbf{C} = \begin{bmatrix} a_1 & 3a_2 & a_3 \\ b_1 & 3b_2 & b_3 \\ c_1 & 3c_2 & c_3 \end{bmatrix},$$

and

$$\mathbf{D} = \begin{bmatrix} a_1 & a_2 & a_3 \\ c_1 & c_2 & c_3 \\ b_1 & b_2 & b_3 \end{bmatrix}.$$

10. If

$$\mathbf{A} = \begin{bmatrix} 1 & -1 & 2 \\ 3 & 4 & 1 \\ 2 & 5 & 1 \end{bmatrix},$$

verify that

$$|\mathbf{A}| = |\mathbf{A}^T|.$$

11. Evaluate:

(a) $\begin{vmatrix} \lambda - 1 & 2 \\ 3 & \lambda - 2 \end{vmatrix}.$

(b) $|\lambda \mathbf{I}_2 - \mathbf{A}|$, where $\mathbf{A} = \begin{bmatrix} 4 & 2 \\ -1 & 1 \end{bmatrix}.$

12. Evaluate:

(a) $\begin{vmatrix} \lambda - 1 & -1 & -2 \\ 0 & \lambda - 2 & 2 \\ 0 & 0 & \lambda - 3 \end{vmatrix}.$

(b) $|\lambda \mathbf{I}_3 - \mathbf{A}|$, where $\mathbf{A} = \begin{bmatrix} -1 & 0 & 1 \\ -2 & 0 & -1 \\ 0 & 0 & 1 \end{bmatrix}.$

13. For each of the matrices in Exercise 11, find all values of λ for which the determinant is 0.

14. For each of the matrices in Exercise 12, find all values of λ for which the determinant is 0.

In Exercises 15 through 18, evaluate the given determinants using the properties of this section.

15. (a) $\begin{vmatrix} 4 & -3 & 5 \\ 5 & 2 & 0 \\ 2 & 0 & 4 \end{vmatrix}.$ (b) $\begin{vmatrix} 2 & 0 & 1 & 4 \\ 3 & 2 & -4 & -2 \\ 2 & 3 & -1 & 0 \\ 11 & 8 & -4 & 6 \end{vmatrix}.$

(c) $\begin{vmatrix} 4 & 1 & 2 \\ 0 & 2 & 3 \\ 0 & 0 & -3 \end{vmatrix}.$

16. (a) $\begin{vmatrix} 4 & 0 & 0 & 0 \\ -1 & 2 & 0 & 0 \\ 1 & 2 & -3 & 0 \\ 1 & 5 & 3 & 5 \end{vmatrix}.$ (b) $\begin{vmatrix} 4 & 1 & 3 \\ 2 & 3 & 0 \\ 1 & 3 & 2 \end{vmatrix}.$

(c) $\begin{vmatrix} 1 & 2 & 3 \\ 2 & 1 & 0 \\ -3 & 1 & 2 \end{vmatrix}.$

17. (a) $\begin{vmatrix} 4 & 2 & 3 & -4 \\ 3 & -2 & 1 & 5 \\ -2 & 0 & 1 & -3 \\ 8 & -2 & 6 & 4 \end{vmatrix}.$ (b) $\begin{vmatrix} 1 & 3 & -4 \\ -2 & 1 & 2 \\ -9 & 15 & 0 \end{vmatrix}.$

(c) $\begin{vmatrix} 1 & 1 & 2 \\ 0 & 2 & -2 \\ 0 & 0 & 3 \end{vmatrix}.$

18. (a) $\begin{vmatrix} 1 & 0 & 1 \\ 1 & 1 & 0 \\ 2 & 1 & 0 \end{vmatrix}.$ (b) $\begin{vmatrix} 2 & 0 & 0 & 0 \\ -5 & 3 & 0 & 0 \\ 3 & 2 & 4 & 0 \\ 4 & 2 & 1 & -5 \end{vmatrix}.$

(c) $\begin{vmatrix} 1 & 2 & -1 \\ 3 & 2 & 0 \\ 1 & 4 & 3 \end{vmatrix}.$

19. Verify that $|\mathbf{AB}| = |\mathbf{A}||\mathbf{B}|$ for the following:

(a) $\mathbf{A} = \begin{bmatrix} 1 & -2 & 3 \\ -2 & 3 & 1 \\ 0 & 1 & 0 \end{bmatrix},$ $\mathbf{B} = \begin{bmatrix} 1 & 0 & 2 \\ 3 & -2 & 5 \\ 2 & 1 & 3 \end{bmatrix}.$

(b) $\mathbf{A} = \begin{bmatrix} 2 & 3 & 6 \\ 0 & 3 & 2 \\ 0 & 0 & -4 \end{bmatrix},$ $\mathbf{B} = \begin{bmatrix} 3 & 0 & 0 \\ 4 & 5 & 0 \\ 2 & 1 & -2 \end{bmatrix}.$

20. If $|\mathbf{A}| = -4$, find
(a) $|\mathbf{A}^2|$.   (b) $|\mathbf{A}^4|$.   (c) $|\mathbf{A}^{-1}|$.

21. If $\mathbf{A}$ and $\mathbf{B}$ are $n \times n$ matrices with $|\mathbf{A}| = 2$ and $|\mathbf{B}| = -3$, calculate $|\mathbf{A}^{-1}\mathbf{B}^T|$.

THEORETICAL EXERCISES

T.1. Prove that if we interchange two numbers in the permutation $j_1 j_2 \cdots j_n$, then the number of inversions is either increased or decreased by an odd number.

(*Hint:* First show that if two adjacent numbers are interchanged, the number of inversions is either increased or decreased by 1. Then prove that an interchange of any two numbers can be achieved by an odd number of successive interchanges of adjacent numbers.)

T.2. Prove Theorem 2.7 for the lower triangular case.

T.3. Show that if c is a real number and $\mathbf{A}$ is $n \times n$, then $|c\mathbf{A}| = c^n |\mathbf{A}|$.

T.4. Prove Corollary 2.1.

T.5. Show that if $|\mathbf{AB}| = 0$, then $|\mathbf{A}| = 0$ or $|\mathbf{B}| = 0$.

T.6. Is $|\mathbf{AB}| = |\mathbf{BA}|$? Justify your answer.

T.7. Show that if $\mathbf{A}$ is a matrix such that in each row and in each column one and only one element is $\neq 0$, then $|\mathbf{A}| \neq 0$.

T.8. Show that if $\mathbf{AB} = \mathbf{I}_n$, then $|\mathbf{A}| \neq 0$ and $|\mathbf{B}| \neq 0$.

T.9. (a) Show that if $\mathbf{A} = \mathbf{A}^{-1}$, then $|\mathbf{A}| = \pm 1$.
(b) Show that if $\mathbf{A}^T = \mathbf{A}^{-1}$, then $|\mathbf{A}| = \pm 1$.

T.10. Show that if $\mathbf{A}$ is a nonsingular matrix such that $\mathbf{A}^2 = \mathbf{A}$, then $|\mathbf{A}| = 1$.

T.11. Show that

$$|\mathbf{A}^T \mathbf{B}^T| = |\mathbf{A}||\mathbf{B}^T| = |\mathbf{A}^T||\mathbf{B}|.$$

T.12. Show that

$$\begin{vmatrix} 1 & a & a^2 \\ 1 & b & b^2 \\ 1 & c & c^2 \end{vmatrix} = (b - a)(c - a)(c - b).$$

This determinant is called a **Vandermonde†
determinant**.

T.13. Let $\mathbf{A} = [a_{ij}]$ be an upper triangular matrix. Prove that $\mathbf{A}$ is nonsingular if and only if $a_{ii} \neq 0$, $1 \leq i \leq n$.

T.14. Show that if $|\mathbf{A}| = 0$, then $|\mathbf{AB}| = 0$.

T.15. Show that if $\mathbf{A}^n = \mathbf{0}$ for some positive integer n, then $|\mathbf{A}| = 0$.

T.16. Show that if $\mathbf{A}$ is $n \times n$ with n odd and skew symmetric ($\mathbf{A}^T = -\mathbf{A}$, see Section 1.3, Exercise T.14), then $|\mathbf{A}| = 0$.

2.2

COFACTOR EXPANSION AND APPLICATIONS

So far, we have been evaluating determinants by using Equation (2) of the preceding section and have been aided by the properties established there. We now develop a different method for evaluating the determinant of an $n \times n$ matrix, which reduces the problem to the evaluation of determinants of matrices of order $n - 1$. We can then repeat the process for these $(n - 1) \times (n - 1)$ matrices until we get to 2×2 matrices.

Definition Let $\mathbf{A} = [a_{ij}]$ be an $n \times n$ matrix. Let $\mathbf{M}_{ij}$ be the $(n - 1) \times (n - 1)$ submatrix of $\mathbf{A}$ obtained by deleting the ith row and jth column of $\mathbf{A}$. The determinant $|\mathbf{M}_{ij}|$ is called the

† Alexandre-Théophile Vandermonde (1735–1796) was born in Paris. His father, a physician, tried to steer him toward a musical career. His published mathematical work consisted of four papers that were presented over a two-year period. He is generally considered the founder of the theory of determinants and also developed formulas for solving general quadratic, cubic, and quartic equations. Vandermonde was a cofounder of the Conservatoire des Arts et Métiers and was its director from 1782. In 1795 he helped to develop a course in political economy. He was an active revolutionary in the French Revolution and was a member of the Commune of Paris and the club of the Jacobins.

minor of a_{ij}. The **cofactor** A_{ij} of a_{ij} is defined as

$$A_{ij} = (-1)^{i+j}|M_{ij}|.$$

Example 1 Let

$$A = \begin{bmatrix} 3 & -1 & 2 \\ 4 & 5 & 6 \\ 7 & 1 & 2 \end{bmatrix}.$$

Then

$$|M_{12}| = \begin{vmatrix} 4 & 6 \\ 7 & 2 \end{vmatrix} = 8 - 42 = -34, \qquad |M_{23}| = \begin{vmatrix} 3 & -1 \\ 7 & 1 \end{vmatrix} = 3 + 7 = 10,$$

and

$$|M_{31}| = \begin{vmatrix} -1 & 2 \\ 5 & 6 \end{vmatrix} = -6 - 10 = -16.$$

Also,

$$A_{12} = (-1)^{1+2}|M_{12}| = (-1)(-34) = 34,$$
$$A_{23} = (-1)^{2+3}|M_{23}| = (-1)(10) = -10,$$

and

$$A_{31} = (-1)^{1+3}|M_{31}| = (1)(-16) = -16.$$

If we think of the sign $(-1)^{i+j}$ as being located in position (i, j) of an $n \times n$ matrix, then the signs form a checkerboard pattern that has a $+$ in the $(1, 1)$ position. The patterns for $n = 3$ and $n = 4$ are as follows:

$$
\begin{matrix}
+ & - & + \\
- & + & - \\
+ & - & +
\end{matrix}
\qquad
\begin{matrix}
+ & - & + & - \\
- & + & - & + \\
+ & - & + & - \\
- & + & - & +
\end{matrix}
$$

$$n = 3 \qquad\qquad n = 4$$

The following theorem, whose proof is omitted, gives a more practical way of evaluating determinants.

Theorem 2.9 *Let $A = [a_{ij}]$ be an $n \times n$ matrix. Then for each $1 \leq i \leq n$,*

$$|A| = a_{i1}A_{i1} + a_{i2}A_{i2} + \cdots + a_{in}A_{in}$$
$$\text{(expansion of } |A| \text{ about the } i\text{th row).} \tag{1}$$

and for each $1 \leq j \leq n$,

$$|A| = a_{1j}A_{1j} + a_{2j}A_{2j} + \cdots + a_{nj}A_{nj}$$
$$\text{(expansion of } |A| \text{ about the } j\text{th column).} \tag{2}$$

Example 2 To evaluate the determinant

$$\begin{vmatrix} 1 & 2 & -3 & 4 \\ -4 & 2 & 1 & 3 \\ 3 & 0 & 0 & -3 \\ 2 & 0 & -2 & 3 \end{vmatrix},$$

we note that it is best to expand about either the second column or the third row because they each have two zeros. Obviously, the optimal course of action is to expand about a row or column having the largest number of zeros because in that case the cofactors A_{ij} of those a_{ij} which are zero do not have to be evaluated, since $a_{ij}A_{ij} = (0)(A_{ij}) = 0$. Thus, expanding about the third row, we have

$$\begin{vmatrix} 1 & 2 & -3 & 4 \\ -4 & 2 & 1 & 3 \\ 3 & 0 & 0 & -3 \\ 2 & 0 & -2 & 3 \end{vmatrix}$$

$$= (-1)^{3+1}(3)\begin{vmatrix} 2 & -3 & 4 \\ 2 & 1 & 3 \\ 0 & -2 & 3 \end{vmatrix} + (-1)^{3+2}(0)\begin{vmatrix} 1 & -3 & 4 \\ -4 & 1 & 3 \\ 2 & -2 & 3 \end{vmatrix} \tag{3}$$

$$+ (-1)^{3+3}(0)\begin{vmatrix} 1 & 2 & 4 \\ -4 & 2 & 3 \\ 2 & 0 & 3 \end{vmatrix} + (-1)^{3+4}(-3)\begin{vmatrix} 1 & 2 & -3 \\ -4 & 2 & 1 \\ 2 & 0 & -2 \end{vmatrix}.$$

We now evaluate

$$\begin{vmatrix} 2 & -3 & 4 \\ 2 & 1 & 3 \\ 0 & -2 & 3 \end{vmatrix}$$

by expanding about the first column, obtaining

$$(-1)^{1+1}(2)\begin{vmatrix} 1 & 3 \\ -2 & 3 \end{vmatrix} + (-1)^{1+2}(2)\begin{vmatrix} -3 & 4 \\ -2 & 3 \end{vmatrix} = (1)(2)(9) + (-1)(2)(-1) = 20.$$

Similarly, we evaluate

$$\begin{vmatrix} 1 & 2 & -3 \\ -4 & 2 & 1 \\ 2 & 0 & -2 \end{vmatrix}$$

by expanding about the third row, obtaining

$$(-1)^{1+3}(2)\begin{vmatrix} 2 & -3 \\ 2 & 1 \end{vmatrix} + (-1)^{3+3}(-2)\begin{vmatrix} 1 & 2 \\ -4 & 2 \end{vmatrix} = (1)(2)(8) + (1)(-2)(10) = -4.$$

Substituting in Equation (3), we find the value of the given determinant as

$$(+1)(3)(20) + 0 + 0 + (-1)(-3)(-4) = 48.$$

On the other hand, expanding about the first column, we have

$$(-1)^{1+1}(1)\begin{vmatrix} 2 & 1 & 3 \\ 0 & 0 & -3 \\ 0 & -2 & 3 \end{vmatrix} + (-1)^{1+2}(-4)\begin{vmatrix} 2 & -3 & 4 \\ 0 & 0 & -3 \\ 0 & -2 & 3 \end{vmatrix}$$

$$+ (-1)^{1+3}(3)\begin{vmatrix} 2 & -3 & 4 \\ 2 & 1 & 3 \\ 0 & -2 & 3 \end{vmatrix} + (-1)^{1+4}(2)\begin{vmatrix} 2 & -3 & 4 \\ 2 & 1 & 3 \\ 0 & 0 & -3 \end{vmatrix}$$

$$= (1)(1)(-12) + (-1)(-4)(-12) + (1)(3)(20) + (-1)(2)(-24) = 48.$$

We can use the properties of Section 2.1 to introduce many zeros in a given row or column and then expand about that row or column. This is illustrated in the following example.

Example 3 Consider the matrix of Example 2. We have

$$\begin{vmatrix} 1 & 2 & -3 & 4 \\ -4 & 2 & 1 & 3 \\ 3 & 0 & 0 & -3 \\ 2 & 0 & -2 & 3 \end{vmatrix}_{c_4 + c_1 \to c_4} = \begin{vmatrix} 1 & 2 & -3 & 5 \\ -4 & 2 & 1 & -1 \\ 3 & 0 & 0 & 0 \\ 2 & 0 & -2 & 5 \end{vmatrix}$$

$$= (-1)^{3+1}(3)\begin{vmatrix} 2 & -3 & 5 \\ 2 & 1 & -1 \\ 0 & -2 & 5 \end{vmatrix}_{r_1 - r_2 \to r_1} = (-1)^4(3)\begin{vmatrix} 0 & -4 & 6 \\ 2 & 1 & -1 \\ 0 & -2 & 5 \end{vmatrix}$$

$$= (-1)^4(3)(-2)(-8) = 48.$$

Here $c_4 + c_1 \to c_4$ means that the fourth column c_4 of the matrix plus the first column c_1 replaces the fourth column.

The Inverse of a Matrix

It is interesting to ask what $a_{i1}A_{k1} + a_{i2}A_{k2} + \cdots a_{in}A_{kn}$ is for $i \ne k$ because, as soon as we answer this question, we shall obtain another method for finding the inverse of a nonsingular matrix.

Theorem 2.10 If $\mathbf{A} = [a_{ij}]$ is an $n \times n$ matrix, then

$$a_{i1}A_{k1} + a_{i2}A_{k2} + \cdots + a_{in}A_{kn} = 0 \quad \text{for } i \ne k; \tag{4}$$

$$a_{1j}A_{1k} + a_{2j}A_{2k} + \cdots + a_{nj}A_{nk} = 0 \quad \text{for } j \ne k. \tag{5}$$

Proof

We only prove the first formula. The second one follows from the first one by Theorem 2.1.

Consider the matrix $\mathbf{B}$ obtained from $\mathbf{A}$ by replacing the kth row of $\mathbf{A}$ by its ith row. Thus $\mathbf{B}$ is a matrix having two identical rows—the ith and kth rows. Then $|\mathbf{B}| = 0$. Now expand $|\mathbf{B}|$ about the kth row. The elements of the kth row of $\mathbf{B}$ are $a_{i1}, a_{i2}, \ldots, a_{in}$. The cofactors of the kth row are $A_{k1}, A_{k2}, \ldots, A_{kn}$. Thus from Equation (1) we have

$$0 = |\mathbf{B}| = a_{i1}A_{k1} + a_{i2}A_{k2} + \cdots + a_{in}A_{kn},$$

which is what we wanted to show.

This theorem says that if we sum the products of the elements of any row (column) times the corresponding cofactors of any other row (column), then we obtain zero.

Example 4 Let

$$\mathbf{A} = \begin{bmatrix} 1 & 2 & 3 \\ -2 & 3 & 1 \\ 4 & 5 & -2 \end{bmatrix}.$$

Then

$$A_{21} = (-1)^{2+1}\begin{vmatrix} 2 & 3 \\ 5 & -2 \end{vmatrix} = 19, \qquad A_{22} = (-1)^{2+2}\begin{vmatrix} 1 & 3 \\ 4 & -2 \end{vmatrix} = -14,$$

$$A_{23} = (-1)^{2+3}\begin{vmatrix} 1 & 2 \\ 4 & 5 \end{vmatrix} = 3.$$

Now

$$a_{31}A_{21} + a_{32}A_{22} + a_{33}A_{23} = (4)(19) + (5)(-14) + (-2)(3) = 0$$

and

$$a_{11}A_{21} + a_{12}A_{22} + a_{13}A_{23} = (1)(19) + (2)(-14) + (3)(3) = 0.$$

We may combine (1) and (4) as

$$a_{i1}A_{k1} + a_{i2}A_{k2} + \cdots + a_{in}A_{kn} = |\mathbf{A}| \qquad \text{if } i = k,$$
$$= 0 \qquad \text{if } i \neq k. \tag{6}$$

Similarly, we may combine (2) and (5) as

$$a_{1j}A_{1k} + a_{2j}A_{2k} + \cdots + a_{nj}A_{nk} = |\mathbf{A}| \qquad \text{if } j = k,$$
$$= 0 \qquad \text{if } j \neq k. \tag{7}$$

Definition Let $\mathbf{A} = [a_{ij}]$ be an $n \times n$ matrix. The $n \times n$ matrix adj $\mathbf{A}$, called the **adjoint** of $\mathbf{A}$, is the matrix whose i, jth element is the cofactor A_{ji} of a_{ji}. Thus

$$\text{adj } \mathbf{A} = \begin{bmatrix} A_{11} & A_{21} & \cdots & A_{n1} \\ A_{12} & A_{22} & \cdots & A_{n2} \\ \vdots & \vdots & & \vdots \\ A_{1n} & A_{2n} & \cdots & A_{nn} \end{bmatrix}.$$

Example 5 Let

$$\mathbf{A} = \begin{bmatrix} 3 & -2 & 1 \\ 5 & 6 & 2 \\ 1 & 0 & -3 \end{bmatrix}.$$

Compute adj $\mathbf{A}$.

Solution

The cofactors of $\mathbf{A}$ are

$$A_{11} = (-1)^{1+1} \begin{vmatrix} 6 & 2 \\ 0 & -3 \end{vmatrix} = -18; \quad A_{12} = (-1)^{1+2} \begin{vmatrix} 5 & 2 \\ 1 & -3 \end{vmatrix} = 17;$$

$$A_{13} = (-1)^{1+3} \begin{vmatrix} 5 & 6 \\ 1 & 0 \end{vmatrix} = -6;$$

$$A_{21} = (-1)^{2+1} \begin{vmatrix} -2 & 1 \\ 0 & -3 \end{vmatrix} = -6; \quad A_{22} = (-1)^{2+2} \begin{vmatrix} 3 & 1 \\ 1 & -3 \end{vmatrix} = -10;$$

$$A_{23} = (-1)^{2+3} \begin{vmatrix} 3 & -2 \\ 1 & 0 \end{vmatrix} = -2;$$

$$A_{31} = (-1)^{3+1} \begin{vmatrix} -2 & 1 \\ 6 & 2 \end{vmatrix} = -10; \quad A_{32} = (-1)^{3+2} \begin{vmatrix} 3 & 1 \\ 5 & 2 \end{vmatrix} = -1;$$

$$A_{33} = (-1)^{3+3} \begin{vmatrix} 3 & -2 \\ 5 & 6 \end{vmatrix} = 28.$$

Then

$$\text{adj } \mathbf{A} = \begin{bmatrix} -18 & -6 & -10 \\ 17 & -10 & -1 \\ -6 & -2 & 28 \end{bmatrix}.$$

Theorem 2.11 *If $\mathbf{A} = [a_{ij}]$ is an $n \times n$ matrix, then*

$$\mathbf{A}(\text{adj } \mathbf{A}) = (\text{adj } \mathbf{A})\mathbf{A} = |\mathbf{A}|\mathbf{I}_n.$$

Proof
We have $\mathbf{A}(\text{adj } \mathbf{A}) =$

$$\begin{bmatrix} a_{11} & a_{12} & \cdots & a_{1n} \\ a_{21} & a_{22} & \cdots & a_{2n} \\ \vdots & \vdots & & \vdots \\ \boxed{a_{i1}} & a_{i2} & \cdots & a_{in} \\ \vdots & \vdots & & \vdots \\ a_{n1} & a_{n2} & \cdots & a_{nn} \end{bmatrix} \begin{bmatrix} A_{11} & A_{21} & \cdots & \boxed{A_{j1}} & \cdots & A_{n1} \\ A_{12} & A_{22} & \cdots & A_{j2} & \cdots & A_{n2} \\ \vdots & \vdots & & \vdots & & \vdots \\ A_{1n} & A_{2n} & \cdots & \boxed{A_{jn}} & \cdots & A_{nn} \end{bmatrix}.$$

The i, jth element in the product matrix $\mathbf{A}(\text{adj } \mathbf{A})$ is, by (6),

$$a_{i1}A_{j1} + a_{i2}A_{j2} + \cdots + a_{in}A_{jn} = |\mathbf{A}| \quad \text{if } i = j$$
$$= 0 \quad \text{if } i \neq j.$$

This means that

$$\mathbf{A}(\text{adj } \mathbf{A}) = \begin{bmatrix} |\mathbf{A}| & 0 & \cdots & 0 \\ 0 & |\mathbf{A}| & & \vdots \\ \vdots & & & \\ 0 & \cdots & 0 & |\mathbf{A}| \end{bmatrix} = |\mathbf{A}|\mathbf{I}_n.$$

The i, jth element in the product matrix $(\text{adj } \mathbf{A})\mathbf{A}$ is, by (7),

$$A_{1i}a_{1j} + A_{2i}a_{2j} + \cdots + A_{ni}a_{nj} = |\mathbf{A}| \quad \text{if } i = j$$
$$= 0 \quad \text{if } i \neq j.$$

Thus $(\text{adj } \mathbf{A})\mathbf{A} = |\mathbf{A}|\mathbf{I}_n$.

Example 6 Consider the matrix of Example 5. Then

$$\begin{bmatrix} 3 & -2 & 1 \\ 5 & 6 & 2 \\ 1 & 0 & -3 \end{bmatrix} \begin{bmatrix} -18 & -6 & -10 \\ 17 & -10 & -1 \\ -6 & -2 & 28 \end{bmatrix} = \begin{bmatrix} -94 & 0 & 0 \\ 0 & -94 & 0 \\ 0 & 0 & -94 \end{bmatrix} = -94\begin{bmatrix} 1 & 0 & 0 \\ 0 & 1 & 0 \\ 0 & 0 & 1 \end{bmatrix}$$

and

$$\begin{bmatrix} -18 & -6 & -10 \\ 17 & -10 & -1 \\ -6 & -2 & 28 \end{bmatrix} \begin{bmatrix} 3 & -2 & 1 \\ 5 & 6 & 2 \\ 1 & 0 & -3 \end{bmatrix} = -94\begin{bmatrix} 1 & 0 & 0 \\ 0 & 1 & 0 \\ 0 & 0 & 1 \end{bmatrix}.$$

We now have a new method for finding the inverse of a nonsingular matrix, and we state this result as the following corollary.

Corollary 2.2 *If* $\mathbf{A}$ *is an* $n \times n$ *matrix and* $|\mathbf{A}| \neq 0$, *then*

$$\mathbf{A}^{-1} = \frac{1}{|\mathbf{A}|}(\text{adj } \mathbf{A}) = \begin{bmatrix} \dfrac{A_{11}}{|\mathbf{A}|} & \dfrac{A_{21}}{|\mathbf{A}|} & \cdots & \dfrac{A_{n1}}{|\mathbf{A}|} \\[2ex] \dfrac{A_{12}}{|\mathbf{A}|} & \dfrac{A_{22}}{|\mathbf{A}|} & \cdots & \dfrac{A_{n2}}{|\mathbf{A}|} \\ \vdots & \vdots & & \vdots \\ \dfrac{A_{1n}}{|\mathbf{A}|} & \dfrac{A_{2n}}{|\mathbf{A}|} & \cdots & \dfrac{A_{nn}}{|\mathbf{A}|} \end{bmatrix}.$$

Proof

By Theorem 2.11, $\mathbf{A}(\text{adj } \mathbf{A}) = |\mathbf{A}|\mathbf{I}_n$, so if $|\mathbf{A}| \neq 0$, then

$$\mathbf{A}\frac{1}{|\mathbf{A}|}(\text{adj } \mathbf{A}) = \frac{1}{|\mathbf{A}|}[\mathbf{A}(\text{adj } \mathbf{A})] = \frac{1}{|\mathbf{A}|}(|\mathbf{A}|\mathbf{I}_n) = \mathbf{I}_n.$$

Hence

$$\mathbf{A}^{-1} = \frac{1}{|\mathbf{A}|}(\text{adj } \mathbf{A}).$$

Example 7 Again consider the matrix of Example 5. Then $|\mathbf{A}| = -94$, and

$$\mathbf{A}^{-1} = \frac{1}{|\mathbf{A}|}(\text{adj } \mathbf{A}) = \begin{bmatrix} \frac{18}{94} & \frac{6}{94} & \frac{10}{94} \\[1ex] -\frac{17}{94} & \frac{10}{94} & \frac{1}{94} \\[1ex] \frac{6}{94} & \frac{2}{94} & -\frac{28}{94} \end{bmatrix}.$$

Theorem 2.12 *A matrix* $\mathbf{A}$ *is nonsingular if and only if* $|\mathbf{A}| \neq 0$.

Proof

If $|\mathbf{A}| \neq 0$, then Corollary 2.2 gives an expression for $\mathbf{A}^{-1}$, so $\mathbf{A}$ is nonsingular.

The converse has already been established in Corollary 2.1, whose proof was left to the reader as Exercise T.4 of Section 2.1. We now prove the converse here. Suppose that $\mathbf{A}$ is nonsingular. Then

$$\mathbf{A}\mathbf{A}^{-1} = \mathbf{I}_n.$$

From Theorem 2.8 we have

$$|\mathbf{A}\mathbf{A}^{-1}| = |\mathbf{A}||\mathbf{A}^{-1}| = |\mathbf{I}_n| = 1,$$

which implies that $|\mathbf{A}| \neq 0$.

This completes the proof.

Corollary 2.3 *If $\mathbf{A}$ is an $n \times n$ matrix, then the homogeneous system $\mathbf{AX} = \mathbf{0}$ has a nontrivial solution if and only if $|\mathbf{A}| = 0$.*

Proof

If $|\mathbf{A}| \neq 0$, then, by Theorem 2.12, $\mathbf{A}$ is nonsingular and thus $\mathbf{AX} = \mathbf{0}$ has only the trivial solution (Theorem 1.12 in Section 1.5).

Conversely, if $|\mathbf{A}| = 0$, then $\mathbf{A}$ is singular (Theorem 2.12). Suppose that $\mathbf{A}$ is row equivalent to a matrix $\mathbf{B}$ in reduced row echelon form. It then follows from Theorem 1.11 in Section 1.5 and Exercise T.7 in Section 1.4 that $\mathbf{B}$ has a row of zeros. The system $\mathbf{BX} = \mathbf{0}$ has the same solutions as the system $\mathbf{AX} = \mathbf{0}$. Let $\mathbf{C}_1$ be the matrix obtained by deleting the zero rows of $\mathbf{B}$. Then the system $\mathbf{BX} = \mathbf{0}$ has the same solutions as the system $\mathbf{C}_1\mathbf{X} = \mathbf{0}$. Since the latter is a homogeneous system of at most $n - 1$ equations in n unknowns, it has a nontrivial solution (Theorem 1.7 in Section 1.4). Hence the given system $\mathbf{AX} = \mathbf{0}$ has a nontrivial solution. We might note that the proof of the converse is essentially the proof of Exercise T.3 in Section 1.5.

In Section 1.5 we developed a practical method for finding $\mathbf{A}^{-1}$. In describing the situation showing that a matrix is singular and has no inverse, we used the fact that if we start with the identity matrix $\mathbf{I}_n$ and use only elementary row operations on $\mathbf{I}_n$, we can never obtain a matrix having a row consisting entirely of zeros. We can now justify this statement as follows. If matrix $\mathbf{B}$ results from $\mathbf{I}_n$ by interchanging two rows of $\mathbf{I}_n$, then $|\mathbf{B}| = -|\mathbf{I}_n| = -1$ (Theorem 2.2); if $\mathbf{C}$ results from $\mathbf{I}_n$ by multiplying a row of $\mathbf{I}_n$ by $c \neq 0$, then $|\mathbf{C}| = c|\mathbf{I}_n| = c$ (Theorem 2.5), and if $\mathbf{D}$ results from $\mathbf{I}_n$ by adding a multiple of a row of $\mathbf{I}_n$ to another row of $\mathbf{I}_n$, then $|\mathbf{D}| = |\mathbf{I}_n| = 1$ (Theorem 2.6). Performing elementary row operations on $\mathbf{I}_n$ will thus never yield a matrix with zero determinant. Now suppose that $\mathbf{F}$ is obtained from $\mathbf{I}_n$ by a sequence of elementary row operations and $\mathbf{F}$ has a row of zeros. Then $|\mathbf{F}| = 0$ (Theorem 2.4). This contradiction justifies the statement used in Section 1.5.

We might note that the method of inverting a nonsingular matrix given in Corollary 2.2 is much less efficient than the method given in Chapter 1. In fact, the computation of $\mathbf{A}^{-1}$ using determinants, as given in Corollary 2.2, becomes too expensive for $n > 4$ from a computing point of view. We discuss these matters in Section 2.3, where we deal with determinants from a computational point of view. However, Corollary 2.2 is still a useful result on other grounds.

We may summarize our results on determinants, homogeneous systems, and nonsingular matrices by noting that the following statements are equivalent:

1. $\mathbf{A}$ is nonsingular.
2. $\mathbf{AX} = \mathbf{0}$ has only the trivial solution.
3. $\mathbf{A}$ is row equivalent to $\mathbf{I}_n$.
4. The linear system $\mathbf{AX} = \mathbf{B}$ has a unique solution for every $n \times 1$ matrix $\mathbf{B}$.
5. $|\mathbf{A}| \neq 0$.

Cramer's[†] Rule

We can use the result of Theorem 2.11 to obtain another method, known as Cramer's rule, for solving a linear system of n equations in n unknowns with a nonsingular coefficient matrix.

Theorem 2.13
(Cramer's
Rule)

Let

$$a_{11}x_1 + a_{12}x_2 + \cdots + a_{1n}x_n = b_1$$
$$a_{21}x_1 + a_{22}x_2 + \cdots + a_{2n}x_n = b_2$$
$$\vdots \qquad \vdots \qquad \qquad \vdots \qquad \vdots$$
$$a_{n1}x_1 + a_{n2}x_2 + \cdots + a_{nn}x_n = b_n$$

be a linear system of n equations in n unknowns and let $\mathbf{A} = [a_{ij}]$ be the coefficient matrix so that we can write the given system as $\mathbf{AX} = \mathbf{B}$, where

$$\mathbf{B} = \begin{bmatrix} b_1 \\ b_2 \\ \vdots \\ b_n \end{bmatrix}.$$

If $|\mathbf{A}| \neq 0$, then the system has the unique solution

$$x_1 = \frac{|\mathbf{A}_1|}{|\mathbf{A}|}, \quad x_2 = \frac{|\mathbf{A}_2|}{|\mathbf{A}|}, \quad \ldots, \quad x_n = \frac{|\mathbf{A}_n|}{|\mathbf{A}|},$$

where $\mathbf{A}_i$ is the matrix obtained from $\mathbf{A}$ by replacing the ith column of $\mathbf{A}$ by $\mathbf{B}$.

Proof

If $|\mathbf{A}| \neq 0$, then, by Theorem 2.12, $\mathbf{A}$ is nonsingular. Hence

$$\mathbf{X} = \begin{bmatrix} x_1 \\ x_2 \\ \vdots \\ x_n \end{bmatrix} = \mathbf{A}^{-1}\mathbf{B} = \begin{bmatrix} \dfrac{A_{11}}{|\mathbf{A}|} & \dfrac{A_{21}}{|\mathbf{A}|} & \cdots & \dfrac{A_{n1}}{|\mathbf{A}|} \\ \dfrac{A_{12}}{|\mathbf{A}|} & \dfrac{A_{22}}{|\mathbf{A}|} & \cdots & \dfrac{A_{n2}}{|\mathbf{A}|} \\ \vdots & \vdots & & \vdots \\ \dfrac{A_{1i}}{|\mathbf{A}|} & \dfrac{A_{2i}}{|\mathbf{A}|} & \cdots & \dfrac{A_{ni}}{|\mathbf{A}|} \\ \vdots & \vdots & & \vdots \\ \dfrac{A_{1n}}{|\mathbf{A}|} & \dfrac{A_{2n}}{|\mathbf{A}|} & \cdots & \dfrac{A_{nn}}{|\mathbf{A}|} \end{bmatrix} \begin{bmatrix} b_1 \\ b_2 \\ \vdots \\ b_n \end{bmatrix}.$$

† Gabriel Cramer (1704–1752) was born in Geneva, Switzerland, and lived there all his life. Remaining single, he traveled extensively, taught at the Académie de Calvin, and participated actively in civic affairs.

The rule for solving systems of linear equations appeared in an appendix to his 1750 book, *Introduction à l'analyse des lignes courbes algébriques*. It was known previously by other mathematicians but was not widely known or very clearly explained until its appearance in Cramer's influential work.

This means that

$$x_i = \frac{A_{1i}}{|\mathbf{A}|} b_1 + \frac{A_{2i}}{|\mathbf{A}|} b_2 + \cdots + \frac{A_{ni}}{|\mathbf{A}|} b_n \qquad (1 \le i \le n).$$

Now let

$$\mathbf{A}_i = \begin{bmatrix} a_{11} & a_{12} & \cdots & a_{1\,i-1} & b_1 & a_{1\,i+1} & \cdots & a_{1n} \\ a_{21} & a_{22} & \cdots & a_{2\,i-1} & b_2 & a_{2\,i+1} & \cdots & a_{2n} \\ \vdots & \vdots & & \vdots & \vdots & \vdots & & \vdots \\ a_{n1} & a_{n2} & \cdots & a_{n\,i-1} & b_n & a_{n\,i+1} & \cdots & a_{nn} \end{bmatrix}.$$

If we evaluate $|\mathbf{A}_i|$ by expanding about the ith column, we find that

$$|\mathbf{A}_i| = A_{1i}b_1 + A_{2i}b_2 + \cdots + A_{ni}b_n.$$

Hence

$$x_i = \frac{|\mathbf{A}_i|}{|\mathbf{A}|}$$

for $i = 1, 2, \ldots, n$. In this expression for x_i, the determinant $|\mathbf{A}_i|$ of $\mathbf{A}_i$ can be calculated by any method. It was only in the derivation of the expression for x_i that we had to evaluate it by expanding about the ith column.

Cramer's rule for solving the linear system $\mathbf{AX} = \mathbf{B}$, where $\mathbf{A}$ is $n \times n$, is as follows:

STEP 1. Compute $|\mathbf{A}|$. If $|\mathbf{A}| = 0$, Cramer's rule is not applicable. Use Gauss–Jordan reduction.

STEP 2. If $|\mathbf{A}| \neq 0$, for each i

$$x_i = \frac{|\mathbf{A}_i|}{|\mathbf{A}|}$$

where $\mathbf{A}_i$ is the matrix obtained from $\mathbf{A}$ by replacing the ith column of $\mathbf{A}$ by $\mathbf{B}$.

Example 8 Consider the following linear system:

$$\begin{aligned} -2x_1 + 3x_2 - x_3 &= 1 \\ x_1 + 2x_2 - x_3 &= 4 \\ -2x_1 - x_2 + x_3 &= -3. \end{aligned}$$

Then

$$|\mathbf{A}| = \begin{vmatrix} -2 & 3 & -1 \\ 1 & 2 & -1 \\ -2 & -1 & 1 \end{vmatrix} = -2.$$

Hence

$$x_1 = \frac{\begin{vmatrix} 1 & 3 & -1 \\ 4 & 2 & -1 \\ -3 & -1 & 1 \end{vmatrix}}{|\mathbf{A}|} = \frac{-4}{-2} = 2;$$

$$x_2 = \frac{\begin{vmatrix} -2 & 1 & -1 \\ 1 & 4 & -1 \\ -2 & -3 & 1 \end{vmatrix}}{|\mathbf{A}|} = \frac{-6}{-2} = 3;$$

$$x_3 = \frac{\begin{vmatrix} -2 & 3 & 1 \\ 1 & 2 & 4 \\ -2 & -1 & -3 \end{vmatrix}}{|\mathbf{A}|} = \frac{-8}{-2} = 4.$$

We note that Cramer's rule is only applicable to the case where we have n equations in n unknowns and where the coefficient matrix $\mathbf{A}$ is nonsingular. Cramer's rule becomes computationally inefficient for $n > 4$, and it is better to use the Gauss–Jordan reduction method discussed in Section 1.4.

APPLICATIONS. At this point the reader may turn to Section 7.1, Lines and Planes, for applications of determinants to lines and planes.

2.2 EXERCISES

1. Let

$$\mathbf{A} = \begin{bmatrix} 1 & 0 & -2 \\ 3 & 1 & 4 \\ 5 & 2 & -3 \end{bmatrix}.$$

Compute all the cofactors.

2. Let

$$\mathbf{A} = \begin{bmatrix} 1 & 0 & 3 & 0 \\ 2 & 1 & 4 & -1 \\ 3 & 2 & 4 & 0 \\ 0 & 3 & -1 & 0 \end{bmatrix}.$$

Compute all the cofactors of the elements in the second row and all the cofactors of the elements in the third column.

In Exercises 3 through 6, evaluate the determinants using Theorem 2.9.

3. (a) $\begin{bmatrix} 1 & 2 & 3 \\ -1 & 5 & 2 \\ 3 & 2 & 0 \end{bmatrix}.$ (b) $\begin{bmatrix} 4 & -4 & 2 & 1 \\ 1 & 2 & 0 & 3 \\ 2 & 0 & 3 & 4 \\ 0 & -3 & 2 & 1 \end{bmatrix}.$

(c) $\begin{bmatrix} 4 & -2 & 0 \\ 0 & 2 & 4 \\ -1 & -1 & -3 \end{bmatrix}.$

4. (a) $\begin{bmatrix} 2 & 2 & -3 & 1 \\ 0 & 1 & 2 & -1 \\ 3 & -1 & 4 & 1 \\ 2 & 3 & 0 & 0 \end{bmatrix}.$ (b) $\begin{bmatrix} 0 & 1 & -2 \\ -1 & 3 & 1 \\ 2 & -2 & 3 \end{bmatrix}.$

(c) $\begin{bmatrix} 2 & 1 & -3 \\ 0 & 1 & 2 \\ -4 & 2 & 1 \end{bmatrix}.$

5. (a) $\begin{bmatrix} 3 & 1 & 2 & -1 \\ 2 & 0 & 3 & -7 \\ 1 & 3 & 4 & -5 \\ 0 & -1 & 1 & -5 \end{bmatrix}$. (b) $\begin{bmatrix} 3 & 1 & 0 \\ 3 & 2 & 1 \\ 0 & 1 & -1 \end{bmatrix}$.

(c) $\begin{bmatrix} 3 & -3 & 0 \\ 2 & 0 & 2 \\ 2 & 1 & -3 \end{bmatrix}$.

6. (a) $\begin{bmatrix} 0 & 0 & -1 & 3 \\ 0 & 1 & 2 & 1 \\ 2 & -2 & 5 & 2 \\ 3 & 3 & 0 & 0 \end{bmatrix}$. (b) $\begin{bmatrix} 4 & 2 & 0 \\ 1 & 1 & 2 \\ -1 & 3 & 4 \end{bmatrix}$.

(c) $\begin{bmatrix} -1 & 2 & -1 \\ 3 & 2 & 1 \\ 1 & 4 & 2 \end{bmatrix}$.

7. Verify Theorem 2.10 for the matrix

$$A = \begin{bmatrix} -2 & 3 & 0 \\ 4 & 1 & -3 \\ 2 & 0 & 1 \end{bmatrix} \text{ by computing}$$

$$a_{11}A_{12} + a_{21}A_{22} + a_{31}A_{32}.$$

8. Let $A = \begin{bmatrix} 2 & 1 & 3 \\ -1 & 2 & 0 \\ 3 & -2 & 1 \end{bmatrix}$.

(a) Find adj A.
(b) Compute $|A|$.
(c) Verify Theorem 2.11; that is, show that

$$A(\text{adj } A) = |A|I_3.$$

9. Let $A = \begin{bmatrix} 6 & 2 & 8 \\ -3 & 4 & 1 \\ 4 & -4 & 5 \end{bmatrix}$.

(a) Find adj A.
(b) Compute $|A|$.
(c) Verify Theorem 2.11; that is, show that

$$(\text{adj } A)A = |A|I_3.$$

In Exercises 10 through 13, compute the inverses of the given matrices, if they exist, using Corollary 2.2.

10. (a) $\begin{bmatrix} 3 & 2 \\ -3 & 4 \end{bmatrix}$. (b) $\begin{bmatrix} 4 & 2 & 2 \\ 0 & 1 & 2 \\ 1 & 0 & 3 \end{bmatrix}$.

(c) $\begin{bmatrix} 2 & 0 & -1 \\ 3 & 7 & 2 \\ 1 & 1 & 0 \end{bmatrix}$.

11. (a) $\begin{bmatrix} 1 & 2 & -3 \\ -4 & -5 & 2 \\ -1 & 1 & -7 \end{bmatrix}$. (b) $\begin{bmatrix} 2 & 3 \\ -1 & 2 \end{bmatrix}$.

(c) $\begin{bmatrix} 4 & 0 & 2 \\ 0 & 3 & 4 \\ 0 & 1 & -2 \end{bmatrix}$.

12. (a) $\begin{bmatrix} 2 & 0 & 1 \\ 3 & 2 & -1 \\ 1 & 0 & 1 \end{bmatrix}$. (b) $\begin{bmatrix} 5 & -1 \\ 2 & -1 \end{bmatrix}$.

(c) $\begin{bmatrix} 1 & 2 & 4 \\ 1 & -5 & 6 \\ 3 & -1 & 2 \end{bmatrix}$.

13. (a) $\begin{bmatrix} -3 & 1 \\ 2 & 0 \end{bmatrix}$. (b) $\begin{bmatrix} 4 & 0 & 0 \\ 0 & -3 & 0 \\ 0 & 0 & 2 \end{bmatrix}$.

(c) $\begin{bmatrix} 0 & 2 & 1 & 3 \\ 2 & -1 & 3 & 4 \\ -2 & 1 & 5 & 2 \\ 0 & 1 & 0 & 2 \end{bmatrix}$.

14. Use Theorem 2.12 to determine which of the following matrices are nonsingular.

(a) $\begin{bmatrix} 1 & 2 & 3 \\ 0 & 1 & 2 \\ 2 & -3 & 1 \end{bmatrix}$. (b) $\begin{bmatrix} 1 & 2 \\ 3 & 4 \end{bmatrix}$.

(c) $\begin{bmatrix} 1 & 3 & 2 \\ 2 & 1 & 4 \\ 1 & -7 & 2 \end{bmatrix}$. (d) $\begin{bmatrix} 1 & 2 & 0 & 5 \\ 3 & 4 & 1 & 7 \\ -2 & 5 & 2 & 0 \\ 0 & 1 & 2 & -7 \end{bmatrix}$.

15. Use Theorem 2.12 to determine which of the following matrices are nonsingular.

(a) $\begin{bmatrix} 4 & 3 & -5 \\ -2 & -1 & 3 \\ 4 & 6 & -2 \end{bmatrix}$. (b) $\begin{bmatrix} 1 & 3 & -1 & 2 \\ 2 & -6 & 4 & 1 \\ 3 & 5 & -1 & 3 \\ 4 & -6 & 5 & 2 \end{bmatrix}$.

(c) $\begin{bmatrix} 2 & 2 & -4 \\ 1 & 5 & 2 \\ 3 & 7 & -2 \end{bmatrix}$. (d) $\begin{bmatrix} 0 & 1 & 2 \\ 1 & 2 & 0 \\ 1 & 3 & 4 \end{bmatrix}$.

16. Find all values of λ for which

(a) $\begin{vmatrix} \lambda - 2 & 2 \\ 3 & \lambda - 3 \end{vmatrix} = 0$.

(b) $|\lambda I_3 - A| = 0$, where $A = \begin{bmatrix} 1 & 0 & -1 \\ 2 & 0 & 1 \\ 0 & 0 & -1 \end{bmatrix}$.

17. Find all values of λ for which

(a) $\begin{vmatrix} \lambda - 1 & -4 \\ 0 & \lambda - 4 \end{vmatrix} = 0$.

(b) $|\lambda I_3 - A| = 0$, where $A = \begin{bmatrix} -3 & -1 & -3 \\ 0 & 3 & 0 \\ -2 & -1 & -2 \end{bmatrix}$.

18. Use Corollary 2.3 to find out whether the following homogeneous systems have nontrivial solutions.

(a) $\begin{aligned} x - 2y + z &= 0 \\ 2x + 3y + z &= 0 \\ 3x + y + 2z &= 0. \end{aligned}$

(b) $\begin{aligned} x + 2y \quad\quad + w &= 0 \\ x + 2y + 3z \quad &= 0 \\ z + 2w &= 0 \\ y + 2z - w &= 0. \end{aligned}$

19. Repeat Exercise 18 for the following homogeneous systems.

(a) $\begin{aligned} x + y - z &= 0 \\ 2x + y + 2z &= 0 \\ 3x - y + z &= 0. \end{aligned}$

(b) $\begin{aligned} x + y + 2z + w &= 0 \\ 2x - y + z - w &= 0 \\ 3x + y + 2z + 3w &= 0 \\ 2x - y - z + w &= 0. \end{aligned}$

In Exercises 20 through 23, if possible, solve the given linear system by Cramer's rule.

20. $\begin{aligned} 2x + 4y + 6z &= 2 \\ x \quad\quad + 2z &= 0 \\ 2x + 3y - z &= -5. \end{aligned}$

21. $\begin{aligned} x + y + z - 2w &= -4 \\ 2y + z + 3w &= 4 \\ 2x + y - z + 2w &= 5 \\ x - y \quad + w &= 4. \end{aligned}$

22. $\begin{aligned} 2x + y + z &= 6 \\ 3x + 2y - 2z &= -2 \\ x + y + 2z &= 4. \end{aligned}$

23. $\begin{aligned} 2x + 3y + 7z &= 2 \\ -2x \quad\quad - 4z &= 0 \\ x + 2y + 4z &= 0. \end{aligned}$

THEORETICAL EXERCISES

T.1. Show by a column (row) expansion that if $A = [a_{ij}]$ is upper (lower) triangular, then $|A| = a_{11}a_{22} \cdots a_{nn}$.

T.2. If $A = [a_{ij}]$ is a 3×3 matrix, develop the general expression for $|A|$ by expanding (a) about the second column, and (b) about the third row. Compare these answers with those obtained for Example 6 in Section 2.1.

T.3. Prove that if A is symmetric, then adj A is also symmetric.

T.4. Prove that if A is a nonsingular upper triangular matrix, then A^{-1} is also upper triangular.

T.5. Prove that

$$A = \begin{bmatrix} a & b \\ c & d \end{bmatrix}$$

is nonsingular if and only if $ad - bc \neq 0$. If this condition is satisfied, use Corollary 2.2 to find A^{-1}.

T.6. Using Corollary 2.2, find the inverse of

$$\mathbf{A} = \begin{bmatrix} 1 & a & a^2 \\ 1 & b & b^2 \\ 1 & c & c^2 \end{bmatrix}.$$

(*Hint:* See Exercise T.12 in Section 2.1, where $|\mathbf{A}|$ was computed.)

T.7. Prove that if $\mathbf{A}$ is singular, then adj $\mathbf{A}$ is singular. [*Hint:* Show that if $\mathbf{A}$ is singular, then $\mathbf{A}(\text{adj } \mathbf{A}) = \mathbf{0}$.]

T.8. Prove that if $\mathbf{A}$ is an $n \times n$ matrix, then $|\text{adj } \mathbf{A}| = |\mathbf{A}|^{n-1}$.

T.9. Do Exercise T.8 in Section 1.4 using determinants.

T.10. Let $\mathbf{AB} = \mathbf{AC}$. Prove that if $|\mathbf{A}| \neq 0$, then $\mathbf{B} = \mathbf{C}$.

T.11. Let $\mathbf{A}$ be an $n \times n$ matrix all of whose entries are integers. Show that if $|\mathbf{A}| = \pm 1$, then all entries of $\mathbf{A}^{-1}$ are integers.

T.12. Show that if $\mathbf{A}$ is nonsingular, then adj $\mathbf{A}$ is nonsingular and

$$(\text{adj } \mathbf{A})^{-1} = \frac{1}{|\mathbf{A}|} \mathbf{A} = \text{adj } (\mathbf{A}^{-1}).$$

2.3

DETERMINANTS FROM A COMPUTATIONAL POINT OF VIEW

In this book we have, by now, developed two methods for solving a linear system of n equations in n unknowns: Gauss–Jordan reduction and Cramer's rule. We also have two methods for finding the inverse of a nonsingular matrix: the method involving determinants and the method discussed in Section 1.4. We must then develop some criteria for choosing one or the other method depending upon our needs.

Most sizable problems in linear algebra are solved on computers so that it is natural to compare two methods by estimating their computing time for the same problem. Since addition is so much faster than multiplication, the number of multiplications is often used as a basis of comparison for two numerical procedures.

Consider the linear system $\mathbf{AX} = \mathbf{B}$, where $\mathbf{A}$ is 25×25. If we find $\mathbf{X}$ by Cramer's rule, we must first obtain $|\mathbf{A}|$. We can find $|\mathbf{A}|$ by cofactor expansion, say $|\mathbf{A}| = a_{11}A_{11} + a_{21}A_{21} + \cdots + a_{n1}A_{n1}$, where we have expanded $|\mathbf{A}|$ about the first column. Note that if each cofactor is available, we require 25 multiplications. Now each cofactor A_{ij} is plus or minus the determinant of a 24×24 matrix, which can be expanded about a given row or column, requiring 24 multiplications. Thus the computation of $|\mathbf{A}|$ requires more than $25 \times 24 \times \cdots \times 2 \times 1 = 25!$ multiplications. Even if we were to use a futuristic computer capable of performing *100 billion* multiplications *per second*, it would take *about 4 million years* to evaluate $|\mathbf{A}|$. However, Gauss–Jordan reduction takes about 25^3 multiplications, and we would find the solution in less than *one second*. Of course, we can compute $|\mathbf{A}|$ in a much more efficient way by using elementary row operations to reduce $\mathbf{A}$ to triangular form and then use Theorem 2.7 (see Example 16 in Section 2.1). When implemented this way, Cramer's rule will require approximately n^4 multiplications. Thus Gauss–Jordan reduction is still much faster.

In general, if we are seeking numerical answers, then any method involving determinants can be used for $n \leq 4$. For $n > 5$, determinant-dependent methods are much less efficient than Gauss–Jordan reduction or the method of Section 1.5 for inverting a matrix.

The importance of determinants obviously does not lie in their computational usage. Note that methods involving determinants enable one to express the inverse of a matrix and the solution to a linear system of n equations in n unknowns by means of expressions or formulas. Gauss–Jordan reduction and the method for finding $\mathbf{A}^{-1}$ given in Section 1.5 do not yield a formula for the answer; we must proceed numerically to obtain the answer. Sometimes we do not need numerical answers but an expression for the answer because we may wish to further manipulate the answer. Another important reason for studying determinants is that they play a key role in the study of eigenvalues and eigenvectors, which will be undertaken in Chapter 5.

KEY IDEAS FOR REVIEW

☐ **Theorem 2.1.** $|\mathbf{A}^T| = |\mathbf{A}|$.

☐ **Theorem 2.2.** If $\mathbf{B}$ results from $\mathbf{A}$ by interchanging two rows (columns) of $\mathbf{A}$, then $|\mathbf{B}| = -|\mathbf{A}|$.

☐ **Theorem 2.3.** If two rows (columns) of $\mathbf{A}$ are equal, then $|\mathbf{A}| = 0$.

☐ **Theorem 2.4.** If a row (column) of $\mathbf{A}$ consists entirely of zeros, then $|\mathbf{A}| = 0$.

☐ **Theorem 2.5.** If $\mathbf{B}$ is obtained from $\mathbf{A}$ by multiplying a row (column) of $\mathbf{A}$ by a real number c, then $|\mathbf{B}| = c|\mathbf{A}|$.

☐ **Theorem 2.6.** If $\mathbf{B}$ is obtained from $\mathbf{A}$ by adding a multiple of a row (column) of $\mathbf{A}$ to another row (column) of $\mathbf{A}$, then $|\mathbf{B}| = |\mathbf{A}|$.

☐ **Theorem 2.7.** If $\mathbf{A} = [a_{ij}]$ is upper (lower) triangular, then $|\mathbf{A}| = a_{11}a_{22} \cdots a_{nn}$.

☐ **Theorem 2.8.** $|\mathbf{AB}| = |\mathbf{A}||\mathbf{B}|$.

☐ **Theorem 2.9. (Cofactor Expansion).** If $\mathbf{A} = [a_{ij}]$ then

$$|\mathbf{A}| = a_{i1}A_{i1} + a_{i2}A_{i2} + \cdots + a_{in}A_{in}$$

and

$$|\mathbf{A}| = a_{1j}A_{1j} + a_{2j}A_{2j} + \cdots + a_{nj}A_{nj}.$$

☐ **Corollary 2.2.** If $|\mathbf{A}| \neq 0$, then

$$\mathbf{A}^{-1} = \begin{bmatrix} \dfrac{A_{11}}{|\mathbf{A}|} & \dfrac{A_{21}}{|\mathbf{A}|} & \cdots & \dfrac{A_{n1}}{|\mathbf{A}|} \\ \dfrac{A_{12}}{|\mathbf{A}|} & \dfrac{A_{22}}{|\mathbf{A}|} & \cdots & \dfrac{A_{n2}}{|\mathbf{A}|} \\ \vdots & \vdots & & \vdots \\ \dfrac{A_{1n}}{|\mathbf{A}|} & \dfrac{A_{2n}}{|\mathbf{A}|} & \cdots & \dfrac{A_{nn}}{|\mathbf{A}|} \end{bmatrix}.$$

☐ **Theorem 2.12.** $\mathbf{A}$ is nonsingular if and only if $|\mathbf{A}| \neq 0$.

☐ **Corollary 2.3.** If $\mathbf{A}$ is an $n \times n$ matrix, then the homogeneous system $\mathbf{AX} = \mathbf{0}$ has a nontrivial solution if and only if $|\mathbf{A}| = 0$.

☐ **Theorem 2.13. (Cramer's Rule).** Let $\mathbf{AX} = \mathbf{B}$ be a linear system of n equations in n unknowns. If $|\mathbf{A}| \neq 0$, then the system has the unique solution

$$x_1 = \frac{|\mathbf{A}_1|}{|\mathbf{A}|}, \quad x_2 = \frac{|\mathbf{A}_2|}{|\mathbf{A}|}, \quad \cdots, \quad x_n = \frac{|\mathbf{A}_n|}{|\mathbf{A}|},$$

where $\mathbf{A}_i$ is the matrix obtained from $\mathbf{A}$ by replacing the ith column of $\mathbf{A}$ by $\mathbf{B}$.

☐ The following statements are equivalent:
  1. $\mathbf{A}$ is nonsingular.
  2. $\mathbf{AX} = \mathbf{0}$ has only the trivial solution.
  3. $\mathbf{A}$ is row equivalent to $\mathbf{I}_n$.
  4. The linear system $\mathbf{AX} = \mathbf{B}$ has a unique solution for every $n \times 1$ matrix $\mathbf{B}$.
  5. $|\mathbf{A}| \neq 0$.

SUPPLEMENTARY EXERCISES

1. Evaluate the following determinants using Equation (2).

(a) $\begin{vmatrix} 0 & 2 & 0 \\ 0 & 0 & -3 \\ 4 & 0 & 0 \end{vmatrix}$. (b) $\begin{vmatrix} 3 & 0 & 0 & 0 \\ 0 & -2 & 0 & 0 \\ 0 & 4 & 1 & 0 \\ 3 & 2 & -1 & -4 \end{vmatrix}$.

2. If

$$\begin{vmatrix} a_1 & a_2 & a_3 \\ b_1 & b_2 & b_3 \\ c_1 & c_2 & c_3 \end{vmatrix} = 5,$$

find the determinants of the following matrices:

(a) $\mathbf{B} = \begin{bmatrix} \frac{1}{2}a_1 & \frac{1}{2}a_2 & \frac{1}{2}a_3 \\ b_1 & b_2 & b_3 \\ c_1 & c_2 & c_3 \end{bmatrix}$.

(b) $\mathbf{C} = \begin{bmatrix} a_1 - b_1 & a_2 - b_2 & a_3 - b_3 \\ 3b_1 & 3b_2 & 3b_3 \\ 2c_1 & 2c_2 & 2c_3 \end{bmatrix}$.

3. Let $\mathbf{A}$ be 4×4 and suppose that $|\mathbf{A}| = 5$. Compute
   (a) $|\mathbf{A}^{-1}|$.   (b) $|2\mathbf{A}|$.   (c) $|2\mathbf{A}^{-1}|$.
   (d) $|(2\mathbf{A})^{-1}|$.

4. Let $|\mathbf{A}| = 3$ and $|\mathbf{B}| = 4$. Compute
   (a) $|\mathbf{AB}|$.   (b) $|\mathbf{ABA}^T|$.   (c) $|\mathbf{B}^{-1}\mathbf{AB}|$.

5. Find all values of λ for which

$$\begin{vmatrix} \lambda + 2 & -1 & 3 \\ 2 & \lambda - 1 & 2 \\ 0 & 0 & \lambda + 4 \end{vmatrix} = 0.$$

6. Evaluate

$$\begin{vmatrix} 2 & -1 & 3 \\ 4 & 1 & 5 \\ -2 & -3 & -2 \end{vmatrix}.$$

7. Evaluate

$$\begin{vmatrix} 3 & 2 & -1 & 1 \\ 4 & 1 & 1 & 0 \\ -1 & 2 & 3 & 4 \\ -2 & 3 & 5 & 1 \end{vmatrix}.$$

8. Compute all the cofactors of

$$\mathbf{A} = \begin{bmatrix} 2 & -1 & 3 \\ 1 & 4 & 5 \\ -3 & -4 & 6 \end{bmatrix}.$$

9. Evaluate

$$\begin{vmatrix} 3 & 2 & -1 & 0 \\ -1 & 0 & 3 & 2 \\ 4 & 1 & 5 & -2 \\ 1 & 3 & 2 & -3 \end{vmatrix}$$

by cofactor expansion.

10. Let

$$\mathbf{A} = \begin{bmatrix} 3 & -1 & 2 \\ 0 & 4 & 5 \\ 1 & 3 & 2 \end{bmatrix}.$$

(a) Find adj $\mathbf{A}$.
(b) Compute $|\mathbf{A}|$.
(c) Show that $\mathbf{A}(\text{adj } \mathbf{A}) = |\mathbf{A}|\mathbf{I}_3$.

11. Compute the inverse of the following matrix, if it exists, using Corollary 2.2.

$$\begin{bmatrix} 2 & -1 & 3 \\ 0 & 1 & 2 \\ -1 & 1 & 2 \end{bmatrix}.$$

12. Find all values of λ for which

$$\begin{bmatrix} \lambda - 3 & 0 & 3 \\ 0 & \lambda + 2 & 0 \\ -5 & 0 & \lambda + 5 \end{bmatrix}$$

is singular.

13. If

$$\mathbf{A} = \begin{bmatrix} \lambda & 0 & 1 \\ 1 & \lambda - 1 & 0 \\ 0 & 0 & \lambda + 1 \end{bmatrix},$$

find all values of λ for which the homogeneous system $\mathbf{AX} = \mathbf{0}$ has only the trivial solution.

14. If possible, solve the following linear system by Cramer's rule.

$$3x + 2y - z = -1$$
$$x - y - z = 0$$
$$2x + y - 2z = 3.$$

15. Using only elementary row or elementary column operations and Theorems 2.2, 2.5, and 2.6 (do not expand the determinants), verify the following.

(a) $\begin{vmatrix} a-b & 1 & a \\ b-c & 1 & b \\ c-a & 1 & c \end{vmatrix} = \begin{vmatrix} a & 1 & b \\ b & 1 & c \\ c & 1 & a \end{vmatrix}.$

(b) $\begin{vmatrix} 1 & a & bc \\ 1 & b & ca \\ 1 & c & ab \end{vmatrix} = \begin{vmatrix} 1 & a & a^2 \\ 1 & b & b^2 \\ 1 & c & c^2 \end{vmatrix}.$

16. Find all values of a for which the linear system

$$2x + ay = 0$$
$$ax + 2y = 0$$

has (a) a unique solution; (b) infinitely many solutions.

17. Find all values of a for which the matrix

$$\begin{bmatrix} a-2 & 2 \\ a-2 & a+2 \end{bmatrix}$$

is nonsingular.

18. Use Cramer's rule to find all values of a for which the linear system

$$x - 2y + 2z = 9$$
$$2x + y = a$$
$$3x - y - z = -10$$

has the solution in which $y = 1$.

THEORETICAL EXERCISES

T.1. Prove that if two rows (columns) of the $n \times n$ matrix **A** are proportional, then $|\mathbf{A}| = 0$.

T.2. Show that if **A** is an $n \times n$ matrix, then $|\mathbf{AA}^T| \geq 0$.

T.3. Let **Q** be an $n \times n$ matrix in which each entry is 1. Show that $|\mathbf{Q} - n\mathbf{I}_n| = 0$.

T.4. Let **P** be an invertible matrix. Prove that if $\mathbf{B} = \mathbf{PAP}^{-1}$, then $|\mathbf{B}| = |\mathbf{A}|$.

T.5. Prove that if **A** is a singular $n \times n$ matrix, then $\mathbf{A}(\text{adj } \mathbf{A}) = \mathbf{0}$. (*Hint*: See Theorem 2.11.)

T.6. Prove that if **A** is a singular $n \times n$ matrix, then **AB** is singular for any $n \times n$ matrix **B**.

T.7. Show that if **A** and **B** are square matrices, then

$$\begin{vmatrix} \mathbf{A} & \mathbf{0} \\ \mathbf{0} & \mathbf{B} \end{vmatrix} = |\mathbf{A}||\mathbf{B}|.$$

T.8. Show that if **A**, **B**, and **C** are square matrices, then

$$\begin{vmatrix} \mathbf{A} & \mathbf{0} \\ \mathbf{C} & \mathbf{B} \end{vmatrix} = |\mathbf{A}||\mathbf{B}|.$$

CHAPTER TEST

1. Evaluate

$$\begin{vmatrix} 1 & 1 & 2 & -1 \\ 0 & 1 & 0 & 3 \\ -1 & 2 & -3 & 4 \\ 0 & 5 & 0 & -2 \end{vmatrix}.$$

2. Let **A** be 3×3 and suppose that $|\mathbf{A}| = 2$. Compute
(a) $|3\mathbf{A}|$. (b) $|3\mathbf{A}^{-1}|$. (c) $|(3\mathbf{A})^{-1}|$.

3. For what value of a is

$$\begin{vmatrix} 2 & 1 & 0 \\ 0 & -1 & 3 \\ 0 & 0 & a \end{vmatrix} + \begin{vmatrix} 0 & a & 1 \\ 1 & 3a & 0 \\ -2 & a & 2 \end{vmatrix} = 14?$$

4. Find all values of a which the matrix

$$\begin{bmatrix} a^2 & 0 & 3 \\ 5 & a & 2 \\ 3 & 0 & 1 \end{bmatrix}$$

   is singular.

5. Solve the following linear system by Cramer's rule.

$$\begin{aligned} x - y + z &= -1 \\ 2x + y - 3z &= 8 \\ x - 2y + 3z &= -5. \end{aligned}$$

6. Answer each of the following as true or false.
   (a) $|\mathbf{A}\mathbf{A}^T| = |\mathbf{A}^2|$.

(b) $|-\mathbf{A}| = -|\mathbf{A}|$.

(c) If $\mathbf{A}^T = \mathbf{A}^{-1}$, then $|\mathbf{A}| = 1$.

(d) If $|\mathbf{A}| = 0$, then $\mathbf{A} = \mathbf{0}$.

(e) If $|\mathbf{A}| = 7$, then $\mathbf{A}\mathbf{X} = \mathbf{0}$ has only the trivial solution.

(f) The sign of the term $a_{15}a_{23}a_{31}a_{42}a_{54}$ in the expansion of the determinant of a 5×5 matrix is $+$.

(g) If $|\mathbf{A}| = 0$, then $|\text{adj } \mathbf{A}| = 0$.

(h) If $\mathbf{B} = \mathbf{P}\mathbf{A}\mathbf{P}^{-1}$, and $\mathbf{P}$ is nonsingular, then $|\mathbf{B}| = |\mathbf{A}|$.

(i) If $\mathbf{A}^4 = \mathbf{I}_n$, then $|\mathbf{A}| = 1$.

(j) If $\mathbf{A}^2 = \mathbf{A}$ and $\mathbf{A} \neq \mathbf{I}_n$, then $|\mathbf{A}| = 0$.

3

Vectors and Vector Spaces

VECTORS IN THE PLANE

Coordinate Systems

In many applications we deal with measurable quantities, such as pressure, mass, and speed, which can be described completely by giving their magnitude. There are also many other measurable quantities, such as velocity, force, and acceleration, which require for their description not only magnitude, but also a direction. These are called **vectors**, and their study comprises this chapter. Vectors will be denoted by boldface capital letters, such as **U**, **V**, **X**, **Y**, and **Z**. The real numbers will be called **scalars** and will be denoted by lowercase italic letters.

We recall that the real number system may be visualized as a straight line L, which is usually taken in a horizontal position. A point O, called the **origin**, is chosen on L; O corresponds to the number 0. A point A is chosen to the right of O, thereby fixing the length of OA as 1 and specifying a positive direction. Thus the positive real numbers lie to the right of O; the negative real numbers lie to the left of O (Figure 3.1).

Figure 3.1

The **absolute value** $|x|$ of the real number x is defined by

$$|x| = \begin{cases} x & \text{if } x \geq 0 \\ -x & \text{if } x < 0. \end{cases}$$

Thus $|3| = 3$, $|-2| = 2$, $|0| = 0$, $|-\frac{2}{3}| = \frac{2}{3}$, and $|-1.82| = 1.82$.

The real number x corresponding to the point P is called the **coordinate** of P and the point P whose coordinate is x is denoted by $P(x)$. The line L is called a **coordinate axis**. If P is to the right of O, then its coordinate is the length of the segment OP. If Q is to the left of O, then its coordinate is the negative of the length of the segment OQ. The distance between the points P and Q with respective coordinates a and b is $|b - a|$.

Example 1 Referring to Figure 3.1, we see that the coordinates of the points B, C, D, and E are, respectively, 3, -3, 1.5, and -4.5. The distance between B and C is $|-3 - 3| = 6$. The distance between A and B is $|3 - 1| = 2$. The distance between C and E is $|-4.5 - (-3)| = 1.5$.

We shall now turn to the analogous situation for the plane. We draw a pair of perpendicular lines intersecting at a point O, called the **origin**. One of the lines, the **x-axis**, is usually taken in a horizontal position. The other line, the **y-axis**, is then taken in a vertical position. We now choose a point on the x-axis to the right of O and a point on the y-axis above O to fix the units of length and positive directions on the x- and y-axes. Frequently, but not always, these points are chosen so that they are both equidistant from O, that is, so that the same unit of length is used for both axes. The x and y axes together are called **coordinate axes** (Figure 3.2). The **projection** of a point P in the plane on a line L is the point Q obtained by intersecting L with the line L' passing through P and perpendicular to L [Figure 3.3(a) and (b)].

Figure 3.2   **Figure 3.3**

Figure 3.4

Figure 3.5

Let P be a point in the plane and let Q be its projection onto the x-axis. The coordinate of Q on the x-axis is called the **x-coordinate** of P. Similarly, let Q' be the projection of P onto the y-axis. The coordinate of Q' is called the **y-coordinate** of P. Thus with every point in the plane we associate an ordered pair (x, y) of real numbers, its coordinates. The point P with coordinates x and y is denoted by $P(x, y)$. Conversely, it is easy to see (Exercise T.1) how we can associate a point in the plane with each ordered pair (x, y) of real numbers (Figure 3.4). The correspondence given above between points in the plane and ordered pairs of real numbers is called a **rectangular coordinate system** or a **Cartesian** (after René Descartes†) **coordinate system**. The set of all points in the plane is denoted by R^2. It is also called **2-space**.

Example 2   In Figure 3.5 we show a number of points and their coordinates.

The coordinates of the origin are $(0, 0)$. The coordinates of the projection of the point $P(x, y)$ on the x-axis are $(x, 0)$ and the coordinates of its projection on the y-axis are $(0, y)$.

† René Descartes (1596–1650) was one of the best-known scientists and philosophers of his day; he was considered by some to be the founder of modern philosophy. After completing a university degree in law, he turned to the private study of mathematics, simultaneously pursuing interests in Parisian night life and in the military, volunteering for brief periods in the Dutch, Bavarian, and French armies. The most productive period of his life was 1628–1648, when he lived in Holland. In 1649 he accepted an invitation from Queen Christina of Sweden to be her private tutor and to establish an Academy of Sciences there. Unfortunately, he had no time for this project, for he died of pneumonia in 1650.

In 1619 Descartes had a dream in which he realized that the method of mathematics is the best way for obtaining truth. However, his only mathematical publication was *La Géométrie*, which appeared as an appendix to his major philosophical work *Discours de la méthode pour bien conduire sa raison, et chercher la vérité dans les sciences*. In *La Géométrie* he proposes the radical idea of doing geometry algebraically. To express a curve algebraically, one chooses any convenient line of reference and, on the line, a point of reference. If y represents the distance from any point of the curve to the reference line and x represents the distance along the line to the reference point, there is an equation relating x and y that represents the curve. The systematic use of "Cartesian" coordinates described above was introduced later in the seventeenth century by authors following up on Descartes' work.

Vectors

Consider the 2×1 matrix

$$\mathbf{X} = \begin{bmatrix} x \\ y \end{bmatrix},$$

where x and y are real numbers. With $\mathbf{X}$ we associate the directed line segment with initial point at the origin $O(0, 0)$ and terminal point at $P(x, y)$. The directed line segment from O to P is denoted by $\overrightarrow{OP}$; O is called its **tail** and P its **head**. We distinguish tail and head by placing an arrow at the head (Figure 3.6).

A directed line segment has a **direction**, which is the angle made with the positive x-axis, indicated by the arrow at its head. The **magnitude** of a directed line segment is its length.

Example 3 Let

$$\mathbf{X} = \begin{bmatrix} 2 \\ 3 \end{bmatrix}.$$

With $\mathbf{X}$ we can associate the directed line segment with tail $O(0, 0)$ and head $P(2, 3)$, shown in Figure 3.7.

Conversely, with a directed line segment $\overrightarrow{OP}$ with tail $O(0, 0)$ and head $P(x, y)$, we can associate the matrix

$$\begin{bmatrix} x \\ y \end{bmatrix}.$$

Example 4 With the directed line segment $\overrightarrow{OP}$ with head $P(4, 5)$, we can associate the matrix

$$\begin{bmatrix} 4 \\ 5 \end{bmatrix}.$$

Figure 3.6

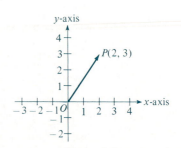

Figure 3.7

Definition A **vector in the plane** is a 2×1 matrix

$$\mathbf{X} = \begin{bmatrix} x \\ y \end{bmatrix},$$

where x and y are real numbers, called the **components of X**. We refer to a vector in the plane merely as a **vector**.

Thus with every vector we can associate a directed line segment, and conversely, with every directed line segment emanating from the origin we can associate a vector. As we have seen, a coordinate system is needed to set up this correspondence. The magnitude and direction of a vector are the magnitude and direction of its associated directed line segment. Frequently, the notions of directed line segment and vector are used interchangeably and a directed line segment is called a **vector**.

Since a vector is a matrix, the vectors

$$\mathbf{X} = \begin{bmatrix} x_1 \\ y_1 \end{bmatrix} \quad \text{and} \quad \mathbf{Y} = \begin{bmatrix} x_2 \\ y_2 \end{bmatrix}$$

are said to be **equal** if $x_1 = x_2$ and $y_1 = y_2$. That is, two vectors are equal if their respective components are equal.

Example 5 The vectors

$$\begin{bmatrix} 1 \\ 0 \end{bmatrix} \quad \text{and} \quad \begin{bmatrix} 1 \\ -2 \end{bmatrix}$$

are not equal, since their respective components are not equal.

With each vector $\mathbf{X} = \begin{bmatrix} x \\ y \end{bmatrix}$ we can also associate in a unique manner the point $P(x, y)$; conversely, with each point $P(x, y)$ we can associate in a unique manner the vector $\begin{bmatrix} x \\ y \end{bmatrix}$.

Thus we also write the vector $\mathbf{X}$ as

$$\mathbf{X} = (x, y).$$

Of course, this association is obtained by means of the directed line segment $\overrightarrow{OP}$, where O is the origin (Figure 3.6).

Thus the plane may be viewed both as the set of all points or as the set of all vectors. For this reason and, depending upon the context, we sometimes take R^2 as the set of all ordered pairs (x, y) and sometimes as the set of all 2×1 matrices $\begin{bmatrix} x \\ y \end{bmatrix}$.

Frequently, in physical applications it is necessary to deal with a directed line segment $\overrightarrow{PQ}$, from the point $P(x, y)$ (not the origin) to the point $Q(x', y')$, as shown in

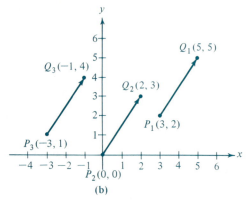

(a)   (b)

Figure 3.8

Figure 3.8(a). Such a directed line segment will also be called a **vector in the plane**, or simply a **vector** with **tail** $P(x, y)$ and **head** $Q(x', y')$. The **components** of such a vector are $x' - x$ and $y' - y$. Thus the vector PQ in Figure 3.8(a) can also be represented by the vector $(x' - x, y' - y)$ with tail O and head $P''(x' - x, y' - y)$. Two such vectors in the plane will be called **equal** if their components are equal. Consider the vectors $\overrightarrow{P_1Q_1}$, $\overrightarrow{P_2Q_2}$, and $\overrightarrow{P_3Q_3}$ joining the points $P_1(3, 2)$ and $Q_1(5, 5)$, $P_2(0, 0)$ and $Q_2(2, 3)$, $P_3(-3, 1)$ and $Q_3(-1, 4)$, respectively, as shown in Figure 3.8(b). Since they all have the same components, they are equal. Moreover, the head $Q_4(x_4', y_4')$ of the vector $\overrightarrow{P_4Q_4} = \begin{bmatrix} 2 \\ 3 \end{bmatrix} = \overrightarrow{P_2Q_2}$, with tail $P_4(-5, 2)$, can be determined as follows. We must have $x_4' - (-5) = 2$ and $y_4' - 2 = 3$ so that $x_4' = 2 - 5 = -3$ and $y_4' = 3 + 2 = 5$. Similarly, the tail $P_5(x_5, y_5)$ of the vector $\overrightarrow{P_5Q_5} = \begin{bmatrix} 2 \\ 3 \end{bmatrix}$ with head $Q_5(8, 6)$ is determined as follows. We must have $8 - x_5 = 2$ and $6 - y_5 = 3$ so that $x_5 = 8 - 2 = 6$ and $y_5 = 6 - 3 = 3$.

Length

By the Pythagorean Theorem (Figure 3.9) the **length** or **magnitude** of the vector $\mathbf{X} = (x, y)$ is

$$\|\mathbf{X}\| = \sqrt{x^2 + y^2}. \tag{1}$$

It also follows, by the Pythagorean Theorem, that the length of the directed line segment with initial point $P_1(x_1, y_1)$ and terminal point $P_2(x_2, y_2)$ is (Figure 3.10)

$$\|\overrightarrow{P_1P_2}\| = \sqrt{(x_2 - x_1)^2 + (y_2 - y_1)^2}. \tag{2}$$

Equation (2) also gives the distance between the points P_1 and P_2.

Figure 3.9

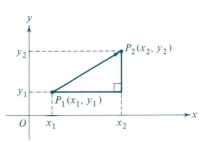

Figure 3.10

Example 6 If $\mathbf{X} = (2, -5)$, then, by Equation (1),
$$\|\mathbf{X}\| = \sqrt{(2)^2 + (-5)^2} = \sqrt{4 + 25} = \sqrt{29}.$$

Example 7 The distance between $P(3, 2)$ and $Q(-1, 5)$, or the length of the directed line segment $\overrightarrow{PQ}$, is, by Equation (2),
$$\|\overrightarrow{PQ}\| = \sqrt{(-1 - 3)^2 + (5 - 2)^2} = \sqrt{4^2 + 3^2} = \sqrt{25} = 5.$$

The length of each vector (directed line segment) $\overrightarrow{P_1Q_1}$, $\overrightarrow{P_2Q_2}$, and $\overrightarrow{P_3Q_3}$ in Figure 3.8(b) is $\sqrt{13}$ (verify).

Two vectors $\mathbf{X}_1 = \begin{bmatrix} x_1 \\ y_1 \end{bmatrix}$ and $\mathbf{X}_2 = \begin{bmatrix} x_2 \\ y_2 \end{bmatrix}$ are said to be **parallel** if $x_1 y_2 = x_2 y_1$. That is, they are parallel if the lines on which they lie are both vertical or have the same slopes. Thus the vectors (directed line segments) $\overrightarrow{P_1Q_1}$, $\overrightarrow{P_2Q_2}$, and $\overrightarrow{P_3Q_3}$ in Figure 3.8(b) are parallel.

Vector Operations

Definition Let $\mathbf{X} = (x_1, y_1)$ and $\mathbf{Y} = (x_2, y_2)$ be two vectors in the plane. The **sum** of the vectors $\mathbf{X}$ and $\mathbf{Y}$ is the vector
$$(x_1 + x_2, y_1 + y_2)$$
and is denoted by $\mathbf{X} + \mathbf{Y}$. Thus vectors are added by adding their components.

Example 8 Let $\mathbf{X} = (1, 2)$ and $\mathbf{Y} = (3, -4)$. Then
$$\mathbf{X} + \mathbf{Y} = (1 + 3, 2 + (-4)) = (4, -2).$$

We can interpret vector addition geometrically as follows. In Figure 3.11, the vector $\mathbf{Z}$, which is parallel to $\mathbf{Y}$, is of the same length as $\mathbf{Y}$ and its tail is the head (x_1, y_1) of $\mathbf{X}$; so

Figure 3.11   Figure 3.12

its head is $(x_1 + x_2, y_1 + y_2)$. Thus the vector with tail O and head $(x_1 + x_2, y_1 + y_2)$ is **X + Y**.

We can also describe **X + Y** as the diagonal of the parallelogram defined by **X** and **Y**, as shown in Figure 3.12.

Example 9 If **X** and **Y** are as in Example 8, then **X + Y** is shown in Figure 3.13.

Definition If $\mathbf{X} = (x, y)$ and c is a scalar (a real number), then the **scalar multiple** $c\mathbf{X}$ of **X** by c is the vector (cx, cy). Thus the scalar multiple $c\mathbf{X}$ of **X** by c is obtained by multiplying each component of **X** by c.

If $c > 0$, then $c\mathbf{X}$ is in the same direction as **X**, whereas if $d < 0$, then $d\mathbf{X}$ is in the opposite direction (Figure 3.14).

Example 10 If $c = 2$, $d = -3$, and $\mathbf{X} = (2, -3)$ then

$$c\mathbf{X} = 2(2, -3) = (4, -6) \qquad \text{and} \qquad d\mathbf{X} = -3(2, -3) = (-6, 9),$$

which are shown in Figure 3.15.

Figure 3.13   Figure 3.14

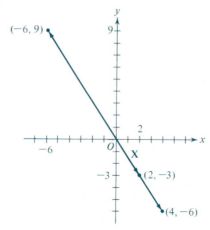

Figure 3.15

The vector

$$(0, 0)$$

is called the **zero vector** and is denoted by **0**. If **X** is any vector, it follows that (Exercise T.2)

$$\mathbf{X} + \mathbf{0} = \mathbf{X}. \tag{3}$$

We can also show (Exercise T.3) that

$$\mathbf{X} + (-1)\mathbf{X} = \mathbf{0}, \tag{4}$$

and we write $(-1)\mathbf{X}$ as $-\mathbf{X}$ and call it the **negative** of **X**. Moreover, we write $\mathbf{X} + (-1)\mathbf{Y}$ as $\mathbf{X} - \mathbf{Y}$ and call it the **difference between X and Y**. The vector $\mathbf{X} - \mathbf{Y}$ is shown in Figure 3.16.

Observe that while vector addition gives one diagonal of a parallelogram, vector subtraction gives the other diagonal.

APPLICATION (VECTORS IN PHYSICS).   When several forces act on a body, we can find a single force, called the **resultant force**, having the equivalent effect. The resultant force can be determined using vectors. The following example illustrates the method.

Figure 3.16

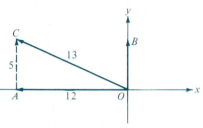

Figure 3.17

Example 11 Suppose that a force of 12 pounds is applied to an object along the negative x-axis and a force of 5 pounds is applied to the object along the positive y-axis. Find the magnitude and direction of the resultant force.

Solution
In Figure 3.17 we have represented the force along the negative x-axis by the vector $\overrightarrow{OA}$ and the force along the positive y-axis by the vector $\overrightarrow{OB}$. The resultant force is the vector $\overrightarrow{OC} = \overrightarrow{OA} + \overrightarrow{OB}$. Thus the magnitude of the resultant force is 13 pounds and its direction is as indicated in the figure.

Vectors are also used in physics to deal with velocity problems, as the following example illustrates.

Example 12 Suppose that a boat is traveling east across a river at the rate of 4 miles per hour while the river's current is flowing south at a rate of 3 miles per hour. Find the resultant velocity of the boat.

Solution
In Figure 3.18, we have represented the velocity of the boat by the vector $\overrightarrow{OA}$ and the velocity of the river's current by the vector $\overrightarrow{OB}$. The resultant velocity is the vector $\overrightarrow{OC} = \overrightarrow{OA} + \overrightarrow{OB}$. Thus the magnitude of the resultant velocity is 5 miles per hour and its direction is as indicated in the figure.

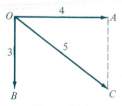

Figure 3.18

Angle Between Two Vectors

The angle between the nonzero vectors $\mathbf{X} = (x_1, y_1)$ and $\mathbf{Y} = (x_2, y_2)$ is the angle θ, $0 \leq \theta \leq \pi$, shown in Figure 3.19. Applying the law of cosines to the triangle in that figure, we obtain

$$\|\mathbf{X} - \mathbf{Y}\|^2 = \|\mathbf{X}\|^2 + \|\mathbf{Y}\|^2 - 2\|\mathbf{X}\| \|\mathbf{Y}\| \cos \theta. \tag{5}$$

From (2),

$$\begin{aligned}
\|\mathbf{X} - \mathbf{Y}\|^2 &= (x_1 - x_2)^2 + (y_1 - y_2)^2 \\
&= x_1^2 + x_2^2 + y_1^2 + y_2^2 - 2(x_1 x_2 + y_1 y_2) \\
&= \|\mathbf{X}\|^2 + \|\mathbf{Y}\|^2 - 2(x_1 x_2 + y_1 y_2).
\end{aligned}$$

If we substitute this expression in (5) and solve for $\cos \theta$ (recall that since $\mathbf{X}$ and $\mathbf{Y}$ are nonzero vectors, then $\|\mathbf{X}\| \neq 0$ and $\|\mathbf{Y}\| \neq 0$), we obtain

$$\cos \theta = \frac{x_1 x_2 + y_1 y_2}{\|\mathbf{X}\| \|\mathbf{Y}\|}. \tag{6}$$

Definition The **inner product** or **dot product** of the vectors $\mathbf{X} = (x_1, y_1)$ and $\mathbf{Y} = (x_2, y_2)$ is defined to be

$$\mathbf{X} \cdot \mathbf{Y} = x_1 x_2 + y_1 y_2.$$

Thus we can rewrite (6) as

$$\cos \theta = \frac{\mathbf{X} \cdot \mathbf{Y}}{\|\mathbf{X}\| \|\mathbf{Y}\|} \qquad (0 \leq \theta \leq \pi). \tag{7}$$

Example 13 If $\mathbf{X} = (2, 4)$ and $\mathbf{Y} = (-1, 2)$, then

$$\mathbf{X} \cdot \mathbf{Y} = (2)(-1) + (4)(2) = 6.$$

Also,

$$\|\mathbf{X}\| = \sqrt{2^2 + 4^2} = \sqrt{20}$$

and

$$\|\mathbf{Y}\| = \sqrt{(-1)^2 + 2^2} = \sqrt{5}.$$

Hence

$$\cos \theta = \frac{6}{\sqrt{20} \cdot \sqrt{5}} = 0.6.$$

We can obtain the approximate angle by using a table of cosines; we find that θ is approximately $53°8'$.

Figure 3.19   **Figure 3.20**

If the nonzero vectors **X** and **Y** are at right angles (Figure 3.20), then the cosine of the angle θ between them is zero. Hence, from (7), we have $\mathbf{X} \cdot \mathbf{Y} = 0$. Conversely, if $\mathbf{X} \cdot \mathbf{Y} = 0$, then $\cos \theta = 0$ and the vectors are at right angles. Thus the nonzero vectors **X** and **Y** are **perpendicular** or **orthogonal** if and only if $\mathbf{X} \cdot \mathbf{Y} = 0$.

Example 14 The vectors $\mathbf{X} = (2, -4)$ and $\mathbf{Y} = (4, 2)$ are orthogonal, since

$$\mathbf{X} \cdot \mathbf{Y} = (2)(4) + (-4)(2) = 0$$

(Figure 3.21).

The dot product satisfies the following properties.

Theorem 3.1 *If* **X**, **Y**, *and* **Z** *are vectors and c is a scalar, then:*
    (a) $\mathbf{X} \cdot \mathbf{X} = \|\mathbf{X}\|^2 \geqslant 0$, *with equality if and only if* $\mathbf{X} = \mathbf{0}$.
    (b) $\mathbf{X} \cdot \mathbf{Y} = \mathbf{Y} \cdot \mathbf{X}$.
    (c) $(\mathbf{X} + \mathbf{Y}) \cdot \mathbf{Z} = \mathbf{X} \cdot \mathbf{Z} + \mathbf{Y} \cdot \mathbf{Z}$.
    (d) $(c\mathbf{X}) \cdot \mathbf{Y} = \mathbf{X} \cdot (c\mathbf{Y}) = c(\mathbf{X} \cdot \mathbf{Y})$.

Proof
Exercise T.7.

Figure 3.21

Unit Vectors

A **unit vector** is a vector whose length is 1. If X is any nonzero vector, then the vector

$$U = \frac{1}{\|X\|} X$$

is a unit vector in the direction of X (Exercise T.5).

Example 15 Let $X = (-3, 4)$. Then

$$\|X\| = \sqrt{(-3)^2 + 4^2} = 5.$$

Hence the vector $U = \frac{1}{5}(-3, 4) = (-\frac{3}{5}, \frac{4}{5})$ is a unit vector, since

$$\|U\| = \sqrt{\left(-\frac{3}{5}\right)^2 + \left(\frac{4}{5}\right)^2} = \sqrt{\frac{9 + 16}{25}} = 1.$$

Also, U lies in the direction of X (Figure 3.22).

There are two unit vectors in R^2 that are of special importance. They are $i = (1, 0)$ and $j = (0, 1)$, the unit vectors along the positive x- and y-axes, respectively (Figure 3.23). If $X = (x, y)$ is any vector in R^2, then we can write X in terms of i and j as

$$X = xi + yj.$$

Example 16 If

$$X = (4, -5),$$

then

$$X = 4i - 5j.$$

Example 17 The vectors i and j are orthogonal (see Exercise 19).

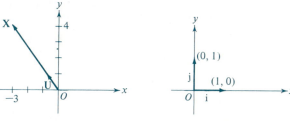

Figure 3.22   Figure 3.23

3.1 EXERCISES

1. Plot the following points in R^2.
    (a) $(2, -1)$ (b) $(-1, 2)$.   (c) $(3, 4)$.
    (d) $(-3, -2)$.   (e) $(0, 2)$.   (f) $(0, -3)$.

2. Sketch a directed line segment in R^2 representing each of the following vectors.

    (a) $\mathbf{X}_1 = \begin{bmatrix} -2 \\ 3 \end{bmatrix}$.   (b) $\mathbf{X}_2 = \begin{bmatrix} 3 \\ 4 \end{bmatrix}$.

    (c) $\mathbf{X}_3 = \begin{bmatrix} -3 \\ -3 \end{bmatrix}$.   (d) $\mathbf{X}_4 = \begin{bmatrix} 0 \\ -3 \end{bmatrix}$.

3. Determine the head of the vector $\begin{bmatrix} -2 \\ 5 \end{bmatrix}$ whose tail is at $(3, 2)$. Make a sketch.

4. Determine the tail of the vector $\begin{bmatrix} 2 \\ 5 \end{bmatrix}$ whose head is at $(1, 2)$. Make a sketch.

5. Find $\mathbf{X} + \mathbf{Y}$, $\mathbf{X} - \mathbf{Y}$, $2\mathbf{X}$, and $3\mathbf{X} - 2\mathbf{Y}$ if
    (a) $\mathbf{X} = (2, 3)$, $\mathbf{Y} = (-2, 5)$.
    (b) $\mathbf{X} = (0, 3)$, $\mathbf{Y} = (3, 2)$.
    (c) $\mathbf{X} = (2, 6)$, $\mathbf{Y} = (3, 2)$.

6. Repeat Exercise 5 for
    (a) $\mathbf{X} = (-1, 3)$, $\mathbf{Y} = (2, 4)$.
    (b) $\mathbf{X} = (-4, -3)$, $\mathbf{Y} = (5, 2)$.
    (c) $\mathbf{X} = (3, 2)$, $\mathbf{Y} = (-2, 0)$.

7. Let $\mathbf{X} = (1, 2)$, $\mathbf{Y} = (-3, 4)$, $\mathbf{Z} = (x, 4)$, and $\mathbf{U} = (-2, y)$. Find x and y so that
    (a) $\mathbf{Z} = 2\mathbf{X}$.   (b) $\frac{3}{2}\mathbf{U} = \mathbf{Y}$.   (c) $\mathbf{Z} + \mathbf{U} = \mathbf{X}$.

8. Let $\mathbf{X} = (-4, 3)$, $\mathbf{Y} = (2, -5)$, and $\mathbf{Z} = (x, y)$. Find x and y so that
    (a) $\mathbf{Z} = 2\mathbf{X} + 3\mathbf{Y}$.   (b) $\mathbf{X} + \mathbf{Z} = 2\mathbf{X} - \mathbf{Y}$.
    (c) $\mathbf{Z} = \frac{5}{2}\mathbf{Y}$.

9. Find the length of the following vectors.
    (a) $(1, 2)$.   (b) $(-3, -4)$.
    (c) $(0, 2)$.   (d) $(-4, 3)$.

10. Find the length of the following vectors.
    (a) $(-2, 3)$.   (b) $(3, 0)$.
    (c) $(-4, -5)$.   (d) $(3, 2)$.

11. Find the distance between the following pairs of points.
    (a) $(2, 3)$, $(3, 4)$.   (b) $(0, 0)$, $(3, 4)$.
    (c) $(-3, 2)$, $(0, 1)$.   (d) $(0, 3)$, $(2, 0)$.

12. Find the distance between the following pairs of points.
    (a) $(4, 2)$, $(1, 2)$.   (b) $(-2, -3)$, $(0, 1)$.
    (c) $(2, 4)$, $(-1, 1)$.   (d) $(2, 0)$, $(3, 2)$.

13. Find a unit vector in the direction of $\mathbf{X}$.
    (a) $\mathbf{X} = (3, 4)$.   (b) $\mathbf{X} = (-2, -3)$.
    (c) $\mathbf{X} = (5, 0)$.

14. Find a unit vector in the direction of $\mathbf{X}$.
    (a) $\mathbf{X} = (2, 4)$.   (b) $\mathbf{X} = (0, -2)$.
    (c) $\mathbf{X} = (-1, -3)$.

15. Find $\mathbf{X} \cdot \mathbf{Y}$.
    (a) $\mathbf{X} = (1, 2)$, $\mathbf{Y} = (2, -3)$.
    (b) $\mathbf{X} = (1, 0)$, $\mathbf{Y} = (0, 1)$.
    (c) $\mathbf{X} = (-3, -4)$, $\mathbf{Y} = (4, -3)$.
    (d) $\mathbf{X} = (2, 1)$, $\mathbf{Y} = (-2, -1)$.

16. Find $\mathbf{X} \cdot \mathbf{Y}$.
    (a) $\mathbf{X} = (0, -1)$, $\mathbf{Y} = (1, 0)$.
    (b) $\mathbf{X} = (2, 2)$, $\mathbf{Y} = (4, -5)$.
    (c) $\mathbf{X} = (2, -1)$, $\mathbf{Y} = (-3, -2)$.
    (d) $\mathbf{X} = (0, 2)$, $\mathbf{Y} = (3, -3)$.

17. Find the cosine of the angle between each pair of vectors $\mathbf{X}$ and $\mathbf{Y}$ in Exercise 15.

18. Find the cosine of the angle between each pair of vectors $\mathbf{X}$ and $\mathbf{Y}$ in Exercise 16.

19. Show that
    (a) $\mathbf{i} \cdot \mathbf{i} = \mathbf{j} \cdot \mathbf{j} = 1$.   (b) $\mathbf{i} \cdot \mathbf{j} = 0$.

20. Which of the vectors $\mathbf{X}_1 = (1, 2)$, $\mathbf{X}_2 = (0, 1)$, $\mathbf{X}_3 = (-2, -4)$, $\mathbf{X}_4 = (-2, 1)$, $\mathbf{X}_5 = (2, 4)$, and $\mathbf{X}_6 = (-6, 3)$ are
    (a) Orthogonal?   (b) In the same direction?
    (c) In opposite directions?

21. Write each of the following vectors in terms of **i** and **j**.
(a) (1, 3).   (b) (−2, −3).
(c) (−2, 0).   (d) (0, 3).

22. Write each of the following vectors as a 2 × 1 matrix.
(a) $3\mathbf{i} - 2\mathbf{j}$.   (b) $2\mathbf{i}$.   (c) $-2\mathbf{i} - 3\mathbf{j}$.

23. A ship is being pushed by a tugboat with a force of 300 pounds along the negative *y*-axis while another tugboat is pushing along the negative *x*-axis with a force of 400 pounds. Find the magnitude and sketch the direction of the resultant force.

24. Suppose that an airplane is flying with an airspeed of 260 kilometers per hour while a wind is blowing to the west at 100 kilometers per hour. Indicate on a figure the approximate direction that the plane must follow to result in a flight directly south. What will be the resultant speed?

THEORETICAL EXERCISES

T.1. Show how we can associate a point in the plane with each ordered pair (*x*, *y*) of real numbers.
T.2. Show that $\mathbf{X} + \mathbf{0} = \mathbf{X}$.
T.3. Show that $\mathbf{X} + (-1)\mathbf{X} = \mathbf{0}$.
T.4. Show that if *c* is a scalar, then $\|c\mathbf{X}\| = |c|\, \|\mathbf{X}\|$.
T.5. Show that if **X** is a nonzero vector, then $\mathbf{U} = (1/\|\mathbf{X}\|)\mathbf{X}$ is a unit vector in the direction of **X**.

T.6. Show that
(a) $1\mathbf{X} = \mathbf{X}$.
(b) $(rs)\mathbf{X} = r(s\mathbf{X})$, where *r* and *s* are scalars.
T.7. Prove Theorem 3.1.
T.8. Show that if **Z** is orthogonal to **X** and **Y**, then **Z** is orthogonal to $r\mathbf{X} + s\mathbf{Y}$, where *r* and *s* are scalars.

3.2

n-Vectors

We shall now generalize the notion of a vector in the plane, discussed in the previous section, to *n*-vectors. For $n = 2$, we obtain the vectors in the plane. The case of $n = 3$ will also be of special interest, and we shall discuss it in some detail.

Definition   An ***n*-vector** is an $n \times 1$ matrix

$$\mathbf{X} = \begin{bmatrix} x_1 \\ x_2 \\ \vdots \\ x_n \end{bmatrix},$$

where $x_1, x_2, \ldots, x_n$ are real numbers, which are called the **components** of **X**.

Since an *n*-vector is a matrix, the *n*-vectors

$$\mathbf{X} = \begin{bmatrix} x_1 \\ x_2 \\ \vdots \\ x_n \end{bmatrix} \quad \text{and} \quad \mathbf{Y} = \begin{bmatrix} y_1 \\ y_2 \\ \vdots \\ y_n \end{bmatrix}$$

are said to be **equal** if $x_i = y_i \ (1 \leq i \leq n)$.

Example 1 The 4-vectors

$$
\begin{bmatrix} 1 \\ -2 \\ 3 \\ 4 \end{bmatrix} \quad \text{and} \quad \begin{bmatrix} 1 \\ -2 \\ 3 \\ -4 \end{bmatrix}
$$

are not equal, since their fourth components are not equal.

The set of all n-vectors is denoted by R^n and is called **n-space**. When the actual value of n need not be specified, we refer to n-vectors simply as **vectors**. The real numbers are called **scalars**. The components of a vector are real numbers and hence the components of a vector are scalars.

Vector Operations

Definition Let

$$
\mathbf{X} = \begin{bmatrix} x_1 \\ x_2 \\ \vdots \\ x_n \end{bmatrix} \quad \text{and} \quad \mathbf{Y} = \begin{bmatrix} y_1 \\ y_2 \\ \vdots \\ y_n \end{bmatrix}
$$

be two vectors in R^n. The **sum** of the vectors $\mathbf{X}$ and $\mathbf{Y}$ is the vector

$$
\begin{bmatrix} x_1 + y_1 \\ x_2 + y_2 \\ \vdots \\ x_n + y_n \end{bmatrix},
$$

and it is denoted by $\mathbf{X} + \mathbf{Y}$.

Example 2 If

$$
\mathbf{X} = \begin{bmatrix} 1 \\ -2 \\ 3 \end{bmatrix} \quad \text{and} \quad \mathbf{Y} \begin{bmatrix} 2 \\ 3 \\ -3 \end{bmatrix}
$$

are vectors in R^3, then

$$
\mathbf{X} + \mathbf{Y} = \begin{bmatrix} 1 + 2 \\ -2 + 3 \\ 3 + (-3) \end{bmatrix} = \begin{bmatrix} 3 \\ 1 \\ 0 \end{bmatrix}.
$$

Definition If

$$\mathbf{X} = \begin{bmatrix} x_1 \\ x_2 \\ \vdots \\ x_n \end{bmatrix}$$

is a vector in R^n and c is a scalar, then the **scalar multiple** $c\mathbf{X}$ of $\mathbf{X}$ by c is the vector

$$\begin{bmatrix} cx_1 \\ cx_2 \\ \vdots \\ cx_n \end{bmatrix}.$$

Example 3 If

$$\mathbf{X} = \begin{bmatrix} 2 \\ 3 \\ -1 \\ 2 \end{bmatrix}$$

is a vector in R^4 and $c = -2$, then

$$c\mathbf{X} = (-2)\begin{bmatrix} 2 \\ 3 \\ -1 \\ 2 \end{bmatrix} = \begin{bmatrix} -4 \\ -6 \\ 2 \\ -4 \end{bmatrix}.$$

The operations of vector addition and scalar multiplication satisfy the following properties.

Theorem 3.2 *Let* $\mathbf{X}$, $\mathbf{Y}$, *and* $\mathbf{Z}$ *be any vectors in* R^n; *let* c *and* d *be any scalars. Then*

   (α) $\mathbf{X} + \mathbf{Y}$ *is a vector in* R^n *(that is,* R^n *is closed under the operation of vector addition).*

   (a) $\mathbf{X} + \mathbf{Y} = \mathbf{Y} + \mathbf{X}$.

   (b) $\mathbf{X} + (\mathbf{Y} + \mathbf{Z}) = (\mathbf{X} + \mathbf{Y}) + \mathbf{Z}$.

   (c) *There is a unique vector* $\mathbf{0}$ *in* R^n,

$$\mathbf{0} = \begin{bmatrix} 0 \\ 0 \\ \vdots \\ 0 \end{bmatrix},$$

   such that $\mathbf{X} + \mathbf{0} = \mathbf{0} + \mathbf{X} = \mathbf{X}$; $\mathbf{0}$ *is called the* **zero vector**.

   (d) *There is a unique vector* $-\mathbf{X}$,

$$-\mathbf{X} = \begin{bmatrix} -x_1 \\ -x_2 \\ \vdots \\ -x_n \end{bmatrix},$$

such that

$$\mathbf{X} + (-\mathbf{X}) = \mathbf{0}.$$

(β) *c*$\mathbf{X}$ *is a vector in* R^n.
   (e) $c(\mathbf{X} + \mathbf{Y}) = c\mathbf{X} + c\mathbf{Y}$.
   (f) $(c + d)\mathbf{X} = c\mathbf{X} + d\mathbf{X}$.
   (g) $c(d\mathbf{X}) = (cd)\mathbf{X}$.
   (h) $1\mathbf{X} = \mathbf{X}$.

Proof

(α) and (β) are immediate from the definitions for vector sum and scalar multiple. We verify (f) here and leave the rest of the proof to the reader (Exercise T.1). Thus

$$(c + d)\mathbf{X} = (c + d)\begin{bmatrix} x_1 \\ x_2 \\ \vdots \\ x_n \end{bmatrix} = \begin{bmatrix} (c + d)x_1 \\ (c + d)x_2 \\ \vdots \\ (c + d)x_n \end{bmatrix} = \begin{bmatrix} cx_1 + dx_1 \\ cx_2 + dx_2 \\ \vdots \\ cx_n + dx_n \end{bmatrix}$$

$$= \begin{bmatrix} cx_1 \\ cx_2 \\ \vdots \\ cx_n \end{bmatrix} + \begin{bmatrix} dx_1 \\ dx_2 \\ \vdots \\ dx_n \end{bmatrix} = c\begin{bmatrix} x_1 \\ x_2 \\ \vdots \\ x_n \end{bmatrix} + d\begin{bmatrix} x_1 \\ x_2 \\ \vdots \\ x_n \end{bmatrix} = c\mathbf{X} + d\mathbf{X}.$$

The vector $-\mathbf{X}$ in (d) is called the **negative** of $\mathbf{X}$. It is easy to verify (Exercise T.2) that

$$-\mathbf{X} = (-1)\mathbf{X}.$$

We shall also write $\mathbf{X} + (-\mathbf{Y})$ as $\mathbf{X} - \mathbf{Y}$.

Example 4 If $\mathbf{X}$ and $\mathbf{Y}$ are as in Example 2, then

$$\mathbf{X} - \mathbf{Y} = \begin{bmatrix} 1 - 2 \\ -2 - 3 \\ 3 - (-3) \end{bmatrix} = \begin{bmatrix} -1 \\ -5 \\ 6 \end{bmatrix}.$$

As in the case of R^2, we shall identify the vector

$$\begin{bmatrix} x_1 \\ x_2 \\ \vdots \\ x_n \end{bmatrix}$$

with the point $(x_1, x_2, \ldots, x_n)$ so that points and vectors can be used interchangeably. Thus we may view R^n as consisting of vectors or of points, and we write

$$\mathbf{X} = (x_1, x_2, \ldots, x_n).$$

Moreover, an *n*-vector is an $n \times 1$ matrix, vector addition is matrix addition, and scalar multiplication is merely the operation of multiplication of a matrix by a real number. Thus R^n can be viewed as the set of all $n \times 1$ matrices with the operations of matrix addition and scalar multiplication. *The important point here is that no matter how we view R^n— as n-vectors, points, or $n \times 1$ matrices—its algebraic behavior is always the same.*

APPLICATION.   Vectors in R^n can be used to handle large amounts of data. Indeed, a number of computer programming languages, notably APL (discussed briefly in the Appendix) and PL/I, make extensive use of vectors. The following example illustrates these ideas.

Example 5 (Inventory Control)   Suppose that a store handles 100 different items. The inventory on hand at the beginning of the week can be described by the inventory vector **A** in R^{100}. The number of items sold at the end of the week can be described by the vector **S**, and the vector

$$\mathbf{A} - \mathbf{S}$$

represents the inventory at the end of the week. If the store receives a new shipment of goods, represented by the vector **B**, then its new inventory would be

$$\mathbf{A} - \mathbf{S} + \mathbf{B}.$$

Visualizing R^3

We cannot draw pictures of R^n for $n > 3$. However, since R^3 is the world we live in, we can visualize it in a manner similar to that used for R^2.

We first fix a **coordinate system** by choosing a point, called the **origin**, and three lines, called the **coordinate axes**, each passing through the origin, so that each line is perpendicular to the other two. These lines are individually called the **x-**, **y-**, and **z-axes**. On each of these axes we choose a point fixing the units of length and positive directions on the coordinate axes. Frequently, but not always, the same unit of length is used for all the coordinate axes. In Figure 3.24(a) and (b) we show two of the many possible coordinate systems.

The coordinate system shown in Figure 3.24(a) is called a **right-handed coordinate system**; the one shown in Figure 3.24(b) is called **left-handed**. A right-handed system is characterized by the following property. If we rotate the *x*-axis counterclockwise toward the *y*-axis, then a right-hand screw will move in the positive *z* direction (Figure 3.25).

The projection of a point P in space on a line L is the point Q obtained by intersecting L with the line L' passing through P and perpendicular to L (Figure 3.26).

The **x-coordinate** of the point P is the number associated with the projection of P on

Figure 3.24

the *x*-axis; similarly, for the **y-** and **z-coordinates**. These three numbers are called the **coordinates** of *P*. Thus with each point in space we associate an ordered triple (x, y, z) of real numbers, and conversely, with each ordered triple of real numbers, we associate a point in space. This correspondence is called a **rectangular coordinate system**. We write $P(x, y, z)$, or simply (x, y, z).

Example 6 In Figure 3.27 (a) and (b) we show two points and their coordinates.

The **xy-plane** is the plane determined by the *x*- and *y*-axes. Similarly, we have the **xz-** and **yz-planes**.

In R^3, the components of a vector **X** are denoted by *x*, *y*, and *z*. Thus $\mathbf{X} = (x, y, z)$.

As in the plane, with the vector $\mathbf{X} = (x, y, z)$ we associate the directed line segment $\overrightarrow{OP}$, whose tail is $O(0, 0, 0)$ and whose head is $P(x, y, z)$ [Figure 3.28(a)]. Again as in the plane, in physical applications we often deal with a directed line segment $\overrightarrow{PQ}$, from the point $P(x, y, z)$ (not the origin) to the point $Q(x', y', z')$, as shown in Figure 3.28(b). Such a directed line segment will also be called a **vector in R^3**, or simply a **vector** with tail at $P(x, y, z)$ and head at $Q(x', y', z')$. The components of such a vector are $x' - x$, $y' - y$, and $z' - z$. Two such vectors in R^3 will be called **equal** if their components are equal. Thus the vector $\overrightarrow{PQ}$ in Figure 3.28(b) can also be represented by the vector $(x' - x, y' - y, z' - z)$ with tail at O and head at $P''(x' - x, y' - y, z' - z)$.

We shall soon define the length of a vector in R^n and the angle between two nonzero vectors in R^n. Once this has been done, it can be shown that two vectors in R^3 are equal if

Figure 3.25   Figure 3.26

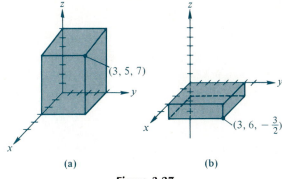

Figure 3.27

and only if any directed line segments representing them are parallel and of the same length.

The sum $\mathbf{X} + \mathbf{Y}$ of the vectors $\mathbf{X} = (x_1, y_1, z_1)$ and $\mathbf{Y} = (x_2, y_2, z_2)$ in R^3 is the diagonal of the parallelogram determined by $\mathbf{X}$ and $\mathbf{Y}$, as shown in Figure 3.29.

The reader will note that Figure 3.29 looks very much like Figure 3.12 in Section 3.1 in that in both R^2 and R^3 the vector $\mathbf{X} + \mathbf{Y}$ is the diagonal of the parallelogram determined by $\mathbf{X}$ and $\mathbf{Y}$.

The scalar multiple in R^3 is shown in Figure 3.30, which looks very much like Figure 3.14 in Section 3.1.

Figure 3.28

Figure 3.29

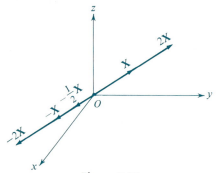

Figure 3.30

Inner Product in R^n

We shall now define the notion of the length of a vector in R^n by generalizing the corresponding idea for R^2.

Definition The **length** or **magnitude** or **norm** of the vector $\mathbf{X} = (x_1, x_2, \ldots, x_n)$ in R^n is

$$\|\mathbf{X}\| = \sqrt{x_1^2 + x_2^2 + \cdots + x_n^2}. \tag{1}$$

We also define the distance from the point $(x_1, x_2, \ldots, x_n)$ to the origin by (1). The **distance** between the points $(x_1, x_2, \ldots, x_n)$ and $(y_1, y_2, \ldots, y_n)$ is then defined as the length of the vector $\mathbf{X} - \mathbf{Y}$, where

$$\mathbf{X} = (x_1, x_2, \ldots, x_n) \quad \text{and} \quad \mathbf{Y} = (y_1, y_2, \ldots, y_n).$$

Thus this distance is given by

$$\|\mathbf{X} - \mathbf{Y}\| = \sqrt{(x_1 - y_1)^2 + (x_2 - y_2)^2 + \cdots + (x_n - y_n)^2}. \tag{2}$$

Example 7 Let $\mathbf{X} = (2, 3, 2, -1)$ and $\mathbf{Y} = (4, 2, 1, 3)$. Then

$$\|\mathbf{X}\| = \sqrt{2^2 + 3^2 + 2^2 + (-1)^2} = \sqrt{18},$$
$$\|\mathbf{Y}\| = \sqrt{4^2 + 2^2 + 1^2 + 3^2} = \sqrt{30}.$$

The distance between the points $(2, 3, 2, -1)$ and $(4, 2, 1, 3)$ is the length of the vector $\mathbf{X} - \mathbf{Y}$. Thus, from Equation (2),

$$\|\mathbf{X} - \mathbf{Y}\| = \sqrt{(2 - 4)^2 + (3 - 2)^2 + (2 - 1)^2 + (-1 - 3)^2} = \sqrt{22}.$$

It should be noted that in R^3, Equations (1) and (2), for the length of a vector and the distance between two points, do not have to be defined. They can easily be established by means of two applications of the Pythagorean Theorem (Exercise T.3).

We shall define the cosine of the angle between two vectors in R^n by generalizing the corresponding formula in R^2. However, first we formulate the notion of inner product in R^n.

Definition If $\mathbf{X} = (x_1, x_2, \ldots, x_n)$ and $\mathbf{Y} = (y_1, y_2, \ldots, y_n)$ are vectors in R^n, then their **inner product** is defined by

$$\mathbf{X} \cdot \mathbf{Y} = x_1 y_1 + x_2 y_2 + \cdots + x_n y_n.$$

This is exactly how the dot product was defined in R^2. The inner product in R^n is also called the **dot product**.

Example 8 If $\mathbf{X}$ and $\mathbf{Y}$ are as in Example 7, then

$$\mathbf{X} \cdot \mathbf{Y} = (2)(4) + (3)(2) + (2)(1) + (-1)(3)$$
$$= 13.$$

Example 9
(Revenue
Monitoring) Consider the store in Example 5. If the vector $\mathbf{P}$ denotes the prices of each of the 100 items, then the dot product

$$\mathbf{S} \cdot \mathbf{P}$$

gives the total revenue received at the end of the week.

The inner product in R^n satisfies the same properties as in R^2. We state these as the following theorem, which reads just like Theorem 3.1.

Theorem 3.3
(Properties
of Inner
Product)

If **X**, **Y**, *and* **Z** *are vectors in* R^n *and* c *is a scalar, then:*

(a) $\mathbf{X} \cdot \mathbf{X} = \|\mathbf{X}\|^2 \geq 0$, *with equality if and only if* $\mathbf{X} = \mathbf{0}$.
(b) $\mathbf{X} \cdot \mathbf{Y} = \mathbf{Y} \cdot \mathbf{X}$.
(c) $(\mathbf{X} + \mathbf{Y}) \cdot \mathbf{Z} = \mathbf{X} \cdot \mathbf{Z} + \mathbf{Y} \cdot \mathbf{Z}$.
(d) $(c\mathbf{X}) \cdot \mathbf{Y} = \mathbf{X} \cdot (c\mathbf{Y}) = c(\mathbf{X} \cdot \mathbf{Y})$.

Proof
Exercise T.4.

We shall now prove a result that will enable us to give a worthwhile definition for the cosine of an angle between two nonzero vectors. This result, called the **Cauchy–Schwarz inequality**, has many important applications in mathematics. The proof of this result, although not difficult, is one that is not too natural but does call for a clever start.

Theorem 3.4
(Cauchy†–
Schwarz‡
Inequality)

If **X** *and* **Y** *are vectors in* R^n, *then*

$$|\mathbf{X} \cdot \mathbf{Y}| \leq \|\mathbf{X}\| \, \|\mathbf{Y}\|. \tag{3}$$

(Observe that $|\ \ |$ *on the left stands for the absolute value of a real number;* $\|\ \ \|$ *on the right denotes the length of a vector.)*

Proof
If $\mathbf{X} = \mathbf{0}$, then $\|\mathbf{X}\| = 0$, and $\mathbf{X} \cdot \mathbf{Y} = 0$, so (3) holds. Now suppose that $\mathbf{X}$ is nonzero. Let r be a scalar and consider the vector $r\mathbf{X} + \mathbf{Y}$. By Theorem 3.3,

$$0 \leq (r\mathbf{X} + \mathbf{Y}) \cdot (r\mathbf{X} + \mathbf{Y}) = r^2 \mathbf{X} \cdot \mathbf{X} + 2r\mathbf{X} \cdot \mathbf{Y} + \mathbf{Y} \cdot \mathbf{Y}$$
$$= ar^2 + 2br + c,$$

where

$$a = \mathbf{X} \cdot \mathbf{X}, \qquad b = \mathbf{X} \cdot \mathbf{Y}, \qquad \text{and} \qquad c = \mathbf{Y} \cdot \mathbf{Y}.$$

---

† Augustin-Louis Cauchy (1789–1857) grew up in a suburb of Paris as a neighbor of several leading mathematicians of the day, attended the École Polytechnique and the École des Ponts et Chaussées, and was for a time a practicing engineer. He was a devout Roman Catholic, with an abiding interest in Catholic charities. He was also strongly devoted to royalty, especially to the Bourbon kings who ruled France after Napoleon's defeat. When Charles X was deposed in 1830, Cauchy voluntarily followed him into exile in Prague.

Cauchy wrote 7 books and more than 700 papers of varying quality, touching all branches of mathematics. He made important contributions to the early theory of determinants, the theory of eigenvalues, the study of ordinary and partial differential equations, the theory of permutation groups, and the foundations of calculus, and he founded the theory of functions of a complex variable.

‡ Hermann Amandus Schwarz (1843–1921) was born in Poland but was educated and taught in Germany. He was a protégé of Karl Weierstrass and Ernst Eduard Kummer, whose daughter he married. His main contributions to mathematics were in the geometric aspects of analysis, such as conformal mappings and minimal surfaces. In connection with the latter he sought certain numbers associated with differential equations, numbers that have since come to be called eigenvalues. The inequality given above was used in the search for these numbers.

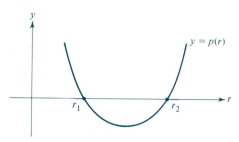

Figure 3.31

Now $p(r) = ar^2 + 2br + c$ is a quadratic polynomial in r (whose graph is a parabola, opening upward, since $a > 0$) that is nonnegative for all values of r. This means that either this polynomial has no real roots, or if it has real roots, then both roots are equal. [If $p(r)$ had two distinct roots r_1 and r_2, then it would be negative for some value of r between r_1 and r_2, as seen in Figure 3.31.]

Recall that the roots of $p(r)$ are given by the quadratic formula as

$$\frac{-2b + \sqrt{4b^2 - 4ac}}{2a} \qquad \text{and} \qquad \frac{-2b - \sqrt{4b^2 - 4ac}}{2a}$$

($a \neq 0$ since $\mathbf{X} \neq \mathbf{0}$). Thus we must have

$$4b^2 - 4ac \leq 0,$$

which means that

$$b^2 \leq ac.$$

Taking square roots of both sides and observing that $\sqrt{a} = \sqrt{\mathbf{X} \cdot \mathbf{X}} = \|\mathbf{X}\|$, $\sqrt{c} = \sqrt{\mathbf{Y} \cdot \mathbf{Y}} = \|\mathbf{Y}\|$, we obtain (3).

Example 10 If $\mathbf{X}$ and $\mathbf{Y}$ are as in Example 7, then from Example 8, $\mathbf{X} \cdot \mathbf{Y} = 13$. Hence

$$|\mathbf{X} \cdot \mathbf{Y}| = 13 \leq \|\mathbf{X}\| \|\mathbf{Y}\| = \sqrt{18}\sqrt{30}.$$

We can now define the angle between two nonzero vectors as follows.

Definition The angle between two nonzero vectors $\mathbf{X}$ and $\mathbf{Y}$ is defined as the unique number θ, $0 \leq \theta \leq \pi$, such that

$$\cos \theta = \frac{\mathbf{X} \cdot \mathbf{Y}}{\|\mathbf{X}\| \|\mathbf{Y}\|} \tag{4}$$

It follows from the Cauchy–Schwarz inequality that

$$\left| \frac{\mathbf{X} \cdot \mathbf{Y}}{\|\mathbf{X}\| \|\mathbf{Y}\|} \right| \leq 1$$

so the expression on the right side of (4) lies between -1 and 1. Thus it makes sense to define the expression on the left side of (4) as $\cos \theta$. In the case of R^3, we can establish by the law of cosines, as was done in R^2, that the cosine of the angle between $\mathbf{X}$ and $\mathbf{Y}$ is given by (4). However, for R^n, $n > 3$, we have to define it by (4).

Example 11 Let $\mathbf{X} = (1, 0, 0, 1)$ and $\mathbf{Y} = (0, 1, 0, 1)$. Then

$$\|\mathbf{X}\| = \sqrt{2}, \qquad \|\mathbf{Y}\| = \sqrt{2}, \qquad \text{and} \qquad \mathbf{X} \cdot \mathbf{Y} = 1.$$

Thus

$$\cos \theta = \tfrac{1}{2}$$

and $\theta = 60°$.

It is very useful to talk about orthogonality and parallelism in R^n, and we accordingly formulate the following definitions.

Definition Two nonzero vectors $\mathbf{X}$ and $\mathbf{Y}$ in R^n are said to be **orthogonal** if $\mathbf{X} \cdot \mathbf{Y} = 0$. If one of the vectors is the zero vector, we agree to say that the vectors are orthogonal. They are said to be **parallel** if $|\mathbf{X} \cdot \mathbf{Y}| = \|\mathbf{X}\| \|\mathbf{Y}\|$. They are in the **same direction** if $\mathbf{X} \cdot \mathbf{Y} = \|\mathbf{X}\| \|\mathbf{Y}\|$. That is, they are orthogonal if $\cos \theta = 0$, parallel if $\cos \theta = \pm 1$, and in the same direction if $\cos \theta = 1$.

Example 12 Consider the vectors $\mathbf{X} = (1, 0, 0, 1)$, $\mathbf{Y} = (0, 1, 1, 0)$, and $\mathbf{Z} = (3, 0, 0, 3)$. Then

$$\mathbf{X} \cdot \mathbf{Y} = 0 \qquad \text{and} \qquad \mathbf{Y} \cdot \mathbf{Z} = 0,$$

which implies that $\mathbf{X}$ and $\mathbf{Y}$ are orthogonal and $\mathbf{Y}$ and $\mathbf{Z}$ are orthogonal. Since

$$\mathbf{X} \cdot \mathbf{Z} = 6, \qquad \|\mathbf{X}\| = \sqrt{2}, \qquad \|\mathbf{Z}\| = \sqrt{18}, \qquad \text{and} \qquad \mathbf{X} \cdot \mathbf{Z} = \|\mathbf{X}\| \|\mathbf{Z}\|,$$

we conclude that $\mathbf{X}$ and $\mathbf{Z}$ are in the same direction.

An easy consequence of the Cauchy–Schwarz inequality is the triangle inequality, which we prove next.

Theorem 3.5 If $\mathbf{X}$ and $\mathbf{Y}$ are vectors in R^n, then
(Triangle
$$\|\mathbf{X} + \mathbf{Y}\| \leq \|\mathbf{X}\| + \|\mathbf{Y}\|.$$
Inequality)

Proof
We have, by Theorem 3.3,

$$
\begin{aligned}
\|\mathbf{X} + \mathbf{Y}\|^2 &= (\mathbf{X} + \mathbf{Y}) \cdot (\mathbf{X} + \mathbf{Y}) \\
&= \mathbf{X} \cdot \mathbf{X} + 2(\mathbf{X} \cdot \mathbf{Y}) + \mathbf{Y} \cdot \mathbf{Y} \\
&= \|\mathbf{X}\|^2 + 2(\mathbf{X} \cdot \mathbf{Y}) + \|\mathbf{Y}\|^2.
\end{aligned}
$$

By the Cauchy–Schwarz inequality we have

$$\|X\|^2 + 2(X \cdot Y) + \|Y\|^2 \leq \|X\|^2 + 2\|X\| \|Y\| + \|Y\|^2$$
$$= (\|X\| + \|Y\|)^2.$$

Taking square roots, we obtain the desired result.

The triangle inequality in R^2 and R^3 merely states that the length of a side of a triangle does not exceed the sum of the lengths of the other two sides (Figure 3.32).

Example 13 For **X** and **Y** as in Example 12, we have

$$\|X + Y\| = \sqrt{4} = 2 < \sqrt{2} + \sqrt{2} = \|X\| + \|Y\|.$$

Another useful result is the Pythagorean Theorem in R^n: If **X** and **Y**, are vectors in R^n, then

$$\|X + Y\|^2 = \|X\|^2 + \|Y\|^2$$

if and only if **X** and **Y** are orthogonal (Exercise T.10).

Definition A **unit vector U** in R^n is a vector of unit length. If **X** is a nonzero vector, then the vector

$$U = \left[\frac{1}{\|X\|} \right] X$$

is a unit vector in the direction of **X**.

Example 14 If **X** is as in Example 12, then since $\|X\| = \sqrt{2}$, the vector $U = (1/\sqrt{2})(1, 0, 0, 1)$ is a unit vector in the direction of **X**.

In the case of R^3 the unit vectors in the positive directions of the x-, y-, and z-axes are denoted by $\mathbf{i} = (1, 0, 0)$, $\mathbf{j} = (0, 1, 0)$, and $\mathbf{k} = (0, 0, 1)$ and are shown in Figure 3.33. If $\mathbf{X} = (x, y, z)$ is any vector in R^3, then we can write **X** in terms of **i**, **j**, and **k** as

$$X = x\mathbf{i} + y\mathbf{j} + z\mathbf{k}.$$

Figure 3.32   Figure 3.33

Example 15   If $X = (2, -1, 3)$, then

$$X = 2i - j + 3k.$$

The n vectors

$$E_1 = (1, 0, \ldots, 0), \qquad E_2 = (0, 1, 0, \ldots, 0), \qquad \cdots \qquad E_n = (0, 0, \ldots, 1)$$

are unit vectors in R^n that are mutually orthogonal, and if $X = (x_1, x_2, \ldots, x_n)$ is any vector in R^n, we have

$$X = x_1 E_1 + x_2 E_2 + \cdots x_n E_n.$$

The vector E_i, $1 \le i \le n$, can be viewed as the ith column of the identity matrix I_n. Thus we see that the columns of I_n form a set of n vectors which are mutually orthogonal. We shall discuss such sets of vectors in more detail in Section 3.8.

3.2 EXERCISES

1. Find $X + Y$, $X - Y$, $2X$, and $3X - 2Y$ if
   (a) $X = (1, 2, -3)$, $Y = (0, 1, -2)$.
   (b) $X = (4, -2, 1, 3)$, $Y = (-1, 2, 5, -4)$.

2. Repeat Exercise 1 for
   (a) $X = (2, 0, -4)$, $Y = (3, 2, 1)$.
   (b) $X = (-3, 5, -3, 0)$, $Y = (2, 1, 5, -2)$.

3. Let $X = (1, -2, 3)$, $Y = (-3, -1, 3)$, $Z = (x, -1, y)$, and $U = (3, u, 2)$. Find x, y, and u so that
   (a) $Z = \frac{1}{2}X$.   (b) $Z + Y = X$.
   (c) $Z + U = Y$.

4. Let $X = (4, -1, -2, 3)$, $Y = (3, -2, -4, 1)$, $Z = (x, -3, -6, y)$, and $U = (2, u, v, 4)$. Find x, y, u, and v so that
   (a) $Z = 3X$.   (b) $Z + U = X$.
   (c) $Z - X = Y$.

5. Let $X = (4, 5, -2, 3)$, $Y = (3, -2, 0, 1)$, $Z = (-3, 2, -5, 3)$, $c = 2$, and $d = 3$. Verify properties (a) through (h) in Theorem 3.2.

6. Plot the following points in R^3.
   (a) $(3, -1, 2)$.   (b) $(1, 0, 2)$.
   (c) $(0, 0, -4)$.   (d) $(1, 0, 0)$.
   (e) $(0, -2, 0)$.

7. Sketch a directed line segment in R^3 representing each of the following vectors.
   (a) $X_1 = (2, -3, -1)$.   (b) $X_2 = (0, 1, 4)$.
   (c) $X_3 = (0, 0, -1)$.

8. For each of the following pairs of points in R^3, determine the vector that is associated with the directed line segment whose tail is the first point and whose head is the second point.
   (a) $(2, 3, -1)$, $(0, 0, 2)$.
   (b) $(1, 1, 0)$, $(0, 1, 1)$.
   (c) $(-1, -2, -3)$, $(3, 4, 5)$.
   (d) $(1, 1, 3)$, $(0, 0, 1)$.

9. Determine the head of the vector $(3, 4, -1)$ whose tail is $(1, -2, 3)$.

10. Find the length of the following vectors.
    (a) $(1, 2, -3)$.   (b) $(2, 3, -1, 4)$.
    (c) $(1, 0, 3)$.   (d) $(0, 0, 3, 4)$.

11. Find the length of the following vectors.
    (a) $(2, 3, 4)$.   (b) $(0, -1, 2, 3)$.
    (c) $(-1, -2, 0)$.   (d) $(1, 2, -3, -4)$.

12. Find the distance between the following pairs of points.
    (a) $(1, -1, 2)$, $(3, 0, 2)$.
    (b) $(4, 2, -1, 5)$, $(2, 3, -1, 4)$.

(c) $(0, 0, 2)$, $(-3, 0, 0)$.

(d) $(1, 0, 0, 2)$, $(3, -1, 5, 2)$.

13. Find the distance between the following pairs of points.

(a) $(1, 1, 0)$, $(2, -3, 1)$.

(b) $(4, 2, -1, 6)$, $(4, 3, 1, 5)$.

(c) $(0, 2, 3)$, $(1, 2, -4)$.

(d) $(3, 4, 0, 1)$, $(2, 2, 1, -1)$.

14. Find $\mathbf{X} \cdot \mathbf{Y}$.

(a) $\mathbf{X} = (1, 2, 3)$, $\mathbf{Y} = (-4, 4, 5)$.

(b) $\mathbf{X} = (0, 2, 3, 1)$, $\mathbf{Y} = (-3, 1, -2, 0)$.

(c) $\mathbf{X} = (0, 0, 1)$, $\mathbf{Y} = (2, 2, 0)$.

(d) $\mathbf{X} = (2, 0, -1, 3)$, $\mathbf{Y} = (-3, -5, 2, -1)$.

15. Find $\mathbf{X} \cdot \mathbf{Y}$.

(a) $\mathbf{X} = (2, 3, 1)$, $\mathbf{Y} = (3, -2, 0)$.

(b) $\mathbf{X} = (1, 2, -1, 3)$, $\mathbf{Y} = (0, 0, -1, -2)$.

(c) $\mathbf{X} = (2, 0, 1)$, $\mathbf{Y} = (2, 2, -1)$.

(d) $\mathbf{X} = (0, 4, 2, 3)$, $\mathbf{Y} = (0, -1, 2, 0)$.

16. Verify Theorem 3.3 for $c = 3$ and $\mathbf{X} = (1, 2, 3)$, $\mathbf{Y} = (1, 2, -4)$, and $\mathbf{Z} = (1, 0, 2)$.

17. Verify Theorem 3.4 for $\mathbf{X}$ and $\mathbf{Y}$ as in Exercise 16.

18. Find the cosine of the angle between each pair of vectors $\mathbf{X}$ and $\mathbf{Y}$ in Exercise 14.

19. Find the cosine of the angle between each pair of vectors $\mathbf{X}$ and $\mathbf{Y}$ in Exercise 15.

20. Show that

(a) $\mathbf{i} \cdot \mathbf{i} = \mathbf{j} \cdot \mathbf{j} = \mathbf{k} \cdot \mathbf{k} = 1$.

(b) $\mathbf{i} \cdot \mathbf{j} = \mathbf{i} \cdot \mathbf{k} = \mathbf{j} \cdot \mathbf{k} = 0$.

21. Which of the vectors $\mathbf{X}_1 = (4, 2, 6, -8)$, $\mathbf{X}_2 = (-2, 3, -1, -1)$, $\mathbf{X}_3 = (-2, -1, -3, 4)$, $\mathbf{X}_4 = (1, 0, 0, 2)$, $\mathbf{X}_5 = (1, 2, 3, -4)$, and $\mathbf{X}_6 = (0, -3, 1, 0)$ are

(a) Orthogonal?   (b) Parallel?

(c) In the same direction?

22. Verify the triangle inequality for $\mathbf{X} = (1, 2, 3, -1)$ and $\mathbf{Y} = (1, 0, -2, 3)$.

23. Find a unit vector in the direction of $\mathbf{X}$.

(a) $\mathbf{X} = (2, -1, 3)$.   (b) $\mathbf{X} = (1, 2, 3, 4)$.

(c) $\mathbf{X} = (0, 1, -1)$.   (d) $\mathbf{X} = (0, -1, 2, -1)$.

24. Find a unit vector in the direction of $\mathbf{X}$.

(a) $\mathbf{X} = (1, 2, -1)$.   (b) $\mathbf{X} = (0, 0, 2, 0)$.

(c) $\mathbf{X} = (-1, 0, -2)$.   (d) $\mathbf{X} = (0, 0, 3, 4)$.

25. Write each of the following vectors in R^3 in terms of $\mathbf{i}$, $\mathbf{j}$, and $\mathbf{k}$.

(a) $(1, 2, -3)$.   (b) $(2, 3, -1)$.

(c) $(0, 1, 2)$.   (d) $(0, 0, -2)$.

26. Write each of the following vectors in R^3 as a 3×1 matrix.

(a) $2\mathbf{i} + 3\mathbf{j} - 4\mathbf{k}$.   (b) $\mathbf{i} + 2\mathbf{j}$.

(c) $-3\mathbf{i}$.   (d) $3\mathbf{i} - 2\mathbf{k}$.

27. Verify that the triangle with vertices $P_1(2, 3, -4)$, $P_2(3, 1, 2)$, and $P_3(-3, 0, 4)$ is isosceles.

28. Verify that the triangle with vertices $P_1(2, 3, -4)$, $P_2(3, 1, 2)$, and $P_3(7, 0, 1)$ is a right triangle.

29. A large steel manufacturer, who has 2000 employees, lists each employee's salary as a component of a vector $\mathbf{S}$ in R^{2000}. If an 8 percent across-the-board salary increase has been approved, find an expression involving $\mathbf{S}$ giving all the new salaries.

30. The vector $\mathbf{A} = (20, 30, 80, 10)$ gives the number of receivers, changers, speakers, and cassette recorders that are on hand in a stereo shop. The vector $\mathbf{B} = (200, 120, 80, 70)$ gives the price (in dollars) of each receiver, changer, speaker, and cassette recorder, respectively. What does the dot product $\mathbf{A} \cdot \mathbf{B}$ tell the shop owner?

31. A brokerage firm records the high and low values of the price of Bethlehem Steel stock each day. The information for a given week is presented in two vectors. $\mathbf{H}$ and $\mathbf{L}$, in R^5, giving the high and low values, respectively. What expression gives the average daily values of the price of Bethlehem Steel stock for the entire 5-day week?

THEORETICAL EXERCISES

T.1. Prove the rest of Theorem 3.2.

T.2. Show that $-\mathbf{X} = (-1)\mathbf{X}$.

T.3. Establish Equations (1) and (2) in R^3, for the length of a vector and the distance between two points, by using the Pythagorean Theorem.

T.4. Prove Theorem 3.3.

T.5. Suppose that $\mathbf{X}$ is orthogonal to both $\mathbf{Y}$ and $\mathbf{Z}$. Show that $\mathbf{X}$ is orthogonal to any vector of the form $r\mathbf{Y} + s\mathbf{Z}$, where r and s are scalars.

T.6. Show that if $\mathbf{X} \cdot \mathbf{Y} = 0$ for all vectors $\mathbf{Y}$, then $\mathbf{X} = \mathbf{0}$.

T.7. Show that $\mathbf{X} \cdot (\mathbf{Y} + \mathbf{Z}) = \mathbf{X} \cdot \mathbf{Y} + \mathbf{X} \cdot \mathbf{Z}$.

T.8. Show that if $\mathbf{X} \cdot \mathbf{Y} = \mathbf{X} \cdot \mathbf{Z}$ for all $\mathbf{X}$, then $\mathbf{Y} = \mathbf{Z}$.

T.9. Show that if c is a scalar, then $\|c\mathbf{X}\| = |c| \, \|\mathbf{X}\|$, where $|c|$ is the absolute value of c.

T.10. (**Pythagorean Theorem in R^n**). Show that $\|\mathbf{X} + \mathbf{Y}\|^2 = \|\mathbf{X}\|^2 + \|\mathbf{Y}\|^2$ if and only if $\mathbf{X} \cdot \mathbf{Y} = 0$.

T.11. If $\mathbf{X}$ and $\mathbf{Y}$ are viewed as $n \times 1$ matrices, show that $\mathbf{X} \cdot \mathbf{Y} = \mathbf{X}^T\mathbf{Y}$.

T.12. Define the **distance** between two vectors $\mathbf{X}$ and $\mathbf{Y}$ in R^n as $d(\mathbf{X}, \mathbf{Y}) = \|\mathbf{X} - \mathbf{Y}\|$. Show that

(a) $d(\mathbf{X}, \mathbf{Y}) \geq 0$.

(b) $d(\mathbf{X}, \mathbf{Y}) = 0$ if and only if $\mathbf{X} = \mathbf{Y}$.

(c) $d(\mathbf{X}, \mathbf{Y}) = d(\mathbf{Y}, \mathbf{X})$.

(d) $d(\mathbf{X}, \mathbf{Z}) \leq d(\mathbf{X}, \mathbf{Y}) + d(\mathbf{Y}, \mathbf{Z})$.

T.13. Prove the **parallelogram law:**

$$\|\mathbf{X} + \mathbf{Y}\|^2 + \|\mathbf{X} - \mathbf{Y}\|^2 = 2\|\mathbf{X}\|^2 + 2\|\mathbf{Y}\|^2.$$

T.14. If $\mathbf{X}$ is a nonzero vector, show that $\mathbf{U} = [1/\|\mathbf{X}\|]\mathbf{X}$ is a unit vector in the direction of $\mathbf{X}$.

3.3

CROSS PRODUCT IN R^3

NOTE: This section is optional but will be needed in Section 7.1.

In this section we discuss an operation that is only meaningful in R^3. Despite this limitation, it has many applications in a number of different situations. We shall consider several of these applications in this section; others will occur in Section 7.1.

Definition If

$$\mathbf{X} = x_1\mathbf{i} + x_2\mathbf{j} + x_3\mathbf{k} \qquad \text{and} \qquad \mathbf{Y} = y_1\mathbf{i} + y_2\mathbf{j} + y_3\mathbf{k}$$

are two vectors in R^3, then their **cross product** is the vector $\mathbf{X} \times \mathbf{Y}$ defined by

$$\mathbf{X} \times \mathbf{Y} = (x_2y_3 - x_3y_2)\mathbf{i} + (x_3y_1 - x_1y_3)\mathbf{j} + (x_1y_2 - x_2y_1)\mathbf{k}. \tag{1}$$

The cross product $\mathbf{X} \times \mathbf{Y}$ can also be written as a "determinant,"

$$\mathbf{X} \times \mathbf{Y} = \begin{vmatrix} \mathbf{i} & \mathbf{j} & \mathbf{k} \\ x_1 & x_2 & x_3 \\ y_1 & y_2 & y_3 \end{vmatrix}. \tag{2}$$

The right side of (2) is not really a determinant, but it is convenient to think of the computation in this manner. If we expand (2) about the first row, we obtain

$$\mathbf{X} \times \mathbf{Y} = \begin{vmatrix} x_2 & x_3 \\ y_2 & y_3 \end{vmatrix} \mathbf{i} - \begin{vmatrix} x_1 & x_3 \\ y_1 & y_3 \end{vmatrix} \mathbf{j} + \begin{vmatrix} x_1 & x_2 \\ y_1 & y_2 \end{vmatrix} \mathbf{k},$$

which is the right side of (1). Observe that the cross product $\mathbf{X} \times \mathbf{Y}$ is a vector while the dot product $\mathbf{X} \cdot \mathbf{Y}$ is a number.

Example 1 Let $\mathbf{X} = 2\mathbf{i} + \mathbf{j} + 2\mathbf{k}$ and $\mathbf{Y} = 3\mathbf{i} - \mathbf{j} - 3\mathbf{k}$. Then

$$\mathbf{X} \times \mathbf{Y} = \begin{vmatrix} \mathbf{i} & \mathbf{j} & \mathbf{k} \\ 2 & 1 & 2 \\ 3 & -1 & -3 \end{vmatrix} = -\mathbf{i} + 12\mathbf{j} - 5\mathbf{k}.$$

Some of the algebraic properties of cross product are described in the following theorem. The proof, which follows easily from the properties of determinants, is left to the reader (Exercise T.1).

Theorem 3.6 *If* $\mathbf{X}$, $\mathbf{Y}$, *and* $\mathbf{Z}$ *are vectors and c is a scalar, then:*
(Properties   (a) $\mathbf{X} \times \mathbf{Y} = -(\mathbf{Y} \times \mathbf{X})$.
of Cross   (b) $\mathbf{X} \times (\mathbf{Y} + \mathbf{Z}) = \mathbf{X} \times \mathbf{Y} + \mathbf{X} \times \mathbf{Z}$.
Product)   (c) $(\mathbf{X} + \mathbf{Y}) \times \mathbf{Z} = \mathbf{X} \times \mathbf{Z} + \mathbf{Y} \times \mathbf{Z}$.
       (d) $c(\mathbf{X} \times \mathbf{Y}) = (c\mathbf{X}) \times \mathbf{Y} = \mathbf{X} \times (c\mathbf{Y})$.
       (e) $\mathbf{X} \times \mathbf{X} = \mathbf{0}$.
       (f) $\mathbf{0} \times \mathbf{X} = \mathbf{X} \times \mathbf{0} = \mathbf{0}$.
       (g) $(\mathbf{X} \times \mathbf{Y}) \times \mathbf{Z} = (\mathbf{Z} \cdot \mathbf{X})\mathbf{Y} - (\mathbf{Z} \cdot \mathbf{Y})\mathbf{X}$.

Example 2 It follows from (1) or (2) that

$$\mathbf{i} \times \mathbf{i} = \mathbf{j} \times \mathbf{j} = \mathbf{k} \times \mathbf{k} = \mathbf{0}; \qquad \mathbf{i} \times \mathbf{j} = \mathbf{k}, \quad \mathbf{j} \times \mathbf{k} = \mathbf{i}, \quad \mathbf{k} \times \mathbf{i} = \mathbf{j}.$$

Also,

$$\mathbf{j} \times \mathbf{i} = -\mathbf{k}, \qquad \mathbf{k} \times \mathbf{j} = -\mathbf{i}, \qquad \mathbf{i} \times \mathbf{k} = -\mathbf{j}.$$

These rules can be remembered by the method indicated in Figure 3.34. Moving around the circle in a clockwise direction, we see that the cross product of two vectors taken in

Figure 3.34

the indicated order is the third vector; moving in a counterclockwise direction, we see that the cross product taken in the indicated order is the negative of the third vector. The cross product of a vector with itself is the zero vector.

Although many of the familiar properties of the real numbers hold for the cross product, it should be noted that two important properties do not hold. The commutative law does not hold, since $\mathbf{X} \times \mathbf{Y} = -(\mathbf{Y} \times \mathbf{X})$. Also, the associative law does not hold, since $\mathbf{i} \times (\mathbf{i} \times \mathbf{j}) = \mathbf{i} \times \mathbf{k} = -\mathbf{j}$, while $(\mathbf{i} \times \mathbf{i}) \times \mathbf{j} = \mathbf{0} \times \mathbf{j} = \mathbf{0}$.

We shall now take a closer look at the geometric properties of the cross product. First, we observe the following two additional properties of the cross product, whose proofs we leave to the reader:

$$(\mathbf{X} \times \mathbf{Y}) \cdot \mathbf{Z} = \mathbf{X} \cdot (\mathbf{Y} \times \mathbf{Z}) \qquad \text{(Exercise T.2)}; \qquad (3)$$
$$\mathbf{X} \times (\mathbf{Y} \times \mathbf{Z}) = (\mathbf{X} \cdot \mathbf{Z})\mathbf{Y} - (\mathbf{X} \cdot \mathbf{Y})\mathbf{Z} \qquad \text{(Exercise T.3)}. \qquad (4)$$

It is also easy to show (Exercise T.4) that

$$(\mathbf{X} \times \mathbf{Y}) \cdot \mathbf{Z} = \begin{vmatrix} x_1 & x_2 & x_3 \\ y_1 & y_2 & y_3 \\ z_1 & z_2 & z_3 \end{vmatrix}.$$

Example 3 Let $\mathbf{X}$ and $\mathbf{Y}$ be as in Example 1, and let $\mathbf{Z} = \mathbf{i} + 2\mathbf{j} + 3\mathbf{k}$. Then

$$\mathbf{X} \times \mathbf{Y} = -\mathbf{i} + 12\mathbf{j} - 5\mathbf{k} \qquad \text{and} \qquad (\mathbf{X} \times \mathbf{Y}) \cdot \mathbf{Z} = 8$$
$$\mathbf{Y} \times \mathbf{Z} = 3\mathbf{i} - 12\mathbf{j} + 7\mathbf{k} \qquad \text{and} \qquad \mathbf{X} \cdot (\mathbf{Y} \times \mathbf{Z}) = 8,$$

which illustrates Equation (3). Also,

$$\mathbf{X} \times (\mathbf{Y} \times \mathbf{Z}) = 31\mathbf{i} - 8\mathbf{j} - 27\mathbf{k},$$
$$\mathbf{X} \cdot \mathbf{Z} = 10, \qquad \mathbf{X} \cdot \mathbf{Y} = -1,$$
$$(\mathbf{X} \cdot \mathbf{Z})\mathbf{Y} = 30\mathbf{i} - 10\mathbf{j} - 30\mathbf{k}, \qquad (\mathbf{X} \cdot \mathbf{Y})\mathbf{Z} = -\mathbf{i} - 2\mathbf{j} - 3\mathbf{k}.$$

Hence

$$(\mathbf{X} \cdot \mathbf{Z})\mathbf{Y} - (\mathbf{X} \cdot \mathbf{Y})\mathbf{Z} = 31\mathbf{i} - 8\mathbf{j} - 27\mathbf{k},$$

which illustrates Equation (4).

From Equation (3) and (a) and (e) of Theorem 3.6, we have

$$(\mathbf{X} \times \mathbf{Y}) \cdot \mathbf{Y} = \mathbf{X} \cdot (\mathbf{Y} \times \mathbf{Y}) = \mathbf{X} \cdot \mathbf{0} = 0, \qquad (5)$$
$$(\mathbf{X} \times \mathbf{Y}) \cdot \mathbf{X} = -(\mathbf{Y} \times \mathbf{X}) \cdot \mathbf{X} = -\mathbf{Y} \cdot (\mathbf{X} \times \mathbf{X}) = -\mathbf{Y} \cdot \mathbf{0} = 0. \qquad (6)$$

From (5) and (6) and the definition for orthogonality of vectors given in Section 3.2, it follows that if $\mathbf{X} \times \mathbf{Y} \neq \mathbf{0}$, then $\mathbf{X} \times \mathbf{Y}$ is orthogonal to both $\mathbf{X}$ and $\mathbf{Y}$ and to the plane determined by them. It can be shown that the direction of $\mathbf{X} \times \mathbf{Y}$ is such that $\mathbf{X}$, $\mathbf{Y}$, and $\mathbf{X} \times \mathbf{Y}$ form a right-handed coordinate system. That is, if θ is the angle between $\mathbf{X}$ and $\mathbf{Y}$, then the direction of $\mathbf{X} \times \mathbf{Y}$ is that in which a right-hand screw perpendicular to the plane of $\mathbf{X}$ and $\mathbf{Y}$ will move if we rotate it through the angle θ from $\mathbf{X}$ to $\mathbf{Y}$ (Figure 3.35).

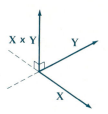

Figure 3.35

The magnitude of $\mathbf{X} \times \mathbf{Y}$ can be determined as follows. From (a) of Theorem 3.3 (Section 3.2) we have

$$\|\mathbf{X} \times \mathbf{Y}\|^2 = (\mathbf{X} \times \mathbf{Y}) \cdot (\mathbf{X} \times \mathbf{Y})$$
$$= \mathbf{X} \cdot [\mathbf{Y} \times (\mathbf{X} \times \mathbf{Y})], \qquad \text{by (3)}$$
$$= \mathbf{X} \cdot [(\mathbf{Y} \cdot \mathbf{Y})\mathbf{X} - (\mathbf{Y} \cdot \mathbf{X})\mathbf{Y}], \qquad \text{by (4)}$$
$$= (\mathbf{X} \cdot \mathbf{X})(\mathbf{Y} \cdot \mathbf{Y}) - (\mathbf{Y} \cdot \mathbf{X})(\mathbf{Y} \cdot \mathbf{X}), \qquad \text{by (d) and (b) of Theorem 3.3}$$
$$= \|\mathbf{X}\|^2 \|\mathbf{Y}\|^2 - (\mathbf{X} \cdot \mathbf{Y})^2, \qquad \text{by (a) of Theorem 3.3.}$$

From Equation (4) of Section 3.2 it follows that

$$\mathbf{X} \cdot \mathbf{Y} = \|\mathbf{X}\| \|\mathbf{Y}\| \cos \theta,$$

where θ is the angle between $\mathbf{X}$ and $\mathbf{Y}$. Hence

$$\|\mathbf{X} \times \mathbf{Y}\|^2 = \|\mathbf{X}\|^2 \|\mathbf{Y}\|^2 - \|\mathbf{X}\|^2 \|\mathbf{Y}\|^2 \cos^2 \theta$$
$$= \|\mathbf{X}\|^2 \|\mathbf{Y}\|^2 (1 - \cos^2 \theta)$$
$$= \|\mathbf{X}\|^2 \|\mathbf{Y}\|^2 \sin^2 \theta.$$

Taking square roots, we obtain

$$\|\mathbf{X} \times \mathbf{Y}\| = \|\mathbf{X}\| \|\mathbf{Y}\| \sin \theta. \tag{7}$$

Note that we do not have $|\sin \theta|$, since $\sin \theta$ is nonnegative for $0 \le \theta \le \pi$. It follows that the vectors $\mathbf{X}$ and $\mathbf{Y}$ are parallel if and only if $\mathbf{X} \times \mathbf{Y} = \mathbf{0}$ (Exercise T.5).

We now consider several applications of cross product.

Area of a Triangle

Consider the triangle with vertices $\mathbf{X}_1$, $\mathbf{X}_2$, and $\mathbf{X}_3$ (Figure 3.36). The area of this triangle is $\frac{1}{2}bh$, where b is the base and h is the height. If we take the segment between $\mathbf{X}_1$ and $\mathbf{X}_2$ to be the base, then

$$b = \|\mathbf{X}_2 - \mathbf{X}_1\|.$$

The height h is given by

$$h = \|\mathbf{X}_3 - \mathbf{X}_1\| \sin \theta.$$

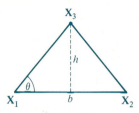

Figure 3.36

Hence, by (7), the area A_T of the triangle is

$$A_T = \tfrac{1}{2}\|\mathbf{X}_2 - \mathbf{X}_1\| \, \|\mathbf{X}_3 - \mathbf{X}_1\| \sin\theta$$
$$= \tfrac{1}{2}\|(\mathbf{X}_2 - \mathbf{X}_1) \times (\mathbf{X}_3 - \mathbf{X}_1)\|.$$

Example 4 To find the area of the triangle with vertices $\mathbf{X}_1 = (2, 2, 4)$, $\mathbf{X}_2 = (-1, 0, 5)$, and $\mathbf{X}_3 = (3, 4, 3)$, we write

$$\mathbf{X}_2 - \mathbf{X}_1 = -3\mathbf{i} - 2\mathbf{j} + \mathbf{k},$$
$$\mathbf{X}_3 - \mathbf{X}_1 = \quad \mathbf{i} + 2\mathbf{j} - \mathbf{k}.$$

Then

$$A_T = \tfrac{1}{2}\|(-3\mathbf{i} - 2\mathbf{j} + \mathbf{k}) \times (\mathbf{i} + 2\mathbf{j} - \mathbf{k})\| = \tfrac{1}{2}\|-2\mathbf{j} - 4\mathbf{k}\|$$
$$= \|-\mathbf{j} - 2\mathbf{k}\| = \sqrt{5}.$$

Area of a Parallelogram

The area A_P of the parallelogram with adjacent sides $\mathbf{X}_2 - \mathbf{X}_1$ and $\mathbf{X}_3 - \mathbf{X}_1$ (Figure 3.37) is $2A_T$, so

$$A_P = \|(\mathbf{X}_2 - \mathbf{X}_1) \times (\mathbf{X}_3 - \mathbf{X}_1)\|.$$

Example 5 If $\mathbf{X}_1$, $\mathbf{X}_2$, and $\mathbf{X}_3$ are as in Example 4, then the area of the parallelogram with adjacent sides $\mathbf{X}_2 - \mathbf{X}_1$ and $\mathbf{X}_3 - \mathbf{X}_1$ is $2\sqrt{5}$.

Figure 3.37

Volume of a Parallelepiped

Consider the parallelepiped with a vertex at the origin and edges $\mathbf{X}$, $\mathbf{Y}$, and $\mathbf{Z}$ (Figure 3.38). The volume v of the parallelepiped is the product of the area of the face containing $\mathbf{Y}$ and $\mathbf{Z}$ and the distance h from this face to the face parallel to it. Now

$$h = \|\mathbf{X}\|\, |\cos\theta|,$$

where θ is the angle between $\mathbf{X}$ and $\mathbf{Y} \times \mathbf{Z}$, and the area of the face determined by $\mathbf{Y}$ and $\mathbf{Z}$ is $\|\mathbf{Y} \times \mathbf{Z}\|$. Hence

$$v = \|\mathbf{Y} \times \mathbf{Z}\|\,\|\mathbf{X}\|\,|\cos\theta| = |\mathbf{X} \cdot (\mathbf{Y} \times \mathbf{Z})|.$$

Example 6 Consider the parallelepiped with a vertex at the origin and edges $\mathbf{X} = \mathbf{i} - 2\mathbf{j} + 3\mathbf{k}$, $\mathbf{Y} = \mathbf{i} + 3\mathbf{j} + \mathbf{k}$, and $\mathbf{Z} = 2\mathbf{i} + \mathbf{j} + 2\mathbf{k}$. Then

$$\mathbf{Y} \times \mathbf{Z} = 5\mathbf{i} - 5\mathbf{k}.$$

Hence $\mathbf{X} \cdot (\mathbf{Y} \times \mathbf{Z}) = -10$. Thus

$$v = |\mathbf{X} \cdot (\mathbf{Y} \times \mathbf{Z})| = |-10| = 10.$$

Figure 3.38

3.3 EXERCISES

In Exercises 1 and 2, compute $\mathbf{X} \times \mathbf{Y}$.

1. (a) $\mathbf{X} = 2\mathbf{i} + 3\mathbf{j} + 4\mathbf{k}$, $\mathbf{Y} = -\mathbf{i} + 3\mathbf{j} - \mathbf{k}$.
    (b) $\mathbf{X} = (1, 0, 1)$, $\mathbf{Y} = (2, 3, -1)$.
    (c) $\mathbf{X} = \mathbf{i} - \mathbf{j} + 2\mathbf{k}$, $\mathbf{Y} = 3\mathbf{i} - 4\mathbf{j} + \mathbf{k}$.
    (d) $\mathbf{X} = (2, -1, 1)$, $\mathbf{Y} = -2\mathbf{X}$.

2. (a) $\mathbf{X} = (1, -1, 2)$, $\mathbf{Y} = (3, 1, 2)$.
    (b) $\mathbf{X} = 2\mathbf{i} + \mathbf{j} - 2\mathbf{k}$, $\mathbf{Y} = \mathbf{i} + 3\mathbf{k}$.
    (c) $\mathbf{X} = 2\mathbf{j} + \mathbf{k}$, $\mathbf{Y} = 3\mathbf{X}$.
    (d) $\mathbf{X} = (4, 0, -2)$, $\mathbf{Y} = (0, 2, -1)$.

3. Let $\mathbf{X} = \mathbf{i} + 2\mathbf{j} - 3\mathbf{k}$, $\mathbf{Y} = 2\mathbf{i} + 3\mathbf{j} + \mathbf{k}$,

$\mathbf{Z} = 2\mathbf{i} - \mathbf{j} + 2\mathbf{k}$, and $c = -3$. Verify properties (a) through (d) of Theorem 3.6.

4. Let $\mathbf{X} = 2\mathbf{i} - \mathbf{j} + 3\mathbf{k}$, $\mathbf{Y} = 3\mathbf{i} + \mathbf{j} - \mathbf{k}$, and $\mathbf{Z} = 3\mathbf{i} + \mathbf{j} + 2\mathbf{k}$.
    (a) Verify Equation (3).
    (b) Verify Equation (4).

5. Let $\mathbf{X} = \mathbf{i} - \mathbf{j} + 2\mathbf{k}$, $\mathbf{Y} = 2\mathbf{i} + 2\mathbf{j} - \mathbf{k}$, and $\mathbf{Z} = \mathbf{i} + \mathbf{j} - \mathbf{k}$.
    (a) Verify Equation (3).
    (b) Verify Equation (4).

6. Verify that each of the cross products $\mathbf{X} \times \mathbf{Y}$ in Exercise 1 is orthogonal to both $\mathbf{X}$ and $\mathbf{Y}$.

7. Verify that each of the cross products $\mathbf{X} \times \mathbf{Y}$ in Exercise 2 is orthogonal to both $\mathbf{X}$ and $\mathbf{Y}$.

8. Verify Equation (7) for the pairs of vectors in Exercise 1.

9. Find the area of the triangle with vertices $\mathbf{X}_1 = (1, -2, 3)$, $\mathbf{X}_2 = (-3, 1, 4)$, and $\mathbf{X}_3 = (0, 4, 3)$.

10. Find the area of the triangle with vertices $\mathbf{X}_1$, $\mathbf{X}_2$,

and $\mathbf{X}_3$, where $\mathbf{X}_2 - \mathbf{X}_1 = 2\mathbf{i} + 3\mathbf{j} - \mathbf{k}$ and $\mathbf{X}_3 - \mathbf{X}_1 = \mathbf{i} + 2\mathbf{j} + 2\mathbf{k}$.

11. Find the area of the parallelogram with adjacent sides $\mathbf{X}_2 - \mathbf{X}_1 = \mathbf{i} + 3\mathbf{j} - 2\mathbf{k}$ and $\mathbf{X}_3 - \mathbf{X}_1 = 3\mathbf{i} - \mathbf{j} - \mathbf{k}$.

12. Find the volume of the parallelepiped with a vertex at the origin and edges $\mathbf{X} = 2\mathbf{i} - \mathbf{j}$, $\mathbf{Y} = \mathbf{i} - 2\mathbf{j} - 2\mathbf{k}$, and $\mathbf{Z} = 3\mathbf{i} - \mathbf{j} + \mathbf{k}$.

13. Repeat Exercise 12 for $\mathbf{X} = \mathbf{i} - 2\mathbf{j} + 4\mathbf{k}$, $\mathbf{Y} = 3\mathbf{i} + 4\mathbf{j} + \mathbf{k}$, and $\mathbf{Z} = -\mathbf{i} + \mathbf{j} + \mathbf{k}$.

THEORETICAL EXERCISES

T.1. Prove Theorem 3.6.

T.2. Show that $(\mathbf{X} \times \mathbf{Y}) \cdot \mathbf{Z} = \mathbf{X} \cdot (\mathbf{Y} \times \mathbf{Z})$.

T.3. Show that $\mathbf{X} \times (\mathbf{Y} \times \mathbf{Z}) = (\mathbf{X} \cdot \mathbf{Z})\mathbf{Y} - (\mathbf{X} \cdot \mathbf{Y})\mathbf{Z}$.

T.4. Show that

$$(\mathbf{X} \times \mathbf{Y}) \cdot \mathbf{Z} = \begin{vmatrix} x_1 & x_2 & x_3 \\ y_1 & y_2 & y_3 \\ z_1 & z_2 & z_3 \end{vmatrix}.$$

T.5. Show that $\mathbf{X}$ and $\mathbf{Y}$ are parallel if and only if $\mathbf{X} \times \mathbf{Y} = \mathbf{0}$.

T.6. Show that $\|\mathbf{X} \times \mathbf{Y}\|^2 + (\mathbf{X} \cdot \mathbf{Y})^2 = \|\mathbf{X}\|^2 \|\mathbf{Y}\|^2$.

T.7. Prove the **Jacobi identity**: $(\mathbf{X} \times \mathbf{Y}) \times \mathbf{Z} + (\mathbf{Y} \times \mathbf{Z}) \times \mathbf{X} + (\mathbf{Z} \times \mathbf{X}) \times \mathbf{Y} = \mathbf{0}$.

3.4

VECTOR SPACES AND SUBSPACES

We have already defined R^n and examined some of its basic properties. We must now study the fundamental structure of R^n. In many applications in mathematics, the sciences, and engineering, the notion of a vector space arises. This idea is merely a carefully constructed generalization of R^n. In studying the properties and structure of a vector space, we can study not only R^n, in particular, but many other important vector spaces. In this section we define the notion of a vector space and a subspace. In the following two sections we study their structure.

Definition 1 A **real vector space** is a set of elements V together with two operations $\oplus$ and $\odot$ satisfying the following properties:

(α) If $\mathbf{X}$ and $\mathbf{Y}$ are any elements of V, then $\mathbf{X} \oplus \mathbf{Y}$ is in V (that is, V is closed under the operation $\oplus$).

(a) $\mathbf{X} \oplus \mathbf{Y} = \mathbf{Y} \oplus \mathbf{X}$, for $\mathbf{X}$ and $\mathbf{Y}$ in V.

(b) $\mathbf{X} \oplus (\mathbf{Y} \oplus \mathbf{Z}) = (\mathbf{X} \oplus \mathbf{Y}) \oplus \mathbf{Z}$, for $\mathbf{X}$, $\mathbf{Y}$, and $\mathbf{Z}$ in V.

(c) There is a unique element **0** in V such that

$$\mathbf{X} \oplus \mathbf{0} = \mathbf{0} \oplus \mathbf{X} = \mathbf{X}, \qquad \text{for all } \mathbf{X} \text{ in } V.$$

(d) For each $\mathbf{X}$ in V, there is a unique element $-\mathbf{X}$ in V such that

$$\mathbf{X} \oplus -\mathbf{X} = \mathbf{0}.$$

(β) If $\mathbf{X}$ is any element of V and c is any real number, then $c \odot \mathbf{X}$ is in V.

(e) $c \odot (\mathbf{X} \oplus \mathbf{Y}) = c \odot \mathbf{X} \oplus c \odot \mathbf{Y}$, for all real numbers c and all $\mathbf{X}$ and $\mathbf{Y}$ in V.

(f) $(c + d) \odot \mathbf{X} = c \odot \mathbf{X} \oplus d \odot \mathbf{X}$, for all real numbers c and d, and all $\mathbf{X}$ in V.

(g) $c \odot (d \odot \mathbf{X}) = (cd) \odot \mathbf{X}$, for all real numbers c and d and all $\mathbf{X}$ in V.

(h) $1 \odot \mathbf{X} = \mathbf{X}$, for all $\mathbf{X}$ in V.

The elements of V are called **vectors**; the real numbers are called **scalars**. The operation $\oplus$ is called **vector addition**; the operation $\odot$ is called **scalar multiplication**. The vector **0** is called the **zero vector**. The vector $-\mathbf{X}$ is called the **negative** of $\mathbf{X}$.

Example 1   Consider the set R^n together with the operations of vector addition and scalar multiplication as defined in Section 3.2. Theorem 3.2 in Section 3.2 established the fact that R^n is a vector space under the above-mentioned operations.

Example 2   Consider the set V of all ordered triples of real numbers of the form $(x, y, 0)$ and define the operations $\oplus$ and $\odot$ by

$$(x, y, 0) \oplus (x', y', 0) = (x + x', y + y', 0)$$
$$c \odot (x, y, 0) = (cx, cy, 0).$$

It is then easy to show (Exercise 3) that V is a vector space, since it satisfies all the properties of Definition 1.

Example 3   Consider the set V of all ordered triples of real numbers (x, y, z) and define the operations $\oplus$ and $\odot$ by

$$(x, y, z) \oplus (x', y', z') = (x + x', y + y', z + z')$$
$$c \odot (x, y, z) = (cx, y, z).$$

It is then easy to verify (Exercise 4) that properties (α), (β), (a), (b), (c), (d), and (e) of Definition 1 hold. Here $\mathbf{0} = (0, 0, 0)$ and the negative of the vector (x, y, z) is the vector $(-x, -y, -z)$. For example, to verify property (e) we proceed as follows. First,

$$c \odot [(x, y, z) \oplus (x', y', z')] = c \odot (x + x', y + y', z + z')$$
$$= (c(x + x'), y + y', z + z').$$

Also,

$$c \odot (x, y, z) \oplus c \odot (x', y', z')$$
$$= (cx, y, z) \oplus (cx', y', z')$$
$$= (cx + cx', y + y', z + z') = (c(x + x'), y + y', z + z').$$

However, property (f) fails to hold. Thus

$$(c + d) \odot (x, y, z) = ((c + d)x, y, z).$$

On the other hand,

$$\begin{aligned}
c \odot (x, y, z) \oplus d \odot (x, y, z) &= (cx, y, z) \oplus (dx, y, z) \\
&= (cx + dx, y + y, z + z) \\
&= ((c + d)x, 2y, 2z).
\end{aligned}$$

Thus V is not a vector space under the prescribed operations. Incidentally, properties (g) and (h) *do* hold for this example.

Example 4 Consider the set V of all 2×3 matrices under the usual operations of matrix addition and scalar multiplication. In Section 1.3 (Theorems 1.1 and 1.3) we have established that the properties in Definition 1 hold, thereby making V into a vector space.

Example 5 Let V be the set of all real-valued functions defined on R^1. If f and g are in V, we define $f \oplus g$ by

$$(f \oplus g)(t) = f(t) + g(t).$$

If f is in V and c is a scalar, we define $c \odot f$ by

$$(c \odot f)(t) = cf(t).$$

Then V is a vector space (Exercise 5).

Example 6 Another source of examples will be sets of polynomials; therefore, we recall some well-known facts about such functions. A **polynomial** (in t) is a function that is expressible as

$$p(t) = a_0 t^n + a_1 t^{n-1} + \cdots + a_{n-1} t + a_n,$$

where $a_0, a_1, \ldots, a_n$ are real numbers. If $a_0 \neq 0$, then $p(t)$ is said to have **degree n**. Thus the degree of a polynomial is the highest power having a nonzero coefficient; $2t + 1$ has degree 1 and the constant polynomial 3 has degree 0. The **zero polynomial** is defined as

$$0t^n + 0t^{n-1} + \cdots + 0t + 0$$

Note that by definition, the zero polynomial has no degree. We now let P_n be the set of all polynomials of degree $\leq n$ together with the zero polynomial. If

$$p(t) = a_0 t^n + a_1 t^{n-1} + \cdots + a_{n-1} t + a_n$$

and

$$q(t) = b_0 t^n + b_1 t^{n-1} + \cdots + b_{n-1} t + b_n,$$

we define $p(t) \oplus q(t)$ as

$$\begin{aligned}
p(t) \oplus q(t) = (a_0 + b_0)t^n \\
+ (a_1 + b_1)t^{n-1} + \cdots + (a_{n-1} + b_{n-1})t + (a_n + b_n).
\end{aligned}$$

If c is a scalar, we also define $c \odot p(t)$ as

$$c \odot p(t) = (ca_0)t^n + (ca_1)t^{n-1} + \cdots + (ca_{n-1})t + (ca_n).$$

We now show that P_n is a vector space.

Let $p(t)$ and $q(t)$ as above be elements of P_n; that is, they are polynomials of degree $\leq n$ or the zero polynomial. Then the definitions above of the operations $\oplus$ and $\odot$ show that $p(t) \oplus q(t)$ and $c \odot p(t)$, for any scalar c, are polynomials of degree $\leq n$ or the zero polynomial. That is, $p(t) \oplus q(t)$ and $c \odot p(t)$ are in P_n so that (α) and (β) in Definition 1 hold. To verify property (a), we observe that

$$q(t) \oplus p(t) = (b_0 + a_0)t^n + (b_1 + a_1)t^{n-1} + \cdots + (b_{n-1} + a_{n-1})t + (b_n + a_n),$$

and since $a_i + b_i = b_i + a_i$ holds for the real numbers, we conclude that $p(t) \oplus q(t) = q(t) \oplus p(t)$. Similarly, we verify property (b). The zero polynomial is the element $\mathbf{0}$ needed in property (c). If $p(t)$ is as given above, then its negative, $-p(t)$, is

$$-a_0t^n - a_1t^{n-1} - \cdots - a_{n-1}t - a_n.$$

We shall now verify property (f) and will leave the verification of the remaining properties to the reader. Thus

$$
\begin{aligned}
(c + d) \odot p(t) &= (c + d)a_0t^n + (c + d)a_1t^{n-1} + \cdots + (c + d)a_{n-1}t + \\
&\quad (c + d)a_n \\
&= ca_0t^n + da_0t^n + ca_1t^{n-1} + da_1t^{n-1} + \cdots + ca_{n-1}t + \\
&\quad da_{n-1}t + ca_n + da_n \\
&= ca_0t^n + ca_1t^{n-1} + \cdots + ca_{n-1}t + ca_n + da_0t^n + \\
&\quad da_1t^{n-1} + \cdots + da_{n-1}t + da_n \\
&= c(a_0t^n + a_1t^{n-1} + \cdots + a_{n-1}t + a_n) + \\
&\quad d(a_0t^n + a_1t^{n-1} + \cdots + a_{n-1}t + a_n) \\
&= c \odot p(t) \oplus d \odot p(t).
\end{aligned}
$$

Example 7 Let V be the set of all real numbers with the operations $\mathbf{X} \oplus \mathbf{Y} = \mathbf{X} - \mathbf{Y}$ ($\oplus$ is ordinary subtraction) and $c \odot \mathbf{X} = c\,\mathbf{X}$ ($\odot$ is ordinary multiplication). Is V a vector space? If it is not, which properties in Definition 1 fail to hold?

Solution
If $\mathbf{X}$ and $\mathbf{Y}$ are in V, and c is a scalar, then $\mathbf{X} \oplus \mathbf{Y}$ and $c \odot \mathbf{X}$ are in V so that (α) and (β) in Definition 1 hold. However, property (a) fails to hold, since

$$\mathbf{X} \oplus \mathbf{Y} = \mathbf{X} - \mathbf{Y}$$

and

$$\mathbf{Y} \oplus \mathbf{X} = \mathbf{Y} - \mathbf{X}$$

Also, property (b) fails to hold (verify). Moreover, property (c) does not hold. Property

Property (d) also does not hold. Properties (e), (g), and (h) hold, but property (f) does not hold:

$$(c + d) \odot \mathbf{X} = (c + d)\mathbf{X} = c\mathbf{X} + d\mathbf{X}$$

while

$$c \odot \mathbf{X} \oplus d \odot \mathbf{X} = c\mathbf{X} \oplus d\mathbf{X} = c\mathbf{X} - d\mathbf{X}.$$

Thus V is not a vector space.

For each natural number n, we have just defined the vector space P_n of all polynomials of degree $\leq n$ together with the zero polynomial. We could also consider the space P of *all* polynomials (of any degree), together with the zero polynomial. Here P is the mathematical union of all the vector spaces P_n. Two polynomials $p(t)$ of degree n and $g(t)$ of degree m are added in P in the same way as they would be added in P_r, where r is the maximum of the two numbers m and n. Then P is a vector space (Exercise 6).

To specify a vector space, we must be given a set V and two operations $\oplus$ and $\odot$ satisfying the properties of Definition 1. The first thing to check is whether (α) and (β) hold, for if either of these fails, we do not have a vector space. We frequently refer to a real vector space simply as a **vector space**. We also write $\mathbf{X} \oplus \mathbf{Y}$ simply as $\mathbf{X} + \mathbf{Y}$ and $c \odot \mathbf{X}$ simply as $c\mathbf{X}$, being careful to keep the particular operation in mind.

Many other important examples of vector spaces occur in numerous areas of mathematics.

The advantage of Definition 1 is that it is not concerned with the question of what a vector is. For example, is a vector in R^3 a point, a directed line segment, or a 3×1 matrix? Definition 1 only deals with the algebraic behavior of the elements in a vector space. In the case of R^3, whichever point of view we take, the algebraic behavior is the same. The mathematician abstracts those features that all such objects have in common (that is, those properties that make them all behave alike) and defines a new structure, called a real vector space. We can now talk about properties of all vector spaces without having to refer to any one vector space in particular. Thus a "vector" is now merely an element of a vector space, and it no longer needs to be associated with a directed line segment. The following theorem presents several useful properties common to all vector spaces.

Theorem 3.7   *If V is a vector space, then:*

(a) $0\mathbf{X} = \mathbf{0}$, *for every* $\mathbf{X}$ *in V.*
(b) $c\mathbf{0} = \mathbf{0}$, *for every scalar c.*
(c) *If* $c\mathbf{X} = \mathbf{0}$, *then* $c = 0$ *or* $\mathbf{X} = \mathbf{0}$.
(d) $(-1)\mathbf{X} = -\mathbf{X}$, *for every* $\mathbf{X}$ *in V.*

Proof

(a) We have

$$0\mathbf{X} = (0 + 0)\mathbf{X} = 0\mathbf{X} + 0\mathbf{X}, \tag{1}$$

by (f) of Definition 1. Adding $-0\mathbf{X}$ to both sides of (1), we obtain by (b), (c), and (d) of Definition 1,

$$\mathbf{0} = 0\mathbf{X} + (-0\mathbf{X}) = (0\mathbf{X} + 0\mathbf{X}) + (-0\mathbf{X})$$
$$= 0\mathbf{X} + [0\mathbf{X} + (-0\mathbf{X})]$$
$$= 0\mathbf{X} + \mathbf{0} = 0\mathbf{X}.$$

(b) Exercise T.1.

(c) Suppose that $c\mathbf{X} = \mathbf{0}$ and $c \neq 0$. Then multiplying both sides by $1/c$, we have $\mathbf{0} = (1/c)\mathbf{0} = (1/c)(c\mathbf{X}) = [(1/c)c]\mathbf{X} = 1\mathbf{X} = \mathbf{X}$, by (b) of this theorem and (g) and (h) of Definition 1.

(d) $(-1)\mathbf{X} + \mathbf{X} = (-1)\mathbf{X} + (1)\mathbf{X} = (-1 + 1)\mathbf{X} = 0\mathbf{X} = \mathbf{0}$ so that

$$(-1)\mathbf{X} = -\mathbf{X}.$$

Subspaces

We now begin to analyze the structure of a vector space. First, it is convenient to have a name for a subset of a given vector space V which is itself a vector space with respect to the same operations as those in V. Thus we have a definition.

Definition Let V be a vector space and W a nonempty subset of V. If W is a vector space with respect to the operations in V, then W is called a **subspace** of V.

Example 8 If V is a vector space, then V is a subspace of itself. Also, $\{\mathbf{0}\}$ is a subspace (Exercise 14) of V. Thus every nonzero vector space has at least two subspaces.

Example 9 Let W be the subset of R^3 consisting of all vectors of the form $(a, b, 0)$, where a and b are any real numbers. To check if W is a subspace of R^3, we first see whether properties (α) and (β) of Definition 1 hold. Thus let $\mathbf{X} = (a_1, b_1, 0)$ and $\mathbf{Y} = (a_2, b_2, 0)$ be vectors in W. Then $\mathbf{X} + \mathbf{Y} = (a_1, b_1, 0) + (a_2, b_2, 0) = (a_1 + a_2, b_1 + b_2, 0)$ is in W, since the third component is zero. Also, if c is a scalar, then $c\mathbf{X} = c(a_1, b_1, 0) = (ca_1, cb_1, 0)$ is in W. Thus properties (α) and (β) of Definition 1 hold. We can also easily verify that properties (a)–(h) hold. Hence W is a subspace of R^3.

We now pause in our listing of subspaces to develop a labor-saving result. We just noted that to verify that a nonempty subset W of a vector space V is a subspace, we must check that (α), (β), and (a)–(h) of Definition 1 hold. However, the following theorem says that it is enough to merely check that (α) and (β) hold.

Theorem 3.8 Let V be a vector space with operations $\oplus$ and $\odot$ and let W be a nonempty subset of V. Then W is a subspace of V if and only if the following conditions hold:
  (α) If $\mathbf{X}$ and $\mathbf{Y}$ are any vectors in W, then $\mathbf{X} \oplus \mathbf{Y}$ is in W.
  (β) If c is any real number and $\mathbf{X}$ is any vector in W, then $c \odot \mathbf{X}$ is in W.

Proof
Exercise T.7.

Example 10 Consider the set W consisting of all 2×3 matrices of the form

$$\begin{bmatrix} a & b & 0 \\ 0 & c & d \end{bmatrix},$$

where a, b, c, and d are arbitrary real numbers. Then W is a subset of the vector space V of Example 4. Show that W is a subspace of V.

Solution
Consider

$$\mathbf{X} = \begin{bmatrix} a_1 & b_1 & 0 \\ 0 & c_1 & d_1 \end{bmatrix} \quad \text{and} \quad \mathbf{Y} = \begin{bmatrix} a_2 & b_2 & 0 \\ 0 & c_2 & d_2 \end{bmatrix} \quad \text{in } W.$$

Then

$$\mathbf{X} + \mathbf{Y} = \begin{bmatrix} a_1 + a_2 & b_1 + b_2 & 0 \\ 0 & c_1 + c_2 & d_1 + d_2 \end{bmatrix} \quad \text{is in } W.$$

so that (α) of Theorem 3.8 is satisfied. Also, if k is a scalar, then

$$k\mathbf{X} = \begin{bmatrix} ka_1 & kb_1 & 0 \\ 0 & kc_1 & kd_1 \end{bmatrix} \quad \text{is in } W$$

so that (β) of Theorem 3.8 is satisfied. Hence W is a subspace of V.

Example 11 Let W be the subset of R^3 consisting of all vectors of the form $(a, b, 1)$, where a and b are any real numbers. To check whether properties (α) and (β) of Theorem 3.8 hold, we let $\mathbf{X} = (a_1, b_1, 1)$ and $\mathbf{Y} = (a_2, b_2, 1)$ be vectors in W. Then $\mathbf{X} + \mathbf{Y} = (a_1, b_1, 1) + (a_2, b_2, 1) = (a_1 + a_2, b_1 + b_2, 2)$, which is not in W, since the third component is 2 and not 1. Since (α) of Theorem 3.8 does not hold, W is not a subspace of R^3.

Example 12 The vector space P_2 is a subset of the vector space P_3. (See Example 6.) It is easy to verify (Exercise 19) that P_2 is a *subspace* of P. In general, the set P_n of all polynomials of degree $\leq n$ and the zero polynomial is a subspace of P. (See Example 6.)

Example 13 Let V be the set of all polynomials of degree exactly $= 2$; V is a *subset* of P_2, but it is not a *subspace* of P_2, since the sum of the polynomials $2t^2 + 3t + 1$ and $-2t^2 + t + 2$, a polynomial of degree 1, is not in V.

Example 14
(Calculus
Required) The sum of two continuous functions is a continuous function; the scalar multiple of a continuous function is a continuous function. Hence the set of all real-valued continuous functions is a subspace of the vector space of all real-valued functions defined in Example 5. This subspace is denoted by $C(-\infty, \infty)$. Similarly, the set of all real-valued functions that are continuous on the interval $[a, b]$ is a subspace, denoted by $C[a, b]$, of the vector space defined in Example 5.

We now come to a very important example of a subspace.

Example 15   Consider the homogeneous system $\mathbf{AX} = \mathbf{0}$, where $\mathbf{A}$ is an $m \times n$ matrix. A solution consists of a vector $\mathbf{X}$ in R^n. Let W be the subset of R^n consisting of all solutions to the homogeneous system. Since $\mathbf{A0} = \mathbf{0}$, we conclude that W is not empty. To check that W is a subspace of R^n, we verify properties (α) and (β) of Theorem 3.8. Thus let $\mathbf{X}$ and $\mathbf{Y}$ be solutions. Then

$$\mathbf{AX} = \mathbf{0} \qquad \text{and} \qquad \mathbf{AY} = \mathbf{0}.$$

Now

$$\mathbf{A}(\mathbf{X} + \mathbf{Y}) = \mathbf{AX} + \mathbf{AY} = \mathbf{0} + \mathbf{0} = \mathbf{0},$$

so $\mathbf{X} + \mathbf{Y}$ is a solution. Also, if c is a scalar, then

$$\mathbf{A}(c\mathbf{X}) = c(\mathbf{AX}) = c\mathbf{0} = \mathbf{0},$$

so $c\mathbf{X}$ is also a solution. Hence W is a subspace of R^n, called the **solution space** of the homogeneous system.

It should be noted that the set of all solutions to the linear system $\mathbf{AX} = \mathbf{B}$, where $\mathbf{A}$ is $m \times n$, is not a subspace of R^n if $\mathbf{B} \neq \mathbf{0}$ (Exercise T.2).

Example 16   A simple way of constructing subspaces in a vector space is as follows. Let $\mathbf{U}_1$ and $\mathbf{U}_2$ be fixed vectors in a vector space V and let W be the set of all vectors in V of the form

$$a_1\mathbf{U}_1 + a_2\mathbf{U}_2,$$

where a_1 and a_2 are any real numbers. To show that W is a subspace of V, we verify properties (α) and (β) of Theorem 3.8. Thus let

$$\mathbf{X} = a_1\mathbf{U}_1 + a_2\mathbf{U}_2 \qquad \text{and} \qquad \mathbf{Y} = b_1\mathbf{U}_1 + b_2\mathbf{U}_2$$

be vectors in W. Then

$$\mathbf{X} + \mathbf{Y} = (a_1\mathbf{U}_1 + a_2\mathbf{U}_2) + (b_1\mathbf{U}_1 + b_2\mathbf{U}_2) = (a_1 + b_1)\mathbf{U}_1 + (a_2 + b_2)\mathbf{U}_2,$$

which is in W. Also, if c is a scalar, then

$$c\mathbf{X} = (ca_1)\mathbf{U}_1 + (ca_2)\mathbf{U}_2$$

is in W. Hence W is a subspace of V.

The construction carried out in Example 16 for two vectors can be easily performed for more than two vectors. Thus if $S = \{\mathbf{X}_1, \mathbf{X}_2, \ldots, \mathbf{X}_k\}$ is a set of vectors in a vector space V, then the set of all vectors in V of the form

$$a_1\mathbf{X}_1 + a_2\mathbf{X}_2 + \cdots + a_k\mathbf{X}_k$$

is a subspace of V (Exercise T.8), which is denoted by

$$\text{span } S \qquad \text{or} \qquad \text{span } \{\mathbf{X}_1, \mathbf{X}_2, \ldots, \mathbf{X}_k\}.$$

3.4 EXERCISES

1. Verify in detail that R^2 is a vector space.

2. Verify in detail that R^3 is a vector space.

3. Verify that the set in Example 2 is a vector space.

4. Verify that all the properties of Definition 1, except property (f), hold for the set in Example 3.

5. Show that the set in Example 5 is a vector space.

6. Show that P is a vector space.

In Exercises 7 through 13, determine whether the given set together with the given operations is a vector space. If it is not a vector space, list the properties of Definition 1 that fail to hold.

7. The set of all ordered triples of real numbers (x, y, z) with the operations

$$(x, y, z) \oplus (x', y', z') = (x', y + y', z')$$

and

$$c \odot (x, y, z) = (cx, cy, cz).$$

8. The set of all ordered triples of real numbers (x, y, z) with the operations

$$(x, y, z) \oplus (x'\ y', z') = (x + x', y + y', z + z')$$

and

$$c \odot (x, y, z) = (x, 1, z).$$

9. The set of all ordered triples of real numbers of the form $(0, 0, z)$ with the operations $(0, 0, z) \oplus (0, 0, z') = (0, 0, z + z')$ and $c \odot (0, 0, z) = (0, 0, cz)$.

10. The set of all real numbers with the usual operations of addition and multiplication.

11. The set of all ordered pairs of real numbers (x, y), where $x \leq 0$, with the usual operations in R^2.

12. The set of all ordered pairs of real numbers (x, y) with the operations $(x, y) \oplus (x', y') = (x + x', y + y')$ and $c \odot (x, y) = (0, 0)$.

13. The set of all positive real numbers X with the operations $X \oplus Y = XY$ and $c \odot X = X^c$.

14. If V is a vector space, show that the subset $\{0\}$ is a subspace.

15. Which of the following subsets of R^3 are subspaces of R^3? The set of all vectors of the form
(a) $(a, b, 2)$.   (b) (a, b, c), where $c = a + b$.
(c) (a, b, c), where $c > 0$.

16. Which of the following subsets of R^3 are subspaces of R^3? The set of all vectors of the form
(a) (a, b, c), where $a = c = 0$.
(b) (a, b, c), where $a = -c$.
(c) (a, b, c), where $b = 2a + 1$.

17. Which of the following subsets of R^4 are subspaces of R^4? The set of all vectors of the form
(a) (a, b, c, d), where $a - b = 2$.
(b) (a, b, c, d), where $c = a + 2b$ and $d = a - 3b$.
(c) (a, b, c, d), where $a = 0$ and $b = -d$.

18. Which of the following subsets of R^4 are subspaces of R^4? The set of all vectors of the form
(a) (a, b, c, d), where $a = b = 0$.
(b) (a, b, c, d), where $a = 1$, $b = 0$, and $a + d = 1$.
(c) (a, b, c, d), where $a > 0$ and $b < 0$.

19. Show that P_2 is a subspace of P.

20. Show that P is a subspace of the vector space defined in Example 5.

21. Let $\mathbf{X} = (1, 2, -3)$ and $\mathbf{Y} = (-2, 3, 0)$ be two vectors in R^3 and let W be the subset of R^3 consisting of all vectors of the form $a\mathbf{X} + b\mathbf{Y}$, where a and b are any real numbers. Verify that W is a subspace of R^3.

22. Let $\mathbf{X} = (2, 0, 3, -4)$ and $\mathbf{Y} = (4, 2, -5, 1)$ be two vectors in R^4 and let W be the subset of R^4 consisting of all vectors of the form $a\mathbf{X} + b\mathbf{Y}$, where a and b are any real numbers. Verify that W is a subspace of R^4.

23. Which of the following subsets of the vector space of all 2×3 matrices defined in Example 4 are subspaces? The set of all matrices of the form

(a) $\begin{bmatrix} a & b & c \\ d & 0 & 0 \end{bmatrix}$, where $b = a + c$.

(b) $\begin{bmatrix} a & b & c \\ d & 0 & 0 \end{bmatrix}$, where $c > 0$.

(c) $\begin{bmatrix} a & b & c \\ d & e & f \end{bmatrix}$, where $a = -2c$ and

$$f = 2e + d.$$

24. Which of the following subsets of the vector space of all 2×3 matrices defined in Example 4 are subspaces? The set of all matrices of the form

(a) $\begin{bmatrix} a & b & c \\ d & e & f \end{bmatrix}$, where $a = 2c + 1$.

(b) $\begin{bmatrix} 0 & 1 & a \\ b & c & 0 \end{bmatrix}$.

(c) $\begin{bmatrix} a & b & c \\ d & e & f \end{bmatrix}$, where $a + c = 0$ and

$$b + d + f = 0.$$

25. (Calculus Required). Which of the following subsets are subspaces of the vector space $C(-\infty, \infty)$ defined in Example 14?
(a) All nonnegative functions.
(b) All constant functions.
(c) All functions f such that $f(0) = 0$.
(d) All functions f such that $f(0) = 5$.
(e) All differentiable functions.

26. (Calculus Required). Which of the following subsets of the vector space $C(-\infty, \infty)$ defined in Example 14 are subspaces?
(a) All integrable functions.
(b) All bounded functions.
(c) All functions that are integrable on $[a, b]$.
(d) All functions that are bounded on $[a, b]$.

THEORETICAL EXERCISES

T.1. Show that $c\mathbf{0} = \mathbf{0}$ for every scalar c.

T.2. Show that the set of all solutions to $\mathbf{AX} = \mathbf{B}$, where $\mathbf{A}$ is $m \times n$, is not a subspace of R^n if $\mathbf{B} \neq \mathbf{0}$.

T.3. Prove that $-(-\mathbf{X}) = \mathbf{X}$.

T.4. Show that a subset W of a vector space V is a subspace of V if and only if the following condition holds: If $\mathbf{X}$ and $\mathbf{Y}$ are any vectors in W and a and b are any scalars, then $a\mathbf{X} + b\mathbf{Y}$ is in W.

T.5. Prove that if $\mathbf{X} + \mathbf{Y} = \mathbf{X} + \mathbf{Z}$, then $\mathbf{Y} = \mathbf{Z}$.

T.6. Show that if $\mathbf{X} \neq \mathbf{0}$ and $a\mathbf{X} = b\mathbf{X}$, then $a = b$.

T.7. Prove Theorem 3.8.

T.8. Let $S = \{\mathbf{X}_1, \mathbf{X}_2, \ldots, \mathbf{X}_k\}$ be a set of vectors in a vector space V. Prove that the set span S consisting of all vectors in V of the form $a_1\mathbf{X}_1 + a_2\mathbf{X}_2 + \cdots + a_k\mathbf{X}_k$ is a subspace of V.

T.9. Let $S = \{\mathbf{X}_1, \mathbf{X}_2, \ldots, \mathbf{X}_k\}$ be a set of vectors in a vector space V, and let W be a subspace of V containing S. Show that W contains span S.

3.5

LINEAR INDEPENDENCE

In this section we turn to a more thorough study of the structure of a vector space and its subspaces. Of the vector spaces studied so far, the only one with a finite number of vectors is the trivial space $\{\mathbf{0}\}$. For if $\mathbf{X} \neq \mathbf{0}$ is in a vector space V, then $c\mathbf{X}$ is also in V, where c is any real number, and so V has infinitely many members. However, in this section and the following one we shall show that each vector space V studied here has a finite number of vectors that completely describe V.

Definition A vector $\mathbf{X}$ in a vector space V is defined to be a **linear combination** of the vectors $\mathbf{X}_1$, $\mathbf{X}_2, \ldots, \mathbf{X}_k$ in V if it can be written as

$$\mathbf{X} = c_1\mathbf{X}_1 + c_2\mathbf{X}_2 + \cdots + c_k\mathbf{X}_k,$$

where $c_1, c_2, \ldots, c_k$ are scalars.

Example 1 In R^4, let $\mathbf{X}_1 = (1, 2, 1, -1)$, $\mathbf{X}_2 = (1, 0, 2, -3)$, and $\mathbf{X}_3 = (1, 1, 0, -2)$. The vector

$$\mathbf{X} = (2, 1, 5, -5)$$

is a linear combination of $\mathbf{X}_1$, $\mathbf{X}_2$, and $\mathbf{X}_3$ if we can find c_1, c_2, and c_3 such that

$$c_1\mathbf{X}_1 + c_2\mathbf{X}_2 + c_3\mathbf{X}_3 = \mathbf{X}.$$

Substituting for $\mathbf{X}_1$, $\mathbf{X}_2$, $\mathbf{X}_3$, and $\mathbf{X}$, we obtain

$$c_1(1, 2, 1, -1) + c_2(1, 0, 2, -3) + c_3(1, 1, 0, -2) = (2, 1, 5, -5)$$

or

$$(c_1 + c_2 + c_3, 2c_1 + 0c_2 + c_3, c_1 + 2c_2 + 0c_3, -c_1 - 3c_2 - 2c_3)$$
$$= (2, 1, 5, -5).$$

Equating corresponding components leads to the linear system

$$
\begin{aligned}
c_1 + c_2 + c_3 &= 2 \\
2c_1 \phantom{{}+ c_2} + c_3 &= 1 \\
c_1 + 2c_2 \phantom{{}+ c_3} &= 5 \\
-c_1 - 3c_2 - 2c_3 &= -5.
\end{aligned}
$$

Solving this linear system by Gauss–Jordan reduction (the method discussed in Section 1.4), we obtain $c_1 = 1$, $c_2 = 2$, $c_3 = -1$, which means that $\mathbf{X}$ is a linear combination of $\mathbf{X}_1$, $\mathbf{X}_2$, and $\mathbf{X}_3$. Thus

$$1(1, 2, 1, -1) + 2(1, 0, 2, -3) - 1(1, 1, 0, -2) = (2, 1, 5, -5).$$

Example 2 In R^3, let $\mathbf{X}_1 = (1, 2, -1)$ and $\mathbf{X}_2 = (1, 0, -1)$. Is the vector $\mathbf{X} = (1, 0, 2)$ a linear combination of $\mathbf{X}_1$ and $\mathbf{X}_2$?

Solution
If $\mathbf{X}$ is to be a linear combination of $\mathbf{X}_1$ and $\mathbf{X}_2$, we must find scalars c_1 and c_2 such that

$$c_1\mathbf{X}_1 + c_2\mathbf{X}_2 = \mathbf{X}.$$

Substituting for $\mathbf{X}_1$, $\mathbf{X}_2$, and $\mathbf{X}$, we obtain

$$c_1(1, 2, -1) + c_2(1, 0, -1) = (1, 0, 2).$$

This leads to the linear system

$$
\begin{aligned}
c_1 + c_2 &= 1 \\
2c_1 \phantom{{}+ c_2} &= 0 \\
-c_1 - c_2 &= 2,
\end{aligned}
$$

which has no solution (verify). Hence $\mathbf{X}$ is not a linear combination of $\mathbf{X}_1$ and $\mathbf{X}_2$.

Example 3　Does $\mathbf{X} = (1, 1, 1)$ belong to span $\{\mathbf{X}_1, \mathbf{X}_2, \mathbf{X}_3\}$, where

$$\mathbf{X}_1 = (1, 0, 1), \qquad \mathbf{X}_2 = (-1, 1, 0), \qquad \text{and} \qquad \mathbf{X}_3 = (0, 0, 1)?$$

Solution

Every vector in span $\{\mathbf{X}_1, \mathbf{X}_2, \mathbf{X}_3\}$ is of the form

$$c_1\mathbf{X}_1 + c_2\mathbf{X}_2 + c_3\mathbf{X}_3.$$

Thus $\mathbf{X}$ belongs to span $\{\mathbf{X}_1, \mathbf{X}_2, \mathbf{X}_3\}$ if we can find c_1, c_2, and c_3 such that

$$c_1\mathbf{X}_1 + c_2\mathbf{X}_2 + c_3\mathbf{X}_3 = \mathbf{X}.$$

Substituting for $\mathbf{X}_1$, $\mathbf{X}_2$, and $\mathbf{X}_3$, we obtain

$$c_1(1, 0, 1) + c_2(-1, 1, 0) + c_3(0, 0, 1) = (1, 1, 1).$$

This leads to the linear system

$$
\begin{aligned}
c_1 - c_2 \quad\quad &= 1 \\
c_2 \quad\quad &= 1 \\
c_1 \quad\quad + c_3 &= 1,
\end{aligned}
$$

which has the solution (verify)

$$c_1 = 2, \qquad c_2 = 1, \qquad c_3 = -1.$$

Hence $\mathbf{X}$ belongs to span $\{\mathbf{X}_1, \mathbf{X}_2, \mathbf{X}_3\}$.

Note that every question of the form "Does $\mathbf{X}$ belong to span $\{\mathbf{X}_1, \mathbf{X}_2, \ldots, \mathbf{X}_k\}$?" results in a question about the consistency of a linear system.

Definition　Let $S = \{\mathbf{X}_1, \mathbf{X}_2, \ldots, \mathbf{X}_k\}$ be a set of vectors in a vector space V. The set S **spans** V, or V **is spanned by** S, if every vector in V is a linear combination of the vectors in S.

The procedure to check if $S = \{\mathbf{X}_1, \mathbf{X}_2, \ldots, \mathbf{X}_k\}$ spans the vector space V is as follows.

STEP 1. Choose an arbitrary vector $\mathbf{X}$ in V. This vector involves some arbitrary parameters.

STEP 2. Determine if $\mathbf{X}$ is a linear combination of the vectors in S for any choice of the parameters involved in $\mathbf{X}$. That is, does the corresponding linear system always have a solution for any choice of the parameters in $\mathbf{X}$? If the linear system has a solution for all choices of the parameters, then S spans V. If there is a solution only for certain choices of the parameters, then S does not span V.

Again we investigate the consistency of a linear system, but this time for an arbitrary right-hand side.

Example 4 Let V be the vector space R^3 and let $S = \{X_1, X_2, X_3\}$, where $X_1 = (1, 2, 1)$, $X_2 = (1, 0, 2)$, and $X_3 = (1, 1, 0)$. Does S span R^3?

Solution

STEP 1. Let $X = (a, b, c)$ be any vector in R^3, where a, b, and c are arbitrary real numbers.

STEP 2. We must find out whether there are constants c_1, c_2, and c_3 such that

$$c_1 X_1 + c_2 X_2 + c_3 X_3 = X.$$

This leads to the linear system

$$
\begin{array}{rcl}
c_1 + c_2 + c_3 &=& a \\
2c_1 \phantom{{}+ c_2} + c_3 &=& b \\
c_1 + 2c_2 \phantom{{}+ c_3} &=& c.
\end{array}
$$

A solution is (verify)

$$
c_1 = \frac{-2a + 2b + c}{3}, \qquad c_2 = \frac{a - b + c}{3}, \qquad c_3 = \frac{4a - b - 2c}{3}.
$$

Since we have obtained a solution for every choice of a, b, and c, we conclude that $S = \{X_1, X_2, X_3\}$ spans R^3. This is equivalent to saying that span $\{X_1, X_2, X_3\} = R^3$.

Example 5 Let V be P_2, the vector space consisting of all polynomials of degree ≤ 2 and the zero polynomial. Let $S = \{p_1(t), p_2(t)\}$, where $p_1(t) = t^2 + 2t + 1$ and $p_2(t) = t^2 + 2$. Does S span P_2?

Solution

STEP 1. Let $p(t) = at^2 + bt + c$ by any polynomial in P_2, where a, b, and c are any real numbers.

STEP 2. We must find out whether there are constants c_1 and c_2 such that

$$p(t) = c_1 p_1(t) + c_2 p_2(t)$$

or

$$at^2 + bt + c = c_1(t^2 + 2t + 1) + c_2(t^2 + 2).$$

Thus

$$(c_1 + c_2)t^2 + (2c_1)t + (c_1 + 2c_2) = at^2 + bt + c.$$

Now two polynomials agree for all values of t only if the coefficients of respective powers of t agree. Thus we get the linear system

$$c_1 + c_2 = a$$
$$2c_1 \quad\quad = b$$
$$c_1 + 2c_2 = c.$$

Using elementary row operations on the augmented matrix of this linear system, we obtain (verify).

$$\begin{bmatrix} 1 & 0 & \vdots & 2a - c \\ 0 & 1 & \vdots & c - a \\ 0 & 0 & \vdots & b - 4a + 2c \end{bmatrix}.$$

If $b - 4a + 2c \neq 0$, then there is no solution. Hence $S = \{p_1(t), p_2(t)\}$ does not span P_2. For example, the polynomial $3t^2 + 2t - 1$ cannot be written as a linear combination of $p_1(t)$ and $p_2(t)$.

Example 6 The vectors $\mathbf{E}_1 = \mathbf{i} = (1, 0)$ and $\mathbf{E}_2 = \mathbf{j} = (0, 1)$, span R^2, for as was observed in Section 3.1, if $\mathbf{X} = (x_1, x_2)$ is any vector in R^2, then $\mathbf{X} = x_1\mathbf{E}_1 + x_2\mathbf{E}_2$. As was noted in Section 3.2, every vector $\mathbf{X}$ in R^3 can be written as a linear combination of the vectors $\mathbf{E}_1 = \mathbf{i} = (1, 0, 0)$, $\mathbf{E}_2 = \mathbf{j} = (0, 1, 0)$, and $\mathbf{E}_3 = \mathbf{k} = (0, 0, 1)$. Thus $\mathbf{E}_1, \mathbf{E}_2$, and $\mathbf{E}_3$ span R^3. Similarly, the vectors $\mathbf{E}_1 = (1, 0, \ldots, 0)$, $\mathbf{E}_2 = (0, 1, 0, \ldots, 0), \ldots,$ $\mathbf{E}_n = (0, 0, \ldots, 0, 1)$ span R^n, since any vector $\mathbf{X} = (x_1, x_2, \ldots, x_n)$ in R^n can be written as

$$\mathbf{X} = x_1\mathbf{E}_1 + x_2\mathbf{E}_2 + \cdots + x_n\mathbf{E}_n.$$

Example 7 The set $S = \{t^n, t^{n-1}, \ldots, t, 1\}$ spans P_n, since every polynomial in P_n is of the form

$$a_0t^n + a_1t^{n-1} + \cdots + a_{n-1}t + a_n,$$

which is a linear combination of the elements in S.

Linear Independence

Definition Let $S = \{\mathbf{X}_1, \mathbf{X}_2, \ldots, \mathbf{X}_k\}$ be a set of vectors in a vector space V. Then S is said to be **linearly dependent** if there exist constants $c_1, c_2, \ldots, c_k$ not all zero such that

$$c_1\mathbf{X}_1 + c_2\mathbf{X}_2 + \cdots + c_k\mathbf{X}_k = \mathbf{0}. \tag{1}$$

Otherwise, S is called **linearly independent**. That is, S is linearly independent if (1) holds only for

$$c_1 = c_2 = \cdots = c_k = 0.$$

It should be emphasized that for any set of vectors $S = \{\mathbf{X}_1, \mathbf{X}_2, \ldots, \mathbf{X}_k\}$, Equation (1) always holds if we choose all the scalars $c_1, c_2, \ldots, c_k$ equal to zero. The important point in this definition is whether or not it is possible to satisfy (1) with at least one of the scalars not equal to zero.

The procedure to determine if $S = \{\mathbf{X}_1, \mathbf{X}_2, \ldots, \mathbf{X}_k\}$ is linearly dependent or linearly independent is as follows.

STEP 1. Form Equation (1), which leads to a homogeneous system.

STEP 2. If the homogeneous system obtained in Step 1 has only the trivial solution, then the set of vectors is linearly independent; if it has a nontrivial solution, then the set of vectors is linearly dependent.

Example 8 In R^4 consider the vectors $\mathbf{X}_1 = (1, 0, 1, 2)$, $\mathbf{X}_2 = (0, 1, 1, 2)$, and $\mathbf{X}_3 = (1, 1, 1, 3)$. To find out whether $S = \{\mathbf{X}_1, \mathbf{X}_2, \mathbf{X}_3\}$ is linearly independent or linearly dependent, we form Equation (1):

$$c_1\mathbf{X}_1 + c_2\mathbf{X}_2 + c_3\mathbf{X}_3 = \mathbf{0},$$

and solve for c_1, c_2, and c_3. The resulting linear system is

$$
\begin{aligned}
c_1 \phantom{{} + c_2} + c_3 &= 0 \\
c_2 + c_3 &= 0 \\
c_1 + c_2 + c_3 &= 0 \\
2c_1 + 2c_2 + 3c_3 &= 0,
\end{aligned}
$$

which has as its only solution $c_1 = c_2 = c_3 = 0$ (verify), showing that S is linearly independent.

Example 9 Consider the vectors

$$\mathbf{X}_1 = (1, 2, -1), \qquad \mathbf{X}_2 = (1, -2, 1), \qquad \mathbf{X}_3 = (-3, 2, -1),$$

and

$$\mathbf{X}_4 = (2, 0, 0) \quad \text{in } R^3.$$

Setting up Equation (1), we are led to the homogeneous system

$$
\begin{aligned}
c_1 + c_2 - 3c_3 + 2c_4 &= 0 \\
2c_1 - 2c_2 + 2c_3 \phantom{{} + 2c_4} &= 0 \\
-c_1 + c_2 - c_3 \phantom{{} + 2c_4} &= 0,
\end{aligned}
$$

a homogeneous system of three equations in four unknowns. By Theorem 1.7, Section 1.4, we are assured of the existence of a nontrivial solution. Hence $\mathbf{X}_1$, $\mathbf{X}_2$, $\mathbf{X}_3$, and $\mathbf{X}_4$ are linearly dependent. In fact, two of the infinitely many solutions are

$$
\begin{aligned}
c_1 = 1, \qquad c_2 = 2, \qquad c_3 = 1, \qquad c_4 = 0; \\
c_1 = 1, \qquad c_2 = 1, \qquad c_3 = 0, \qquad c_4 = -1.
\end{aligned}
$$

Example 10 The vectors $\mathbf{E}_1$ and $\mathbf{E}_2$ in R^2, defined in Example 6, are linearly independent, since

$$c_1(1, 0) + c_2(0, 1) = (0, 0)$$

can hold only if $c_1 = c_2 = 0$. Similarly, the vectors $\mathbf{E}_1$, $\mathbf{E}_2$, and $\mathbf{E}_3$ in R^3, and more generally, the vectors $\mathbf{E}_1$, $\mathbf{E}_2$, . . . , $\mathbf{E}_n$ in R^n are linearly independent (Exercise T.1).

Corollary 3.3 in Section 3.7, to follow, gives another way of testing whether n given vectors in R^n are linearly dependent or linearly independent. We form the matrix $\mathbf{A}$, whose columns are the given n vectors. Then the given vectors are linearly independent if and only if $|\mathbf{A}| \neq 0$. Thus, in Example 10,

$$\mathbf{A} = \begin{bmatrix} 1 & 0 \\ 0 & 1 \end{bmatrix}$$

and $|\mathbf{A}| = 1$ so that $\mathbf{E}_1$ and $\mathbf{E}_2$ are linearly independent.

Example 11 Consider the vectors

$$p_1(t) = t^2 + t + 2, \qquad p_2(t) = 2t^2 + t, \qquad p_3(t) = 3t^2 + 2t + 2.$$

To find out whether $S = \{p_1(t), p_2(t), p_3(t)\}$ is linearly dependent or linearly independent, we set up Equation (1) and solve for c_1, c_2, and c_3. The resulting homogeneous system is (verify)

$$\begin{aligned} c_1 + 2c_2 + 3c_3 &= 0 \\ c_1 + c_2 + 2c_3 &= 0 \\ 2c_1 + \phantom{2c_2 + {}}2c_3 &= 0, \end{aligned}$$

which has infinitely many solutions (verify). A particular solution is $c_1 = 1$, $c_2 = 1$, $c_3 = -1$, so

$$p_1(t) + p_2(t) - p_3(t) = 0.$$

Hence S is linearly dependent.

Example 12 If $S = \{\mathbf{X}_1, \mathbf{X}_2, . . . , \mathbf{X}_k\}$ are k vectors in any vector space and $\mathbf{X}_i$ is the zero vector, Equation (1) holds by letting $c_i = 1$ and $c_j = 0$ for $j \neq i$. Thus S is linearly dependent. Hence every set of vectors containing the zero vector is linearly dependent.

Let S_1 and S_2 be finite subsets of a vector space and let S_1 be a subset of S_2. Then (a) if S_1 is linearly dependent, so is S_2; and (b) if S_2 is linearly independent, so is S_1 (Exercise T.2).

We consider next the meaning of linear dependence in R^2 and R^3. Suppose that $\{\mathbf{X}_1, \mathbf{X}_2\}$ is linearly dependent in R^2. Then there exist scalars c_1 and c_2 not both zero such that $c_1\mathbf{X}_1 + c_2\mathbf{X}_2 = \mathbf{0}$. If $c_1 \neq 0$, then $\mathbf{X}_1 = -(c_2/c_1)\mathbf{X}_2$. If $c_2 \neq 0$, then $\mathbf{X}_2 = -(c_1/c_2)\mathbf{X}_1$. Thus one of the vectors is a multiple of the other. Conversely, suppose that $\mathbf{X}_1 = c\mathbf{X}_2$. Then $1\mathbf{X}_1 - c\mathbf{X}_2 = \mathbf{0}$, and since the coefficients of $\mathbf{X}_1$ and $\mathbf{X}_2$ are not both zero, it follows that $\mathbf{X}_1$ and $\mathbf{X}_2$ are linearly dependent. Thus $\{\mathbf{X}_1, \mathbf{X}_2\}$ is linearly dependent in R^2 if and only if one of the vectors is a multiple of the other [Figure 3.39(a)]. Hence two vectors in R^2 are linearly dependent if and only if they both lie on the same line passing through the origin [Figure 3.39(a) and (b)].

(a) Linearly dependent.   (b) Linearly independent.

Figure 3.39

Suppose now that $\{\mathbf{X}_1, \mathbf{X}_2, \mathbf{X}_3\}$ is a linearly dependent set of vectors in R^3. Then we can write

$$c_1\mathbf{X}_1 + c_2\mathbf{X}_2 + c_3\mathbf{X}_3 = \mathbf{0},$$

where c_1, c_2, and c_3 are not all zero; say that $c_2 \neq 0$. Then

$$\mathbf{X}_2 = -\frac{c_1}{c_2}\mathbf{X}_1 - \frac{c_3}{c_2}\mathbf{X}_3,$$

which means that $\mathbf{X}_2$ is in span $\{\mathbf{X}_1, \mathbf{X}_3\}$.

Now span $\{\mathbf{X}_1, \mathbf{X}_3\}$ is either a plane through the origin (when $\{\mathbf{X}_1, \mathbf{X}_3\}$ is linearly independent) or a line through the origin (when $\{\mathbf{X}_1, \mathbf{X}_3\}$ is linearly dependent). Since a line through the origin always lies in a plane through the origin, we conclude that $\mathbf{X}_1$, $\mathbf{X}_2$, and $\mathbf{X}_3$ all lie in the same plane through the origin. Conversely, suppose that $\mathbf{X}_1$, $\mathbf{X}_2$, and $\mathbf{X}_3$ all lie in the same plane through the origin. Then either all three vectors are the zero vectors, or all three vectors lie on the same line through the origin, or all three vectors lie in a plane through the origin spanned by two vectors, say $\mathbf{X}_1$ and $\mathbf{X}_3$. Thus in all these cases $\mathbf{X}_2$ is a linear combination of $\mathbf{X}_1$ and $\mathbf{X}_3$:

$$\mathbf{X}_2 = a_1\mathbf{X}_1 + a_3\mathbf{X}_3.$$

Then

$$a_1\mathbf{X}_1 - 1\mathbf{X}_2 + a_3\mathbf{X}_3 = \mathbf{0},$$

which means that $\{\mathbf{X}_1, \mathbf{X}_2, \mathbf{X}_3\}$ is linearly dependent. Hence three vectors in R^3 are linearly dependent if and only if they all lie in the same plane passing through the origin [Figure 3.40(a) and (b)].

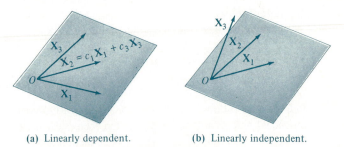

(a) Linearly dependent.   (b) Linearly independent.

Figure 3.40

Theorem 3.9 *Let* $S = \{\mathbf{X}_1, \mathbf{X}_2, \ldots, \mathbf{X}_k\}$ *be a set of nonzero vectors in a vector space V. Then S is linearly dependent if and only if one of the vectors* $\mathbf{X}_j$ *is a linear combination of the preceding vectors in S.*

Proof

Suppose that $\mathbf{X}_j$ is a linear combination of the preceding vectors,

$$\mathbf{X}_j = c_1\mathbf{X}_1 + c_2\mathbf{X}_2 + \cdots + c_{j-1}\mathbf{X}_{j-1}.$$

Then

$$c_1\mathbf{X}_1 + c_2\mathbf{X}_2 + \cdots + c_{j-1}\mathbf{X}_{j-1} + (-1)\mathbf{X}_j + 0\mathbf{X}_{j+1} + \cdots + 0\mathbf{X}_k = \mathbf{0}. \quad (2)$$

Since at least one coefficient, -1, is not zero in (2), we conclude that S is linearly dependent.

Conversely, suppose that S is linearly dependent. Then there exist scalars $c_1, c_2, \ldots,$ c_k, not all zero such that (1) holds. Now let j be the largest subscript for which $c_j \neq 0$. If $j > 1$, then

$$\mathbf{X}_j = -\left(\frac{c_1}{c_j}\right)\mathbf{X}_1 - \left(\frac{c_2}{c_j}\right)\mathbf{X}_2 - \cdots - \left(\frac{c_{j-1}}{c_j}\right)\mathbf{X}_{j-1}.$$

If $j = 1$, then $c_j\mathbf{X}_j = \mathbf{0}$, which implies that $\mathbf{X}_j = \mathbf{0}$, a contradiction to the hypothesis that none of the vectors in S is the zero vector.

Example 13 If $\mathbf{X}_1, \mathbf{X}_2, \mathbf{X}_3$, and $\mathbf{X}_4$ are as in Example 9, then $\mathbf{X}_1 + 2\mathbf{X}_2 + \mathbf{X}_3 + 0\mathbf{X}_4 = \mathbf{0}$, which implies that $\mathbf{X}_3 = -\mathbf{X}_1 - 2\mathbf{X}_2$. Also, $\mathbf{X}_1 + \mathbf{X}_2 + 0\mathbf{X}_3 - \mathbf{X}_4 = \mathbf{0}$, which implies that $\mathbf{X}_4 = \mathbf{X}_1 + \mathbf{X}_2$.

We observe that Theorem 3.9 does not say that *every* vector $\mathbf{X}_j$ is a linear combination of the preceding vectors in S. Thus, in Example 13, $\mathbf{X}_1 + 2\mathbf{X}_2 + \mathbf{X}_3 + 0\mathbf{X}_4 = \mathbf{0}$. We cannot solve for $\mathbf{X}_4$ as a linear combination of $\mathbf{X}_1, \mathbf{X}_2$, and $\mathbf{X}_3$, since its coefficient, in this equation, is zero. However, in this same example, we have $\mathbf{X}_1 + \mathbf{X}_2 + 0\mathbf{X}_3 - \mathbf{X}_4$ $= \mathbf{0}$. In this equation we *can* solve for $\mathbf{X}_4$, since its coefficient is nonzero.

We can also show that if $S = \{\mathbf{X}_1, \mathbf{X}_2, \ldots, \mathbf{X}_k\}$ is a set of vectors in a vector space, then S is linearly dependent if and only if one of the vectors in S is a linear combination of all the other vectors in S (Exercise T.3). Thus, in Example 13.

$$\mathbf{X}_1 = -\mathbf{X}_3 - 2\mathbf{X}_2, \quad \mathbf{X}_2 = -\tfrac{1}{2}\mathbf{X}_1 - \tfrac{1}{2}\mathbf{X}_3, \quad \mathbf{X}_1 = \mathbf{X}_4 - \mathbf{X}_2, \quad \text{and} \quad \mathbf{X}_2 = \mathbf{X}_4 - \mathbf{X}_1.$$

Again, this result does not say that *every* vector in S is a linear combination of all the other vectors in S. In Equation (1),

$$c_1\mathbf{X}_1 + c_2\mathbf{X}_2 + \cdots + c_k\mathbf{X}_k = \mathbf{0}$$

we cannot solve for a vector whose coefficient is zero.

3.5 EXERCISES

1. Which of the following vectors are linear combinations of $X_1 = (4, 2, -3)$, $X_2 = (2, 1, -2)$, and $X_3 = (-2, -1, 0)$?
   (a) $(1, 1, 1)$.   (b) $(4, 2, -6)$.
   (c) $(-2, -1, 1)$.   (d) $(-1, 2, 3)$.

2. Which of the following vectors are linear combinations of $X_1 = (1, 2, 1, 0)$, $X_2 = (4, 1, -2, 3)$, $X_3 = (1, 2, 6, -5)$, and $X_4 = (-2, 3, -1, 2)$?
   (a) $(3, 6, 3, 0)$.   (b) $(1, 0, 0, 0)$.
   (c) $(3, 6, -2, 5)$.   (d) $(0, 0, 0, 1)$.

3. Which of the following vectors are linear combinations of

   $$p_1(t) = t^2 + 2t + 1, \qquad p_2(t) = t^2 + 3,$$

   and

   $$p_3(t) = t - 1?$$

   (a) $t^2 + t + 2$.   (b) $2t^2 + 2t + 3$.
   (c) $-t^2 + t - 4$.   (d) $-2t^2 + 3t + 1$.

4. Which of the following vectors are linear combinations of

   $$A_1 = \begin{bmatrix} 1 & -1 \\ 0 & 3 \end{bmatrix}, \qquad A_2 = \begin{bmatrix} 1 & 1 \\ 0 & 2 \end{bmatrix},$$

   $$A_3 = \begin{bmatrix} 2 & 2 \\ -1 & 1 \end{bmatrix}?$$

   (a) $\begin{bmatrix} 5 & 1 \\ -1 & 9 \end{bmatrix}$.   (b) $\begin{bmatrix} -3 & -1 \\ 3 & 2 \end{bmatrix}$.

   (c) $\begin{bmatrix} 3 & -2 \\ 3 & 2 \end{bmatrix}$.   (d) $\begin{bmatrix} 1 & 0 \\ 2 & 1 \end{bmatrix}$.

5. In each part, determine whether the given vector X belongs to span $\{X_1, X_2, X_3\}$, where

   $$X_1 = (1, 0, 0, 1), X_2 = (1, -1, 0, 0),$$

   and

   $$X_3 = (0, 1, 2, 1).$$

   (a) $X = (-1, 4, 2, 2)$.   (b) $X = (1, 2, 0, 1)$.
   (c) $X = (-1, 1, 4, 3)$.   (d) $X = (0, 1, 1, 0)$.

6. In each part, determine whether the given vector $p(t)$ belongs to span $\{p_1(t), p_2(t), p_3(t)\}$, where $p_1(t) = t^2 - t$, $p_2(t) = t^2 - 2t + 1$, and $p_3(t) = -t^2 + 1$.
   (a) $p(t) = 3t^2 - 3t + 1$.
   (b) $p(t) = t^2 - t + 1$.
   (c) $p(t) = t + 1$.
   (d) $p(t) = 2t^2 - t - 1$.

7. Which of the following sets of vectors span R^2?
   (a) $(1, 2), (-1, 1)$.
   (b) $(0, 0), (1, 1), (-2, -2)$.
   (c) $(1, 3), (2, -3), (0, 2)$.
   (d) $(2, 4), (-1, 2)$.

8. Which of the following sets of vectors span R^3?
   (a) $(1, -1, 2), (0, 1, 1)$.
   (b) $(1, 2, -1), (6, 3, 0), (4, -1, 2), (2, -5, 4)$.
   (c) $(2, 2, 3), (-1, -2, 1), (0, 1, 0)$.
   (d) $(1, 0, 0), (0, 1, 0), (0, 0, 1), (1, 1, 1)$.

9. Which of the following sets of vectors span R^4?
   (a) $(1, 0, 0, 1)$, $(0, 1, 0, 0)$, $(1, 1, 1, 1)$, $(1, 1, 1, 0)$.
   (b) $(1, 2, 1, 0), (1, 1, -1, 0), (0, 0, 0, 1)$.
   (c) $(6, 4, -2, 4)$, $(2, 0, 0, 1)$, $(3, 2, -1, 2)$, $(5, 6, -3, 2), (0, 4, -2, -1)$.
   (d) $(1, 1, 0, 0)$, $(1, 2, -1, 1)$, $(0, 0, 1, 1)$, $(2, 1, 2, 1)$.

10. Which of the following sets of polynomials span P_2?
    (a) $t^2 + 1, t^2 + t, t + 1$.
    (b) $t^2 + 1, t - 1, t^2 + t$.
    (c) $t^2 + 2, 2t^2 - t + 1, t + 2, t^2 + t + 4$.
    (d) $t^2 + 2t - 1, t^2 - 1$.

11. Do the polynomials $t^3 + 2t + 1$, $t^2 - t + 2$, $t^3 + 2$, $-t^3 + t^2 - 5t + 2$ span P_3?

12. Which of the following sets of vectors in R^3 are linearly dependent? For those that are, express one vector as a linear combination of the rest.
    (a) $X_1 = (4, 2, 1)$, $X_2 = (2, 6, -5)$, $X_3 = (1, -2, 3)$.

(b) $\mathbf{X}_1 = (1, 2, -1)$, $\mathbf{X}_2 = (3, 2, 5)$.

(c) $\mathbf{X}_1 = (1, 1, 0)$, $\mathbf{X}_2 = (0, 2, 3)$, $\mathbf{X}_3 = (1, 2, 3)$, $\mathbf{X}_4 = (3, 6, 6)$.

(d) $\mathbf{X}_1 = (1, 2, 3)$, $\mathbf{X}_2 = (1, 1, 1)$, $\mathbf{X}_3 = (1, 0, 1)$.

13. Consider the vector space R^4. Follow the directions of Exercise 12.

(a) $\mathbf{X}_1 = (1, 1, 2, 1)$, $\mathbf{X}_2 = (1, 0, 0, 2)$, $\mathbf{X}_3 = (4, 6, 8, 6)$, $\mathbf{X}_4 = (0, 3, 2, 1)$.

(b) $\mathbf{X}_1 = (1, 1, 1, 1)$, $\mathbf{X}_2 = (1, 0, 0, 2)$, $\mathbf{X}_3 = (0, 1, 0, 2)$.

(c) $\mathbf{X}_1 = (1, 1, 1, 1)$, $\mathbf{X}_2 = (2, 3, 1, 2)$, $\mathbf{X}_3 = (3, 1, 2, 1)$, $\mathbf{X}_4 = (2, 2, 1, 1)$.

(d) $\mathbf{X}_1 = (4, 2, -1, 3)$, $\mathbf{X}_2 = (6, 5, -5, 1)$, $\mathbf{X}_3 = (2, -1, 3, 5)$.

14. Consider the vector space P_2. Follow the directions of Exercise 12.

(a) $p_1(t) = t^2 + 1$, $p_2(t) = t - 2$, $p_3(t) = t + 3$.

(b) $p_1(t) = 2t^2 + 1$, $p_2(t) = t^2 + 3$, $p_3(t) = t$.

(c) $p_1(t) = 3t + 1$, $p_2(t) = 3t^2 + 1$, $p_3(t) = 2t^2 + t + 1$.

(d) $p_1(t) = t^2 - 4$, $p_2(t) = 5t^2 - 5t - 6$, $p_3(t) = 3t^2 - 5t + 2$.

15. For what values of λ is the set of vectors $\{t + 3, 2t + \lambda^2 + 2\}$ in P_1 linearly independent?

THEORETICAL EXERCISES

T.1. Prove that the vectors $\mathbf{E}_1, \mathbf{E}_2, \ldots, \mathbf{E}_n$ in R^n are linearly independent.

T.2. Let S_1 and S_2 be finite subsets of a vector space and let S_1 be a subset of S_2. Prove:

(a) If S_1 is linearly dependent, so is S_2.

(b) If S_2 is linearly independent, so is S_1.

T.3. Let $S = \{\mathbf{X}_1, \mathbf{X}_2, \ldots, \mathbf{X}_k\}$ be a set of vectors in a vector space. Prove that S is linearly dependent if and only if one of the vectors in S is a linear combination of all the other vectors in S.

T.4. Suppose that $S = \{\mathbf{X}_1, \mathbf{X}_2, \mathbf{X}_3\}$ is a linearly independent set of vectors in a vector space V. Prove that $T = \{\mathbf{Y}_1, \mathbf{Y}_2, \mathbf{Y}_3\}$ is also linearly independent where $\mathbf{Y}_1 = \mathbf{X}_1 + \mathbf{X}_2 + \mathbf{X}_3$, $\mathbf{Y}_2 = \mathbf{X}_2 + \mathbf{X}_3$, and $\mathbf{Y}_3 = \mathbf{X}_3$.

T.5. Let $\mathbf{X}_1$, $\mathbf{X}_2$, and $\mathbf{X}_3$ be vectors in a vector space such that $\{\mathbf{X}_1, \mathbf{X}_2\}$ is linearly independent. Show that if $\mathbf{X}_3$ does not belong to span $\{\mathbf{X}_1, \mathbf{X}_2\}$, then $\{\mathbf{X}_1, \mathbf{X}_2, \mathbf{X}_3\}$ is linearly independent.

T.6. Let $\mathbf{A}$ be an $m \times n$ matrix in reduced row echelon form. Prove that the nonzero rows of $\mathbf{A}$, viewed as vectors in R^n, form a linearly independent set of vectors.

T.7. Let $S = \{\mathbf{X}_1, \mathbf{X}_2, \ldots, \mathbf{X}_k\}$ be a set of vectors in a vector space, and let $T = \{\mathbf{Y}_1, \mathbf{Y}_2, \ldots, \mathbf{Y}_m\}$, where each $\mathbf{Y}_i$, $i = 1, 2, \ldots, m$, is a linear combination of the vectors in S. Prove that

$$\mathbf{Z} = b_1\mathbf{Y}_1 + b_2\mathbf{Y}_2 + \cdots + b_m\mathbf{Y}_m$$

is a linear combination of the vectors in S.

T.8. Let S_1 and S_2 be finite subsets of a vector space V and let S_1 be a subset of S_2. If S_2 is linearly dependent, show by examples that S_1 may be either linearly dependent or linearly independent.

T.9. Let S_1 and S_2 be finite subsets of a vector space and let S_1 be a subset of S_2. If S_1 is linearly independent, show by examples that S_2 may be either linearly dependent or linearly independent.

3.6

BASIS AND DIMENSION

In this section we continue our study of the structure of a vector space V by determining a set of vectors in V that completely describes V.

Definition A set of vectors $S = \{\mathbf{X}_1, \mathbf{X}_2, \ldots, \mathbf{X}_n\}$ in a vector space V is called a **basis** for V if S spans V and S is linearly independent.

Example 1 The vectors $\mathbf{E}_1 = (1, 0)$ and $\mathbf{E}_2 = (0, 1)$ form a basis for R^2, the vectors $\mathbf{E}_1$, $\mathbf{E}_2$, and $\mathbf{E}_3$ form a basis for R^3 and, in general, the vectors $\mathbf{E}_1$, $\mathbf{E}_2$, $\ldots$, $\mathbf{E}_n$ form a basis for R^n. Each of these sets of vectors is called the **natural basis** for R^2, R^3, and R^n, respectively.

Example 2 Show that the set $S = \{\mathbf{X}_1, \mathbf{X}_2, \mathbf{X}_3, \mathbf{X}_4\}$, where $\mathbf{X}_1 = (1, 0, 1, 0)$, $\mathbf{X}_2 = (0, 1, -1, 2)$, $\mathbf{X}_3 = (0, 2, 2, 1)$, and $\mathbf{X}_4 = (1, 0, 0, 1)$ is a basis for R^4.

Solution
To show that S is linearly independent, we form the equation

$$c_1\mathbf{X}_1 + c_2\mathbf{X}_2 + c_3\mathbf{X}_3 + c_4\mathbf{X}_4 = \mathbf{0}$$

and solve for c_1, c_2, c_3, and c_4. Substituting for $\mathbf{X}_1$, $\mathbf{X}_2$, $\mathbf{X}_3$, and $\mathbf{X}_4$, we obtain the linear system

$$
\begin{aligned}
c_1 \quad\quad\quad\quad + c_4 &= 0 \\
c_2 + 2c_3 \quad\quad &= 0 \\
c_1 - c_2 + 2c_3 \quad\quad &= 0 \\
2c_2 + c_3 + c_4 &= 0,
\end{aligned}
$$

which has as its only solution $c_1 = c_2 = c_3 = c_4 = 0$ (verify), showing that S is linearly independent.

To show that S spans R^4, we let $\mathbf{X} = (a, b, c, d)$ be any vector in R^4. We now seek constants k_1, k_2, k_3, and k_4 such that

$$k_1\mathbf{X}_1 + k_2\mathbf{X}_2 + k_3\mathbf{X}_3 + k_4\mathbf{X}_4 = \mathbf{X}.$$

Substituting for $\mathbf{X}_1$, $\mathbf{X}_2$, $\mathbf{X}_3$, $\mathbf{X}_4$, and $\mathbf{X}$, we find a solution (verify) for k_1, k_2, k_3, and k_4 to the resulting linear system for any a, b, c, and d. Hence S spans R^4 and is a basis for R^4.

Example 3 In Example 7 of Section 3.5 we observed that the set of polynomials $S = \{t^n, t^{n-1}, \ldots, t, 1\}$ spans P_n. Moreover, it is easy to show that S is linearly independent as follows. Set up Equation (1) of Section 3.5,

$$c_1 t^n + c_2 t^{n-1} + \cdots + c_n t + c_{n+1} = 0, \tag{1}$$

which holds for all values of t. In other words, every real number t is a root of the polynomial $p(t) = c_1 t^n + c_2 t^{n-1} + \cdots + c_n t + c_{n+1}$. But a nonzero polynomial can have only a finite number of roots, so we conclude that Equation (1) can hold only if

$$c_1 = c_2 = \cdots = c_n = c_{n+1} = 0.$$

Thus S is a basis for P_n.

Example 4 The set

$$S = \left\{ \begin{bmatrix} 1 & 0 \\ 0 & 0 \end{bmatrix}, \begin{bmatrix} 0 & 1 \\ 0 & 0 \end{bmatrix}, \begin{bmatrix} 0 & 0 \\ 1 & 0 \end{bmatrix}, \begin{bmatrix} 0 & 0 \\ 0 & 1 \end{bmatrix} \right\}$$

is a basis for the vector space V of all 2×2 matrices. To verify that S is linearly independent, we form Equation (1) of Section 3.5:

$$c_1 \begin{bmatrix} 1 & 0 \\ 0 & 0 \end{bmatrix} + c_2 \begin{bmatrix} 0 & 1 \\ 0 & 0 \end{bmatrix} + c_3 \begin{bmatrix} 0 & 0 \\ 1 & 0 \end{bmatrix} + c_4 \begin{bmatrix} 0 & 0 \\ 0 & 1 \end{bmatrix} = \begin{bmatrix} 0 & 0 \\ 0 & 0 \end{bmatrix}.$$

This gives

$$\begin{bmatrix} c_1 & c_2 \\ c_3 & c_4 \end{bmatrix} = \begin{bmatrix} 0 & 0 \\ 0 & 0 \end{bmatrix},$$

which implies that $c_1 = c_2 = c_3 = c_4 = 0$. Hence S is linearly independent. To verify that S spans V, we take any vector $\begin{bmatrix} a & b \\ c & d \end{bmatrix}$ in V and we must find scalars c_1, c_2, c_3, and c_4 such that

$$\begin{bmatrix} a & b \\ c & d \end{bmatrix} = c_1 \begin{bmatrix} 1 & 0 \\ 0 & 0 \end{bmatrix} + c_2 \begin{bmatrix} 0 & 1 \\ 0 & 0 \end{bmatrix} + c_3 \begin{bmatrix} 0 & 0 \\ 1 & 0 \end{bmatrix} + c_4 \begin{bmatrix} 0 & 0 \\ 0 & 1 \end{bmatrix}.$$

We find that $c_1 = a$, $c_2 = b$, $c_3 = c$, and $c_4 = d$ so that S spans V.

Theorem 3.10 *If $S = \{X_1, X_2, \ldots, X_n\}$ is a basis for a vector space V, then every vector in V can be written in one and only one way as a linear combination of the vectors in S.*

Proof

First, every vector X in V can be written as a linear combination of the vectors in S because S spans V. Now let

$$X = c_1 X_1 + c_2 X_2 + \cdots + c_n X_n \tag{2}$$

and

$$X = d_1 X_1 + d_2 X_2 + \cdots + d_n X_n. \tag{3}$$

Subtracting (3) from (2), we obtain

$$0 = (c_1 - d_1)X_1 + (c_2 - d_2)X_2 + \cdots + (c_n - d_n)X_n.$$

Since S is linearly independent, it follows that $c_i - d_i = 0$, $1 \leq i \leq n$, so $c_i = d_i$, $1 \leq i \leq n$.

Even though a nonzero vector space contains an infinite number of elements, a vector space with a basis is in a sense completely described by a finite number of elements, namely, those elements in a basis.

The following useful result, whose proof is left as an exercise (Exercise T.1), will be used in proving Theorem 3.11, which yields an algorithm for finding a basis for a vector space spanned by a given set of vectors.

Suppose that $S = \{X_1, X_2, \ldots, X_n\}$ spans a vector space V and X_j is a linear combination of the preceding vectors in S. Then the set $S_1 = \{X_1, X_2, \ldots, X_{j-1}, X_{j+1}, \ldots, X_n\}$, consisting of S with X_j deleted, also spans V.

Theorem 3.11 *Let $S = \{X_1, X_2, \ldots, X_n\}$ be a set of nonzero vectors in a vector space V and let $W = span\ S$. Then some subset of S is a basis for W.*

Proof

If S is linearly independent, then since S already spans W, we conclude that S is a basis for W. To see if S is linearly dependent or is linearly independent, we solve

$$c_1X_1 + c_2X_2 + \cdots + c_nX_n = 0. \tag{4}$$

for $c_1, c_2, \ldots, c_n$. Equation (4) leads to a homogeneous system in the n unknowns c_1, $c_2, \ldots, c_n$, which is solved by first transforming the augmented matrix of the system to reduced row echelon form. Suppose that S is linearly dependent so that the system has a nontrivial solution. This means that the unknowns corresponding to the leading 1's in the reduced row echelon form matrix can be solved for in terms of other unknowns. The latter unknowns can be assigned arbitrary real values. Without loss of generality, we may assume that $c_1, c_2, \ldots, c_k$ are the unknowns corresponding to the leading 1's. Thus c_1, $c_2, \ldots, c_k$ can be expressed in terms of $c_{k+1}, c_{k+2}, \ldots, c_n$; for example,

$$c_1 = b_1c_{k+1} + b_2c_{k+2} + \cdots + b_{n-k}c_n. \tag{5}$$

Letting $c_{k+1} = 1, c_{k+2} = 0, \ldots, c_n = 0$ in Equation (4), we see that X_{k+1} is a linear combination of $X_1, X_2, \ldots, X_k$. By the remark made just before the statement of this theorem, the set of vectors obtained from S by deleting X_{k+1} spans W. Similarly, letting $c_{k+1} = 0, c_{k+2} = 1, c_{k+3} = 0. \ldots, c_n = 0$, we find that X_{k+2} is a linear combination of $X_1, X_2, \ldots, X_k$ and the set of vectors obtained from S by deleting X_{k+1} and X_{k+2} spans W. Continuing in this manner, $X_{k+3}, X_{k+4}, \ldots, X_n$ are linear combinations of $X_1, X_2, \ldots, X_k$, so it follows that $\{X_1, X_2, \ldots, X_k\}$ spans W.

We next show that $\{X_1, X_2, \ldots, X_k\}$ is linearly independent. Consider the equation

$$c_1X_1 + c_2X_2 + \cdots + c_kX_k = 0. \tag{6}$$

If we can find values for $c_1, c_2, \ldots, c_k$ satisfying Equation (6), then these values will also satisfy the following equation:

$$c_1X_1 + c_2X_2 + \cdots + c_kX_k + 0X_{k+1} + 0X_{k+2} + \cdots + 0X_n = 0. \tag{7}$$

On the other hand, observe that every solution to Equation (7) is a solution to Equation (4) in which the $n - k$ arbitrary constants $c_{k+1}, c_{k+2}, \ldots, c_n$ have been set equal to zero. Since each of the constants $c_1, c_2, \ldots, c_k$ can be written in terms of $c_{k+1}, c_{k+2}, \ldots, c_n$ [see Equation (5)] and these have been chosen to be zero to satisfy Equation (4), it follows from Equation (5) that $c_1 = 0, c_2 = 0, \ldots, c_k = 0$. Thus the only choice of constants that satisfies Equation (6) is $c_1, c_2, \ldots, c_k$ all zero; hence $\{X_1, X_2, \ldots, X_k\}$ is linearly independent and is therefore a basis for W.

The proof of Theorem 3.11 leads to a simple procedure for finding a subset of a spanning set that is a basis.

The procedure for finding a subset of $S = \{X_1, X_2, \ldots, X_n\}$ that is a basis for $W = \text{span } S$ is as follows.

STEP 1. Form Equation (4)

$$c_1X_1 + c_2X_2 + \cdots + c_nX_n = \mathbf{0}.$$

STEP 2. Construct the augmented matrix associated with the homogeneous system of Equation (4), and transform it to reduced row echelon form.

STEP 3. The vectors corresponding to the columns containing the leading 1's form a basis for $W = \text{span } S$. Thus, if $S = \{X_1, X_2, \ldots, X_6\}$ and the leading 1's occur in columns 1, 3, and 4, then $\{X_1, X_3, X_4\}$ is a basis for span S.

REMARK In step 2 of the procedure above, it is sufficient to transform the augmented matrix to row echelon form (see Section 1.4).

Example 5 Let $S = \{X_1, X_2, X_3, X_4\}$ be a set of vectors in R^4, where $X_1 = (1, 2, -2, 1)$, $X_2 = (-3, 0, -4, 3)$, $X_3 = (2, 1, 1, -1)$, and $X_4 = (-3, 3, -9, 6)$. Find a subset of S that is a basis for $W = \text{span } S$.

Solution

STEP 1. From Equation (4)

$$c_1 (1, 2, -2, 1) + c_2 (-3, 0, -4, 3) + c_3 (2, 1, 1, -1)$$
$$+ c_4 (-3, 3, -9, 6) = (0, 0, 0, 0)$$

STEP 2. Equating corresponding components, we obtain the homogeneous system

$$\begin{aligned} c_1 - 3c_2 + 2c_3 - 3c_4 &= 0 \\ 2c_1 \qquad + c_3 + 3c_4 &= 0 \\ -2c_1 - 4c_2 + c_3 - 9c_4 &= 0 \\ c_1 + 3c_2 - c_3 + 6c_4 &= 0 \end{aligned}$$

The reduced row echelon form of the associated augmented matrix is (verify)

$$\left[\begin{array}{cccc:c} 1 & 0 & \frac{1}{2} & \frac{3}{2} & 0 \\ 0 & 1 & -\frac{1}{2} & \frac{3}{2} & 0 \\ 0 & 0 & 0 & 0 & 0 \\ 0 & 0 & 0 & 0 & 0 \end{array} \right].$$

STEP 3. The leading 1's appear in columns 1 and 2, so $\{\mathbf{X}_1, \mathbf{X}_2\}$ is a basis for $W = $ span S.

REMARK   In the proof of Theorem 3.11, the order of the vector in the original spanning set S determines which basis for W is obtained. If, for example, we consider Example 5, where $S = \{\mathbf{Y}_1, \mathbf{Y}_2, \mathbf{Y}_3, \mathbf{Y}_4\}$ with $\mathbf{Y}_1 = \mathbf{X}_4$, $\mathbf{Y}_2 = \mathbf{X}_3$, $\mathbf{Y}_3 = \mathbf{X}_2$, and $\mathbf{Y}_4 = \mathbf{X}_1$, then the reduced row echelon form of the augmented matrix is (verify)

$$\left[\begin{array}{cccc:c} 1 & 0 & \frac{1}{3} & \frac{1}{3} & 0 \\ 0 & 1 & -1 & 1 & 0 \\ 0 & 0 & 0 & 0 & 0 \\ 0 & 0 & 0 & 0 & 0 \end{array}\right].$$

It now follows that $\{\mathbf{Y}_1, \mathbf{Y}_2\} = \{\mathbf{X}_4, \mathbf{X}_3\}$ is a basis for $W = $ span S.

We are now about to establish a main result (Corollary 3.1) of this section, which will tell us about the number of vectors in two different bases. First, observe that if $\{\mathbf{X}_1, \mathbf{X}_2, \dots, \mathbf{X}_n\}$ is a basis for a vector space V, then $\{c\mathbf{X}_1, \mathbf{X}_2, \dots, \mathbf{X}_n\}$ is also a basis if $c \neq 0$. Thus a vector space always has infinitely many bases.

Theorem 3.12 *If $S = \{\mathbf{X}_1, \mathbf{X}_2, \dots, \mathbf{X}_n\}$ is a basis for a vector space V and $T = \{\mathbf{Y}_1, \mathbf{Y}_2, \dots, \mathbf{Y}_r\}$ is a linearly independent set of vectors in V, then $r \leq n$.*

Proof

Let $T_1 = \{\mathbf{Y}_1, \mathbf{X}_1, \dots, \mathbf{X}_n\}$. Since S spans V, so does T_1. Since $\mathbf{Y}_1$ is a linear combination of the vectors in S, we find that T_1 is linearly dependent. Then, by Theorem 3.9, some $\mathbf{X}_j$ is a linear combination of the preceding vectors in T_1. Delete that particular vector $\mathbf{X}_j$.

Let $S_1 = \{\mathbf{Y}_1, \mathbf{X}_1, \dots, \mathbf{X}_{j-1}, \mathbf{X}_{j+1}, \dots, \mathbf{X}_n\}$. Note that S_1 spans V. Next, let $T_2 = \{\mathbf{Y}_2, \mathbf{Y}_1, \mathbf{X}_1, \dots, \mathbf{X}_{j-1}, \mathbf{X}_{j+1}, \dots, \mathbf{X}_n\}$. Then T_2 is linearly dependent and some vector in T_2 is a linear combination of the preceding vectors in T_2. Since T is linearly independent, this vector cannot be $\mathbf{Y}_1$, so it is $\mathbf{X}_i$, $i \neq j$. Repeat this process over and over. If the $\mathbf{X}$ vectors are all eliminated before we run out of $\mathbf{Y}$ vectors, then the resulting set of $\mathbf{Y}$ vectors, a subset of T, is linearly dependent, which implies that T is also linearly dependent. Since we have reached a contradiction, we conclude that the number r of $\mathbf{Y}$ vectors must be no greater than the number n of $\mathbf{X}$ vectors. That is, $r \leq n$.

Corollary 3.1 *If $S = \{\mathbf{X}_1, \mathbf{X}_2, \dots, \mathbf{X}_n\}$ and $T = \{\mathbf{Y}_1, \mathbf{Y}_2, \dots, \mathbf{Y}_m\}$ are bases for a vector space, then $n = m$.*

Proof

Since T is a linearly independent set of vectors, Theorem 3.12 implies that $m \leq n$. Similarly, $n \leq m$ because S is linearly independent. Hence $n = m$.

Thus, although a vector space has many bases, all bases have the same number of vectors. We can then make the following definition.

Definition   The **dimension** of a nonzero vector space V is the number of vectors in a basis for V. We often write **dim** V for the dimension of V. Since the set $\{\mathbf{0}\}$ is linearly dependent, it is natural to say that the vector space $\{\mathbf{0}\}$ has dimension **zero**.

Example 6   The dimension of R^2 is 2; the dimension of R^3 is 3; and in general, the dimension of R^n is n.

Example 7   The dimension of P_2 is 3; the dimension of P_3 is 4; and in general, the dimension of P_n is $n + 1$.

Almost all the vector spaces considered in this book are **finite dimensional**; that is, their dimension is a finite number. However, we point out that there are many vector spaces of infinite dimension that are extremely important in mathematics and physics; their study lies beyond the scope of this book. The vector space P and the vector space $C(-\infty, \infty)$ defined in Example 14 of Section 3.4 are not finite dimensional. All finite-dimensional vector spaces of the same dimension differ only in the nature of their elements; their algebraic properties are identical.

It is easy to show that if V is a finite-dimensional vector space, then every nonzero subspace W of V has a finite basis and dim $W \leq$ dim V (Exercise T.3).

Example 8   The subspace W of R^4 considered in Example 5 has dimension 2.

We might also consider the subspaces of R^2 [recall that R^2 can be viewed as the (x, y)-plane]. First, we have $\{\mathbf{0}\}$ and R^2, the trivial subspaces of dimension 0 and 2, respectively. Now the subspace V of R^2 spanned by a vector $\mathbf{X} \neq \mathbf{0}$ is a one-dimensional subspace of R^2; V is represented by a line through the origin. Thus the subspaces of R^2 are $\{\mathbf{0}\}$, R^2, and all the lines through the origin.

It can be shown that if a vector space V has dimension n, then any set of $n + 1$ vectors in V is necessarily linearly dependent (Exercise T.4). Any set of more than n vectors in R^n is linearly dependent. Thus the four vectors in R^3 considered in Example 9 of Section 3.5 are linearly dependent. Also, if a vector space V is of dimension n, then no set of $n - 1$ vectors in V can span V (Exercise T.5). Thus in Example 5 of Section 3.5 polynomials $p_1(t)$ and $p_2(t)$ do not span P_2, which is of dimension 3.

We now come to a theorem that we shall have occasion to use several times in constructing a basis containing a given set of linearly independent vectors. We shall leave the proof as an exercise (Exercise T.6). The example following the theorem completely imitates the proof.

Theorem 3.13   *If S is a linearly independent set of vectors in a finite-dimensional vector space V, then there is a basis T for V, which contains S.*

Theorem 3.13 says that a linearly independent set of vectors in a vector space V can be extended to a basis for V.

Example 9 Suppose that we wish to find a basis for R^4 that contains the vectors $\mathbf{X}_1 = (1, 0, 1, 0)$ and $\mathbf{X}_2 = (-1, 1, -1, 0)$.

We use Theorem 3.13 as follows. First, let $\{\mathbf{E}_1, \mathbf{E}_2, \mathbf{E}_3, \mathbf{E}_4\}$ be the natural basis for R^4, where

$$\mathbf{E}_1 = (1, 0, 0, 0), \qquad \mathbf{E}_2 = (0, 1, 0, 0), \qquad \mathbf{E}_3 = (0, 0, 1, 0),$$

and

$$\mathbf{E}_4 = (0, 0, 0, 1).$$

Form the set $S_1 = \{\mathbf{X}_1, \mathbf{X}_2, \mathbf{E}_1, \mathbf{E}_2, \mathbf{E}_3, \mathbf{E}_4\}$. Since S_1 spans R^4, it contains, by Theorem 3.11, a basis for R^4. The basis is obtained by deleting from S_1 every vector that is a linear combination of the preceding vectors. We check whether $\mathbf{E}_1$ is a linear combination of $\mathbf{X}_1$ and $\mathbf{X}_2$. Since the answer is no, we retain $\mathbf{E}_1$. Now we check whether $\mathbf{E}_2$ is a linear combination of $\mathbf{X}_1$, $\mathbf{X}_2$, and $\mathbf{E}_1$. Since the answer is yes, we delete $\mathbf{E}_2$. Next, we check whether $\mathbf{E}_3$ is a linear combination of $\mathbf{X}_1$, $\mathbf{X}_2$, and $\mathbf{E}_1$. Since the answer is yes, we delete $\mathbf{E}_3$. We now check whether $\mathbf{E}_4$ is a linear combination of $\mathbf{X}_1$, $\mathbf{X}_2$, and $\mathbf{E}_1$. The answer is no, so our basis consists of the vectors $\mathbf{X}_1$, $\mathbf{X}_2$, $\mathbf{E}_1$, and $\mathbf{E}_4$.

From the definition of a basis, a set of vectors in a vector space V is a basis for V if it spans V and is linearly independent. However, if we are given the additional information that the dimension of V is n, we need only verify one of the two conditions. This is the content of the following theorem.

Theorem 3.14 *Let V be an n-dimensional vector space, and let $S = \{\mathbf{X}_1, \mathbf{X}_2, \ldots , \mathbf{X}_n\}$ be a set of n vectors in V.*

(a) *If S is linearly independent, then it is a basis for V.*
(b) *If S spans V, then it is a basis for V.*

Proof
Exercise T.7.

We have already observed in Example 15 of Section 3.4 that the set of all solutions to the homogeneous system $\mathbf{AX} = \mathbf{0}$, where $\mathbf{A}$ is $m \times n$, is a subspace of R^n. An extremely important problem, which will occur repeatedly in Chapter 5, is that of finding a basis for this solution space.

The procedure for finding a basis for the solution space of a homogeneous system $\mathbf{AX} = \mathbf{0}$ is as follows.

STEP 1. Solve the given homogeneous system by Gauss–Jordan reduction. The solution may involve arbitrary constants.

STEP 2. Write the solution as a linear combination of vectors, using any arbitrary constants as coefficients. The set of vectors S thus determined spans the solution space W.

STEP 3. Find a subset of S that is a basis for W.

Example 10 Find a basis for the solution space W of the homogeneous system

$$\begin{bmatrix} 1 & 2 & 0 & 3 & 1 \\ 2 & 3 & 0 & 3 & 1 \\ 1 & 1 & 2 & 2 & 1 \\ 3 & 5 & 0 & 6 & 2 \\ 2 & 3 & 2 & 5 & 2 \end{bmatrix} \begin{bmatrix} x_1 \\ x_2 \\ x_3 \\ x_4 \\ x_5 \end{bmatrix} = \begin{bmatrix} 0 \\ 0 \\ 0 \\ 0 \\ 0 \end{bmatrix}.$$

Solution

STEP 1. To solve the given system by the Gauss–Jordan reduction method, we transform the augmented matrix to reduced row echelon form, obtaining

$$\begin{bmatrix} 1 & 0 & 0 & -3 & -1 & \vdots & 0 \\ 0 & 1 & 0 & 3 & 1 & \vdots & 0 \\ 0 & 0 & 1 & 1 & \frac{1}{2} & \vdots & 0 \\ 0 & 0 & 0 & 0 & 0 & \vdots & 0 \\ 0 & 0 & 0 & 0 & 0 & \vdots & 0 \end{bmatrix}.$$

Every solution is of the form (verify)

$$\mathbf{X} = \begin{bmatrix} 3s + t \\ -3s - t \\ -s - \frac{1}{2}t \\ s \\ t \end{bmatrix}, \tag{8}$$

where s and t are any real numbers.

STEP 2. Every vector in W is a solution and is then of the form given by Equation (8). We can then write every vector in W as

$$\mathbf{X} = s \begin{bmatrix} 3 \\ -3 \\ -1 \\ 1 \\ 0 \end{bmatrix} + t \begin{bmatrix} 1 \\ -1 \\ -\frac{1}{2} \\ 0 \\ 1 \end{bmatrix}. \tag{9}$$

Since s and t can take on any values, we first let $s = 1$, $t = 0$, and then let $s = 0$, $t = 1$, obtaining as solutions

$$\mathbf{X}_1 = \begin{bmatrix} 3 \\ -3 \\ -1 \\ 1 \\ 0 \end{bmatrix} \quad \text{and} \quad \mathbf{X}_2 = \begin{bmatrix} 1 \\ -1 \\ -\frac{1}{2} \\ 0 \\ 1 \end{bmatrix}.$$

That is, $\mathbf{X}_1$ and $\mathbf{X}_2$ belong to W. From Equation (9) we see that $S = \{\mathbf{X}_1, \mathbf{X}_2\}$ spans W.

STEP 3. Since $S = \{\mathbf{X}_1, \mathbf{X}_2\}$ is linearly independent, it is a basis for W. Thus the dimension of W is 2.

APPLICATIONS. At this point the reader may turn to Section 7.1, Lines and Planes.

3.6 EXERCISES

1. Which of the following sets of vectors are bases for R^2?
   (a) $(1, 3), (1, -1)$.
   (b) $(0, 0), (1, 2), (2, 4)$.
   (c) $(1, 2), (2, -3), (3, 2)$.
   (d) $(1, 3), (-2, 6)$.

2. Which of the following sets of vectors are bases for R^3?
   (a) $(1, 2, 0), (0, 1, -1)$.
   (b) $(1, 1, -1), (2, 3, 4), (4, 1, -1), (0, 1, -1)$.
   (c) $(3, 2, 2), (-1, 2, 1), (0, 1, 0)$.
   (d) $(1, 0, 0), (0, 2, -1), (3, 4, 1), (0, 1, 0)$.

3. Which of the following sets of vectors are bases for R^4?
   (a) $(1, 0, 0, 1),\quad (0, 1, 0, 0),\quad (1, 1, 1, 1),$ $(0, 1, 1, 1)$.
   (b) $(1, -1, 0, 2), (3, -1, 2, 1), (1, 0, 0, 1)$.
   (c) $(-2, 4, 6, 4),\quad (0, 1, 2, 0),\quad (-1, 2, 3, 2),$ $(-3, 2, 5, 6), (-2, -1, 0, 4)$.
   (d) $(0, 0, 1, 1),\quad (-1, 1, 1, 2),\quad (1, 1, 0, 0),$ $(2, 1, 2, 1)$.

4. Which of the following sets of vectors are bases for P_2?
   (a) $-t^2 + t + 2, 2t^2 + 2t + 3, 4t^2 - 1$.
   (b) $t^2 + 2t - 1, 2t^2 + 3t - 2$.
   (c) $t^2 + 1,\quad 3t^2 + 2t,\quad 3t^2 + 2t + 1,\quad 6t^2 +$ $6t + 3$.
   (d) $3t^2 + 2t + 1, t^2 + t + 1, t^2 + 1$.

5. Which of the following sets of vectors are bases for P_3?
   (a) $t^3 + 2t^2 + 3t, 2t^3 + 1, 6t^3 + 8t^2 + 6t + 4,$ $t^3 + 2t^2 + t + 1$.
   (b) $t^3 + t^2 + 1, t^3 - 1, t^3 + t^2 + t$.
   (c) $t^3 + t^2 + t + 1,\quad t^3 + 2t^2 + t + 3, 2t^3 +$ $t^2 + 3t + 2, t^3 + t^2 + 2t + 2$.
   (d) $t^3 - t, t^3 + t^2 + 1, t - 1$.

6. Show that the matrices
$$\begin{bmatrix} 1 & 1 \\ 0 & 0 \end{bmatrix}, \begin{bmatrix} 0 & 0 \\ 1 & 1 \end{bmatrix}, \begin{bmatrix} 1 & 0 \\ 0 & 1 \end{bmatrix}, \begin{bmatrix} 0 & 1 \\ 1 & 1 \end{bmatrix}$$
form a basis for the vector space of all 2×2 matrices.

7. Which of the followings sets of vectors form a basis for R^3? Express the vector $\mathbf{X} = (2, 1, 3)$ as a linear combination of the vectors in each set that is a basis.
   (a) $\mathbf{X}_1 = (1, 1, 1),\quad \mathbf{X}_2 = (1, 2, 3),\quad \mathbf{X}_3 = (0, 1, 0)$.
   (b) $\mathbf{X}_1 = (1, 2, 2),\quad \mathbf{X}_2 = (2, 1, 3),\quad \mathbf{X}_3 = (0, 0, 0)$.
   (c) $\mathbf{X}_1 = (-2, 1, 3),\quad \mathbf{X}_2 = (-1, 2, 3),\quad \mathbf{X}_3 = (-1, -4, -3)$.

8. Which of the following subsets form a basis for P_2? Express $5t^2 - 3t + 8$ as a linear combination of the vectors in each subset that is a basis.
   (a) $\{t^2 + t, t - 1, t + 1\}$.
   (b) $\{t^2 + 1, t - 1\}$.
   (c) $\{t^2 + t, t^2, t^2 + 1\}$.
   (d) $\{t^2 + 1, t^2 - t + 1\}$.

9. Let $S = \{\mathbf{X}_1, \mathbf{X}_2, \mathbf{X}_3, \mathbf{X}_4\}$, where $\mathbf{X}_1 = (1, 2, 2)$, $\mathbf{X}_2 = (3, 2, 1),\quad \mathbf{X}_3 = (11, 10, 7),\quad$ and $\quad \mathbf{X}_4 = (7, 6, 4)$. Find a basis for the subspace $W = $ span S of R^3. What is dim W?

10. Let $\quad S = \{\mathbf{X}_1, \mathbf{X}_2, \mathbf{X}_3, \mathbf{X}_4, \mathbf{X}_5\}$, where $\mathbf{X}_1 = (1, 1, 0, -1),\quad \mathbf{X}_2 = (0, 1, 2, 1),\quad \mathbf{X}_3 = (1, 0, 1, -1), \mathbf{X}_4 = (1, 1, -6, -3),$ and $\mathbf{X}_5 = (-1, -5, 1, 0)$. Find a basis for the subspace $W = $ span S of R^4. What is dim W?

11. Consider the following subset of P_3:

$$S = \{t^3 + t^2 - 2t + 1, \ t^2 + 1, \ t^3 - 2t,$$
$$2t^3 + 3t^2 - 4t + 3\}.$$

Find a basis for the subspace $W = \text{span } S$. What is dim W?

12. Consider the following subset of the vector space of all real-valued functions

$$S = \{\cos^2 t, \ \sin^2 t, \ \cos 2t\}.$$

Find a basis for the subspace $W = \text{span } S$. What is dim W?

13. Find a basis for the following subspaces of R^3.
   (a) All vectors of the form (a, b, c), where $b = a + c$.
   (b) All vectors of the form (a, b, c), where $b = a$.
   (c) All vectors of the form (a, b, c), where $a = 0$.
   (d) All vectors of the form $(a - b, b + c, 2a - b + c)$.

14. Find the dimensions of the following subspaces of R^4.
   (a) All vectors of the form (a, b, c, d), where $d = a + b$.
   (b) All vectors of the form (a, b, c, d), where $c = a - b$ and $d = a + b$.
   (c) All vectors of the form (a, b, c, d), where $a = b$.
   (d) All vectors of the form $(a + c, a - b, b + c, -a + b)$.

15. Find the dimensions of the subspaces of R^2 spanned by the vectors in Exercise 1.

16. Find the dimensions of the subspaces of R^3 spanned by the vectors in Exercise 2.

17. Find the dimensions of the subspaces of R^4 spanned by the vectors in Exercise 3.

18. Find a basis for R^3 that includes the vectors
   (a) $(1, 0, 2)$.   (b) $(1, 0, 2)$ and $(0, 1, 3)$.

19. Find a basis for R^4 that includes the vectors $(1, 0, 1, 0)$ and $(0, 1, -1, 0)$.

20. Find a basis for the solution space W of

$$\begin{bmatrix} 1 & 2 & 1 & 2 & 1 \\ 1 & 2 & 2 & 1 & 2 \\ 2 & 4 & 3 & 3 & 3 \\ 0 & 0 & 1 & -1 & -1 \end{bmatrix} \begin{bmatrix} x_1 \\ x_2 \\ x_3 \\ x_4 \\ x_5 \end{bmatrix} = \begin{bmatrix} 0 \\ 0 \\ 0 \\ 0 \end{bmatrix}.$$

What is the dimension of W?

21. Find the dimension of the solution space of

$$\begin{bmatrix} 1 & 0 & 2 \\ 2 & 1 & 3 \\ 3 & 1 & 2 \end{bmatrix} \begin{bmatrix} x_1 \\ x_2 \\ x_3 \end{bmatrix} = \begin{bmatrix} 0 \\ 0 \\ 0 \end{bmatrix}.$$

22. Find a basis for the solution space W of

$$\begin{bmatrix} 1 & 2 & 2 & -1 & 1 \\ 0 & 2 & 2 & -2 & -1 \\ 2 & 6 & 2 & -4 & 1 \\ 1 & 4 & 0 & -3 & 0 \end{bmatrix} \begin{bmatrix} x_1 \\ x_2 \\ x_3 \\ x_4 \\ x_5 \end{bmatrix} = \begin{bmatrix} 0 \\ 0 \\ 0 \\ 0 \end{bmatrix}.$$

What is the dimension of W?

In Exercises 23 and 24, find a basis for the solution space of the homogeneous system $(\lambda I_3 - A)X = 0$ for the given scalar λ and given matrix A.

23. $\lambda = 1$;

$$A = \begin{bmatrix} 0 & 0 & 1 \\ 1 & 0 & -3 \\ 0 & 1 & 3 \end{bmatrix}.$$

24. $\lambda = 3$;

$$A = \begin{bmatrix} 1 & 1 & -2 \\ -1 & 2 & 1 \\ 0 & 1 & -1 \end{bmatrix}.$$

THEORETICAL EXERCISES

T.1. Suppose that $S = \{\mathbf{X}_1, \mathbf{X}_2, \ldots, \mathbf{X}_n\}$ spans a vector space V and $\mathbf{X}_j$ is a linear combination of the preceding vectors in S. Prove that the set $S_1 = \{\mathbf{X}_1, \mathbf{X}_2, \ldots, \mathbf{X}_{j-1}, \mathbf{X}_{j+1}, \ldots, \mathbf{X}_n\}$, consisting of S with $\mathbf{X}_j$ deleted, also spans V.

T.2 Suppose that in the nonzero vector space V, the largest number of vectors in a linearly independent set is m. Show that any set of m linearly independent vectors in V is a basis for V.

T.3. Show that if V is a finite-dimensional vector space, then every nonzero subspace W of V has a finite basis and dim $W \leq$ dim V.

T.4. Prove that if dim $V = n$, then any $n + 1$ vectors in V are linearly dependent.

T.5. Prove that if dim $V = n$, then no set of $n - 1$ vectors in V can span V.

T.6. Prove Theorem 3.13.

T.7. Prove Theorem 3.14.

T.8. Prove that if W is a subspace of a finite-dimensional vector space V and dim $W =$ dim V, then $W = V$.

T.9. Classify all the subspaces of R^3.

T.10. Prove that if $\{\mathbf{X}_1, \mathbf{X}_2, \ldots, \mathbf{X}_n\}$ is a basis for a vector space V and $c \neq 0$, then

$$\{c\mathbf{X}_1, \mathbf{X}_2, \ldots, \mathbf{X}_n\}$$

is also a basis for V.

T.11. Let $S = \{\mathbf{X}_1, \mathbf{X}_2, \mathbf{X}_3\}$ be a basis for vector space V. Show that $T = \{\mathbf{Y}_1, \mathbf{Y}_2, \mathbf{Y}_3\}$, where $\mathbf{Y}_1 = \mathbf{X}_1 + \mathbf{X}_2 + \mathbf{X}_3$, $\mathbf{Y}_2 = \mathbf{X}_2 + \mathbf{X}_3$, and $\mathbf{Y}_3 = \mathbf{X}_3$, is also a basis for V.

3.7

THE RANK OF A MATRIX AND APPLICATIONS

In this section we obtain another effective method for finding a basis for a vector space V spanned by a given set of vectors $S = \{\mathbf{X}_1, \mathbf{X}_2, \ldots, \mathbf{X}_n\}$. In the proof to Theorem 3.11 we develop a technique for choosing a basis for V that is a subset of S. The method to be developed in this section produces a basis for V that is not guaranteed to be a subset of S. We also obtain information about the existence of nontrivial solutions to a homogeneous system.

Definition Let

$$\mathbf{A} = \begin{bmatrix} a_{11} & a_{12} & \cdots & a_{1n} \\ a_{21} & a_{22} & \cdots & a_{2n} \\ \vdots & \vdots & & \vdots \\ a_{m1} & a_{m2} & \cdots & a_{mn} \end{bmatrix}$$

be an $m \times n$ matrix. The rows of $\mathbf{A}$,

$$\mathbf{X}_1 = (a_{11}, a_{12}, \ldots, a_{1n})$$
$$\mathbf{X}_2 = (a_{21}, a_{22}, \ldots, a_{2n})$$
$$\vdots$$
$$\mathbf{X}_m = (a_{m1}, a_{m2}, \ldots, a_{mn})$$

considered as vectors in R^n, span a subspace of R^n, called the **row space** of $\mathbf{A}$. Similarly, the columns of $\mathbf{A}$,

$$Y_1 = \begin{bmatrix} a_{11} \\ a_{21} \\ \vdots \\ a_{m1} \end{bmatrix}, \quad Y_2 = \begin{bmatrix} a_{12} \\ a_{22} \\ \vdots \\ a_{m2} \end{bmatrix}, \quad \ldots, \quad Y_n = \begin{bmatrix} a_{1n} \\ a_{2n} \\ \vdots \\ a_{mn} \end{bmatrix}$$

considered as vectors in R^m, span a subspace of R^m, called the **column space** of **A**.

Theorem 3.15   *If* **A** *and* **B** *are two* $m \times n$ *row equivalent matrices, then the row spaces of* **A** *and* **B** *are equal.*

Proof

If **A** and **B** are row equivalent, then the rows of **B** are obtained from those of **A** by a finite number of the three elementary row operations. Thus each row of **B** is a linear combination of the rows of **A**. Hence the row space of **B** is contained in the row space of **A**. Similarly, **A** can be obtained from **B** by a finite number of the three elementary row operations, so the row space of **A** is contained in the row space of **B**. Hence the row spaces of **A** and **B** are equal.

It follows from Theorem 3.15 that if we take a given matrix **A** and transform it to reduced row echelon form **B**, then the row spaces of **A** and **B** are equal. Furthermore, it is not difficult to show (Exercise T.1) that the nonzero rows of a matrix that is in reduced row echelon form are a basis for its row space. We can use this method to find a basis for a vector space spanned by a given set of vectors in R^n, as illustrated in Example 1.

Example 1   Let

$$S = \{(1, -2, 0, 3, -4), (3, 2, 8, 1, 4), (2, 3, 7, 2, 3), (-1, 2, 0, 4, -3)\},$$

and let V be the subspace of R^5 given by $V = \text{span } S$. Find a basis for V.

Solution

Note that V is the row space of the matrix **A** whose rows are the given vectors

$$A = \begin{bmatrix} 1 & -2 & 0 & 3 & -4 \\ 3 & 2 & 8 & 1 & 4 \\ 2 & 3 & 7 & 2 & 3 \\ -1 & 2 & 0 & 4 & -3 \end{bmatrix}.$$

Using elementary row operations, we find that **A** is row equivalent to the matrix (verify)

$$B = \begin{bmatrix} 1 & 0 & 2 & 0 & 1 \\ 0 & 1 & 1 & 0 & 1 \\ 0 & 0 & 0 & 1 & -1 \\ 0 & 0 & 0 & 0 & 0 \end{bmatrix},$$

which is in reduced row echelon form. The row spaces of $\mathbf{A}$ and $\mathbf{B}$ are identical, and a basis for the row space of $\mathbf{B}$ consists of the nonzero rows of $\mathbf{B}$. Hence

$$(1, 0, 2, 0, 1), \qquad (0, 1, 1, 0, 1), \qquad \text{and} \qquad (0, 0, 0, 1, -1)$$

form a basis for V.

We may now summarize the method used in Example 1 to find a basis for the subspace V of R^n given by $V = \text{span } S$, where S is a set of vectors in R^n.

The procedure for finding a basis for the subspace V of R^n given by $V = \text{span } S$, where $S = \{\mathbf{X}_1, \mathbf{X}_2, \ldots, \mathbf{X}_k\}$ is a set of vectors in R^n that are given in row form, is as follows.

STEP 1. Form the matrix

$$\mathbf{A} = \begin{bmatrix} \mathbf{X}_1 \\ \mathbf{X}_2 \\ \vdots \\ \mathbf{X}_k \end{bmatrix}$$

whose rows are the given vectors in S.

STEP 2. Transform $\mathbf{A}$ to reduced row echelon form, obtaining the matrix $\mathbf{B}$.

STEP 3. The nonzero rows of $\mathbf{B}$ form a basis for V.

Of course, the basis that has been obtained by the procedure used in Example 1 produced a basis that is not a subset of the given spanning set. The method used in Example 5 of Section 3.6 does produce a basis that is a subset of the spanning set. However, the basis for a subspace V of R^n that is obtained by the procedure used in Example 1 is analogous, in its simplicity, to the standard basis for R^n. Thus if $\mathbf{X} = (a_1, a_2, \ldots, a_u)$ is a vector in V and $\{\mathbf{X}_1, \mathbf{X}_2, \ldots, \mathbf{X}_k\}$ is a basis for V obtained by the method of Example 1 where the leading 1s occur in columns $j_1, j_2, \ldots, j_k$, then it can be shown that

$$\mathbf{X} = a_{j_1}\mathbf{X}_1 + a_{j_2}\mathbf{X}_2 + \cdots + a_{j_k}\mathbf{X}_k.$$

Thus, let V be the subspace of Example 1 and suppose that the vector $\mathbf{X} = (5, 4, 14, 6, 3)$ is in V. To write $\mathbf{X}$ as a linear combination of the basis determined in Example 1, we have $j_1 = 1, j_2 = 2$, and $j_3 = 4$, so

$$\mathbf{X} = (5, 4, 14, 6, 3) = 5(1, 0, 2, 0, 1) + 4(0, 1, 1, 0, 1) + 6(0, 0, 0, 1, -1).$$

Definition   The dimension of the row space of $\mathbf{A}$ is called the **row rank** of $\mathbf{A}$, and the dimension of the column space of $\mathbf{A}$ is called the **column rank** of $\mathbf{A}$.

Example 2 The matrix **A** in Example 1 has row rank equal to 3.

Example 3 Find the column rank of the matrix **A** in Example 1.

Solution

We must find the dimension of the column space of **A**. That is, we must find the dimension of the subspace of R^4 spanned by the columns of **A**. If we write the columns of **A** as row vectors, we obtain the matrix

$$\mathbf{A}^T = \begin{bmatrix} 1 & 3 & 2 & -1 \\ -2 & 2 & 3 & 2 \\ 0 & 8 & 7 & 0 \\ 3 & 1 & 2 & 4 \\ -4 & 4 & 3 & -3 \end{bmatrix}.$$

Transforming $\mathbf{A}^T$ to reduced row echelon form, we obtain (verify)

$$\mathbf{C} = \begin{bmatrix} 1 & 0 & 0 & \frac{11}{24} \\ 0 & 1 & 0 & -\frac{49}{24} \\ 0 & 0 & 1 & \frac{7}{3} \\ 0 & 0 & 0 & 0 \\ 0 & 0 & 0 & 0 \end{bmatrix}.$$

Thus the vectors $(1, 0, 0, \frac{11}{24})$, $(0, 1, 0, -\frac{49}{24})$, and $(0, 0, 1, \frac{7}{3})$ form a basis for the row space of $\mathbf{A}^T$. Hence the vectors

$$\begin{bmatrix} 1 \\ 0 \\ 0 \\ \frac{11}{24} \end{bmatrix}, \quad \begin{bmatrix} 0 \\ 1 \\ 0 \\ -\frac{49}{24} \end{bmatrix}, \quad \text{and} \quad \begin{bmatrix} 0 \\ 0 \\ 1 \\ \frac{7}{3} \end{bmatrix}$$

form a basis for the column space of **A** and we conclude that the column rank of **A** is 3.

REMARK The method used in the solution of Example 3 shows how to modify the procedure above for finding a basis for the subspace $V = \text{span } S$ when S is a set of vectors in R^n that are given in column form.

Example 4 Let

$$S = \left\{ \begin{bmatrix} 1 \\ -2 \\ 1 \end{bmatrix}, \begin{bmatrix} -1 \\ 1 \\ 1 \end{bmatrix}, \begin{bmatrix} 1 \\ -3 \\ 3 \end{bmatrix}, \begin{bmatrix} 3 \\ -5 \\ 1 \end{bmatrix}, \begin{bmatrix} 1 \\ -4 \\ 5 \end{bmatrix} \right\},$$

and let V be the subspace of R^3 given by $V = \text{span } S$. Find a basis for V.

Solution

STEP 1. Form the matrix

$$\mathbf{A}^T = \begin{bmatrix} 1 & -2 & 1 \\ -1 & 1 & 1 \\ 1 & -3 & 3 \\ 3 & -5 & 1 \\ 1 & -4 & 5 \end{bmatrix}$$

whose columns are the given vectors in S written in row form.

STEP 2. Transforming $\mathbf{A}^T$ to reduced row echelon form, we obtain (verify) the matrix

$$\mathbf{B}^T = \begin{bmatrix} 1 & 0 & -3 \\ 0 & 1 & -2 \\ 0 & 0 & 0 \\ 0 & 0 & 0 \\ 0 & 0 & 0 \end{bmatrix}.$$

STEP 3. The nonzero rows of $\mathbf{B}^T$, written as columns

$$\begin{bmatrix} 1 \\ 0 \\ -3 \end{bmatrix} \quad \text{and} \quad \begin{bmatrix} 0 \\ 1 \\ -2 \end{bmatrix},$$

form a basis for V.

From Examples 1 through 4 we observe that the row and column ranks of $\mathbf{A}$ are equal. This is always true, and is a very important result in linear algebra. We now turn to the proof of this theorem.

Theorem 3.16   *The row rank and column rank of the $m \times n$ matrix $\mathbf{A} = [a_{ij}]$ are equal.*

Proof

Let $\mathbf{X}_1, \mathbf{X}_2, \ldots, \mathbf{X}_m$ be the row vectors of $\mathbf{A}$, where

$$\mathbf{X}_i = [a_{i1} \quad a_{i2} \quad \cdots \quad a_{in}] \quad (1 \le i \le m).$$

Let row rank $\mathbf{A} = k$ and let the set of vectors $\{\mathbf{Y}_1, \mathbf{Y}_2, \ldots, \mathbf{Y}_k\}$ form a basis for the row space of $\mathbf{A}$, where $\mathbf{Y}_i = [b_{i1} \quad b_{i2} \quad \cdots \quad b_{in}]$ for $i = 1, 2, \ldots, k$.

Now each of the row vectors is a linear combination of $\mathbf{Y}_1, \mathbf{Y}_2, \ldots, \mathbf{Y}_k$:

$$\mathbf{X}_1 = r_{11}\mathbf{Y}_1 + r_{12}\mathbf{Y}_2 + \cdots + r_{1k}\mathbf{Y}_k$$
$$\mathbf{X}_2 = r_{21}\mathbf{Y}_1 + r_{22}\mathbf{Y}_2 + \cdots + r_{2k}\mathbf{Y}_k$$
$$\vdots \qquad \vdots \qquad \vdots \qquad \qquad \vdots$$
$$\mathbf{X}_m = r_{m1}\mathbf{Y}_1 + r_{m2}\mathbf{Y}_2 + \cdots + r_{mk}\mathbf{Y}_k,$$

where the r_{ij} are uniquely determined real numbers. Recalling that two matrices are equal if their corresponding entries are equal, we equate the entries of these vector equations to get

$$
\begin{aligned}
a_{1j} &= r_{11}b_{1j} + r_{12}b_{2j} + \cdots + r_{1k}b_{kj} \\
a_{2j} &= r_{21}b_{1j} + r_{22}b_{2j} + \cdots + r_{2k}b_{kj} \\
&\ \ \vdots \qquad\quad \vdots \qquad\quad \vdots \qquad\qquad \vdots \\
a_{mj} &= r_{m1}b_{1j} + r_{m2}b_{2j} + \cdots + r_{mk}b_{kj}
\end{aligned}
$$

or

$$
\begin{bmatrix} a_{1j} \\ a_{2j} \\ \vdots \\ a_{mj} \end{bmatrix} = b_{1j}\begin{bmatrix} r_{11} \\ r_{21} \\ \vdots \\ r_{m1} \end{bmatrix} + b_{2j}\begin{bmatrix} r_{12} \\ r_{22} \\ \vdots \\ r_{m2} \end{bmatrix} + \cdots + b_{kj}\begin{bmatrix} r_{1k} \\ r_{2k} \\ \vdots \\ r_{mk} \end{bmatrix}
$$

for $j = 1, 2, \ldots, n$.

Since every column of $\mathbf{A}$ is a linear combination of k vectors, the dimension of the column space of $\mathbf{A}$ is at most k, or column rank $\mathbf{A} \leq k =$ row rank $\mathbf{A}$. Similarly, we get row rank $\mathbf{A} \leq$ column rank $\mathbf{A}$. Hence the row and column rank of $\mathbf{A}$ are equal.

Definition We now merely refer to the **rank** of an $m \times n$ matrix, and write rank $\mathbf{A}$.

We next summarize the procedure for computing the rank of a matrix,

---

The procedure for computing the rank of the matrix $\mathbf{A}$ is as follows.

STEP 1. Using elementary row operations, transform $\mathbf{A}$ to a matrix $\mathbf{B}$ in reduced row echelon form.

STEP 2. Rank $\mathbf{A} =$ the number of nonzero rows of $\mathbf{B}$.

For a square matrix, its rank can be used to determine whether the matrix is singular or nonsingular, as the following theorem shows.

Theorem 3.17 *An $n \times n$ matrix is nonsingular if and only if rank $\mathbf{A} = n$.*

Proof
Suppose that $\mathbf{A}$ is nonsingular. Then $\mathbf{A}$ is row equivalent to $\mathbf{I}_n$ (Theorem 1.11, Section 1.5), and so rank $\mathbf{A} = n$.

Conversely, if rank $\mathbf{A} = n$, then $\mathbf{A}$ is row equivalent to $\mathbf{I}_n$ and so $\mathbf{A}$ is nonsingular (Theorem 1.11, Section 1.5).

From a practical point of view, this result is not too useful, since most of the time we want to know not only whether $\mathbf{A}$ is nonsingular, but also its inverse. The method developed in Chapter 1 enables us to find $\mathbf{A}^{-1}$, if it exists, and tells us if it does not exist. Thus we do not have to learn first if $\mathbf{A}^{-1}$ exists, and then go through another procedure to obtain it.

An immediate consequence of Theorem 3.17 is the following corollary, which is a criterion for the rank of an $n \times n$ matrix to be n.

Corollary 3.2 *If $\mathbf{A}$ is an $n \times n$ matrix, then rank $\mathbf{A} = n$ if and only if $|\mathbf{A}| \neq 0$.*

Proof
Exercise T.2.

The following corollary gives another method of testing whether n given vectors in R^n are linearly dependent or linearly independent.

Corollary 3.3 *Let $S = \{\mathbf{X}_1, \mathbf{X}_2, \ldots, \mathbf{X}_n\}$ be a set of n vectors in R^n and let $\mathbf{A}$ be the matrix whose rows (columns) are the vectors in S. Then S is linearly independent if and only if $|\mathbf{A}| \neq 0$.*

Proof
Exercise T.3.

For $n > 4$, the method of Corollary 3.3 for testing linear dependence is not as efficient as the direct method of Section 3.5, calling for the solution of a linear system.

Corollary 3.4 *The homogeneous system $\mathbf{AX} = \mathbf{0}$ of n linear equations in n unknowns has a nontrivial solution if and only if rank $\mathbf{A} < n$.*

Proof
This follows from Corollary 3.2 and from the fact that $\mathbf{AX} = \mathbf{0}$ has a nontrivial solution if and only if $\mathbf{A}$ is singular (Theorem 1.12, Section 1.5).

Example 5 Let

$$\mathbf{A} = \begin{bmatrix} 1 & 2 & 0 \\ 0 & 1 & 3 \\ 2 & 1 & 3 \end{bmatrix}.$$

If we transform $\mathbf{A}$ to reduced row echelon form $\mathbf{B}$, we find that $\mathbf{B} = \mathbf{I}_3$ (verify). Thus rank $\mathbf{A} = 3$ and $\mathbf{A}$ is nonsingular. Moreover, the homogeneous system $\mathbf{AX} = \mathbf{0}$ has only the trivial solution.

Example 6 Let

$$\mathbf{A} = \begin{bmatrix} 1 & 2 & 0 \\ 1 & 1 & -3 \\ 1 & 3 & 3 \end{bmatrix}.$$

Then $\mathbf{A}$ is row equivalent to

$$\begin{bmatrix} 1 & 0 & -6 \\ 0 & 1 & 3 \\ 0 & 0 & 0 \end{bmatrix},$$

a matrix in reduced row echelon form. Hence rank $\mathbf{A} < 3$, and $\mathbf{A}$ is singular. Moreover, $\mathbf{AX} = \mathbf{0}$ has a nontrivial solution.

We may summarize our results on homogeneous systems, nonsingular matrices, and rank by noting that the following statements are equivalent for an $n \times n$ matrix $\mathbf{A}$.

1. $\mathbf{A}$ is nonsingular.
2. $\mathbf{AX} = \mathbf{0}$ has only the trivial solution.
3. $\mathbf{A}$ is row equivalent to $\mathbf{I}_n$.
4. The linear system $\mathbf{AX} = \mathbf{B}$ has a unique solution for every $n \times 1$ matrix $\mathbf{B}$.
5. $|\mathbf{A}| \neq 0$.
6. $\mathbf{A}$ has rank n.
7. The rows (columns) of $\mathbf{A}$ form a linearly independent set of n vectors in R^n.

Our final application of rank is to linear systems. If we consider the linear system $\mathbf{AX} = \mathbf{B}$ of m linear equations in n unknowns, where $\mathbf{A} = [a_{ij}]$, then we observe that the system can also be written as the equation

$$x_1 \begin{bmatrix} a_{11} \\ a_{21} \\ \vdots \\ a_{m1} \end{bmatrix} + x_2 \begin{bmatrix} a_{12} \\ a_{22} \\ \vdots \\ a_{m2} \end{bmatrix} + \cdots + x_n \begin{bmatrix} a_{1n} \\ a_{2n} \\ \vdots \\ a_{mn} \end{bmatrix} = \begin{bmatrix} b_1 \\ b_2 \\ \vdots \\ b_m \end{bmatrix}.$$

Thus $\mathbf{AX} = \mathbf{B}$ has a solution if and only if $\mathbf{B}$ is a linear combination of the columns of $\mathbf{A}$; that is, if and only if $\mathbf{B}$ belongs to the column space of $\mathbf{A}$. This means that if $\mathbf{AX} = \mathbf{B}$ has a solution, then rank $\mathbf{A} = $ rank $[\mathbf{A} \mid \mathbf{B}]$, where $[\mathbf{A} \mid \mathbf{B}]$ is the augmented matrix of the system. Conversely, if rank $[\mathbf{A} \mid \mathbf{B}] = $ rank $\mathbf{A}$, then $\mathbf{B}$ is in the column space of $\mathbf{A}$, which means that the system has a solution. Therefore, $\mathbf{AX} = \mathbf{B}$ has a solution if and only if rank $\mathbf{A} = $ rank $[\mathbf{A} \mid \mathbf{B}]$. This result, while of interest, is not of great computational value, since we usually are interested in finding a solution rather than in knowing whether or not a solution exists.

Example 7 Consider the linear system

$$\begin{bmatrix} 2 & 1 & 3 \\ 1 & -2 & 2 \\ 0 & 1 & 3 \end{bmatrix} \begin{bmatrix} x_1 \\ x_2 \\ x_3 \end{bmatrix} = \begin{bmatrix} 1 \\ 2 \\ 3 \end{bmatrix}.$$

Since rank $\mathbf{A} = $ rank $[\mathbf{A} \mid \mathbf{B}] = 3$, the linear system has a solution.

Example 8 The linear system

$$\begin{bmatrix} 1 & 2 & 3 \\ 1 & -3 & 4 \\ 2 & -1 & 7 \end{bmatrix} \begin{bmatrix} x_1 \\ x_2 \\ x_3 \end{bmatrix} = \begin{bmatrix} 4 \\ 5 \\ 6 \end{bmatrix}$$

has no solution, because rank $\mathbf{A} = 2$ and rank $[\mathbf{A} \mid \mathbf{B}] = 3$.

3.7 EXERCISES

1. Let $S = \{\mathbf{X}_1, \mathbf{X}_2, \mathbf{X}_3, \mathbf{X}_4, \mathbf{X}_5\}$, where $\mathbf{X}_1 = (1, 2, 3)$, $\mathbf{X}_2 = (2, 1, 4)$, $\mathbf{X}_3 = (-1, -1, 2)$, $\mathbf{X}_4 = (0, 1, 2)$, and $\mathbf{X}_5 = (1, 1, 1)$. Find a basis for the subspace $V = \text{span } S$ of R^3.

2. Let $S = \{\mathbf{X}_1, \mathbf{X}_2, \mathbf{X}_3, \mathbf{X}_4, \mathbf{X}_5\}$, where $\mathbf{X}_1 = (1, 1, 2, 1)$, $\mathbf{X}_2 = (1, 0, -3, 1)$, $\mathbf{X}_3 = (0, 1, 1, 2)$, $\mathbf{X}_4 = (0, 0, 1, 1)$, and $\mathbf{X}_5 = (1, 0, 0, 1)$. Find a basis for the subspace $V = \text{span } S$ of R^4.

3. Let $S = \{\mathbf{X}_1, \mathbf{X}_2, \mathbf{X}_3, \mathbf{X}_4, \mathbf{X}_5\}$, where $\mathbf{X}_1 = \begin{bmatrix} 1 \\ 2 \\ 1 \\ 2 \end{bmatrix}$,

$$\mathbf{X}_2 = \begin{bmatrix} 2 \\ 1 \\ 2 \\ 1 \end{bmatrix}, \mathbf{X}_3 = \begin{bmatrix} 3 \\ 2 \\ 3 \\ 2 \end{bmatrix}, \mathbf{X}_4 = \begin{bmatrix} 3 \\ 3 \\ 3 \\ 3 \end{bmatrix},$$

and

$$\mathbf{X}_5 = \begin{bmatrix} 5 \\ 3 \\ 5 \\ 3 \end{bmatrix}.$$

Find a basis for the subspace $V = \text{span } S$ of R^4.

4. Let $S = \{\mathbf{X}_1, \mathbf{X}_2, \mathbf{X}_3, \mathbf{X}_4, \mathbf{X}_5\}$, where $\mathbf{X}_1 = \begin{bmatrix} 1 \\ 2 \\ 1 \\ 1 \end{bmatrix}$,

$$\mathbf{X}_2 = \begin{bmatrix} 2 \\ 1 \\ 3 \\ 1 \end{bmatrix}, \mathbf{X}_3 = \begin{bmatrix} 0 \\ 2 \\ 1 \\ 2 \end{bmatrix}, \mathbf{X}_4 = \begin{bmatrix} 3 \\ 2 \\ 1 \\ 4 \end{bmatrix},$$

and

$$\mathbf{X}_5 = \begin{bmatrix} 5 \\ 0 \\ 0 \\ -1 \end{bmatrix}.$$

Find a basis for the subspace $V = \text{span } S$ of R^4.

5. Find the row and column ranks of

$$\mathbf{A} = \begin{bmatrix} 1 & 2 & 3 & 2 & 1 \\ 3 & 1 & -5 & -2 & 1 \\ 7 & 8 & -1 & 2 & 5 \end{bmatrix}.$$

6. Find the row and column ranks of

$$\mathbf{A} = \begin{bmatrix} 1 & 3 & 2 & 0 & 0 & 1 \\ 2 & 1 & -5 & 1 & 2 & 0 \\ 3 & 2 & 5 & 1 & -2 & 1 \\ 5 & 8 & 9 & 1 & -2 & 2 \\ 9 & 9 & 4 & 2 & 0 & 2 \end{bmatrix}.$$

In Exercises 7 through 11, compute the rank of each matrix.

7. $\begin{bmatrix} 1 & 2 & 1 & 3 \\ 2 & 1 & -4 & -5 \\ 7 & 8 & -5 & -1 \\ 10 & 14 & -2 & 8 \end{bmatrix}.$

8. $\begin{bmatrix} 1 & 2 & 1 & 3 \\ 2 & 1 & -4 & -5 \\ 1 & 1 & 0 & 0 \\ 0 & 0 & 1 & 1 \end{bmatrix}.$

9. $\begin{bmatrix} 1 & 2 & 3 \\ -1 & 2 & 1 \\ 3 & 1 & 2 \end{bmatrix}.$

10. $\begin{bmatrix} 1 & -2 & -1 \\ 2 & -1 & 3 \\ 7 & -8 & 3 \end{bmatrix}$.

11. $\begin{bmatrix} 1 & -2 & -1 \\ 2 & -1 & 3 \\ 7 & -8 & 3 \\ 5 & -7 & 0 \end{bmatrix}$.

12. If A is a 3×4 matrix, what is the largest possible value for rank A?

13. If A is a 4×6 matrix, show that the columns of A are linearly dependent.

14. If A is a 5×3 matrix, show that the rows of A are linearly dependent.

In Exercises 15 through 17, use Theorem 3.17 to determine whether each matrix is singular or nonsingular.

15. $\begin{bmatrix} 1 & 2 & -3 \\ -1 & 2 & 3 \\ 0 & 8 & 0 \end{bmatrix}$.

16. $\begin{bmatrix} 1 & 2 & -3 \\ -1 & 2 & 3 \\ 0 & 1 & 1 \end{bmatrix}$.

17. $\begin{bmatrix} 1 & 1 & 4 & -1 \\ 1 & 2 & 3 & 2 \\ -1 & 3 & 2 & 1 \\ -2 & 6 & 12 & -4 \end{bmatrix}$.

Use Corollary 3.3 to do Exercises 18 and 19.

18. Is

$$S = \left\{ \begin{bmatrix} 2 \\ 2 \\ 3 \end{bmatrix}, \begin{bmatrix} 1 \\ 0 \\ 2 \end{bmatrix}, \begin{bmatrix} 0 \\ 1 \\ 3 \end{bmatrix} \right\}$$

a linearly independent set of vectors in R^3?

19. Is

$$S = \left\{ \begin{bmatrix} 4 \\ 1 \\ 2 \end{bmatrix}, \begin{bmatrix} 2 \\ 5 \\ -5 \end{bmatrix}, \begin{bmatrix} 2 \\ -1 \\ 3 \end{bmatrix} \right\}$$

a linear independent set of vectors in R^3?

In Exercises 20 through 22, find which homogeneous systems have a nontrivial solution for the given matrix A by using Corollary 3.4.

20. $A = \begin{bmatrix} 1 & 1 & 2 & -1 \\ 1 & 3 & -1 & 2 \\ 1 & 1 & 1 & 3 \\ 1 & 2 & 1 & 1 \end{bmatrix}$.

21. $A = \begin{bmatrix} 1 & 2 & 3 \\ 0 & 1 & 0 \\ 1 & 0 & 3 \end{bmatrix}$.

22. $A = \begin{bmatrix} 1 & 2 & -1 \\ 2 & -1 & 3 \\ 5 & -4 & 3 \end{bmatrix}$.

In Exercises 23 through 26, determine which of the linear systems have a solution by comparing the ranks of the coefficient and augmented matrices.

23. $\begin{bmatrix} 1 & 2 & 5 & -2 \\ 2 & 3 & -2 & 4 \\ 5 & 1 & 0 & 2 \end{bmatrix} \begin{bmatrix} x_1 \\ x_2 \\ x_3 \\ x_4 \end{bmatrix} = \begin{bmatrix} 0 \\ 0 \\ 0 \end{bmatrix}$.

24. $\begin{bmatrix} 1 & 2 & 5 & -2 \\ 2 & 3 & -2 & 4 \\ 5 & 1 & 0 & 2 \end{bmatrix} \begin{bmatrix} x_1 \\ x_2 \\ x_3 \\ x_4 \end{bmatrix} = \begin{bmatrix} 1 \\ -13 \\ 3 \end{bmatrix}$.

25. $\begin{bmatrix} 1 & -2 & -3 & 4 \\ 4 & -1 & -5 & 6 \\ 2 & 3 & 1 & -2 \end{bmatrix} \begin{bmatrix} x_1 \\ x_2 \\ x_3 \\ x_4 \end{bmatrix} = \begin{bmatrix} 1 \\ 2 \\ 2 \end{bmatrix}$.

26. $\begin{bmatrix} 1 & 1 & 1 \\ 1 & -1 & 1 \\ 5 & 1 & 5 \end{bmatrix} \begin{bmatrix} x_1 \\ x_2 \\ x_3 \end{bmatrix} = \begin{bmatrix} 6 \\ 2 \\ 5 \end{bmatrix}$.

THEORETICAL EXERCISES _____

T.1. Show that the nonzero rows of a matrix that is in reduced row echelon form are a basis for its row space.

T.2. Prove Corollary 3.2.

T.3. Prove Corollary 3.3.

T.4. Let **A** be an $n \times n$ matrix. Show that the homogeneous system $\mathbf{AX} = \mathbf{0}$ has a nontrivial solution if and only if the columns of **A** are linearly dependent.

T.5. Let **A** be an $n \times n$ matrix. Show that
$$\text{rank } \mathbf{A} = n,$$
if and only if the columns of **A** are linearly independent.

T.6. Let **A** be an $n \times n$ matrix. Prove that the rows of **A** are linearly independent if and only if the columns of **A** span R^n.

T.7. Let **A** be an $m \times n$ matrix. Show that the linear system $\mathbf{AX} = \mathbf{B}$ has a solution for every $m \times 1$ matrix **B** if and only if rank $\mathbf{A} = m$.

3.8 _____

ORTHONORMAL BASES IN R^n

A given subspace W of R^n need not contain any of the natural basis vectors, but we want to show that it has a basis with the same properties. That is, we want to show that W contains a basis S such that every vector in S is of unit length and every two vectors in S are orthogonal. The method used to obtain such a basis is the Gram–Schmidt process.

Definition A set $S = \{\mathbf{X}_1, \mathbf{X}_2, \ldots, \mathbf{X}_k\}$ in R^n is called **orthogonal** if any two distinct vectors in S are orthogonal, that is, if $\mathbf{X}_i \cdot \mathbf{X}_j = 0$ for $i \neq j$. An **orthonormal** set of vectors is an orthogonal set of unit vectors.

Example 1 If $\mathbf{X}_1 = (1, 0, 2)$, $\mathbf{X}_2 = (-2, 0, 1)$, and $\mathbf{X}_3 = (0, 1, 0)$, then $\{\mathbf{X}_1, \mathbf{X}_2, \mathbf{X}_3\}$ is an orthogonal set in R^3. The vectors

$$\mathbf{Y}_1 = \left(\frac{1}{\sqrt{5}}, 0, \frac{2}{\sqrt{5}} \right) \quad \text{and} \quad \mathbf{Y}_2 = \left(\frac{-2}{\sqrt{5}}, 0, \frac{1}{\sqrt{5}} \right)$$

are unit vectors in the directions of $\mathbf{X}_1$ and $\mathbf{X}_2$, respectively. Since $\mathbf{X}_3$ is also of unit length, it follows that $\{\mathbf{Y}_1, \mathbf{Y}_2, \mathbf{X}_3\}$ is an orthonormal set. Also, span $\{\mathbf{X}_1, \mathbf{X}_2, \mathbf{X}_3\}$ is the same as span $\{\mathbf{Y}_1, \mathbf{Y}_2, \mathbf{X}_3\}$.

Example 2 The natural basis

$$\{(1, 0, 0), (0, 1, 0), (0, 0, 1)\}$$

is an orthonormal set in R^3. More generally, the natural basis in R^n is an orthonormal set.

An important result about orthogonal sets is the following theorem.

Theorem 3.18 *Let $S = \{\mathbf{X}_1, \mathbf{X}_2, \ldots, \mathbf{X}_k\}$ be an orthogonal set of nonzero vectors in R^n. Then S is linearly independent.*

Proof

Consider the equation

$$c_1\mathbf{X}_1 + c_2\mathbf{X}_2 + \cdots + c_k\mathbf{X}_k = \mathbf{0}. \tag{1}$$

Taking the inner product of both sides of (1) with $\mathbf{X}_i$, $1 \leq i \leq k$, we have

$$(c_1\mathbf{X}_1 + c_2\mathbf{X}_2 + \cdots + c_k\mathbf{X}_k) \cdot \mathbf{X}_i = \mathbf{0} \cdot \mathbf{X}_i. \tag{2}$$

By properties (c) and (d) of Theorem 3.3, Section 3.2, the left side of (2) is

$$c_1(\mathbf{X}_1 \cdot \mathbf{X}_i) + c_2(\mathbf{X}_2 \cdot \mathbf{X}_i) + \cdots + c_k(\mathbf{X}_k \cdot \mathbf{X}_i),$$

and the right side is 0. Since $\mathbf{X}_j \cdot \mathbf{X}_i = 0$ if $i \neq j$, (2) becomes

$$0 = c_i(\mathbf{X}_i \cdot \mathbf{X}_i) = c_i\|\mathbf{X}_i\|^2. \tag{3}$$

By (a) of Theorem 3.3, Section 3.2, $\|\mathbf{X}_i\| \neq 0$, since $\mathbf{X}_i \neq \mathbf{0}$. Hence (3) implies that $c_i = 0$, $1 \leq i \leq k$, and S is linearly independent.

Corollary 3.5 *An orthonormal set of vectors in R^n is linearly independent.*

Proof

Exercise T.2.

Definition From Theorem 3.14 of Section 3.6, it follows that an orthonormal set of n vectors in R^n is a basis for R^n (Exercise T.3). An **orthogonal (orthonormal) basis** for a vector space is a basis that is an orthogonal (orthonormal) set.

Theorem 3.19 *Let W be a nonzero subspace of R^n with basis $S = \{\mathbf{X}_1, \mathbf{X}_2, \ldots, \mathbf{X}_m\}$. Then there exists*
(Gram†– *an orthonormal basis $T = \{\mathbf{Z}_1, \mathbf{Z}_2, \ldots, \mathbf{Z}_m\}$ for W.*
Schmidt‡

Process) Proof

The proof is constructive; that is, we develop the desired basis T. However, we first find an orthogonal basis $T^* = \{\mathbf{Y}_1, \mathbf{Y}_2, \ldots, \mathbf{Y}_m\}$ for W.

First, we pick any one of the vectors in S, say $\mathbf{X}_1$, and call it $\mathbf{Y}_1$. Thus $\mathbf{Y}_1 = \mathbf{X}_1$. We now look for a vector $\mathbf{Y}_2$ in the subspace W_1 of W spanned by $\{\mathbf{X}_1, \mathbf{X}_2\}$, which is orthogonal to $\mathbf{Y}_1$. Since $\mathbf{Y}_1 = \mathbf{X}_1$, W_1 is also the subspace spanned by $\{\mathbf{Y}_1, \mathbf{X}_2\}$. Thus

$$\mathbf{Y}_2 = c_1\mathbf{Y}_1 + c_2\mathbf{X}_2.$$

---

† Jörgen Pederson Gram (1850–1916) was a Danish actuary.
‡ Erhard Schmidt (1876–1959) taught at several leading German Universities and was a student of both Hermann Amandus Schwarz and David Hilbert. He made important contributions to the study of integral equations and partial differential equations and, as part of this study, he introduced the method for finding an orthonormal basis in 1907. In 1908 he wrote a paper on infinitely many linear equations in infinitely many unknowns, in which he founded the theory of Hilbert spaces and in which he again used his method.

We try to determine c_1 and c_2 so that $\mathbf{Y}_1 \cdot \mathbf{Y}_2 = 0$. Now

$$\begin{aligned} 0 = \mathbf{Y}_2 \cdot \mathbf{Y}_1 &= (c_1 \mathbf{Y}_1 + c_2 \mathbf{X}_2) \cdot \mathbf{Y}_1 \\ &= c_1(\mathbf{Y}_1 \cdot \mathbf{Y}_1) + c_2(\mathbf{X}_2 \cdot \mathbf{Y}_1). \end{aligned} \tag{4}$$

Since $\mathbf{Y}_1 \neq \mathbf{0}$ (why?), $\mathbf{Y}_1 \cdot \mathbf{Y}_1 \neq 0$, and solving for c_1 and c_2 in (4), we have

$$c_1 = -c_2 \frac{\mathbf{X}_2 \cdot \mathbf{Y}_1}{\mathbf{Y}_1 \cdot \mathbf{Y}_1}.$$

We may assign an arbitrary nonzero value to c_2. Thus, letting $c_2 = 1$, we obtain

$$c_1 = -\frac{\mathbf{X}_2 \cdot \mathbf{Y}_1}{\mathbf{Y}_1 \cdot \mathbf{Y}_1}.$$

Hence

$$\mathbf{Y}_2 = c_1 \mathbf{Y}_1 + c_2 \mathbf{X}_2 = \mathbf{X}_2 - \left(\frac{\mathbf{X}_2 \cdot \mathbf{Y}_1}{\mathbf{Y}_1 \cdot \mathbf{Y}_1} \right) \mathbf{Y}_1.$$

Notice that at this point we have an orthogonal subset $\{\mathbf{Y}_1, \mathbf{Y}_2\}$ of W (see Figure 3.41).

Next, we look for a vector $\mathbf{Y}_3$ in the subspace W_2 of W spanned by $\{\mathbf{X}_1, \mathbf{X}_2, \mathbf{X}_3\}$, which is orthogonal to both $\mathbf{Y}_1$ and $\mathbf{Y}_2$. Of course, W_2 is also the subspace spanned by $\{\mathbf{Y}_1, \mathbf{Y}_2, \mathbf{X}_3\}$ (why?). Thus

$$\mathbf{Y}_3 = d_1 \mathbf{Y}_1 + d_2 \mathbf{Y}_2 + d_3 \mathbf{X}_3.$$

We let $d_3 = 1$ and try to find d_1 and d_2 so that

$$\mathbf{Y}_3 \cdot \mathbf{Y}_1 = 0 \qquad \text{and} \qquad \mathbf{Y}_3 \cdot \mathbf{Y}_2 = 0.$$

Now

$$0 = \mathbf{Y}_3 \cdot \mathbf{Y}_1 = (d_1 \mathbf{Y}_1 + d_2 \mathbf{Y}_2 + \mathbf{X}_3) \cdot \mathbf{Y}_1 = d_1(\mathbf{Y}_1 \cdot \mathbf{Y}_1) + \mathbf{X}_3 \cdot \mathbf{Y}_1, \tag{5}$$

$$0 = \mathbf{Y}_3 \cdot \mathbf{Y}_2 = (d_1 \mathbf{Y}_1 + d_2 \mathbf{Y}_2 + \mathbf{X}_3) \cdot \mathbf{Y}_2 = d_2(\mathbf{Y}_2 \cdot \mathbf{Y}_2) + \mathbf{X}_3 \cdot \mathbf{Y}_2, \tag{6}$$

In obtaining the right sides of (5) and (6), we have used the fact that $\mathbf{Y}_1 \cdot \mathbf{Y}_2 = 0$. Now observe that $\mathbf{Y}_2 \neq \mathbf{0}$ (why?). Solving (5) and (6) for d_1 and d_2, respectively, we obtain

Figure 3.41

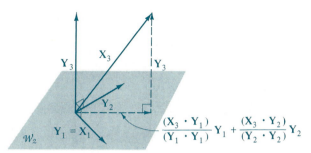

Figure 3.42

$$d_1 = -\frac{\mathbf{X}_3 \cdot \mathbf{Y}_1}{\mathbf{Y}_1 \cdot \mathbf{Y}_1} \quad \text{and} \quad d_2 = -\frac{\mathbf{X}_3 \cdot \mathbf{Y}_2}{\mathbf{Y}_2 \cdot \mathbf{Y}_2}.$$

Thus

$$\mathbf{Y}_3 = \mathbf{X}_3 - \left(\frac{\mathbf{X}_3 \cdot \mathbf{Y}_1}{\mathbf{Y}_1 \cdot \mathbf{Y}_1}\right)\mathbf{Y}_1 - \left(\frac{\mathbf{X}_3 \cdot \mathbf{Y}_2}{\mathbf{Y}_2 \cdot \mathbf{Y}_2}\right)\mathbf{Y}_2.$$

Notice that at this point we have an orthogonal subset $\{\mathbf{Y}_1, \mathbf{Y}_2, \mathbf{Y}_3\}$ of W (see Figure 3.42).

The Gram–Schmidt process for computing an orthonormal basis $T = \{\mathbf{Z}_1, \mathbf{Z}_2, \ldots, \mathbf{Z}_m\}$ for a nonzero subspace W of R^n with basis $S = \{\mathbf{X}_1, \mathbf{X}_2, \ldots, \mathbf{X}_m\}$ is as follows.

STEP 1. Let $\mathbf{Y}_1 = \mathbf{X}_1$.

STEP 2. Compute the vectors $\mathbf{Y}_2, \mathbf{Y}_3, \ldots, \mathbf{Y}_m$, successively, one at a time, by the formula

$$\mathbf{Y}_i = \mathbf{X}_i - \left(\frac{\mathbf{X}_i \cdot \mathbf{Y}_1}{\mathbf{Y}_1 \cdot \mathbf{Y}_1}\right)\mathbf{Y}_1 - \left(\frac{\mathbf{X}_i \cdot \mathbf{Y}_2}{\mathbf{Y}_2 \cdot \mathbf{Y}_2}\right)\mathbf{Y}_2 - \cdots - \left(\frac{\mathbf{X}_i \cdot \mathbf{Y}_{i-1}}{\mathbf{Y}_{i-1} \cdot \mathbf{Y}_{i-1}}\right)\mathbf{Y}_{i-1},$$

The set of vectors $T^* = \{\mathbf{Y}_1, \mathbf{Y}_2, \ldots, \mathbf{Y}_m\}$ is an orthogonal set.

STEP 3. Let

$$\mathbf{Z}_i = \frac{\mathbf{Y}_i}{\|\mathbf{Y}_i\|}, \qquad 1 \leq i \leq m$$

Then $T = \{\mathbf{Z}_1, \mathbf{Z}_2, \ldots, \mathbf{Z}_m\}$ is an orthonormal basis for W.

We next seek a vector $\mathbf{Y}_4$ in the subspace W_3 of W spanned by $\{\mathbf{X}_1, \mathbf{X}_2, \mathbf{X}_3 \, \mathbf{X}_4\}$, and thus by $\{\mathbf{Y}_1, \mathbf{Y}_2, \mathbf{Y}_3, \mathbf{X}_4\}$, which is orthogonal to $\mathbf{Y}_1, \mathbf{Y}_2,$ and $\mathbf{Y}_3$. We can then write

$$\mathbf{Y}_4 = \mathbf{X}_4 - \left(\frac{\mathbf{X}_4 \cdot \mathbf{Y}_1}{\mathbf{Y}_1 \cdot \mathbf{Y}_1} \right) \mathbf{Y}_1 - \left(\frac{\mathbf{X}_4 \cdot \mathbf{Y}_2}{\mathbf{Y}_2 \cdot \mathbf{Y}_2} \right) \mathbf{Y}_2 - \left(\frac{\mathbf{X}_4 \cdot \mathbf{Y}_3}{\mathbf{Y}_3 \cdot \mathbf{Y}_3} \right) \mathbf{Y}_3.$$

We continue in this manner until we have an orthogonal set $T^* = \{\mathbf{Y}_1, \mathbf{Y}_2, \ldots, \mathbf{Y}_m\}$ of m vectors. It then follows that T^* is a basis for W. If we now normalize the $\mathbf{Y}_i$, that is, let

$$\mathbf{Z}_i = \frac{\mathbf{Y}_i}{\|\mathbf{Y}_i\|} \qquad (1 \le i \le m),$$

then $T = \{\mathbf{Z}_1, \mathbf{Z}_2, \ldots, \mathbf{Z}_m\}$ is an orthonormal basis for W.

Example 3 Consider the basis $S = \{\mathbf{X}_1, \mathbf{X}_2, \mathbf{X}_3\}$ for R^3, where

$$\mathbf{X}_1 = (1, 1, 1), \qquad \mathbf{X}_2 = (-1, 0, -1), \qquad \text{and} \qquad \mathbf{X}_3 = (-1, 2, 3).$$

Use the Gram–Schmidt process to transform S to an orthogonal basis for R^3.

Solution

STEP 1. Let $\mathbf{Y}_1 = \mathbf{X}_1 = (1, 1, 1)$.

STEP 2. We now compute $\mathbf{Y}_2$ and $\mathbf{Y}_3$:

$$\mathbf{Y}_2 = \mathbf{X}_2 - \left(\frac{\mathbf{X}_2 \cdot \mathbf{Y}_1}{\mathbf{Y}_1 \cdot \mathbf{Y}_1} \right) \mathbf{Y}_1 = (-1, 0, -1) + \frac{2}{3} (1, 1, 1) = \left(-\frac{1}{3}, \frac{2}{3}, -\frac{1}{3} \right)$$

and

$$\mathbf{Y}_3 = \mathbf{X}_3 - \left(\frac{\mathbf{X}_3 \cdot \mathbf{Y}_1}{\mathbf{Y}_1 \cdot \mathbf{Y}_1} \right) \mathbf{Y}_1 - \left(\frac{\mathbf{X}_3 \cdot \mathbf{Y}_2}{\mathbf{Y}_2 \cdot \mathbf{Y}_2} \right) \mathbf{Y}_2$$

$$= (-1, 2, 3) - \left(\frac{4}{3} \right) (1, 1, 1) - \left(\frac{\frac{8}{3}}{\frac{6}{9}} \right) \left(-\frac{1}{3}, \frac{2}{3}, -\frac{1}{3} \right)$$

$$= (-2, 0, 2).$$

Then $T^* = \{\mathbf{Y}_1, \mathbf{Y}_2, \mathbf{Y}_3\}$ is an orthogonal basis for R^3.

STEP 3. Let

$$\mathbf{Z}_1 = \frac{\mathbf{Y}_1}{\|\mathbf{Y}_1\|} = \frac{1}{\sqrt{3}} (1, 1, 1)$$

$$\mathbf{Z}_2 = \frac{\mathbf{Y}_2}{\|\mathbf{Y}_2\|} = \frac{1}{\sqrt{\frac{6}{9}}} \left(-\frac{1}{3}, \frac{2}{3}, -\frac{1}{3} \right)$$

$$\mathbf{Z}_3 = \frac{\mathbf{Y}_3}{\|\mathbf{Y}_3\|} = \frac{1}{\sqrt{8}} (-2, 0, 2).$$

Then

$$T = \{\mathbf{Z}_1, \mathbf{Z}_2, \mathbf{Z}_3\}$$

$$= \left\{ \left(\frac{1}{\sqrt{3}}, \frac{1}{\sqrt{3}}, \frac{1}{\sqrt{3}} \right), \left(-\frac{1}{\sqrt{6}}, \frac{2}{\sqrt{6}}, -\frac{1}{\sqrt{6}} \right), \left(-\frac{1}{\sqrt{2}}, 0, \frac{1}{\sqrt{2}} \right) \right\}.$$

is an orthonormal basis for R^3.

REMARK   At the *end* of Step 2 in the Gram–Schmidt process we have an orthogonal basis for W. Clearing the fractions in each $\mathbf{Y}_i$ by multiplying by a proper scalar, we obtain a resulting set of vectors that is also an orthogonal basis for W. Normalizing each of the vectors in this orthogonal basis, with no cumbersome fractions, yields the desired orthonormal basis T^*. Thus, in Example 3, we have

$$T^* = \left\{ (1, 1, 1), \left(-\frac{1}{3}, \frac{2}{3}, -\frac{1}{3} \right), (-2, 0, 2) \right\}.$$

Clearing fractions in each of these vectors, we obtain the following orthogonal basis for W:

$$\{(1, 1, 1), (-1, 2, -1), (-2, 0, 2)\}.$$

Normalizing each of the vectors in this orthogonal basis, we obtain the orthonormal basis T as in Example 3.

Example 4   Let W be the subspace of R^4 with basis $S = \{\mathbf{X}_1, \mathbf{X}_2\}$ where $\mathbf{X}_1 = (1, -2, 0, 1)$ and $\mathbf{X}_2 = (-1, 0, 0, -1)$. Use the Gram–Schmidt process to transform S to an orthonormal basis.

Solution

STEP 1.   Let $\mathbf{Y}_1 = \mathbf{X}_1 = (1, -2, 0, 1)$.

STEP 2.   We now compute $\mathbf{Y}_2$ and $\mathbf{Y}_3$:

$$\mathbf{Y}_2 = \mathbf{X}_2 - \left(\frac{\mathbf{X}_2 \cdot \mathbf{Y}_1}{\mathbf{Y}_1 \cdot \mathbf{Y}_1} \right) \mathbf{Y}_1 = (-1, 0, 0, -1) - \left(-\frac{2}{6} \right)(1, -2, 0, 1)$$

$$= \left(-\frac{2}{3}, -\frac{2}{3}, 0, -\frac{2}{3} \right).$$

Then

$$T^* = \{\mathbf{Y}_1, \mathbf{Y}_2\} = \left\{ (1, -2, 0, 1), \left(-\frac{2}{3}, -\frac{2}{3}, 0, -\frac{2}{3} \right) \right\}$$

is an orthogonal basis for W. Clearing fractions in each $\mathbf{Y}_i$ in T^*, we find that

$$\{(1, -2, 0, 1), (-2, -2, 0, -2)\}$$

is also an orthogonal basis for W.

STEP 3. Normalizing these vectors, we obtain the following orthonormal basis for W:

$$\left\{\left(\frac{1}{\sqrt{6}}, -\frac{2}{\sqrt{6}}, 0, \frac{1}{\sqrt{6}}\right), \left(-\frac{1}{\sqrt{3}}, -\frac{1}{\sqrt{3}}, 0, -\frac{1}{\sqrt{3}}\right)\right\}.$$

In the proof of Theorem 3.19 we have also established the following result. At each stage of the Gram–Schmidt process, the ordered set $\{Z_1, Z_2, \ldots, Z_k\}$ is an orthonormal basis for the subspace span $\{X_1, X_2, \ldots, X_k\}$, $1 \le k \le m$. Also, the final orthonormal basis T depends upon the order of the vectors in the given basis S. Thus, if we change the order of the vectors in S, we might obtain a different orthonormal basis T for W.

We make one final observation with regard to the Gram–Schmidt process. In our proof of Theorem 3.19, we first obtained an orthogonal basis T^* and then normalized all the vectors in T^* to obtain the orthonormal basis T. Of course, an alternative course of action is to normalize each vector as soon as we produce it.

Suppose that $S = \{X_1, X_2, \ldots, X_n\}$ is a basis for R^n. In Section 3.5 we had to solve a linear system of n equations in n unknowns to write a given vector X of R^n as a linear combination of the vectors in S. However, if S is orthonormal, we can obtain the same result with much less work. This is the content of the following theorem.

Theorem 3.20 *Let $S = \{X_1, X_2, \ldots, X_n\}$ be an orthonormal basis for R^n and X any vector in R^n. Then*

$$X = c_1 X_1 + c_2 X_2 + \cdots + c_n X_n,$$

where

$$c_i = X \cdot X_i \qquad (1 \le i \le n).$$

Proof
Exercise T.4.

Example 5 Consider the orthonormal basis $T = \{Z_1, Z_2, Z_3\}$ obtained in Example 3, where

$$Z_1 = \left(\frac{1}{\sqrt{3}}, \frac{1}{\sqrt{3}}, \frac{1}{\sqrt{3}}\right), \qquad Z_2 = \left(-\frac{1}{\sqrt{6}}, \frac{2}{\sqrt{6}}, -\frac{1}{\sqrt{6}}\right),$$

$$Z_3 = \left(-\frac{1}{\sqrt{2}}, 0, \frac{1}{\sqrt{2}}\right).$$

Let $X = (4, 3, -1)$. To write X as a linear combination of the vectors in S, we compute (verify)

$$c_1 = \mathbf{X} \cdot \mathbf{Z}_1 = \frac{6}{\sqrt{3}}, \qquad c_2 = \mathbf{X} \cdot \mathbf{Z}_2 = \frac{3}{\sqrt{6}}, \qquad c_3 = \mathbf{X} \cdot \mathbf{Z}_3 = -\frac{5}{\sqrt{2}}.$$

Then

$$\mathbf{X} = c_1\mathbf{Z}_1 + c_2\mathbf{Z}_2 + c_3\mathbf{Z}_3 = \frac{6}{\sqrt{3}}\mathbf{Z}_1 + \frac{3}{\sqrt{6}}\mathbf{Z}_2 - \frac{5}{\sqrt{2}}\mathbf{Z}_3.$$

OPTIONAL APPLICATION (Projection). If W is an m-dimensional subspace of R^n, then we know that the Gram–Schmidt process enables us to construct an orthonormal basis for W. We shall now discuss an important application of this result.

Theorem 3.21   *Let W be an m-dimensional subspace of R^n with orthonormal basis $\{\mathbf{X}_1, \mathbf{X}_2, \ldots, \mathbf{X}_m\}$. Then every vector $\mathbf{X}$ in R^n can be uniquely written as*

$$\mathbf{X} = \mathbf{Z} + \mathbf{Y},$$

where $\mathbf{Z}$ is in W and $\mathbf{Y}$ is orthogonal to every vector in W.

Proof
Let

$$\mathbf{Z} = (\mathbf{X} \cdot \mathbf{X}_1)\mathbf{X}_1 + (\mathbf{X} \cdot \mathbf{X}_2)\mathbf{X}_2 + \cdots + (\mathbf{X} \cdot \mathbf{X}_m)\mathbf{X}_m \qquad (7)$$

and

$$\mathbf{Y} = \mathbf{X} - \mathbf{Z}.$$

Since $\mathbf{Z}$ is a linear combination of $\mathbf{X}_1, \mathbf{X}_2, \ldots, \mathbf{X}_m$, we conclude that $\mathbf{Z}$ belongs to W. We next show that $\mathbf{Y}$ is orthogonal to every vector in W. First note that

$$\begin{aligned} \mathbf{Y} \cdot \mathbf{X}_i &= (\mathbf{X} - \mathbf{Z}) \cdot \mathbf{X}_i \\ &= \mathbf{X} \cdot \mathbf{X}_i - \mathbf{Z} \cdot \mathbf{X}_i. \end{aligned}$$

Now

$$\begin{aligned} \mathbf{Z} \cdot \mathbf{X}_i &= (\mathbf{X} \cdot \mathbf{X}_1)(\mathbf{X}_1 \cdot \mathbf{X}_i) + (\mathbf{X} \cdot \mathbf{X}_2)(\mathbf{X}_2 \cdot \mathbf{X}_i) + \cdots \\ &\quad + (\mathbf{X} \cdot \mathbf{X}_i)(\mathbf{X}_i \cdot \mathbf{X}_i) + \cdots + (\mathbf{X} \cdot \mathbf{X}_m)(\mathbf{X}_m \cdot \mathbf{X}_i) \\ &= \mathbf{X} \cdot \mathbf{X}_i. \end{aligned}$$

Thus $\mathbf{Y} \cdot \mathbf{X}_i = 0$ for $1 \le i \le m$.
   Now if

$$\mathbf{Z}' = c_1\mathbf{X}_1 + c_2\mathbf{X}_2 + \cdots + c_m\mathbf{X}_m$$

is an arbitrary vector in W, then

$$\begin{aligned} \mathbf{Y} \cdot \mathbf{Z}' &= \mathbf{Y} \cdot (c_1\mathbf{X}_1 + c_2\mathbf{X}_2 + \cdots + c_m\mathbf{X}_m) \\ &= c_1(\mathbf{Y} \cdot \mathbf{X}_1) + c_2(\mathbf{Y} \cdot \mathbf{X}_2) + \cdots + c_m(\mathbf{Y} \cdot \mathbf{X}_m) \\ &= 0. \end{aligned}$$

We omit the proof that the vectors $\mathbf{Z}$ and $\mathbf{Y}$ are unique.

Figure 3.43

Definition The vector $\mathbf{Z}$ obtained in Equation (7) is called **orthogonal projection** of $\mathbf{X}$ on W and is denoted by $\text{proj}_W \mathbf{X}$. In Figure 3.43, we illustrate Theorem 3.21 when W is a two-dimensional subspace of R^3 (a plane through the origin).

Example 6 Let W be the two-dimensional subspace of R^3 with orthonormal basis $\{\mathbf{X}_1, \mathbf{X}_2\}$, where $\mathbf{X}_1 = (2/3, 1/3, -2/3)$ and $\mathbf{X}_2 = (1/\sqrt{2}, 0, 1/\sqrt{2})$. Write the vector $\mathbf{X} = (2, 1, 3)$ as $\mathbf{Z} + \mathbf{Y}$, with $\mathbf{Z}$ and $\mathbf{Y}$ as in Theorem 3.21.

Solution
We have

$$\text{proj}_W \mathbf{X} = \mathbf{Z} = (\mathbf{X} \cdot \mathbf{X}_1)\mathbf{X}_1 + (\mathbf{X} \cdot \mathbf{X}_2)\mathbf{X}_2$$

$$= -\frac{1}{3}\left(\frac{2}{3}, \frac{1}{3}, -\frac{2}{3}\right) + \frac{5}{\sqrt{2}}\left(\frac{1}{\sqrt{2}}, 0, \frac{1}{\sqrt{2}}\right)$$

$$= \left(\frac{41}{18}, -\frac{1}{9}, \frac{49}{18}\right),$$

and

$$\mathbf{Y} = \mathbf{X} - \mathbf{Z}$$

$$= (2, 1, 3) - \left(\frac{41}{18}, -\frac{1}{9}, \frac{49}{18}\right)$$

$$= \left(-\frac{5}{18}, \frac{10}{9}, \frac{5}{18}\right).$$

Then

$$\mathbf{X} = \mathbf{Z} + \mathbf{Y}.$$

It is clear from Figure 3.43 that the distance from $\mathbf{X}$ to the plane W is given by the length of the vector $\mathbf{Y} = \mathbf{X} - \mathbf{Z}$, that is, by

$$\|\mathbf{X} - \text{proj}_W \mathbf{X}\|.$$

Example 7 Let W be the subspace of R^3 defined in Example 6, and let $\mathbf{X} = (1, 1, 0)$. Find the distance from $\mathbf{X}$ to W.

Solution

We first compute

$$\text{proj}_W \mathbf{X} = (\mathbf{X} \cdot \mathbf{X}_1)\mathbf{X}_1 + (\mathbf{X} \cdot \mathbf{X}_2)\mathbf{X}_2$$

$$= 1\left(\frac{2}{3}, \frac{1}{3}, -\frac{2}{3}\right) + \frac{1}{\sqrt{2}}\left(\frac{1}{\sqrt{2}}, 0, \frac{1}{\sqrt{2}}\right)$$

$$= \left(\frac{5}{6}, \frac{1}{3}, -\frac{1}{6}\right).$$

Then

$$\mathbf{X} - \text{proj}_W \mathbf{X} = (1, 1, 0) - \left(\frac{5}{6}, \frac{1}{3}, -\frac{1}{6}\right)$$

$$= \left(\frac{1}{6}, \frac{2}{3}, \frac{1}{6}\right).$$

Finally,

$$\|\mathbf{X} - \text{proj}_W \mathbf{X}\| = \sqrt{\frac{1}{36} + \frac{4}{9} + \frac{1}{36}}$$

$$= \sqrt{2},$$

so the distance from $\mathbf{X}$ to W is $\sqrt{2}$.

3.8 EXERCISES

1. Which of the following are orthogonal sets of vectors?
   (a) $(1, -1, 2)$, $(0, 2, -1)$, $(-1, 1, 1)$.
   (b) $(1, 2, -1, 1)$, $(0, -1, -2, 0)$, $(1, 0, 0, -1)$.
   (c) $(0, 1, 0, -1)$, $(1, 0, 1, 1)$, $(-1, 1, -1, 2)$.

2. Which of the following are orthonormal sets of vectors?
   (a) $(\frac{1}{3}, \frac{2}{3}, \frac{2}{3})$, $(\frac{2}{3}, \frac{1}{3}, -\frac{2}{3})$, $(\frac{2}{3}, -\frac{2}{3}, \frac{1}{3})$.
   (b) $(1/\sqrt{2}, 0, -1/\sqrt{2})$, $(1/\sqrt{3}, 1/\sqrt{3}, 1/\sqrt{3})$, $(0, 1, 0)$.
   (c) $(0, 2, 2, 1)$, $(1, 1, -2, 2)$, $(0, -2, 1, 2)$.

3. Use the Gram–Schmidt process to find an orthonormal basis for the subspace of R^3 with basis $\{(1, -1, 0), (2, 0, 1)\}$.

4. Use the Gram–Schmidt process to find an orthonormal basis for the subspace of R^3 with basis $\{(1, 0, 2), (-1, 1, 0)\}$.

5. Use the Gram–Schmidt process to find an ortho-normal basis for the subspace of R^4 with basis $\{(1, -1, 0, 1), (2, 0, 0, -1), (0, 0, 1, 0)\}$.

6. Use the Gram–Schmidt process to find an orthonormal basis for the subspace of R^4 with basis $\{(1, 1, -1, 0), (0, 2, 0, 1), (-1, 0, 0, 1)\}$.

7. Use the Gram–Schmidt process to transform the basis $\{(1, 2), (-3, 4)\}$ for R^2 into (a) an orthogonal basis; (b) an orthonormal basis.

8. Use the Gram–Schmidt process to transform the basis $\{(1, 1, 1), (0, 1, 1), (1, 2, 3)\}$ for R^3 into an orthonormal basis for R^3.

9. Find an orthonormal basis for R^3 containing the vectors $(\frac{2}{3}, -\frac{2}{3}, \frac{1}{3})$ and $(\frac{2}{3}, \frac{1}{3}, -\frac{2}{3})$.

10. Use the Gram–Schmidt process to construct an orthonormal basis for the subspace W of R^3 *spanned* by $\{(1, 1, 1), (2, 2, 2), (0, 0, 1), (1, 2, 3)\}$.

11. Use the Gram–Schmidt process to construct an orthonormal basis of the subspace W of R^4 *spanned* by $\{(1, 1, 0, 0), (2, -1, 0, 1), (3, -3, 0, -2),$ $(1, -2, 0, -3)\}$.

12. Consider the orthonormal basis

$$S = \left\{ \left(\frac{1}{\sqrt{2}}, \frac{1}{\sqrt{2}} \right), \left(-\frac{1}{\sqrt{2}}, \frac{1}{\sqrt{2}} \right) \right\}$$

for R^2. Using Theorem 3.20, write the vector $(2, 3)$ as a linear combination of the vectors in S.

13. Consider the orthonormal basis

$$S = \left\{ \left(\frac{1}{\sqrt{5}}, 0, \frac{2}{\sqrt{5}} \right), \right.$$

$$\left. \left(-\frac{2}{\sqrt{5}}, 0, \frac{1}{\sqrt{5}} \right), (0, 1, 0) \right\}$$

for R^3. Using Theorem 3.20, write the vector $(2, -3, 1)$ as a linear combination of the vectors in S.

14. (Optional). Let W be the subspace of R^3 with orthonormal basis $\{X_1, X_2\}$, where $X_1 = (0, 1, 0)$ and $X_2 = (1/\sqrt{5}, 0, 2/\sqrt{5})$. Write the vector $X = (1, 2, -1)$ as $Z + Y$, with Z and Y as in Theorem 3.21.

15. (Optional). Let W be the subspace of R^4 with orthonormal basis $\{X_1, X_2, X_3\}$, where $X_1 = (1/\sqrt{2}, 0, 0, -1/\sqrt{2})$, $X_2 = (0, 0, 1, 0)$, and $X_3 = (1/\sqrt{2}, 0, 0, 1/\sqrt{2})$. Write the vector $(1, 0, 2, 3)$ as $Z + Y$, with Z and Y as in Theorem 3.21.

16. (Optional). Let W be the subspace of R^3 defined in Exercise 14, and let $X = (-1, 0, 1)$. Find the distance from X to W.

17. (Optional). Let W be the subspace of R^4 defined in Exercise 15, and let $X = (1, 2, -1, 0)$. Find the distance from X to W.

THEORETICAL EXERCISES

T.1. Verify that the natural basis for R^n is an orthonormal set.

T.2. Prove Corollary 3.5.

T.3. Show that an orthonormal set of n vectors in R^n is a basis for R^n.

T.4. Prove Theorem 3.20.

T.5. Let $X, Y_1, Y_2, \ldots, Y_n$ be vectors in R^n. Show that if X is orthogonal to $Y_1, Y_2, \ldots, Y_n$, then X is orthogonal to every vector in span $\{Y_1, Y_2, \ldots, Y_n\}$.

T.6. Let X be a fixed vector in R^n. Prove that the set of all vectors in R^n that are orthogonal to X is a subspace of R^n.

KEY IDEAS FOR REVIEW

☐ **Theorem 3.2.** Properties of vector addition and scalar multipliction in R^n (see page 115).

☐ **Theorem 3.3 (Properties of Inner Product).** See page 122.

☐ **Theorem 3.4 (Cauchy-Schwarz Inequality)**

$$|X \cdot Y| \le \|X\| \, \|Y\|$$

☐ **Vector space.** See page 134.

☐ **Theorem 3.8.** Let V be a vector space with operations $\oplus$ and $\odot$ and let W be a nonempty subset of V. Then W is a subspace of V if and only if

(α) If X and Y are any vectors in W, then $X \oplus Y$ is in W.

(β) If X is in W and c is a scalar, then cX is in W.

☐ $S = \{X_1, X_2, \ldots, X_k\}$ is linearly dependent if and only if one of the vectors in S is a linear combination of all the other vectors in S.

☐ **Theorem 3.10.** If $S = \{X_1, X_2, \ldots, X_n\}$ is a basis for a vector space V, then every vector in V can be written in one and only one way as a linear combination of the vectors in S.

☐ **Theorem 3.12.** If $S = \{X_1, X_2, \ldots, X_n\}$ is a basis for a vector space V and $T = \{Y_1, Y_2, \ldots, Y_r\}$ is a linearly independent set of vectors in V, then $r \leq n$.

☐ **Corollary 3.1.** All bases for a vector space V must have the same number of vectors.

☐ **Theorem 3.14.** Let V be an n-dimensional vector space, and let $S = \{X_1, X_2, \ldots, X_n\}$ be a set of n vectors in V.

(a) If S is linearly independent, then it is a basis for V.

(b) If S spans V, then it is a basis for V.

☐ **Basis for solution space of AX = 0.** See page 160.

☐ **Theorem 3.16** Row rank $\mathbf{A}$ = column rank $\mathbf{A}$.

☐ **Theorem 3.17.** An $n \times n$ matrix is nonsingular if and only if rank $\mathbf{A} = n$.

☐ **Corollary 3.2** If $\mathbf{A}$ is an $n \times n$ matrix, then

rank $\mathbf{A} = n$ if and only if $|\mathbf{A}| \neq 0$.

☐ **Corollary 3.4.** If $\mathbf{A}$ is $n \times n$, then $\mathbf{AX} = \mathbf{0}$ has a nontrivial solution if and only if rank $\mathbf{A} < n$.

☐ The following statements are equivalent for an $n \times n$ matrix $\mathbf{A}$.

1. $\mathbf{A}$ is nonsingular.
2. $\mathbf{AX} = \mathbf{0}$ has only the trivial solution.
3. $\mathbf{A}$ is row equivalent to $\mathbf{I}_n$.
4. The linear system $\mathbf{AX} = \mathbf{B}$ has a unique solution for every $n \times 1$ matrix $\mathbf{B}$.
5. $|\mathbf{A}| \neq 0$.
6. $\mathbf{A}$ has rank n.
7. The rows (columns) of $\mathbf{A}$ form a linearly independent set of n vectors in R^n.

☐ **Theorem 3.18.** An orthogonal set of nonzero vectors in R^n is linearly independent.

☐ **Theorem 3.19 (Gram–Schmidt Process)** See page 177.

☐ **Theorem 3.20.** If $\{X_1, X_2, \ldots, X_n\}$ is an orthonormal basis for R^n and $\mathbf{X}$ is a vector in R^n, then $\mathbf{X} = (\mathbf{X} \cdot \mathbf{X}_1)\mathbf{X}_1 + (\mathbf{X} \cdot \mathbf{X}_2)\mathbf{X}_2 + \cdots + (\mathbf{X} \cdot \mathbf{X}_n)\mathbf{X}_n$.

SUPPLEMENTARY EXERCISES

1. Let $\mathbf{X} = (1, -1, 2, 3)$, and $\mathbf{Y} = (2, 3, 1, -2)$. Compute
   (a) $\|\mathbf{X}\|$.   (b) $\|\mathbf{Y}\|$.
   (c) $\|\mathbf{X} - \mathbf{Y}\|$.
   (d) $\mathbf{X} \cdot \mathbf{Y}$.
   (e) Cosine of the angle between $\mathbf{X}$ and $\mathbf{Y}$.

2. If $\mathbf{X} = (x, y)$ is any vector in R^2, show that the vector $\mathbf{Y} = (-y, x)$ is orthogonal to $\mathbf{X}$.

3. Find all values of c for which $\|c(1, -2, 2, 0)\| = 9$.

4. Consider the set W of all vectors in R^4 of the form (a, b, c, d), where $a = b + c$ and $d = a + 1$. Is W a subspace of R^4?

5. Is the vector $(4, 2, 1)$ a linear combination of the vectors $(1, 2, 1)$, $(-1, 1, 2)$, and $(-3, -3, 0)$?

6. Do the vectors

   $(-1, 2, 1), (2, 1, -1),$ and $(0, 5, -1)$ span R^3?

7. Is the set of vectors

   $\{t^2 + 2t + 2, 2t^2 + 3t + 1, -t - 3\}$

   linearly dependent or linearly independent? If it is linearly dependent, write one of the vectors as a linear combination of the other two.

8. Does the set of vectors

   $\{t - 1, t^2 + 2t + 1, t^2 + t - 2\}$

   form a basis for P_2?

9. Find a basis for the subspace of R^4 consisting of all vectors of the form $(a + b, b + c, a - b - 2c, b + c)$. What is the dimension of this subspace?

10. Find the dimension of the subspace of P_2 consisting of all polynomials $a_0 t^2 + a_1 t + a_2$, where $a_2 = 0$.

11. Find a basis for the solution space of the homogeneous system

$$x_1 + 2x_2 - x_3 + x_4 + 2x_5 = 0$$
$$x_1 + x_2 + 2x_3 - 3x_4 + x_5 = 0.$$

What is the dimension of the solution space? Find a basis for $V = $ span S.

12. Let $S = \{(1, 2, -1, 2), (0, -1, 3, -6),$
$(2, 3, 1, -2), (3, 2, 1, -4), (1, -1, 0, -2)\}$.
Find a basis for $V = $ span S.

13. For what value(s) of λ is the set

$$\{t + 3, 2t + \lambda^2 + 2\}$$

linearly independent?

14. Let

$$A = \begin{bmatrix} \lambda^2 & 0 & 1 \\ -1 & \lambda & -2 \\ -1 & 0 & -1 \end{bmatrix}.$$

For what value(s) of λ does the homogeneous system $AX = 0$ have only the trivial solution?

15. For what values of a is the vector $(a^2, a, 1)$ in span $\{(1, 2, 3), (1, 1, 1), (0, 1, 2)\}$?

16. For what values of a is the vector $(a^2, -3a, -2)$ in span $\{(1, 2, 3), (0, 1, 1) (1, 3, 4)\}$?

17. Let X_1, X_2, and X_3 be vectors in a vector space. Verify that

span $\{X_1, X_2, X_3\} = $
    span $\{X_1 + X_2, X_1 - X_3, X_1 + X_3\}$.

18. For what values of k will the set S form a basis for R^6?

$S = \{(1, 2, -2, 1, 1, 1), (0, 2, 0, 0, 0, 0),$
$(0, 5, k, 1, 0, 0), (0, 2, 1, k, 0, 0),$
$(0, 1, -2, 4, -3, 1), (0, 3, 1, 1, 1, 0)\}$.

19. Consider vector space R^2.
   (a) For what values of m and b will all vectors of the form $(x, mx + b)$ be a subspace of R^2?
   (b) For what value of r will the set of all vectors of the form (x, rx^2) be a subspace of R^2?

20. Let A be a fixed $n \times n$ matrix and let the set of all $n \times n$ matrices B such that $AB = BA$ be denoted by $C(A)$. Is $C(A)$ a subspace of the vector space of all $n \times n$ matrices?

21. Prove that a subspace W of R^3 coincides with R^3 if and only if it contains the vectors $(1, 0, 0)$, $(0, 1, 0)$, and $(0, 0, 1)$.

22. Let A be a fixed $m \times n$ matrix and define W to be the subset of all $m \times 1$ matrices B in R^m for which the linear system $AX = B$ has a solution.
   (a) Is W a subspace of R^m?
   (b) What is the relationship between W and the column space of A?

23. **(Calculus Required)** Let $C[a, b]$ denote the set of all real-valued continuous functions defined on $[a, b]$. If f and g are in $C[a, b]$, we define $f \oplus g$ by $(f \oplus g)(t) = f(t) + g(t)$, for t in $[a, b]$. If f is in $C[a, b]$ and c is a scalar, we define $c \odot f$ by $(c \odot f)(t) = c f(t)$, for t in $[a, b]$.
   (a) Show that $C[a, b]$ is a real vector space.
   (b) Let $W(k)$ be the set of all functions in $C[a, b]$ with $f(a) = k$. For what values of k will $W(k)$ be a subspace of $C[a, b]$?
   (c) Let $t_1, t_2, \ldots, t_n$ be a fixed set of points in $[a, b]$. Show that the subset of all functions f in $C[a, b]$ that have roots at $t_1, t_2, \ldots, t_n$, that is, $f(t_i) = 0$ for $i = 1, 2, \ldots, n$, forms a subspace.

24. Let $V = $ span $\{X_1, X_2\}$, where

$$X_1 = (1, 0, 2) \quad \text{and} \quad X_2 = (1, 1, 1).$$

Find a basis S for R^3 that includes X_1 and X_2. (*Hint:* Use the technique developed in the proof of Theorem 3.11.)

25. Find the rank of the matrix

$$\begin{bmatrix} 1 & 2 & -1 & 3 & 1 \\ 0 & 1 & -3 & 2 & 3 \\ 2 & 3 & 1 & 4 & -1 \\ -1 & 2 & 2 & 2 & -5 \\ 3 & 1 & -1 & 2 & 4 \end{bmatrix}.$$

26. Let A be an 8×5 matrix. How large can the dimension of the row space be? Explain.

27. Exercise T.6 of Section 3.8 proves that the set of all vectors in R^n that are orthogonal to a fixed vector X is a subspace of R^n. For $X = (1, -2, 1)$ find an orthogonal basis for the subspace of vectors in R^3 that are orthogonal to X. [*Hint:* Solve the linear system $X \cdot Y = 0$, when $Y = (y_1, y_2, y_3)$.]

28. Find an orthonormal basis for the set of vectors X such that $AX = 0$ when

(a) $A = \begin{bmatrix} 1 & 0 & 5 & -2 \\ 0 & 1 & -2 & -5 \end{bmatrix}$.

(b) $A = \begin{bmatrix} 1 & 0 & 5 & -2 \\ 0 & 1 & -2 & 4 \end{bmatrix}$.

29. Given the orthonormal basis

$$S = \left\{ \left(\frac{1}{\sqrt{2}}, 0, -\frac{1}{\sqrt{2}} \right), (0, 1, 0), \right.$$
$$\left. \left(\frac{1}{\sqrt{2}}, 0, \frac{1}{\sqrt{2}} \right) \right\}$$

for R^3, write the vector $(1, 2, 3)$ as a linear combination of the vectors in S.

30. Use the Gram–Schmidt process to find an orthonormal basis for the subspace of R^4 with basis $\{(1, 0, 0, -1), (1, -1, 0, 0), (0, 1, 0, 1)\}$.

31. Let $W = \text{span } \{(1, 0, 1), (0, 1, 0)\}$ in R^3.
(a) Find a basis for the othogonal complement of W (see Supplementary Exercise T.8 below).
(b) Show that vectors $(1, 0, 1)$, $(0, 1, 0)$ and the basis for the orthogonal complement of W from part (a) form a basis for R^3.
(c) Find vectors Y and Z as defined in Supplementary Exercise T.8 below, so that $X = Z + Y$ for each of the following vectors.

(i) $X = (1, 0, 0)$ (ii) $X = (1, 2, 3)$

THEORETICAL EXERCISES

T.1. Let A be an $n \times n$ matrix. Prove that A is nonsingular if and only if the dimension of the solution space of the homogeneous system $AX = 0$ is zero.

T.2. Let $\{X_1, X_2, \ldots, X_n\}$ be a basis for an n-dimensional vector space V. Prove that if A is a nonsingular $n \times n$ matrix, then $\{AX_1, AX_2, \ldots, AX_n\}$ is also a basis for V.

T.3. Let X and Y be vectors in R^n. Show that $X \cdot Y = 0$ if and only if $\|X + Y\| = \|X - Y\|$.

T.4. Let $S = \{X_1, X_2, \ldots, X_m\}$ be a basis for a subspace of R^n. Show that if X is a vector in R^n that is orthogonal to each vector in S, then $X = 0$.

T.5. Show that rank $A =$ rank A^T, for any $m \times n$ matrix A.

T.6. Let A and B be $m \times n$ matrices that are row equivalent.
(a) Prove that rank $A =$ rank B.
(b) Prove that for X in R^n, $AX = 0$ if and only if $BX = 0$.

T.7. Let A be $m \times n$ and B be $n \times k$.
(a) Prove that rank $(AB) \leq \min \{\text{rank } A, \text{rank } B\}$.
(b) Find A and B such that

rank $(AB) < \min \{\text{rank } A, \text{rank } B\}$.

(c) If $k = n$ and B is nonsingular, prove that rank $(AB) =$ rank A.
(d) If $m = n$ and A is nonsingular, prove that rank $(AB) =$ rank B.
(e) For nonsingular matrices P and Q, what is rank (PAQ)?

T.8. If W is a subspace of R^n then the set of all vectors in R^n that are orthogonal to every vector in W is called the **orthogonal complement** of W.
(a) Prove that the orthogonal complement of W is a subspace of R^n.
(b) Prove that every vector X in R^n is uniquely expressible in the form $Z + Y$ where Z is in W and Y is in the orthogonal complement of W.

CHAPTER TEST

1. Consider the set W of all vectors in R^3 of the form (a, b, c), where $a + b + c = 0$. Is W a subspace of R^3?

2. Find a basis for the solution space of the homogeneous system

$$\begin{aligned} x_1 + 3x_2 + 3x_3 - x_4 + 2x_5 &= 0 \\ x_1 + 2x_2 + 2x_3 - 2x_4 + 2x_5 &= 0 \\ x_1 + x_2 + x_3 - 3x_4 + 2x_5 &= 0. \end{aligned}$$

3. Does the set of vectors $\{(1, -1, 1), (1, -3, 1), (1, -2, 2)\}$ form a basis for R^3?

4. For what value(s) of λ is the set of vectors $\{(\lambda^2 - 5, 1, 0), (2, -2, 3), (2, 3, -3)\}$ linearly dependent?

5. Use the Gram–Schmidt process to find an orthonormal basis for the subspace of R^4 with basis $\{(1, 0, -1, 0), (1, -1, 0, 0), (3, 1, 0, 0)\}$.

6. Answer each of the following as true or false.
   (a) All vectors of the form $(a, 0, -a)$ form a subspace of R^3.
   (b) In R^n, $\|c\mathbf{X}\| = c\|\mathbf{X}\|$.
   (c) Every set of vectors in R^3 containing two vectors is linearly independent.
   (d) The solution space of the homogeneous system $\mathbf{AX} = \mathbf{0}$ is spanned by the columns of $\mathbf{A}$.
   (e) If the columns of an $n \times n$ matrix form a basis for R^n, so do the rows.
   (f) If $\mathbf{A}$ is an 8×8 matrix such that the homogeneous system $\mathbf{AX} = \mathbf{0}$ has only the trivial solution, then rank $\mathbf{A} < 8$.
   (g) Every orthonormal set of five vectors in R^5 is a basis for R^5.
   (h) Every linearly independent set of vectors in R^3 contains three vectors.
   (i) If $\mathbf{A}$ is an $n \times n$ symmetric matrix, then

   $$\text{rank } \mathbf{A} = n.$$

   (j) Every set of vectors spanning R^3 contains at least three vectors.

4

Linear Transformations and Matrices

Note. Chapter 5 can be covered without first covering Chapter 4. The several places in Chapter 5 where Sections 4.1 and 4.3 are used have been clearly marked. These discussions can be omitted without any loss of continuity.

4.1

DEFINITION AND EXAMPLES

Functions occur in almost every application of mathematics. In this chapter we shall study the properties of certain functions called linear transformations, mapping one vector space into another vector space. Linear transformations play an important role in many areas of mathematics, as well as in numerous applied problems in the physical sciences, the social sciences, and economics.

Definition Let V and W be vector spaces. A **linear transformation L** of V into W is a function assigning a unique vector $\mathbf{L(X)}$ in W to each $\mathbf{X}$ in V such that:
  (a) $\mathbf{L(X + Y) = L(X) + L(Y)}$, for every $\mathbf{X}$ and $\mathbf{Y}$ in V.
  (b) $\mathbf{L(}c\mathbf{X) = }c\mathbf{L(X)}$, for every $\mathbf{X}$ in V and every scalar c.

In the definition above observe that in (a) the + in $\mathbf{X} + \mathbf{Y}$ refers to the addition operation in V, whereas the + in $\mathbf{L}(\mathbf{X}) + \mathbf{L}(\mathbf{Y})$ refers to the addition operation in W. Similarly, in (b) the scalar product $c\mathbf{X}$ is in V, while the scalar product $c\mathbf{L}(\mathbf{X})$ is in W.

We shall write the fact that $\mathbf{L}$ maps V into W, even if it is not a linear transformation, as

$$\mathbf{L}\colon V \to W.$$

If $V = W$, the linear transformation $\mathbf{L}\colon V \to V$ is also called a **linear operator** on V.

Example 1 Let $\mathbf{L}\colon R^3 \to R^2$ be defined by

$$\mathbf{L}(x, y, z) = (x, y).$$

To verify that $\mathbf{L}$ is a linear transformation, we let

$$\mathbf{X} = (x_1, y_1, z_1) \qquad \text{and} \qquad \mathbf{Y} = (x_2, y_2, z_2).$$

Then

$$\begin{aligned}
\mathbf{L}(\mathbf{X} + \mathbf{Y}) &= \mathbf{L}((x_1, y_1, z_1) + (x_2, y_2, z_2)) \\
&= \mathbf{L}(x_1 + x_2, y_1 + y_2, z_1 + z_2) = (x_1 + x_2, y_1 + y_2) \\
&= (x_1, y_1) + (x_2, y_2) = \mathbf{L}(\mathbf{X}) + \mathbf{L}(\mathbf{Y}).
\end{aligned}$$

Also, if c is a real number, then

$$\begin{aligned}
\mathbf{L}(c\mathbf{X}) &= \mathbf{L}(cx_1, cy_1, cz_1) = (cx_1, cy_1) = c(x_1, y_1) \\
&= c\mathbf{L}(\mathbf{X}).
\end{aligned}$$

Hence $\mathbf{L}$ is a linear transformation, which is called a **projection**. It is a simple and helpful matter to describe geometrically the effect of $\mathbf{L}$. The image under $\mathbf{L}$ of a vector $\mathbf{X} = (a, b, c)$ in R^3 is found by drawing a line through the end point $P(a, b, c)$ of $\mathbf{X}$ and perpendicular to R^2, the x, y-plane. We obtain the point of intersection $Q(a, b)$ of this line with the x, y-plane. The vector $\mathbf{Y} = (a, b)$ in R^2 with end point Q is the image of $\mathbf{X}$ under $\mathbf{L}$ (Figure 4.1).

Example 2 Let $\mathbf{L}\colon R^3 \to R^3$ be defined by

$$\mathbf{L}\left(\begin{bmatrix} x \\ y \\ z \end{bmatrix}\right) = \begin{bmatrix} x + 1 \\ 2y \\ z \end{bmatrix}.$$

Figure 4.1

To determine whether $\mathbf{L}$ is a linear transformation, let

$$\mathbf{X} = \begin{bmatrix} x_1 \\ y_1 \\ z_1 \end{bmatrix} \quad \text{and} \quad \mathbf{Y} = \begin{bmatrix} x_2 \\ y_2 \\ z_2 \end{bmatrix}.$$

Then

$$\mathbf{L}(\mathbf{X} + \mathbf{Y}) = \mathbf{L}\left(\begin{bmatrix} x_1 \\ y_1 \\ z_1 \end{bmatrix} + \begin{bmatrix} x_2 \\ y_2 \\ z_2 \end{bmatrix} \right) = \mathbf{L}\left(\begin{bmatrix} x_1 + x_2 \\ y_1 + y_2 \\ z_1 + z_2 \end{bmatrix} \right) = \begin{bmatrix} (x_1 + x_2) + 1 \\ 2(y_1 + y_2) \\ z_1 + z_2 \end{bmatrix}.$$

On the other hand,

$$\mathbf{L}(\mathbf{X}) + \mathbf{L}(\mathbf{Y}) = \begin{bmatrix} x_1 + 1 \\ 2y_1 \\ z_1 \end{bmatrix} + \begin{bmatrix} x_2 + 1 \\ 2y_2 \\ z_2 \end{bmatrix} = \begin{bmatrix} (x_1 + x_2) + 2 \\ 2(y_1 + y_2) \\ z_1 + z_2 \end{bmatrix}.$$

Since $\mathbf{L}(\mathbf{X} + \mathbf{Y}) \neq \mathbf{L}(\mathbf{X}) + \mathbf{L}(\mathbf{Y})$, we conclude that the function $\mathbf{L}$ is not a linear transformation.

Example 3 Let $\mathbf{L}: R^3 \rightarrow R^3$ be defined by

$$\mathbf{L}(\mathbf{X}) = r\mathbf{X}$$

where r is a real number. It is easy to verify that $\mathbf{L}$ is a linear transformation. If $r > 1$, $\mathbf{L}$ is called a **dilation**; if $0 < r < 1$, $\mathbf{L}$ is called a **contraction**. In Figure 4.2(a), we show the vector $\mathbf{L}_1(\mathbf{X}) = 2\mathbf{X}$, and in Figure 4.2(b) the vector $\mathbf{L}_2(\mathbf{X}) = \frac{1}{2}\mathbf{X}$. Thus dilation stretches a vector, and contraction shrinks it.

Example 4 Let $\mathbf{L}: R^2 \rightarrow R^3$ be defined by

$$\mathbf{L}\left(\begin{bmatrix} x \\ y \end{bmatrix} \right) = \begin{bmatrix} 1 & 0 \\ 0 & 1 \\ 1 & -1 \end{bmatrix} \begin{bmatrix} x \\ y \end{bmatrix}.$$

Then $\mathbf{L}$ is a linear transformation (verify).

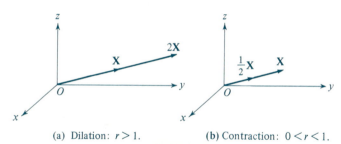

(a) Dilation: $r > 1$.   (b) Contraction: $0 < r < 1$.

Figure 4.2

Example 5 Let **L**: $R^n \to R^m$ be defined by

$$\mathbf{L(X) = AX}$$

for **X** in R^n where **A** is a fixed $m \times n$ matrix. It then follows that **L** is a linear transformation (Exercise T.1).

Example 6 *(Calculus Required)* In calculus we rotated every point in the x, y-plane counterclockwise through an angle ϕ about the origin of a Cartesian coordinate system. Thus, if P has coordinates (x, y), then after rotating, we get P' with coordinates (x', y'), where

$$x' = x \cos \phi - y \sin \phi \quad \text{and} \quad y' = x \sin \phi + y \cos \phi,$$

or

$$x = x' \cos \phi + y' \sin \phi, \quad \text{and} \quad y = -x' \sin \phi + y' \cos \phi.$$

We did this to simplify the general equation of second degree $ax^2 + bxy + cy^2 + dx + ey + f = 0$. Substituting for x and y in terms of x' and y', we obtain $a'x'^2 + b'x'y' + c'y'^2 + d'x' + e'y' + f' = 0$. The key point is to choose ϕ so that $b' = 0$. Once this is done (we might now have to perform a translation of coordinates), we identify the general equation of second degree as a circle, ellipse, hyperbola, parabola, or a degenerate form of these. Note that we may perform this change of coordinates by considering the map **L**: $R^2 \to R^2$ defined by

$$\mathbf{L}\left(\begin{bmatrix} x \\ y \end{bmatrix} \right) = \begin{bmatrix} \cos \phi & -\sin \phi \\ \sin \phi & \cos \phi \end{bmatrix} \begin{bmatrix} x \\ y \end{bmatrix}.$$

If $\mathbf{X} = \begin{bmatrix} x \\ y \end{bmatrix}$ is the vector from the origin to $P(x, y)$, then **L(X)**, the vector obtained by rotating **X** counterclockwise through the angle ϕ (Figure 4.3), is the vector from the origin to $P'(x', y')$. Then **L** is a linear transformation (verify) called a **rotation**.

Example 7 *(Production)* A book publisher publishes a book in three different editions: trade, book club, and deluxe. Each book requires a certain amount of paper and canvas (for the cover). The requirements are given (in grams) by the matrix

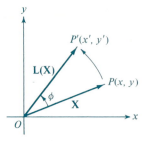

Figure 4.3

$$\mathbf{A} = \begin{array}{c} \text{Trade} \quad \begin{array}{c}\text{Book}\\\text{Club}\end{array} \quad \text{Deluxe} \\ \begin{bmatrix} 300 & 500 & 800 \\ 40 & 50 & 60 \end{bmatrix} \begin{array}{c}\text{Paper}\\\text{Canvas}\end{array} \end{array}$$

Let

$$\mathbf{X} = \begin{bmatrix} x_1 \\ x_2 \\ x_3 \end{bmatrix}$$

denote the production vector, where x_1, x_2, and x_3 are the number of trade, book club, and deluxe books, respectively, that are published. The linear transformation $\mathbf{L}: R^3 \to R^2$ defined by $\mathbf{L}(\mathbf{X}) = \mathbf{AX}$ gives the vector $\begin{bmatrix} y_1 \\ y_2 \end{bmatrix}$, where y_1 is the total amount of paper required and y_2 is the total amount of canvas required.

Example 8 *(Cryptography)* Cryptography is the technique of coding and decoding messages; it goes back to the time of the ancient Greeks. One method of coding and decoding messages uses matrices. We first associate a different number with every letter in the alphabet. For example,

$$\begin{array}{ccccccccc} \text{A} & \text{B} & \text{C} & \text{D} & \cdots & \text{X} & \text{Y} & \text{Z} \\ \updownarrow & \updownarrow & \updownarrow & \updownarrow & & \updownarrow & \updownarrow & \updownarrow \\ 1 & 2 & 3 & 4 & \cdots & 24 & 25 & 26. \end{array}$$

Suppose that we now want to transmit the message STUDY MATH. Using the preceding substitution scheme, our message is 19, 20, 21, 4, 25, 13, 1, 20, 8. A code of this type could be cracked without too much difficulty by a number of techniques, including the analysis of frequency of letters. To make it difficult to crack the code, we proceed as follows. Break the message into three vectors in R^3 (if this cannot be done, we can add extra letters). Our message is now

$$\begin{bmatrix} 19 \\ 20 \\ 21 \end{bmatrix}, \quad \begin{bmatrix} 4 \\ 25 \\ 13 \end{bmatrix}, \quad \begin{bmatrix} 1 \\ 20 \\ 8 \end{bmatrix}$$

Choose a nonsingular 3×3 matrix such as

$$\mathbf{A} = \begin{bmatrix} 1 & 2 & 3 \\ 1 & 1 & 2 \\ 0 & 1 & 2 \end{bmatrix},$$

which is known to *both* sender and receiver. Now define the linear transformation $\mathbf{L}: R^3 \to R^3$ by $\mathbf{L}(\mathbf{X}) = \mathbf{AX}$. Thus our message becomes

$$\mathbf{A} \begin{bmatrix} 19 \\ 20 \\ 21 \end{bmatrix} = \begin{bmatrix} 122 \\ 81 \\ 62 \end{bmatrix}, \quad \mathbf{A} \begin{bmatrix} 4 \\ 25 \\ 13 \end{bmatrix} = \begin{bmatrix} 93 \\ 55 \\ 51 \end{bmatrix}, \quad \mathbf{A} \begin{bmatrix} 1 \\ 20 \\ 8 \end{bmatrix} = \begin{bmatrix} 65 \\ 37 \\ 36 \end{bmatrix},$$

and we transmit 122, 81, 62, 93, 55, 51, 65, 37, 36. Suppose now that we receive back from our correspondent the message

$$114, \ 81, \ 58, \ 104, \ 69, \ 55,$$

which he or she encoded with the same key matrix **A** above.

To decode it, proceed as follows. Break the message into two vectors of R^3:

$$\begin{bmatrix} 114 \\ 81 \\ 58 \end{bmatrix}, \quad \begin{bmatrix} 104 \\ 69 \\ 55 \end{bmatrix}.$$

We want to solve the equation

$$\mathbf{L}(\mathbf{X}_1) = \begin{bmatrix} 114 \\ 81 \\ 58 \end{bmatrix} = \mathbf{A}\mathbf{X}_1$$

for $\mathbf{X}_1$. Since **A** is nonsingular,

$$\mathbf{X}_1 = \mathbf{A}^{-1} \begin{bmatrix} 114 \\ 81 \\ 58 \end{bmatrix} = \begin{bmatrix} 0 & 1 & -1 \\ 2 & -2 & -1 \\ -1 & 1 & 1 \end{bmatrix} \begin{bmatrix} 114 \\ 81 \\ 58 \end{bmatrix} = \begin{bmatrix} 23 \\ 8 \\ 25 \end{bmatrix}.$$

Similarly,

$$\mathbf{X}_2 = \mathbf{A}^{-1} \begin{bmatrix} 104 \\ 69 \\ 55 \end{bmatrix} = \begin{bmatrix} 14 \\ 15 \\ 20 \end{bmatrix}.$$

Using our correspondence between letters and numbers, we have received the message WHY NOT.

Additional material on cryptography may be found in the references given in Further Reading at the end of this section.

Example 9 Let **L**: $R^2 \to R^2$ be defined by

$$\mathbf{L}(x, y) = (x, -y).$$

Then **L** is a linear transformation, which is shown in Figure 4.4. This linear transformation is called a **reflection with respect to the x-axis**. The reader may consider a **reflection with respect to the y-axis**.

Example 10 Let W be the vector space of all real-valued functions and let V be the subspace of all
(Calculus differentiable functions. Let **L**: $V \to W$ be defined by
Required)

$$\mathbf{L}(f) = f',$$

where f' is the derivative of f. It is easy to show (Exercise 9), from the properties of differentiation, that **L** is a linear transformation.

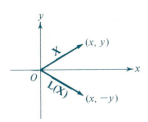

Figure 4.4

Example 11 Let $V = C[0, 1]$ be the vector space of all real-valued functions that are continuous on the
(Calculus interval $[0, 1]$. Let $W = R^1$. Define $\mathbf{L}: V \to W$ by
Required)

$$\mathbf{L}(f) = \int_0^1 f(x) \, dx.$$

It is easy to show (Exercise 10), from the properties of integration, that $\mathbf{L}$ is a linear
transformation.

Theorem 4.1 *If $\mathbf{L}: V \to W$ is a linear transformation, then*

$$\mathbf{L}(c_1\mathbf{X}_1 + c_2\mathbf{X}_2 + \cdots + c_k\mathbf{X}_k) = c_1\mathbf{L}(\mathbf{X}_1) + c_2\mathbf{L}(\mathbf{X}_2) + \cdots + c_k\mathbf{L}(\mathbf{X}_k)$$

for any vectors $\mathbf{X}_1, \mathbf{X}_2, \ldots, \mathbf{X}_k$ in V and any scalars $c_1, c_2, \ldots, c_k$.

Proof
Exercise T.2.

The following theorem gives some basic properties of linear transformations.

Theorem 4.2 *Let $\mathbf{L}: V \to W$ be a linear transformation. Then:*
(a) $\mathbf{L}(\mathbf{0}_V) = \mathbf{0}_W$, where $\mathbf{0}_V$ and $\mathbf{0}_W$ are the zero vectors in V and W, respectively.
(b) $\mathbf{L}(\mathbf{X} - \mathbf{Y}) = \mathbf{L}(\mathbf{X}) - \mathbf{L}(\mathbf{Y})$.

Proof
(a) We have

$$\mathbf{0}_V = \mathbf{0}_V + \mathbf{0}_V.$$

Then

$$\mathbf{L}(\mathbf{0}_V) = \mathbf{L}(\mathbf{0}_V + \mathbf{0}_V) = \mathbf{L}(\mathbf{0}_V) + \mathbf{L}(\mathbf{0}_V). \tag{1}$$

Subtracting $\mathbf{L}(\mathbf{0}_V)$ from both sides of (1), we obtain

$$\mathbf{L}(\mathbf{0}_V) = \mathbf{0}_W.$$

(b) Exercise T.3.

A function f mapping a set V into a set W can be specified by a formula that assigns to
every member of V a unique element of W. On the other hand, we can also specify a
function by listing next to each member of V its assigned element of W. An example of

this would be provided by listing the names of all charge account customers of a department store along with their charge account number. At first glance, it appears impossible to describe a linear transformation $L: V \rightarrow W$ of a vector space $V \neq \{0\}$ into a vector space W in this latter manner, since V has infinitely many members in it. However, the following very useful theorem tells us that once we know what L does to a basis for V, then we have completely determined L. Thus for a finite-dimensional vector space V, it is possible to describe L by giving only the images of a finite number of vectors in V.

Theorem 4.3 *Let* $L: V \rightarrow W$ *be a linear transformation of an n-dimensional vector space V into a vector space W. Also let* $S = \{X_1, X_2, \ldots, X_n\}$ *be a basis for V. If X is any vector in V, then* $L(X)$ *is completely determined by* $\{L(X_1), L(X_2), \ldots, L(X_n)\}$.

Proof

Since X is in V, we can write

$$X = c_1 X_1 + c_2 X_2 + \cdots + c_n X_n, \qquad (2)$$

where $c_1, c_2, \ldots, c_n$ are uniquely determined real numbers. Then

$$L(X) = L(c_1 X_1 + c_2 X_2 + \cdots + c_n X_n)$$
$$= c_1 L(X_1) + c_2 L(X_2) + \cdots + c_n L(X_n),$$

by Theorem 4.1. Thus $L(X)$ has been completely determined by the elements $L(X_1)$, $L(X_2), \ldots, L(X_n)$.

It might be noted that in the proof of Theorem 4.3, the real numbers c_i determined in Equation (2) depend on the basis vectors in S. Thus, if we change S, then we may change the c_i's.

Example 12 Let $L: R^2 \rightarrow R^2$ be a linear transformation for which we know that $L(1, 0) = (2, -2)$ and $L(0, 1) = (4, 1)$.
(a) What is $L(3, -2)$?
(b) What is $L(a, b)$?

Solution

(a) The set $\{(1, 0), (0, 1)\}$ is a basis for R^2, and

$$(3, -2) = 3(1, 0) + 2(0, 1).$$

Then, by Theorem 4.3,

$$L(3, -2) = L(3(1, 0) + 2(0, 1))$$
$$= 3L(1, 0) - 2L(0, 1)$$
$$= 3(2, -2) - 2(4, 1)$$
$$= (-2, -8)$$

(b) Proceeding as in part (a),

$$(a, b) = a(1, 0) + b(0, 1)$$

so by Theorem 4.3,

$$\begin{aligned}
\mathbf{L}(a, b) &= \mathbf{L}(a(1, 0) + b(0, 1)) \\
&= a\mathbf{L}(1, 0) + b\mathbf{L}(0, 1) \\
&= a(2, -2) + b(4, 1) \\
&= (2a + 4b, -2a + b)
\end{aligned}$$

FURTHER READING IN CRYPTOGRAPHY

Elementary presentation
Kohn, Bernice. *Secret Codes and Ciphers.* Englewood Cliffs, N.J.: Prentice-Hall, Inc., 1968 (63 pages).

Advanced presentation
Fisher, James L. *Applications-Oriented Algebra.* New York: T. Harper & Row, Publishers, 1977 (Chapter 9, "Coding Theory").
Kahn, David. *The Codebreakers.* New York: The New American Library Inc., 1973.

4.1 EXERCISES

1. Which of the following are linear transformations?
(a) $\mathbf{L}(x, y) = (x + 1, y, x + y)$.

(b) $\mathbf{L}\left(\begin{bmatrix} x \\ y \\ z \end{bmatrix}\right) = \begin{bmatrix} x + y \\ y \\ x - z \end{bmatrix}$.

(c) $\mathbf{L}(x, y) = (x^2 + x, y - y^2)$.

2. Which of the following are linear transformations?
(a) $\mathbf{L}(x, y, z) = (x - y, x^2, 2z)$.

(b) $\mathbf{L}\left(\begin{bmatrix} x \\ y \\ z \end{bmatrix}\right) = \begin{bmatrix} 2x - 3y \\ 3y - 2z \\ 2z \end{bmatrix}$.

(c) $\mathbf{L}(x, y) = (x - y, 2x + 2)$.

3. Which of the following are linear transformations?
(a) $\mathbf{L}(x, y, z) = (x + y, 0, 2x - z)$.

(b) $\mathbf{L}\left(\begin{bmatrix} x \\ y \end{bmatrix}\right) = \begin{bmatrix} x^2 - y^2 \\ x^2 + y^2 \end{bmatrix}$.

(c) $\mathbf{L}(x, y) = (x - y, 0, 2x + 3)$.

4. Let $\mathbf{L}: R^3 \to R^2$ be a linear transformation for which we know that $\mathbf{L}(1, 0, 0) = (2, -4)$, $\mathbf{L}(0, 1, 0) = (3, -5)$, $\mathbf{L}(0, 0, 1) = (2, 3)$.
(a) What is $\mathbf{L}(1, -2, 3)$?
(b) What is $\mathbf{L}(a, b, c)$?

5. Let $\mathbf{L}: R^2 \to R^2$ be a linear transformation for which we know that

$$\mathbf{L}\left(\begin{bmatrix} 1 \\ 1 \end{bmatrix}\right) = \begin{bmatrix} 2 \\ -3 \end{bmatrix} \quad \text{and} \quad \mathbf{L}\left(\begin{bmatrix} 0 \\ 1 \end{bmatrix}\right) = \begin{bmatrix} 1 \\ 2 \end{bmatrix}.$$

(a) What is $\mathbf{L}\left(\begin{bmatrix} 3 \\ -2 \end{bmatrix}\right)$?

(b) What is $\mathbf{L}\left(\begin{bmatrix} a \\ b \end{bmatrix}\right)$?

6. Describe the following linear transformations geometrically.
(a) $\mathbf{L}(x, y) = (-x, y)$.
(b) $\mathbf{L}(x, y) = (-x, -y)$.
(c) $\mathbf{L}(x, y) = (-y, x)$.

7. Verify that the functions in Examples 3, 4, and 9 are linear transformations.

8. Let $\mathbf{L}: P_2 \to P_3$ be a linear transformation for which we know that $\mathbf{L}(1) = 1$, $\mathbf{L}(t) = t^2$, and $\mathbf{L}(t^2) = t^3 + t$.
(a) Find $\mathbf{L}(2t^2 - 5t + 3)$.
(b) Find $\mathbf{L}(at^2 + bt + c)$.

9. (Calculus Required). Verify that the function in Example 10 is a linear transformation.

10. **(Calculus Required).** Verify that the function in Example 11 is a linear transformation.

11. Use the substitution and matrix **A** of Example 8.
    (a) Code the message SEND HIM MONEY.
    (b) Decode the message 67, 44, 41, 49, 39, 19, 113, 76, 62, 104, 69, 55.

12. Use the substitution scheme of Example 8 and the matrix

$$\mathbf{A} = \begin{bmatrix} 5 & 3 \\ 2 & 1 \end{bmatrix}.$$

   (a) Code the message WORK HARD.
   (b) Decode the message 93, 36, 60, 21, 159, 60, 110, 43.

THEORETICAL EXERCISES

T.1. Verify that the function in Example 5 is a linear transformation.

T.2. Prove Theorem 4.1.

T.3. Prove (b) of Theorem 4.2.

T.4. Prove that $\mathbf{L}: V \rightarrow W$ is a linear transformation if and only if

$$\mathbf{L}(a\mathbf{X} + b\mathbf{Y}) = a\mathbf{L}(\mathbf{X}) + b\mathbf{L}(\mathbf{Y}),$$

for any scalars a and b and any vectors $\mathbf{X}$ and $\mathbf{Y}$ in V.

T.5. Let V be the vector space of all $n \times n$ matrices and consider the following function Tr: $V \rightarrow R^1$ (the **trace**): If $\mathbf{A} = [a_{ij}]$ is in V, then $\text{Tr}(\mathbf{A}) = a_{11} + a_{22} + \cdots + a_{nn}$. Prove that Tr is a linear trans-

formation (see Supplementary Exercise T.1 in Chapter 1).

T.6. Let V be the vector space of all $n \times n$ matrices and let $\mathbf{L}: V \rightarrow V$ be the function defined by $\mathbf{L}(\mathbf{A}) = \mathbf{A}^T$ for **A** in V. Is **L** a linear transformation? Justify your answer.

T.7. Let V be the vector space of all $n \times n$ matrices and let $\mathbf{L}: V \rightarrow V$ be the function defined by

$$\mathbf{L}(\mathbf{A}) = \begin{cases} \mathbf{A}^{-1} & \text{if } \mathbf{A} \text{ is nonsingular} \\ \mathbf{0} & \text{if } \mathbf{A} \text{ is singular} \end{cases}$$

for **A** in V. Is **L** a linear transformation? Justify your answer.

4.2 _____

THE KERNEL AND RANGE OF A LINEAR TRANSFORMATION

In this section we study special types of linear transformations; we formulate the notions of one-to-one linear transformations and onto linear transformations. We also develop methods for determining when a linear transformation is one-to-one or onto.

Definition A linear transformation $\mathbf{L}: V \rightarrow W$ is said to be **one-to-one** if for all $\mathbf{X}_1$, $\mathbf{X}_2$, in V, $\mathbf{X}_1 \neq \mathbf{X}_2$ implies that $\mathbf{L}(\mathbf{X}_1) \neq \mathbf{L}(\mathbf{X}_2)$. An equivalent statement is that **L** is one-to-one if for all $\mathbf{X}_1$, $\mathbf{X}_2$, in V, $\mathbf{L}(\mathbf{X}_1) = \mathbf{L}(\mathbf{X}_2)$ implies that $\mathbf{X}_1 = \mathbf{X}_2$.

This definition says that **L** is one-to-one if $\mathbf{L}(\mathbf{X}_1)$ and $\mathbf{L}(\mathbf{X}_2)$ are distinct whenever $\mathbf{X}_1$ and $\mathbf{X}_2$ are distinct (Figure 4.5).

Example 1 Let $\mathbf{L}: R^2 \rightarrow R^2$ be defined by

$$\mathbf{L}(x, y) = (x + y, x - y).$$

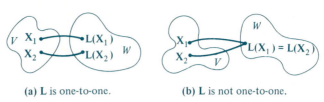

(a) **L** is one-to-one.   (b) **L** is not one-to-one.

Figure 4.5

To determine whether **L** is one-to-one, we let

$$\mathbf{X}_1 = (x_1, y_1) \quad \text{and} \quad \mathbf{X}_2 = (x_2, y_2).$$

Then if

$$\mathbf{L}(\mathbf{X}_1) = \mathbf{L}(\mathbf{X}_2),$$

we have

$$x_1 + y_1 = x_2 + y_2$$
$$x_1 - y_1 = x_2 - y_2.$$

Adding these equations, we obtain $2x_1 = 2x_2$, or $x_1 = x_2$, which implies that $y_1 = y_2$. Hence, $\mathbf{X}_1 = \mathbf{X}_2$ and **L** is one-to-one.

Example 2 Let **L**: $R^3 \to R^2$ be the linear transformation defined in Example 1 of Section 4.1 (the projection function) by

$$\mathbf{L}(x, y, z) = (x, y).$$

Since $(1, 3, 3) \neq (1, 3, -2)$, but

$$\mathbf{L}(1, 3, 3) = \mathbf{L}(1, 3, -2) = (1, 3),$$

we conclude that **L** is not one-to-one.

We shall now develop some more efficient ways of determining whether or not a linear transformation is one-to-one.

Definition Let **L**: $V \to W$ be a linear transformation. The **kernel** of **L**, ker **L**, is the subset of V consisting of all vectors **X** such that $\mathbf{L}(\mathbf{X}) = \mathbf{0}_W$.

We observe that property (a) of Theorem 4.2 of Section 4.1 assures us that ker **L** is never an empty set, since $\mathbf{0}_V$ is in ker **L**.

Example 3 Let **L**: $R^3 \to R^2$ be as defined in Example 2. The vector $(0, 0, 2)$ is in ker **L**, since $\mathbf{L}(0, 0, 2) = (0, 0)$. However, the vector $(2, -3, 4)$ is not in ker **L**, since $\mathbf{L}(2, -3, 4) = (2, -3)$. To find ker **L**, we must determine all **X** in R^3 so that $\mathbf{L}(\mathbf{X}) = \mathbf{0}$. That is, we seek $\mathbf{X} = (x_1, x_2, x_3)$ so that $\mathbf{L}(\mathbf{X}) = \mathbf{L}(x_1, x_2, x_3) = \mathbf{0} = (0, 0)$. However, $\mathbf{L}(\mathbf{X}) = (x_1, x_2)$.

Thus $(x_1, x_2) = (0, 0)$, so $x_1 = 0$, $x_2 = 0$, and x_3 can be any real number. Hence, ker $\mathbf{L}$ consists of all vectors in R^3 of the form $(0, 0, r)$, where r is any real number. It is clear that ker $\mathbf{L}$ consists of the z-axis in x, y, z three-dimensional space R^3.

Example 4 If $\mathbf{L}$ is as defined in Example 1, then ker $\mathbf{L}$ consists of all vectors $\mathbf{X}$ in R^2 such that $\mathbf{L}(\mathbf{X}) = \mathbf{0}$. Thus we must solve the linear system

$$x + y = 0$$
$$x - y = 0$$

for x and y. The only solution is $\mathbf{X} = \mathbf{0}$, so ker $\mathbf{L} = \{\mathbf{0}\}$.

Example 5 If $\mathbf{L}: R^4 \rightarrow R^2$ is defined by

$$\mathbf{L}\left(\begin{bmatrix} x \\ y \\ z \\ w \end{bmatrix}\right) = \begin{bmatrix} x + y \\ z + w \end{bmatrix},$$

then ker $(\mathbf{L})$ consists of all vectors $\mathbf{X}$ in R^4 such that $\mathbf{L}(\mathbf{X}) = \mathbf{0}$. This leads to the linear system

$$x + y = 0$$
$$z + w = 0.$$

Thus ker $\mathbf{L}$ consists of all vectors of the form

$$\begin{bmatrix} r \\ -r \\ s \\ -s \end{bmatrix},$$

where r and s are any real numbers.

Theorem 4.4 *If* $\mathbf{L}: V \rightarrow W$ *is a linear transformation, then* ker $\mathbf{L}$ *is a subspace of V.*

Proof
First, observe that ker $\mathbf{L}$ is not an empty set, since $\mathbf{0}_V$ is in ker $\mathbf{L}$. Also, let $\mathbf{X}$ and $\mathbf{Y}$ be in ker $\mathbf{L}$. Then since $\mathbf{L}$ is a linear transformation,

$$\mathbf{L}(\mathbf{X} + \mathbf{Y}) = \mathbf{L}(\mathbf{X}) + \mathbf{L}(\mathbf{Y}) = \mathbf{0}_W + \mathbf{0}_W = \mathbf{0}_W,$$

so $\mathbf{X} + \mathbf{Y}$ is in ker $\mathbf{L}$. Also, if c is a scalar, then since $\mathbf{L}$ is a linear transformation,

$$\mathbf{L}(c\mathbf{X}) = c\mathbf{L}(\mathbf{X}) = c\mathbf{0}_W = \mathbf{0}_W,$$

so $c\mathbf{X}$ is in ker $\mathbf{L}$. Hence ker $\mathbf{L}$ is a subspace of V.

Example 6 If L is as in Examples 1 and 4, then ker L is the subspace $\{0\}$; its dimension is zero.

Example 7 If L is as in Examples 2 and 3, then a basis for ker L is

$$\{(0, 0, 1)\}$$

and dim (ker L) = 1. Thus ker L consists of the z-axis in x, y, z three-dimensional space R^3.

Example 8 If L is as in Example 5, then a basis for ker L consists of the vectors

$$\begin{bmatrix} 1 \\ -1 \\ 0 \\ 0 \end{bmatrix} \quad \text{and} \quad \begin{bmatrix} 0 \\ 0 \\ 1 \\ -1 \end{bmatrix};$$

thus dim (ker L) = 2.

If $L: R^n \to R^m$ is a linear transformation defined by $L(X) = AX$, where A is an $m \times n$ matrix, then the kernel of L is the solution space of the homogeneous system $AX = 0$.

An examination of the elements in ker L allows us to decide whether L is or is not one-to-one.

Theorem 4.5 *A linear transformation* $L: V \to W$ *is one-to-one if and only if* ker $L = \{0_V\}$.

Proof
Let L be one-to-one. We show that ker $L = \{0_V\}$. Let X be in ker L. Then $L(X) = 0_W$. Also, we already know that $L(0_V) = 0_W$. Thus $L(X) = L(0_V)$. Since L is one-to-one, we conclude that $X = 0_V$. Hence ker $L = \{0_V\}$.

Conversely, suppose that ker $L = \{0_V\}$. We wish to show that L is one-to-one. Assume that $L(X_1) = L(X_2)$, for X_1 and X_2 in V. Then

$$L(X_1) - L(X_2) = 0_W,$$

so $L(X_1 - X_2) = 0_W$, which means that $X_1 - X_2$ is in ker L. Therefore, $X_1 - X_2 = 0_V$, so $X_1 = X_2$. Thus L is one-to-one.

Example 9 The linear transformation of Examples 1 and 4 is one-to-one; the one in Examples 2 and 3 is not.

In Section 4.3 we shall prove that for every linear transformation $L: R^n \to R^m$, we can find a unique $m \times n$ matrix A so that if X is in R^n, then $L(X) = AX$. It then follows that to find ker L, we need to find the solution space of the homogeneous system $AX = 0$. Hence to find ker L we need only use techniques with which we are already familiar.

Definition   If $L: V \rightarrow W$ is a linear transformation, then the **range** of L, denoted by range L, is the set of all vectors in W that are images, under L, of vectors in V. Thus a vector Y is in range L if we can find some vector X in V such that $L(X) = Y$. If range $L = W$, we say that L is **onto**.

Theorem 4.6   *If $L: V \rightarrow W$ is a linear transformation, then range L is a subspace of W.*

Proof

First, observe that range L is not an empty set, since $0_W = L(0_V)$, so 0_W is in range L. Let Y_1 and Y_2 be in range L. Then $Y_1 = L(X_1)$ and $Y_2 = L(X_2)$ for some X_1 and X_2 in V. Now

$$Y_1 + Y_2 = L(X_1) + L(X_2) = L(X_1 + X_2),$$

which implies that $Y_1 + Y_2$ is in range L. Also, if c is a scalar, then $cY_1 = cL(X_1) = L(cX_1)$, so cY_1 is in range L. Hence range L is a subspace of W.

Example 10   Let L be the linear transformation defined in Example 2. To find out whether L is onto, we choose any vector $Y = (x_2, y_2)$ in R^2 and seek a vector $X = (x_1, y_1, z_1)$ in R^3 such that $L(X) = Y$. Since $L(X) = (x_1, y_1)$, we find that if $x_1 = x_2$ and $y_1 = y_2$, then $L(X) = Y$. Therefore, L is onto and dimension of range L is 2.

Example 11   Let $L: R^3 \rightarrow R^3$ be defined by

$$L\left(\begin{bmatrix} x_1 \\ x_2 \\ x_3 \end{bmatrix}\right) = \begin{bmatrix} 1 & 0 & 1 \\ 1 & 1 & 2 \\ 2 & 1 & 3 \end{bmatrix}\begin{bmatrix} x_1 \\ x_2 \\ x_3 \end{bmatrix}.$$

Is L onto? Given any $Y = \begin{bmatrix} y_1 \\ y_2 \\ y_3 \end{bmatrix}$ in R^3, where y_1, y_2, and y_3 are any real numbers,

can we find $X = \begin{bmatrix} x_1 \\ x_2 \\ x_3 \end{bmatrix}$ in R^3 so that $L(X) = Y$? We are seeking a solution to the linear system

$$\begin{bmatrix} 1 & 0 & 1 \\ 1 & 1 & 2 \\ 2 & 1 & 3 \end{bmatrix}\begin{bmatrix} x_1 \\ x_2 \\ x_3 \end{bmatrix} = \begin{bmatrix} y_1 \\ y_2 \\ y_3 \end{bmatrix}.$$

The reduced row echelon form of the augmented matrix is (verify)

$$\begin{bmatrix} 1 & 0 & 1 & \vdots & y_1 \\ 0 & 1 & 1 & \vdots & y_2 - y_1 \\ 0 & 0 & 0 & \vdots & y_3 - y_2 - y_1 \end{bmatrix}.$$

Thus a solution exists only for $y_3 - y_2 - y_1 = 0$, and so **L** is not onto.

To find a basis for range **L**, we note that

$$\mathbf{L}\left(\begin{bmatrix} x_1 \\ x_2 \\ x_3 \end{bmatrix}\right) = \begin{bmatrix} 1 & 0 & 1 \\ 1 & 1 & 2 \\ 2 & 1 & 3 \end{bmatrix}\begin{bmatrix} x_1 \\ x_2 \\ x_3 \end{bmatrix} = \begin{bmatrix} x_1 + x_3 \\ x_1 + x_2 + 2x_3 \\ 2x_1 + x_2 + 3x_3 \end{bmatrix}$$

$$= x_1\begin{bmatrix} 1 \\ 1 \\ 2 \end{bmatrix} + x_2\begin{bmatrix} 0 \\ 1 \\ 1 \end{bmatrix} + x_3\begin{bmatrix} 1 \\ 2 \\ 3 \end{bmatrix}.$$

Since

$$\mathbf{Y}_1 = \begin{bmatrix} 1 \\ 1 \\ 2 \end{bmatrix} = \mathbf{L}\left(\begin{bmatrix} 1 \\ 0 \\ 0 \end{bmatrix}\right), \qquad \mathbf{Y}_2 = \begin{bmatrix} 0 \\ 1 \\ 1 \end{bmatrix} = \mathbf{L}\left(\begin{bmatrix} 0 \\ 1 \\ 0 \end{bmatrix}\right),$$

and

$$\mathbf{Y}_3 = \begin{bmatrix} 1 \\ 2 \\ 3 \end{bmatrix} = \mathbf{L}\left(\begin{bmatrix} 0 \\ 0 \\ 1 \end{bmatrix}\right),$$

$\mathbf{Y}_1$, $\mathbf{Y}_2$, and $\mathbf{Y}_3$ belong to range **L**. From the last equation it follows that $\{\mathbf{Y}_1, \mathbf{Y}_2, \mathbf{Y}_3\}$ spans range **L**. That is, range **L** is the subspace of R^3 spanned by the columns of the matrix defining **L**. Now $\{\mathbf{Y}_1, \mathbf{Y}_2\}$ is linearly independent, while $\mathbf{Y}_3 = \mathbf{Y}_1 + \mathbf{Y}_2$. Thus $\mathbf{Y}_1$ and $\mathbf{Y}_2$ form a basis for range **L**, and dim (range **L**) = 2.

It is also interesting to find dim (ker **L**) for this example; that is, we wish to find all **X** in R^3 so that $\mathbf{L}(\mathbf{X}) = \mathbf{0}$. From the reduced echelon form of the augmented matrix we find that $x_1 = -x_3$ and $x_2 = -x_3$. Thus ker **L** consists of all vectors of the form $\begin{bmatrix} -r \\ -r \\ r \end{bmatrix}$, where r is any real number. A basis for ker **L** is given by the vector $\begin{bmatrix} -1 \\ -1 \\ 1 \end{bmatrix}$; **L** is not one-to-one and dim (ker **L**) = 1.

A basis for range **L** can be obtained by the method discussed in Section 3.7 as shown in the following example.

Example 12 Let $\mathbf{L}: R^4 \rightarrow R^3$ be defined by

$$\mathbf{L}(a_1, a_2, a_3, a_4) = (a_1 + a_2, a_3 + a_4, a_1 + a_3).$$

Find a basis for range **L**.

Solution

We have

$$\mathbf{L}(a_1, a_2, a_3, a_4) = a_1(1, 0, 1) + a_2(1, 0, 0)$$
$$+ a_3(0, 1, 1) + a_4(0, 1, 0).$$

Thus
$$\{(1, 0, 1), (1, 0, 0), (0, 1, 1), (0, 1, 0)\}$$

spans range **L**. Proceeding as in Section 3.7, we form the matrix whose rows are the given vectors

$$\begin{bmatrix} 1 & 0 & 1 \\ 1 & 0 & 0 \\ 0 & 1 & 1 \\ 0 & 1 & 0 \end{bmatrix},$$

which we transform to reduced row echelon form, obtaining (verify)

$$\begin{bmatrix} 1 & 0 & 0 \\ 0 & 1 & 0 \\ 0 & 0 & 1 \\ 0 & 0 & 0 \end{bmatrix}.$$

Hence $\{(1, 0, 0), (0, 1, 0), (0, 0, 1)\}$ is a basis for range **L**.

To find out if a linear transformation is one-to-one or onto, we must solve a linear system of equations. This is one further demonstration of the frequency with which linear systems must be solved to answer many questions in linear algebra.

We also observe that in Example 11 we have

$$\dim (\ker \mathbf{L}) + \dim (\text{range } \mathbf{L}) = \dim V.$$

This very important result is always true, and we now prove it in the following theorem.

Theorem 4.7 *If* **L**: $V \rightarrow W$ *is a linear transformation, then*

$$\dim (\ker \mathbf{L}) + \dim (\text{range } \mathbf{L}) = \dim V. \qquad (1)$$

Proof

Let $n = \dim V$ and $k = \dim (\ker \mathbf{L})$. If $k = n$, then $\ker \mathbf{L} = V$ (Exercise T.8, Section 3.6), which implies that $\mathbf{L(X)} = \mathbf{0}$ for every **X** in V. Hence range $\mathbf{L} = \{\mathbf{0}\}$, dim (range **L**) = 0, and the conclusion holds. Suppose that $1 \leq k < n$. We shall prove that dim (range **L**) = $n - k$. Let $\{\mathbf{X}_1, \mathbf{X}_2, \ldots, \mathbf{X}_k\}$ be a basis for ker **L**. By Theorem 3.13 of Section 3.6 we can extend this basis to a basis $S = \{\mathbf{X}_1, \mathbf{X}_2, \ldots, \mathbf{X}_k, \mathbf{X}_{k+1}, \ldots, \mathbf{X}_n\}$ for V. We now prove that the set $T = \{\mathbf{L(X}_{k+1}), \mathbf{L(X}_{k+2}), \ldots, \mathbf{L(X}_n)\}$ is a basis for range **L**, which will prove that (1) holds.

First, we show that T spans range **L**. Let **Y** be any vector in range **L**. Then $\mathbf{Y} = \mathbf{L(X)}$ for some **X** in V. Since S is a basis for V, we can write

$$\mathbf{X} = c_1\mathbf{X}_1 + c_2\mathbf{X}_2 + \cdots + c_n\mathbf{X}_n,$$

where $c_1, c_2, \ldots, c_n$ are real numbers. Then

$$\mathbf{Y} = \mathbf{L(X)}$$
$$= \mathbf{L}(c_1\mathbf{X}_1 + c_2\mathbf{X}_2 + \cdots + c_k\mathbf{X}_k + c_{k+1}\mathbf{X}_{k+1} + \cdots + c_n\mathbf{X}_n)$$

$$= c_1 \mathbf{L}(\mathbf{X}_1) + c_2 \mathbf{L}(\mathbf{X}_2) + \cdots + c_k \mathbf{L}(\mathbf{X}_k)$$
$$+ c_{k+1} \mathbf{L}(\mathbf{X}_{k+1}) + \cdots + c_n \mathbf{L}(\mathbf{X}_n)$$
$$= c_{k+1} \mathbf{L}(\mathbf{X}_{k+1}) + \cdots + c_n \mathbf{L}(\mathbf{X}_n),$$

because $\mathbf{X}_1, \mathbf{X}_2, \ldots, \mathbf{X}_k$ are in ker $\mathbf{L}$. Hence T spans range $\mathbf{L}$.

Now we show that T is linearly independent. Suppose that

$$c_{k+1} \mathbf{L}(\mathbf{X}_{k+1}) + c_{k+2} \mathbf{L}(\mathbf{X}_{k+2}) + \cdots + c_n \mathbf{L}(\mathbf{X}_n) = \mathbf{0}_W; \qquad (2)$$

then

$$\mathbf{L}(c_{k+1} \mathbf{X}_{k+1} + c_{k+2} \mathbf{X}_{k+2} + \cdots + c_n \mathbf{X}_n) = \mathbf{0}_W.$$

Hence the vector $c_{k+1} \mathbf{X}_{k+1} + c_{k+2} \mathbf{X}_{k+2} + \cdots + c_n \mathbf{X}_n$ is in ker $\mathbf{L}$, and we can write it as a linear combination of the vectors in the basis for ker $\mathbf{L}$:

$$c_{k+1} \mathbf{X}_{k+1} + c_{k+2} \mathbf{X}_{k+2} + \cdots + c_n \mathbf{X}_n = d_1 \mathbf{X}_1 + d_2 \mathbf{X}_2 + \cdots + d_k \mathbf{X}_k,$$

where $d_1, d_2, \ldots, d_k$ are real numbers. Then

$$d_1 \mathbf{X}_1 + d_2 \mathbf{X}_2 + \cdots + d_k \mathbf{X}_k - c_{k+1} \mathbf{X}_{k+1} - c_{k+2} \mathbf{X}_{k+2} - \cdots - c_n \mathbf{X}_n = \mathbf{0}_V.$$

Since S is linearly independent, we conclude that

$$d_1 = d_2 = \cdots = d_k = c_{k+1} = \cdots = c_n = 0.$$

Referring back to Equation (2), we find that this means that T is linearly independent and is a basis for range $\mathbf{L}$.

If $k = 0$, then ker $\mathbf{L}$ has no basis: we let $\{\mathbf{X}_1, \mathbf{X}_2, \ldots, \mathbf{X}_n\}$ be a basis for V. The proof now proceeds with minor modifications as above.

The dimension of ker $\mathbf{L}$ is also called the **nullity** of $\mathbf{L}$.

We have seen that a linear transformation may be one-one and onto or onto and not one-one. However, the following corollary shows that each of these properties implies the other if the vector spaces V and W have the same dimensions.

Corollary 4.1   *Let* $\mathbf{L}: V \to W$ *be a linear transformation and let* dim V = dim W.
   (a) *If* $\mathbf{L}$ *is one-to-one, then it is onto.*
   (b) *If* $\mathbf{L}$ *is onto, then it is one-to-one.*

Proof
Exercise T.1.

Consider now the homogeneous system

$$\mathbf{AX} = \mathbf{0}, \qquad (3)$$

where $\mathbf{A}$ is $m \times n$. The solution space W is a subspace of R^n; we now study the dimension of W. We examine the linear transformation $\mathbf{L}: R^n \to R^m$ defined by

$$\mathbf{L}(\mathbf{X}) = \mathbf{AX},$$

for X in R^n. It is easy to show (Exercise T.2) that the column space of A is range L. Also, as mentioned above, the solution space W of (3) is the kernel of L. By Equation (1)

$$n = \dim (\ker L) + \dim (\text{range } L).$$

Since $\dim (\text{range } L) = \dim (\text{column space of } A) = \text{rank } A$ (see Section 3.7), we conclude that

$$\dim W = \dim (\ker L) = n - \text{rank } A.$$

Thus the dimension of the solution space of (3) is the number of columns of A minus its rank.

Example 13 Consider the homogeneous system of Example 11 of Section 1.4. The coefficient matrix is

$$A = \begin{bmatrix} 1 & 1 & 1 & 1 \\ 1 & 0 & 0 & 1 \\ 1 & 2 & 1 & 0 \end{bmatrix}.$$

which is row equivalent to (verify)

$$B = \begin{bmatrix} 1 & 0 & 0 & 1 \\ 0 & 1 & 0 & -1 \\ 0 & 0 & 1 & 1 \end{bmatrix}.$$

The matrix B is in reduced row echelon form and rank $A = 3$. Hence the dimension of the solution space is $4 - 3 = 1$. In Example 11 of Section 1.4 we found that the solution to the system is of the form

$$X = \begin{bmatrix} -r \\ r \\ -r \\ r \end{bmatrix},$$

where r is any real number. Thus a basis for the solution space is the vector

$$\begin{bmatrix} -1 \\ 1 \\ -1 \\ 1 \end{bmatrix},$$

which confirms the fact that the dimension of the solution space is 1.

If $L: R^n \to R^n$ is a linear transformation defined by $L(X) = AX$, where A is an $n \times n$ matrix, using Theorem 4.5, Equation (1), and Corollary 3.2, we can show (Exercise T.3) that L is one-to-one if and only if $|A| \neq 0$.

4.2 EXERCISES

1. Let $L: R^2 \rightarrow R^2$ be the linear transformation defined by $L(a_1, a_2) = (a_1, 0)$.
   (a) Is $(0, 2)$ in ker L?   (b) Is $(2, 2)$ in ker L?
   (c) Is $(3, 0)$ in range L? (d) Is $(3, 2)$ in range L?
   (e) Find ker L.
   (f) Find range L.

2. Let $L: R^2 \rightarrow R^2$ be the linear transformation defined by $L\left(\begin{bmatrix} a_1 \\ a_2 \end{bmatrix}\right) = \begin{bmatrix} 1 & 2 \\ 2 & 4 \end{bmatrix}\begin{bmatrix} a_1 \\ a_2 \end{bmatrix}$.

   (a) Is $\begin{bmatrix} 1 \\ 2 \end{bmatrix}$ in ker L?   (b) Is $\begin{bmatrix} 2 \\ -1 \end{bmatrix}$ in ker L?

   (c) Is $\begin{bmatrix} 3 \\ 6 \end{bmatrix}$ in range L?   (d) Is $\begin{bmatrix} 2 \\ 3 \end{bmatrix}$ in range L?

   (e) Find ker L.
   (f) Find a set of vectors spanning range L.

3. Let $L: R^2 \rightarrow R^3$ be defined by
$$L(x, y) = (x, x + y, y).$$
   (a) Find ker L.
   (b) Is L one-to-one?
   (c) Is L onto?

4. Let $L: R^4 \rightarrow R^3$ be defined by
$$L(x, y, z, w) = (x + y, z + w, x + z).$$
   (a) Find a basis for ker L.
   (b) Find a basis for range L.
   (c) Verify Theorem 4.7.

5. Let $L: R^5 \rightarrow R^4$ be defined by
$$L\left(\begin{bmatrix} x_1 \\ x_2 \\ x_3 \\ x_4 \\ x_5 \end{bmatrix}\right) = \begin{bmatrix} 1 & 0 & -1 & 3 & -1 \\ 1 & 0 & 0 & 2 & -1 \\ 2 & 0 & -1 & 5 & -1 \\ 0 & 0 & -1 & 1 & 0 \end{bmatrix}\begin{bmatrix} x_1 \\ x_2 \\ x_3 \\ x_4 \\ x_5 \end{bmatrix}.$$
   (a) Find a basis for ker L.
   (b) Find a basis for range L.
   (c) Verify Theorem 4.7.

6. Let $L: R^3 \rightarrow R^3$ be defined by
$$L\left(\begin{bmatrix} x \\ y \\ z \end{bmatrix}\right) = \begin{bmatrix} 4 & 2 & 2 \\ 2 & 3 & -1 \\ -1 & 1 & -2 \end{bmatrix}\begin{bmatrix} x \\ y \\ z \end{bmatrix}.$$
   (a) Is L one-to-one?
   (b) Find the dimension of range L.

7. Let $L: R^4 \rightarrow R^3$ be defined by
$$L\left(\begin{bmatrix} x \\ y \\ z \\ w \end{bmatrix}\right) = \begin{bmatrix} x + y \\ y - z \\ z - w \end{bmatrix}.$$
   (a) Is L onto?
   (b) Find the dimension of ker L.
   (c) Verify Theorem 4.7.

8. Let $L: R^3 \rightarrow R^3$ be defined by
$$L(x, y, z) = (x - y, x + 2y, z).$$
   (a) Find a basis for ker L.
   (b) Find a basis for range L.
   (c) Verify Theorem 4.7.

9. Verify Theorem 4.7 for the following linear transformations.
   (a) $L(x, y) = (x + y, y)$.
   (b) $L\left(\begin{bmatrix} x \\ y \\ z \end{bmatrix}\right) = \begin{bmatrix} 4 & -1 & -1 \\ 2 & 2 & 3 \\ 2 & -3 & -4 \end{bmatrix}\begin{bmatrix} x \\ y \\ z \end{bmatrix}.$
   (c) $L(x, y, z) = (x + y - z, x + y, y + z)$.

10. Let $L: R^4 \rightarrow R^4$ be defined by
$$L\left(\begin{bmatrix} x \\ y \\ z \\ w \end{bmatrix}\right) = \begin{bmatrix} 1 & 2 & 1 & 3 \\ 2 & 1 & -1 & 2 \\ 1 & 0 & 0 & -1 \\ 4 & 1 & -1 & 0 \end{bmatrix}\begin{bmatrix} x \\ y \\ z \\ w \end{bmatrix}.$$
   (a) Find a basis for ker L.
   (b) Find a basis for range L.
   (c) Verify Theorem 4.7.

11. Let $L: P_2 \rightarrow P_2$ be the linear transformation defined by $L(at^2 + bt + c) = (a + c)t^2 + (b + c)t$.
    (a) Is $t^2 - t - 1$ in ker L?
    (b) Is $t^2 + t - 1$ in ker L?
    (c) Is $2t^2 - t$ in range L?
    (d) Is $t^2 - t + 2$ in range L?
    (e) Find a basis for ker L.
    (f) Find a basis for range L.

12. Let $L: P_3 \rightarrow P_3$ be the linear transformation defined by $L(at^3 + bt^2 + ct + d) = (a - b)t^3 + (c - d)t$.
    (a) Is $t^3 + t^2 + t - 1$ in ker L?
    (b) Is $t^3 - t^2 + t - 1$ in ker L?
    (c) Is $3t^3 + t$ in range L?
    (d) Is $3t^3 - t^2$ in range L?
    (e) Find a basis for ker L.
    (f) Find a basis for range L.

13. (Calculus Required). Let $L: P_2 \rightarrow P_1$ be the linear transformation defined by $L(p(t)) = p'(t)$.
    (a) Find a basis for ker L.
    (b) Find a basis for range L.

14. (Calculus Required). Let $L: P_2 \rightarrow R^1$ be the linear transformation defined by $L(p(t)) = \int_0^1 p(t)\, dt$.

(a) Find a basis for ker L.
(b) Find a basis for range L.

15. Let $L: R^4 \rightarrow R^6$ be a linear transformation.
    (a) If dim (ker L) = 2, what is dim (range L)?
    (b) If dim (range L) = 3, what is dim (ker L)?

16. Let $L: V \rightarrow R^5$ be a linear transformation.
    (a) If L is onto and dim (ker L) = 2, what is dim V?
    (b) If L is one-to-one and onto, what is dim V?

In Exercises 17 through 19, determine the dimension of the solution space of the homogeneous system $AX = 0$ for the given matrix A.

17. $A = \begin{bmatrix} 1 & 1 & -2 \\ 1 & 2 & 3 \\ 0 & 1 & 3 \end{bmatrix}$.

18. $A = \begin{bmatrix} 1 & 1 & -2 & 0 & 0 \\ 1 & 2 & 3 & 6 & 7 \\ 2 & 1 & 3 & 6 & 5 \end{bmatrix}$.

19. $A = \begin{bmatrix} 1 & 1 & -2 \\ 1 & 2 & 3 \\ 3 & 4 & -1 \end{bmatrix}$.

THEORETICAL EXERCISES

T.1. Prove Corollary 4.1.

T.2. Let A be an $m \times n$ matrix and let $L: R^n \rightarrow R^m$ be defined by $L(X) = AX$ for X in R^n. Show that the column space of A is the range of L.

T.3. Let $L: R^n \rightarrow R^n$ be a linear transformation defined by $L(X) = AX$, where A is an $n \times n$ matrix. Show that L is one-to-one if and only if $|A| \neq 0$. [Hint: Use Theorem 4.5, Equation (1), and Corollary 3.2.]

T.4. Let $L: V \rightarrow W$ be a linear transformation. If $\{X_1, X_2, \ldots, X_k\}$ spans V, show that $\{L(X_1), L(X_2), \ldots, L(X_k)\}$ spans range L.

T.5. Let $L: V \rightarrow W$ be a linear transformation.
    (a) Show that dim (range L) $\leq$ dim V.
    (b) Prove that if L is onto, then dim $W \leq$ dim V.

T.6. Let $L: V \rightarrow W$ be a linear transformation, and let

$S = \{X_1, X_2, \ldots, X_n\}$ be a set of vectors in S. Prove that if $T = \{L(X_1), L(X_2), \ldots, L(X_n)\}$ is linearly independent, then so is S. (Hint: Assume that S is linearly dependent. What can we say about T?)

T.7. Let $L: V \rightarrow W$ be a linear transformation. Then prove that L is one-to-one if and only if dim (range L) = dim V.

T.8. Let $L: V \rightarrow W$ be a linear transformation. Show that L is one-to-one if and only if the image of every linearly independent set of vectors in V is a linearly independent set of vectors in W.

T.9. Let $L: V \rightarrow W$ be a linear transformation, and let dim V = dim W. Show that L is one-to-one if and only if the image under L of a basis for V is a basis for W.

4.3

THE MATRIX OF A LINEAR TRANSFORMATION

In Example 5 of Section 4.1 we saw that if A is an $m \times n$ matrix, then we can define a linear transformation $L: R^n \to R^m$ by $L(X) = AX$, for X in R^n. Now let $L: V \to W$ be a linear transformation of a finite-dimensional vector space V into a finite-dimensional vector space W. In this section we shall show how to attach a unique matrix to L that will enable us to find $L(X)$, for X in V, by merely performing matrix multiplication.

Example 1 Suppose that $L: R^2 \to R^3$ is the linear transformation defined by

$$L\left(\begin{bmatrix} x \\ y \end{bmatrix}\right) = \begin{bmatrix} x + y \\ x - y \\ 2x + 3y \end{bmatrix}.$$

If $X = \begin{bmatrix} x \\ y \end{bmatrix}$ is any vector in R^2, then

$$X = x\begin{bmatrix} 1 \\ 0 \end{bmatrix} + y\begin{bmatrix} 0 \\ 1 \end{bmatrix}$$

so that

$$L(X) = L\left(x\begin{bmatrix} 1 \\ 0 \end{bmatrix} + y\begin{bmatrix} 0 \\ 1 \end{bmatrix}\right)$$

$$= xL\left(\begin{bmatrix} 1 \\ 0 \end{bmatrix}\right) + yL\left(\begin{bmatrix} 0 \\ 1 \end{bmatrix}\right)$$

$$= x\begin{bmatrix} 1 \\ 1 \\ 2 \end{bmatrix} + y\begin{bmatrix} 1 \\ -1 \\ 3 \end{bmatrix}$$

$$= \begin{bmatrix} 1 & 1 \\ 1 & -1 \\ 2 & 3 \end{bmatrix}\begin{bmatrix} x \\ y \end{bmatrix}.$$

Thus L has been expressed as $L(X) = AX$, where

$$A = \begin{bmatrix} 1 & 1 \\ 1 & -1 \\ 2 & 3 \end{bmatrix}.$$

In Example 1 we obtained the matrix A under the assumption that vectors in R^2 and R^3 are always represented in terms of the natural bases. Examples 1, 2, and 3 in Section 4.4

use natural bases to illustrate computer graphics. We now extend this procedure to allow for arbitrary bases for R^n and R^m and also to deal with arbitrary vector spaces V and W. We first discuss the notion of a coordinate vector.

Coordinate Vectors

Definition Let V be an n-dimensional vector space with basis $S = \{\mathbf{X}_1, \mathbf{X}_2, \ldots, \mathbf{X}_n\}$. If

$$\mathbf{X} = a_1\mathbf{X}_1 + a_2\mathbf{X}_2 + \cdots + a_n\mathbf{X}_n \tag{1}$$

is any vector in V, then the vector

$$[\mathbf{X}]_S = \begin{bmatrix} a_1 \\ a_2 \\ \vdots \\ a_n \end{bmatrix}$$

in R^n is called the **coordinate vector of X with respect to the basis** S. The components of $[\mathbf{X}]_S$ are called the **coordinates of X with respect to** S.

Note that the coordinate vector $[\mathbf{X}]_S$ depends on the order in which the vectors in S are listed; a change in the order of this listing may change the coordinates of $\mathbf{X}$ with respect to S.

Example 2 Let $S = \{\mathbf{X}_1, \mathbf{X}_2, \mathbf{X}_3\}$ be a basis for R^3, where

$$\mathbf{X}_1 = \begin{bmatrix} 1 \\ -1 \\ 0 \end{bmatrix}, \qquad \mathbf{X}_2 = \begin{bmatrix} 0 \\ 1 \\ 1 \end{bmatrix}, \qquad \text{and} \qquad \mathbf{X}_3 = \begin{bmatrix} 1 \\ -1 \\ 1 \end{bmatrix}.$$

If $\mathbf{X} = \begin{bmatrix} 1 \\ 2 \\ 2 \end{bmatrix}$, then to find $[\mathbf{X}]_S$ we must find c_1, c_2, and c_3 [see Equation (1)] such that

$$\begin{bmatrix} 1 \\ 2 \\ 2 \end{bmatrix} = c_1\mathbf{X}_1 + c_2\mathbf{X}_2 + c_3\mathbf{X}_3.$$

This leads to the linear system

$$\begin{bmatrix} 1 & 0 & 1 \\ -1 & 1 & -1 \\ 0 & 1 & 1 \end{bmatrix} \begin{bmatrix} c_1 \\ c_2 \\ c_3 \end{bmatrix} = \begin{bmatrix} 1 \\ 2 \\ 2 \end{bmatrix}.$$

The solution is (verify)

$$c_1 = 2, \qquad c_2 = 3, \qquad \text{and} \qquad c_3 = -1.$$

Thus

$$[\mathbf{X}]_S = \begin{bmatrix} 2 \\ 3 \\ -1 \end{bmatrix}.$$

Example 3 Let S be the natural basis for R^n and let $\mathbf{X}$ be a vector in R^n. Then

$$[\mathbf{X}]_S = \mathbf{X}.$$

Example 4 Let $S = \{\mathbf{X}_1, \mathbf{X}_2, \ldots, \mathbf{X}_n\}$ be a basis for an n-dimensional vector space V. Then since

$$\mathbf{X}_1 = 1\mathbf{X}_1 + 0\mathbf{X}_2 + \cdots + 0\mathbf{X}_n,$$

we have

$$[\mathbf{X}_1]_S = \begin{bmatrix} 1 \\ 0 \\ \vdots \\ 0 \end{bmatrix}.$$

Similarly,

$$[\mathbf{X}_j]_S = \mathbf{E}_j \qquad (1 \leq j \leq n),$$

where $\{\mathbf{E}_1, \mathbf{E}_2, \ldots, \mathbf{E}_n\}$ is the natural basis for R^n.

Example 5 Consider the vector space P_1 of all polynomials of degree ≤ 1 and the zero polynomial. Let $S = \{t, 1\}$ be a basis for P_1. If $p(t) = 5t - 2$, then

$$[p(t)]_S = \begin{bmatrix} 5 \\ -2 \end{bmatrix}.$$

On the other hand, if $T = \{t + 1, t - 1\}$ is a basis for P_1, we have

$$5t - 2 = \tfrac{3}{2}(t + 1) + \tfrac{7}{2}(t - 1),$$

which implies that

$$[p(t)]_T = \begin{bmatrix} \tfrac{3}{2} \\ \tfrac{7}{2} \end{bmatrix}.$$

The Matrix of a Linear Transformation

Theorem 4.8 *Let $\mathbf{L}: V \to W$ be a linear transformation of an n-dimensional vector space V into an m-dimensional vector space W ($n \neq 0$ and $m \neq 0$) and let $S = \{\mathbf{X}_1, \mathbf{X}_2, \ldots, \mathbf{X}_n\}$ and $T = \{\mathbf{Y}_1, \mathbf{Y}_2, \ldots, \mathbf{Y}_m\}$ be bases for V and W, respectively. Then the $m \times n$ matrix $\mathbf{A}$, whose jth column is the coordinate vector $[\mathbf{L}(\mathbf{X}_j)]_T$ of $\mathbf{L}(\mathbf{X}_j)$ with respect to T, is associ-*

ated with **L** and has the following property: If **Y** = **L**(**X**) for some **X** in V, then

$$[\mathbf{Y}]_T = \mathbf{A}[\mathbf{X}]_S, \tag{2}$$

where $[\mathbf{X}]_S$ and $[\mathbf{Y}]_T$ are the coordinate vectors of **X** and **Y** with respect to the respective bases S and T. Moreover, **A** is the only matrix with this property.

Proof

The proof is a constructive one; that is, we show how to construct the matrix **A**. Consider the vector $\mathbf{X}_j$ in V for j = 1, 2, . . . , n. Then $\mathbf{L}(\mathbf{X}_j)$ is a vector in W, and since T is a basis for W we can express this vector as a linear combination of the vectors in T in a unique manner. Thus

$$\mathbf{L}(\mathbf{X}_j) = c_{1j}\mathbf{Y}_1 + c_{2j}\mathbf{Y}_2 + \cdots + c_{mj}\mathbf{Y}_m \qquad (1 \leq j \leq n). \tag{3}$$

This means that the coordinate vector of $\mathbf{L}(\mathbf{X}_j)$ with respect to T is

$$[\mathbf{L}(\mathbf{X}_j)]_T = \begin{bmatrix} c_{1j} \\ c_{2j} \\ \vdots \\ c_{mj} \end{bmatrix}.$$

We now define the $m \times n$ matrix **A** by choosing $[\mathbf{L}(\mathbf{X}_j)]_T$ as the jth column of **A** and show that this matrix satisfies the properties stated in the theorem. We leave the rest of the proof as Exercise T.1 and amply illustrate it in the examples below.

We now summarize the procedure given in Theorem 4.8.

The procedure for computing the matrix of a linear transformation **L**: $V \rightarrow W$ with respect to the bases $S = \{\mathbf{X}_1, \mathbf{X}_2, . . . , \mathbf{X}_n\}$ and $T = \{\mathbf{Y}_1, \mathbf{Y}_2, . . . , \mathbf{Y}_m\}$ for V and W, respectively, is as follows.

STEP 1. Compute $\mathbf{L}(\mathbf{X}_j)$ for j = 1, 2, . . . , n.

STEP 2. Find the coordinate vector $[\mathbf{L}(\mathbf{X}_j)]_T$ of $\mathbf{L}(\mathbf{X}_j)$ with respect to the basis T. This means that we have to express $\mathbf{L}(\mathbf{X}_j)$ as a linear combination of the vectors in T (See Equation (3).).

STEP 3. The matrix **A** of **L** with respect to S and T is formed by choosing $[\mathbf{L}(\mathbf{X}_j)]_T$ as the jth column of **A**.

Example 6 Let **L**: $R^3 \rightarrow R^2$ be defined by

$$\mathbf{L}\left(\begin{bmatrix} x \\ y \\ z \end{bmatrix}\right) = \begin{bmatrix} x + y \\ y - z \end{bmatrix}. \tag{4}$$

Let

$$S = \{\mathbf{X}_1, \mathbf{X}_2, \mathbf{X}_3\} \qquad \text{and} \qquad T = \{\mathbf{Y}_1, \mathbf{Y}_2\}$$

be bases for R^3 and R^2, respectively, where

$$\mathbf{X}_1 = \begin{bmatrix} 1 \\ 0 \\ 0 \end{bmatrix}, \qquad \mathbf{X}_2 = \begin{bmatrix} 0 \\ 1 \\ 0 \end{bmatrix}, \qquad \mathbf{X}_3 = \begin{bmatrix} 0 \\ 0 \\ 1 \end{bmatrix},$$

$$\mathbf{Y}_1 = \begin{bmatrix} 1 \\ 0 \end{bmatrix}, \qquad \text{and} \qquad \mathbf{Y}_2 = \begin{bmatrix} 0 \\ 1 \end{bmatrix}.$$

We now find the matrix $\mathbf{A}$ associated with $\mathbf{L}$. We have

$$\mathbf{L}(\mathbf{X}_1) = \begin{bmatrix} 1 + 0 \\ 0 - 0 \end{bmatrix} = \begin{bmatrix} 1 \\ 0 \end{bmatrix} = 1\mathbf{Y}_1 + 0\mathbf{Y}_2, \qquad \text{so } [\mathbf{L}(\mathbf{X}_1)]_T = \begin{bmatrix} 1 \\ 0 \end{bmatrix};$$

$$\mathbf{L}(\mathbf{X}_2) = \begin{bmatrix} 0 + 1 \\ 1 - 0 \end{bmatrix} = \begin{bmatrix} 1 \\ 1 \end{bmatrix} = 1\mathbf{Y}_1 + 1\mathbf{Y}_2, \qquad \text{so } [\mathbf{L}(\mathbf{X}_2)]_T = \begin{bmatrix} 1 \\ 1 \end{bmatrix};$$

$$\mathbf{L}(\mathbf{X}_3) = \begin{bmatrix} 0 + 0 \\ 0 - 1 \end{bmatrix} = \begin{bmatrix} 0 \\ -1 \end{bmatrix} = 0\mathbf{Y}_1 - 1\mathbf{Y}_2, \qquad \text{so } [\mathbf{L}(\mathbf{X}_3)]_T = \begin{bmatrix} 0 \\ -1 \end{bmatrix}.$$

Hence

$$\mathbf{A} = \begin{bmatrix} 1 & 1 & 0 \\ 0 & 1 & -1 \end{bmatrix}.$$

Example 7 Let $\mathbf{L}: R^3 \to R^2$ be defined as in Example 6. Now let

$$S = \{\mathbf{X}_1, \mathbf{X}_2, \mathbf{X}_3\} \qquad \text{and} \qquad T = \{\mathbf{Y}_1, \mathbf{Y}_2\}$$

be bases for R^3 and R^2, respectively, where

$$\mathbf{X}_1 = \begin{bmatrix} 1 \\ 0 \\ 1 \end{bmatrix}, \qquad \mathbf{X}_2 = \begin{bmatrix} 0 \\ 1 \\ 1 \end{bmatrix}, \qquad \mathbf{X}_3 = \begin{bmatrix} 1 \\ 1 \\ 1 \end{bmatrix},$$

$$\mathbf{Y}_1 = \begin{bmatrix} 1 \\ 2 \end{bmatrix}, \qquad \text{and} \qquad \mathbf{Y}_2 = \begin{bmatrix} -1 \\ 1 \end{bmatrix}.$$

Then

$$\mathbf{L}(\mathbf{X}_1) = \begin{bmatrix} 1 \\ -1 \end{bmatrix} = 0\mathbf{Y}_1 - 1\mathbf{Y}_2, \qquad \text{so } [\mathbf{L}(\mathbf{X}_1)]_T = \begin{bmatrix} 0 \\ -1 \end{bmatrix};$$

$$\mathbf{L}(\mathbf{X}_2) = \begin{bmatrix} 1 \\ 0 \end{bmatrix} = \tfrac{1}{3}\mathbf{Y}_1 - \tfrac{2}{3}\mathbf{Y}_2, \qquad \text{so } [\mathbf{L}(\mathbf{X}_2)]_T = \begin{bmatrix} \tfrac{1}{3} \\ -\tfrac{2}{3} \end{bmatrix};$$

$$\mathbf{L}(\mathbf{X}_3) = \begin{bmatrix} 2 \\ 0 \end{bmatrix} = \tfrac{2}{3}\mathbf{Y}_1 - \tfrac{4}{3}\mathbf{Y}_2, \qquad \text{so } [\mathbf{L}(\mathbf{X}_3)]_T = \begin{bmatrix} \tfrac{2}{3} \\ -\tfrac{4}{3} \end{bmatrix}.$$

Hence the matrix $\mathbf{A}$ associated with $\mathbf{L}$ is

$$\mathbf{A} = \begin{bmatrix} 0 & \frac{1}{3} & \frac{2}{3} \\ -1 & -\frac{2}{3} & -\frac{4}{3} \end{bmatrix}.$$

Equation (2) is then

$$[\mathbf{L}(\mathbf{X})]_T = \begin{bmatrix} 0 & \frac{1}{3} & \frac{2}{3} \\ -1 & -\frac{2}{3} & -\frac{4}{3} \end{bmatrix} [\mathbf{X}]_S. \tag{5}$$

To illustrate this equation, let

$$\mathbf{X} = \begin{bmatrix} 1 \\ 6 \\ 3 \end{bmatrix}.$$

Then from the definition of $\mathbf{L}$, Equation (4), we have

$$\mathbf{L}(\mathbf{X}) = \begin{bmatrix} 1 + 6 \\ 6 - 3 \end{bmatrix} = \begin{bmatrix} 7 \\ 3 \end{bmatrix}.$$

Now (verify)

$$[\mathbf{X}]_S = \begin{bmatrix} -3 \\ 2 \\ 4 \end{bmatrix}.$$

Then from (5),

$$[\mathbf{L}(\mathbf{X})]_T = \mathbf{A}[\mathbf{X}]_S = \begin{bmatrix} \frac{10}{3} \\ -\frac{11}{3} \end{bmatrix}.$$

Hence

$$\mathbf{L}(\mathbf{X}) = \frac{10}{3} \begin{bmatrix} 1 \\ 2 \end{bmatrix} - \frac{11}{3} \begin{bmatrix} -1 \\ 1 \end{bmatrix} = \begin{bmatrix} 7 \\ 3 \end{bmatrix},$$

which agrees with the previous value for $\mathbf{L}(\mathbf{X})$.

Notice that the matrices obtained in Examples 6 and 7 are different even though $\mathbf{L}$ is the same in both cases. It can be shown that there is a relationship between these two matrices. The study of this relationship is beyond the scope of this book.

Definition   The matrix $\mathbf{A}$ of Theorem 4.8 is called the **matrix of L with respect to the bases S and T**. Equation (2) is called the **representation of L with respect to S and T**. We also say that Equation (2) **represents L with respect to S and T**.

We observe that having $\mathbf{A}$ enables us to replace $\mathbf{L}$ by $\mathbf{A}$, $\mathbf{X}$ by $[\mathbf{X}]_S$, and $\mathbf{Y}$ by $[\mathbf{Y}]_T$ in the expression $\mathbf{L}(\mathbf{X}) = \mathbf{Y}$ to obtain $\mathbf{A}[\mathbf{X}]_S = [\mathbf{Y}]_T$. We can thus work with matrices rather than with linear transformations. Physicists and others who deal at great length with linear transformations perform most of their computations with the matrices of the linear transformations.

Example 8 Let $\mathbf{L}: R^3 \rightarrow R^2$ be as defined in Example 7. Now let

$$S = \{\mathbf{X}_3, \mathbf{X}_2, \mathbf{X}_1\} \qquad \text{and} \qquad T = \{\mathbf{Y}_1, \mathbf{Y}_2\},$$

where $\mathbf{X}_1$, $\mathbf{X}_2$, $\mathbf{X}_3$, $\mathbf{Y}_1$, and $\mathbf{Y}_2$ are as in Example 7. Then the matrix of $\mathbf{L}$ is

$$\mathbf{A} = \begin{bmatrix} \frac{2}{3} & \frac{1}{3} & 0 \\ -\frac{4}{3} & -\frac{2}{3} & -1 \end{bmatrix}.$$

Note that if we change the order of the vectors in the bases S and T, then the matrix $\mathbf{A}$ of $\mathbf{L}$ may change.

Example 9 Let $\mathbf{L}: R^3 \rightarrow R^2$ be defined by

$$\mathbf{L}\left(\begin{bmatrix} x \\ y \\ z \end{bmatrix}\right) = \begin{bmatrix} 1 & 1 & 1 \\ 1 & 2 & 3 \end{bmatrix}\begin{bmatrix} x \\ y \\ z \end{bmatrix}.$$

Let

$$S = \{\mathbf{X}_1, \mathbf{X}_2, \mathbf{X}_3\} \qquad \text{and} \qquad T = \{\mathbf{Y}_1, \mathbf{Y}_2\}$$

be the natural bases for R^3 and R^2, respectively. Now

$$\mathbf{L}(\mathbf{X}_1) = \begin{bmatrix} 1 & 1 & 1 \\ 1 & 2 & 3 \end{bmatrix}\begin{bmatrix} 1 \\ 0 \\ 0 \end{bmatrix} = \begin{bmatrix} 1 \\ 1 \end{bmatrix} = 1\mathbf{Y}_1 + 1\mathbf{Y}_2, \qquad \text{so } [\mathbf{L}(\mathbf{X}_1)]_T = \begin{bmatrix} 1 \\ 1 \end{bmatrix};$$

$$\mathbf{L}(\mathbf{X}_2) = \begin{bmatrix} 1 & 1 & 1 \\ 1 & 2 & 3 \end{bmatrix}\begin{bmatrix} 0 \\ 1 \\ 0 \end{bmatrix} = \begin{bmatrix} 1 \\ 2 \end{bmatrix} = 1\mathbf{Y}_1 + 2\mathbf{Y}_2, \qquad \text{so } [\mathbf{L}(\mathbf{X}_2)]_T = \begin{bmatrix} 1 \\ 2 \end{bmatrix}.$$

Also,

$$[\mathbf{L}(\mathbf{X}_3)]_T = \begin{bmatrix} 1 \\ 3 \end{bmatrix} \qquad \text{(verify)}.$$

Then the matrix of $\mathbf{L}$ is

$$\mathbf{A} = \begin{bmatrix} 1 & 1 & 1 \\ 1 & 2 & 3 \end{bmatrix}.$$

Of course, the reason that $\mathbf{A}$ is the same matrix as the one involved in the definition of $\mathbf{L}$ is that the natural bases are being used for R^3 and R^2.

Example 10 Let $\mathbf{L}: R^3 \rightarrow R^2$ be defined as in Example 9. Now let

$$S = \{\mathbf{X}_1, \mathbf{X}_2, \mathbf{X}_3\} \qquad \text{and} \qquad T = \{\mathbf{Y}_1, \mathbf{Y}_2\},$$

where

$$\mathbf{X}_1 = \begin{bmatrix} 1 \\ 1 \\ 0 \end{bmatrix}, \qquad \mathbf{X}_2 = \begin{bmatrix} 0 \\ 1 \\ 1 \end{bmatrix}, \qquad \mathbf{X}_3 = \begin{bmatrix} 0 \\ 0 \\ 1 \end{bmatrix},$$

$$\mathbf{Y}_1 = \begin{bmatrix} 1 \\ 2 \end{bmatrix}, \qquad \text{and} \qquad \mathbf{Y}_2 = \begin{bmatrix} 1 \\ 3 \end{bmatrix}.$$

Then the matrix $\mathbf{A}$ of $\mathbf{L}$ is (verify)

$$\mathbf{A} = \begin{bmatrix} 3 & 1 & 0 \\ -1 & 1 & 1 \end{bmatrix}.$$

This matrix is, of course, far different from the one that defined $\mathbf{L}$. Thus, although a matrix $\mathbf{A}$ may be involved in the definition of a linear transformation $\mathbf{L}$, we cannot conclude that it is necessarily the matrix representing $\mathbf{L}$ with respect to two given bases S and T.

Example 11 Let $\mathbf{L}: P_1 \rightarrow P_2$ be defined by

$$\mathbf{L}(p(t)) = tp(t).$$

(a) Find the matrix of $\mathbf{L}$ with respect to the bases $S = \{t, 1\}$ and $T = \{t^2, t, 1\}$ for P_1 and P_2, respectively.

(b) If $p(t) = 3t - 2$, compute $\mathbf{L}(p(t))$ directly and using the matrix obtained in (a).

Solution

(a) We have

$$\mathbf{L}(t) = t \cdot t = t^2 = 1(t^2) + 0(t) + 0(1), \qquad \text{so } [\mathbf{L}(t)]_T = \begin{bmatrix} 1 \\ 0 \\ 0 \end{bmatrix}.$$

$$\mathbf{L}(1) = t \cdot 1 = t = 0(t^2) + 1(t) + 0(1), \qquad \text{so } [\mathbf{L}(1)]_T = \begin{bmatrix} 0 \\ 1 \\ 0 \end{bmatrix}.$$

Hence the matrix of $\mathbf{L}$ is

$$\mathbf{A} = \begin{bmatrix} 1 & 0 \\ 0 & 1 \\ 0 & 0 \end{bmatrix}.$$

(b) Computing $\mathbf{L}(p(t))$ directly, we have

$$\mathbf{L}(p(t)) = t(p(t)) = t(3t - 2) = 3t^2 - 2t.$$

To compute $\mathbf{L}(p(t))$ using $\mathbf{A}$, we first write

$$p(t) = 3 \cdot t + (-2)1, \qquad \text{so } [p(t)]_S = \begin{bmatrix} 3 \\ -2 \end{bmatrix}.$$

Then

$$[\mathbf{L}(p(t))]_T = \mathbf{A}[p(t)]_S = \begin{bmatrix} 1 & 0 \\ 0 & 1 \\ 0 & 0 \end{bmatrix} \begin{bmatrix} 3 \\ -2 \end{bmatrix} = \begin{bmatrix} 3 \\ -2 \\ 0 \end{bmatrix}.$$

Hence

$$\mathbf{L}(p(t)) = 3t^2 + (-2)t + 0(1)$$
$$= 3t^2 - 2t.$$

Example 12　Let $\mathbf{L} \colon P_1 \to P_2$ be as defined in Example 11.
(a) Find the matrix of $\mathbf{L}$ with respect to the bases $S = \{t, 1\}$ and $T = \{t^2, t - 1, t + 1\}$ for P_1 and P_2, respectively.
(b) If $p(t) = 3t - 2$, compute $\mathbf{L}(p(t))$ using the matrix obtained in (a).

Solution
(a) We have

$$\mathbf{L}(t) = t^2 = 1t^2 + 0(t - 1) + 0(t + 1), \qquad \text{so } [\mathbf{L}(t)]_T = \begin{bmatrix} 1 \\ 0 \\ 0 \end{bmatrix}.$$

$$\mathbf{L}(1) = t = 0t^2 + \tfrac{1}{2}(t - 1) + \tfrac{1}{2}(t + 1), \qquad \text{so } [\mathbf{L}(1)]_T = \begin{bmatrix} 0 \\ \tfrac{1}{2} \\ \tfrac{1}{2} \end{bmatrix}.$$

Then the matrix of $\mathbf{L}$ is

$$\mathbf{A} = \begin{bmatrix} 1 & 0 \\ 0 & \tfrac{1}{2} \\ 0 & \tfrac{1}{2} \end{bmatrix}.$$

(b) We have

$$[\mathbf{L}(p(t))]_T = \mathbf{A}[p(t)]_S = \begin{bmatrix} 1 & 0 \\ 0 & \tfrac{1}{2} \\ 0 & \tfrac{1}{2} \end{bmatrix} \begin{bmatrix} 3 \\ -2 \end{bmatrix} = \begin{bmatrix} 3 \\ -1 \\ -1 \end{bmatrix}.$$

Hence

$$\mathbf{L}(p(t)) = 3t^2 + (-1)(t - 1) + (-1)(t + 1)$$
$$= 3t^2 - 2t.$$

Suppose that $L: V \to W$ is a linear transformation and that $\mathbf{A}$ is the matrix of L with respect to ordered bases for V and W. Then the problem of finding ker L reduces to the problem of finding the solution space $\mathbf{AX} = \mathbf{0}$. Moreover, the problem of finding range L reduces to the problem of finding the column space of $\mathbf{A}$.

If $L: V \to V$ is a linear operator and V is an n-dimensional vector space, then to obtain a representation of L, we fix bases S and T for V and obtain the matrix of L with respect to S and T. However, it is often convenient in this case to choose $S = T$. To avoid redundancy in this case, we refer to $\mathbf{A}$ as the **matrix of L with respect to S**. We refer to the representation as the **representation of L with respect to S**.

It is, of course, clear that if $\mathbf{I}: S \to S$ is the identity linear transformation defined by $\mathbf{I}(\mathbf{X}) = \mathbf{X}$, for every $\mathbf{X}$ in V, then the matrix of $\mathbf{I}$ with respect to any basis S for V is the identity matrix $\mathbf{I}_n$ (Exercise T.2), where $n = \dim V$.

4.3 EXERCISES

1. Let
$$S = \left\{ \begin{bmatrix} 1 \\ -1 \end{bmatrix}, \begin{bmatrix} 2 \\ 3 \end{bmatrix} \right\}$$
be a basis for R^2. Find the coordinate vectors of the following vectors with respect to S.

(a) $\begin{bmatrix} -3 \\ -7 \end{bmatrix}$.   (b) $\begin{bmatrix} 3 \\ 7 \end{bmatrix}$.   (c) $\begin{bmatrix} 12 \\ 13 \end{bmatrix}$.   (d) $\begin{bmatrix} 2 \\ 3 \end{bmatrix}$

2. Let
$$S = \left\{ \begin{bmatrix} 1 \\ -1 \\ 2 \end{bmatrix}, \begin{bmatrix} 0 \\ 2 \\ 1 \end{bmatrix}, \begin{bmatrix} 1 \\ 0 \\ 0 \end{bmatrix} \right\}$$
be a basis for R^3. Find the coordinate vectors of the following vectors with respect to S.

(a) $\begin{bmatrix} 1 \\ 4 \\ 2 \end{bmatrix}$.   (b) $\begin{bmatrix} 3 \\ -4 \\ 3 \end{bmatrix}$.   (c) $\begin{bmatrix} 2 \\ -5 \\ 5 \end{bmatrix}$.   (d) $\begin{bmatrix} 1 \\ -2 \\ 4 \end{bmatrix}$.

3. Repeat Exercise 2 for
$$S = \left\{ \begin{bmatrix} 1 \\ 1 \\ 3 \end{bmatrix}, \begin{bmatrix} -1 \\ 0 \\ 1 \end{bmatrix}, \begin{bmatrix} 1 \\ 1 \\ 0 \end{bmatrix} \right\}.$$

(a) $\begin{bmatrix} 2 \\ 2 \\ 3 \end{bmatrix}$.   (b) $\begin{bmatrix} -2 \\ 0 \\ -1 \end{bmatrix}$.   (c) $\begin{bmatrix} -1 \\ 0 \\ 1 \end{bmatrix}$.   (d) $\begin{bmatrix} 0 \\ 1 \\ -5 \end{bmatrix}$.

4. Let
$$S = \{t + 2, t - 1\}$$
be a basis for P_1. Find the coordinate vectors of the following vectors with respect to S.

(a) $3t + 3$.   (b) $-2t - 1$.
(c) $t + 2$.   (d) $2t - 5$.

5. Let
$$S = \{t^2 + 1, t - 1, t\}$$
be a basis for P_2. Find the coordinate vectors of the following vectors with respect to S.

(a) $t^2 + 2$.   (b) $2t - 1$.
(c) $t^2 - t + 3$.   (d) $-t + 1$.

6. The set $S = \left\{ \begin{bmatrix} 1 \\ -1 \\ 1 \end{bmatrix}, \begin{bmatrix} 0 \\ 0 \\ 1 \end{bmatrix}, \begin{bmatrix} 1 \\ 1 \\ 0 \end{bmatrix} \right\}$ is a

basis for R^3. Find $\mathbf{X}$ if $[\mathbf{X}]_S$ is

(a) $\begin{bmatrix} 2 \\ -1 \\ 3 \end{bmatrix}$.   (b) $\begin{bmatrix} 0 \\ 0 \\ 1 \end{bmatrix}$.   (c) $\begin{bmatrix} 0 \\ 2 \\ 1 \end{bmatrix}$.

7. The set $S = \{t^2 + 1, t + 1, t^2 + t\}$ is a basis for P_2. Find $p(t)$ if $[p(t)]_S$ is

(a) $\begin{bmatrix} 3 \\ -1 \\ -2 \end{bmatrix}$.   (b) $\begin{bmatrix} 1 \\ 0 \\ 0 \end{bmatrix}$.   (c) $\begin{bmatrix} 2 \\ 0 \\ -1 \end{bmatrix}$.

8. Let $L: R^2 \rightarrow R^2$ be defined by

$$L\left(\begin{bmatrix} x \\ y \end{bmatrix}\right) = \begin{bmatrix} x + 2y \\ 2x - y \end{bmatrix}.$$

Let S be the natural basis for R^2 and let

$$T = \left\{\begin{bmatrix} -1 \\ 2 \end{bmatrix}, \begin{bmatrix} 2 \\ 0 \end{bmatrix}\right\}$$ be another basis for R^2. Find

the matrix of L with respect to
(a) S.   (b) S and T.   (c) T and S.   (d) T.

(e) Compute $L\left(\begin{bmatrix} 1 \\ 2 \end{bmatrix}\right)$ using the definition of L and
also using the matrices obtained in (a), (b), (c), and (d).

⋆9. Let $L: R^2 \rightarrow R^3$ be defined by

$$L\left(\begin{bmatrix} x \\ y \end{bmatrix}\right) = \begin{bmatrix} x - 2y \\ 2x + y \\ x + y \end{bmatrix}.$$

Let S and T be the natural bases for R^2 and R^3, respectively. Also, let

$$S' = \left\{\begin{bmatrix} 1 \\ -1 \end{bmatrix}, \begin{bmatrix} 0 \\ 1 \end{bmatrix}\right\}$$

and

$$T' = \left\{\begin{bmatrix} 1 \\ 1 \\ 0 \end{bmatrix}, \begin{bmatrix} 0 \\ 1 \\ 1 \end{bmatrix}, \begin{bmatrix} 1 \\ -1 \\ 1 \end{bmatrix}\right\}$$

be bases for R^2 and R^3, respectively. Find the matrix of L with respect to
(a) S and T.   (b) S' and T'.

(c) Compute $L\left(\begin{bmatrix} 1 \\ 2 \end{bmatrix}\right)$ using the matrices obtained
in (a) and (b).

10. Let $L: R^3 \rightarrow R^3$ be defined by

$$L\left(\begin{bmatrix} x \\ y \\ z \end{bmatrix}\right) = \begin{bmatrix} x + 2y + z \\ 2x - y \\ 2y + z \end{bmatrix}.$$

Let S be the natural basis for R^3 and let

$$T = \left\{\begin{bmatrix} 1 \\ 0 \\ 1 \end{bmatrix}, \begin{bmatrix} 0 \\ 1 \\ 1 \end{bmatrix}, \begin{bmatrix} 0 \\ 0 \\ 1 \end{bmatrix}\right\}$$ be another basis for R^3.

Find the matrix of L with respect to
(a) S.   (b) S and T.   (c) T and S.   (d) T.

(e) Compute $L\left(\begin{bmatrix} 1 \\ 1 \\ -2 \end{bmatrix}\right)$ using the representations

obtained in (a), (b), (c), and (d).

⋆11. Let $L: R^3 \rightarrow R^2$ be defined by

$$L\left(\begin{bmatrix} x \\ y \\ z \end{bmatrix}\right) = \begin{bmatrix} x + y \\ y - z \end{bmatrix}.$$

Let S and T be the natural bases for R^3 and R^2, respectively. Also, let

$$S' = \left\{\begin{bmatrix} 1 \\ 1 \\ 0 \end{bmatrix}, \begin{bmatrix} 0 \\ 1 \\ 0 \end{bmatrix}, \begin{bmatrix} -1 \\ 1 \\ 1 \end{bmatrix}\right\}$$

and

$$T' = \left\{\begin{bmatrix} -1 \\ 1 \end{bmatrix}, \begin{bmatrix} 1 \\ 2 \end{bmatrix}\right\}$$

be bases for R^3 and R^2, respectively.
Find the representation of L with respect to
(a) S and T.   (b) S' and T'.

(c) Compute $L\left(\begin{bmatrix} 1 \\ 2 \\ 3 \end{bmatrix}\right)$ using both representations.

12. Let $L: R^2 \rightarrow R^3$ be defined by

$$L\left(\begin{bmatrix} x \\ y \end{bmatrix}\right) = \begin{bmatrix} 1 & 1 \\ 1 & -1 \\ 1 & 2 \end{bmatrix} \begin{bmatrix} x \\ y \end{bmatrix}.$$

(a) Represent L with respect to the natural bases for R^2 and R^3.
(b) Represent L with respect to the bases S' and T' of Exercise 9.

(c) Compute $L\left(\begin{bmatrix} 2 \\ -3 \end{bmatrix}\right)$ using both representations.

13. Let **L**: $P_1 \rightarrow P_3$ be defined by $\mathbf{L}(p(t)) = t^2 p(t)$. Let

$$S = \{t, 1\} \quad \text{and} \quad S' = \{t, t + 1\}$$

be bases for P_1. Let

$$T = \{t^3, t^2, t, 1\}$$

and

$$T' = \{t^3, t^2 - 1, t, t + 1\}$$

be bases for P_3. Find the matrix of **L** with respect to
(a) S and T.   (b) S' and T'.

14. Let **L**: $P_1 \rightarrow P_2$ be defined by $\mathbf{L}(p(t)) = tp(t) + p(0)$. Let

$$S = \{t, 1\} \quad \text{and} \quad S' = \{t + 1, t - 1\}$$

be bases for P_1. Let

$$T = \{t^2, t, 1\}$$

and

$$T' = \{t^2 + 1, t - 1, t + 1\}$$

be bases for P_2. Find the matrix of **L** with respect to
(a) S and T.   (b) S' and T'.
(c) Find $\mathbf{L}(-3t + 3)$ using the definition of **L** and the matrices obtained in (a) and (b).

15. Let **L**: $R^2 \rightarrow R^2$ be a linear transformation. Suppose that the matrix of **L** with respect to the basis

$$S = \{\mathbf{X}_1, \mathbf{X}_2\}$$

is

$$\mathbf{A} = \begin{bmatrix} 2 & -3 \\ -1 & 4 \end{bmatrix},$$

where

$$\mathbf{X}_1 = \begin{bmatrix} 1 \\ 2 \end{bmatrix} \quad \text{and} \quad \mathbf{X}_2 = \begin{bmatrix} 1 \\ -1 \end{bmatrix}.$$

(a) Compute $[\mathbf{L}(\mathbf{X}_1)]_S$ and $[\mathbf{L}(\mathbf{X}_2)]_S$.
(b) Compute $\mathbf{L}(\mathbf{X}_1)$ and $\mathbf{L}(\mathbf{X}_2)$.
(c) Compute $\mathbf{L}\begin{bmatrix} -2 \\ 3 \end{bmatrix}$.

16. Let the matrix of **L**: $R^2 \rightarrow R^2$ with respect to the basis

$$S = \left\{ \begin{bmatrix} 1 \\ -1 \end{bmatrix}, \begin{bmatrix} 0 \\ 1 \end{bmatrix} \right\}$$

be

$$\begin{bmatrix} 1 & 2 \\ -2 & 3 \end{bmatrix}.$$

Find the matrix of **L** with respect to the natural basis for R^2.

[*Hint:* Let $\mathbf{X}_1 = \begin{bmatrix} 1 \\ -1 \end{bmatrix}$, $\mathbf{X}_2 = \begin{bmatrix} 0 \\ 1 \end{bmatrix}$. Find $\mathbf{L}(\mathbf{X}_1)$ and $\mathbf{L}(\mathbf{X}_2)$.]

17. Let **L**: $R^3 \rightarrow R^3$ be defined by

$$\mathbf{L}\left(\begin{bmatrix} 1 \\ 0 \\ 0 \end{bmatrix} \right) = \begin{bmatrix} 1 \\ 1 \\ 0 \end{bmatrix},$$

$$\mathbf{L}\left(\begin{bmatrix} 0 \\ 1 \\ 0 \end{bmatrix} \right) = \begin{bmatrix} 2 \\ 0 \\ 1 \end{bmatrix},$$

$$\mathbf{L}\left(\begin{bmatrix} 0 \\ 0 \\ 1 \end{bmatrix} \right) = \begin{bmatrix} 1 \\ 0 \\ 1 \end{bmatrix}.$$

(a) Find the matrix of **L** with respect to the natural basis S for R^3.

(b) Find $\mathbf{L}\left(\begin{bmatrix} 1 \\ 2 \\ 3 \end{bmatrix} \right)$ using the definition of **L** and also using the matrix obtained in (a).

18. Let the matrix of **L**: $R^3 \rightarrow R^2$ with respect to the bases

$$S = \{\mathbf{X}_1, \mathbf{X}_2, \mathbf{X}_3\} \quad \text{and} \quad T = \{\mathbf{Y}_1, \mathbf{Y}_2\}$$

be

$$\mathbf{A} = \begin{bmatrix} 1 & 2 & 1 \\ -1 & 1 & 0 \end{bmatrix},$$

where

$$\mathbf{X}_1 = \begin{bmatrix} -1 \\ 1 \\ 0 \end{bmatrix}, \quad \mathbf{X}_2 = \begin{bmatrix} 0 \\ 1 \\ 1 \end{bmatrix}, \quad \text{and} \quad \mathbf{X}_3 = \begin{bmatrix} 1 \\ 0 \\ 0 \end{bmatrix}$$

and

$$\mathbf{Y}_1 = \begin{bmatrix} 1 \\ 2 \end{bmatrix}, \quad \mathbf{Y}_2 = \begin{bmatrix} 1 \\ -1 \end{bmatrix}.$$

(a) Compute $[\mathbf{L}(\mathbf{X}_1)]_T$, $[\mathbf{L}(\mathbf{X}_2)]_T$, and $[\mathbf{L}(\mathbf{X}_3)]_T$.
(b) Compute $\mathbf{L}(\mathbf{X}_1)$, $\mathbf{L}(\mathbf{X}_2)$, and $\mathbf{L}(\mathbf{X}_3)$.

(c) Compute $\mathbf{L}\left(\begin{bmatrix} 2 \\ 1 \\ -1 \end{bmatrix} \right)$.

19. **(Calculus Required).** Let V be the vector space with basis $S = \{\sin t, \cos t\}$, and let $T = \{\sin t - \cos t, \sin t + \cos t\}$ be another basis for V. Find the matrix of the linear operator $\mathbf{L}: V \to V$ defined by $\mathbf{L}(f) = f'$ with respect to
(a) S.   (b) T.   (c) S and T.   (d) T and S.

20. **(Calculus Required).** Let V be the vector space with basis $S = \{e^t, e^{-t}\}$. Find the matrix of the linear operator $\mathbf{L}: V \to V$ defined by $\mathbf{L}(f) = f'$ with respect to S.

THEORETICAL EXERCISES

T.1. Complete the proof of Theorem 4.8.
T.2. If $\mathbf{I}: V \to V$ is the identity transformation defined by $\mathbf{I}(\mathbf{X}) = \mathbf{X}$ for any $\mathbf{X}$ in V, show that the matrix of $\mathbf{I}$ with respect to any basis S for V is the identity matrix $\mathbf{I}_n$, where $n = \dim V$.
T.3. Let $\mathbf{0}: V \to W$ be the zero linear transformation, defined by $\mathbf{0}(\mathbf{X}) = \mathbf{0}_W$, for any $\mathbf{X}$ in V. Prove that

the matrix of $\mathbf{0}$ with respect to any bases for V and W is the $m \times n$ zero matrix (where $n = \dim V$, $m = \dim W$).
T.4. Let $\mathbf{L}: V \to V$ be a linear operator defined by $\mathbf{L}(\mathbf{X}) = c\mathbf{X}$, where c is a fixed constant. Prove that the matrix of $\mathbf{L}$ with respect to any basis for V is a scalar matrix (see Section 1.2).

4.4

COMPUTER GRAPHICS (OPTIONAL)

In this section we give illustrations of linear operators $\mathbf{L}: R^2 \to R^2$ that are useful in two-dimensional computer graphics. We will assume that the natural basis is used in R^2 to obain a matrix representation of $\mathbf{L}$. In each of the following three examples we describe a geometric transformation on a set of points in R^2 and show the construction of the matrix representing the corresponding linear operator $\mathbf{L}: R^2 \to R^2$.

Example 1   A reflection with respect to the x-axis of a vector $\mathbf{X}$ in R^2 is defined by the linear operator

$$\mathbf{L}(\mathbf{X}) = \mathbf{L}\left(\begin{bmatrix} x \\ y \end{bmatrix} \right) = \begin{bmatrix} x \\ -y \end{bmatrix}$$

(see Example 9 in Section 4.1). Then

$$\mathbf{L}\left(\begin{bmatrix} 1 \\ 0 \end{bmatrix} \right) = \begin{bmatrix} 1 \\ 0 \end{bmatrix} \quad \text{and} \quad \mathbf{L}\left(\begin{bmatrix} 0 \\ 1 \end{bmatrix} \right) = \begin{bmatrix} 0 \\ -1 \end{bmatrix}.$$

Hence the matrix representation of **L** with respect to the natural basis is

$$A = \begin{bmatrix} 1 & 0 \\ 0 & -1 \end{bmatrix}.$$

Thus we have

$$L(X) = AX = \begin{bmatrix} x \\ -y \end{bmatrix}.$$

To illustrate a reflection with respect to the x-axis in computer graphics, let the triangle T in Figure 4.6(a) have vertices

$$(-1, 4), \qquad (3, 1), \qquad \text{and} \qquad (2, 6).$$

To reflect T with respect to the x-axis, we let

$$X_1 = \begin{bmatrix} -1 \\ 4 \end{bmatrix}, \qquad X_2 = \begin{bmatrix} 3 \\ 1 \end{bmatrix}, \qquad \text{and} \qquad X_3 = \begin{bmatrix} 2 \\ 6 \end{bmatrix}$$

and compute the images $L(X_1)$, $L(X_2)$, $L(X_3)$ by forming the products

$$AX_1 = \begin{bmatrix} 1 & 0 \\ 0 & -1 \end{bmatrix}\begin{bmatrix} -1 \\ 4 \end{bmatrix} = \begin{bmatrix} -1 \\ -4 \end{bmatrix}$$

$$AX_2 = \begin{bmatrix} 1 & 0 \\ 0 & -1 \end{bmatrix}\begin{bmatrix} 3 \\ 1 \end{bmatrix} = \begin{bmatrix} 3 \\ -1 \end{bmatrix}$$

$$AX_3 = \begin{bmatrix} 1 & 0 \\ 0 & -1 \end{bmatrix}\begin{bmatrix} 2 \\ 6 \end{bmatrix} = \begin{bmatrix} 2 \\ -6 \end{bmatrix}$$

Figure 4.6a

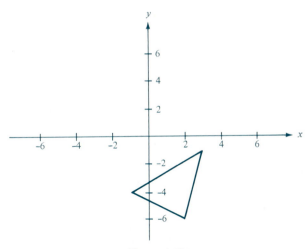

Figure 4.6b

Thus the image of T has vertices

$$(-1, -4), \qquad (3, -1), \qquad \text{and} \qquad (2, -6)$$

and is displayed in Figure 4.6(b).

Example 2 The reflection with respect to line $y = -x$ of a vector $\mathbf{X}$ in R^2 is defined by the linear operator

$$\mathbf{L(X)} = \mathbf{L}\left(\begin{bmatrix} x \\ y \end{bmatrix}\right) = \begin{bmatrix} -y \\ -x \end{bmatrix}.$$

Then

$$\mathbf{L}\left(\begin{bmatrix} 1 \\ 0 \end{bmatrix}\right) = \begin{bmatrix} 0 \\ -1 \end{bmatrix} \qquad \text{and} \qquad \mathbf{L}\left(\begin{bmatrix} 0 \\ 1 \end{bmatrix}\right) = \begin{bmatrix} -1 \\ 0 \end{bmatrix}.$$

Hence the matrix representation of $\mathbf{L}$ with respect to the natural basis is

$$\mathbf{B} = \begin{bmatrix} 0 & -1 \\ -1 & 0 \end{bmatrix}.$$

To illustrate the reflection with respect to the line $y = -x$, we use triangle T as defined in Example 1 and compute the products

$$\mathbf{BX_1} = \begin{bmatrix} 0 & -1 \\ -1 & 0 \end{bmatrix}\begin{bmatrix} -1 \\ 4 \end{bmatrix} = \begin{bmatrix} -4 \\ 1 \end{bmatrix}$$

$$\mathbf{BX_2} = \begin{bmatrix} 0 & -1 \\ -1 & 0 \end{bmatrix}\begin{bmatrix} 3 \\ 1 \end{bmatrix} = \begin{bmatrix} -1 \\ -3 \end{bmatrix}$$

$$\mathbf{BX_3} = \begin{bmatrix} 0 & -1 \\ -1 & 0 \end{bmatrix}\begin{bmatrix} 2 \\ 6 \end{bmatrix} = \begin{bmatrix} -6 \\ -2 \end{bmatrix}$$

Thus the image of T has vertices

$$(-4, 1), \qquad (-1, -3), \qquad \text{and} \qquad (-6, -2)$$

and is displayed in Figure 4.7.

To perform a reflection with respect to the x-axis on triangle T of Example 1 followed by a reflection with respect to the line $y = -x$, we compute

$$\mathbf{B(AX_1)}, \quad \mathbf{B(AX_2)}, \quad \mathbf{B(AX_3)}$$

It is easy to show that reversing the order of these linear operators produces a different image (verify). Thus the order in which graphics transformations are performed is important.

Example 3 Rotations in a plane are defined in Example 6 of Section 4.1. A plane figure is rotated counterclockwise through an angle ϕ using the linear operator whose matrix representation with respect to the natural basis in R^2 is given by

$$\mathbf{A} = \begin{bmatrix} \cos\phi & -\sin\phi \\ \sin\phi & \cos\phi \end{bmatrix}.$$

Suppose that we wish to rotate the parabola $y = x^2$ through 50°. We start by choosing a sample of points from the parabola, say

$$(-2, 4), \qquad (-1, 1), \qquad (0, 0), \qquad (\tfrac{1}{2}, \tfrac{1}{4}), \qquad \text{and} \qquad (3, 9).$$

See Figure 4.8(a).

Figure 4.7

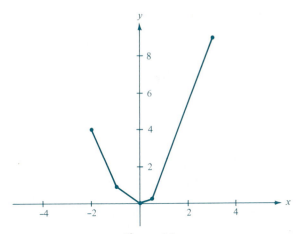

Figure 4.8a

We then compute the images of this sample of points. Thus, letting

$$\mathbf{X_1} = (-2, 4), \qquad \mathbf{X_2} = (-1, 1), \qquad \mathbf{X_3} = (0, 0), \qquad \mathbf{X_4} = (\tfrac{1}{2}, \tfrac{1}{4}), \qquad \text{and}$$
$$\mathbf{X_5} = (3, 9),$$

we compute the products (to four decimal places)

$$\mathbf{AX_1} = \begin{bmatrix} -4.3498 \\ 1.0391 \end{bmatrix}, \qquad \mathbf{AX_2} = \begin{bmatrix} -1.4088 \\ -0.1233 \end{bmatrix}, \qquad \mathbf{AX_3} = \begin{bmatrix} 0 \\ 0 \end{bmatrix},$$

$$\mathbf{AX_4} = \begin{bmatrix} 0.1299 \\ 0.5437 \end{bmatrix}, \qquad \text{and} \qquad \mathbf{AX_5} = \begin{bmatrix} -4.9660 \\ 8.0832 \end{bmatrix}.$$

These image points are plotted as shown in Figure 4.8(b) and successive points are connected showing the image of the parabola.

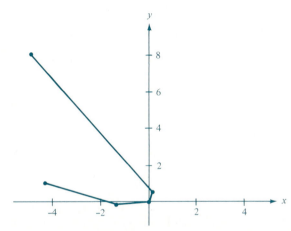

Figure 4.8b

Other linear operators used in two-dimensional computer graphics are considered in the exercises at the end of this section. For a detailed discussion of computer graphics, the reader is referred to the book by Rogers and Adams listed in Further Reading at the end of this section.

An arbitrary 2×2 matrix **A** defines a linear operator **L**: $R^2 \to R^2$ by

$$\mathbf{L(X) = AX}$$

for a vector **X** in R^2. Using computer graphics, it is interesting to view the image of a closed figure, like the triangle T in Figure 4.6(a), that results from using such a linear operator. The original figure and its image can be compared in a number of ways. We present one such comparison in the next example.

Example 4 The area of a triangle with vertices (x_1, y_1), (x_2, y_2), and (x_3, y_3) is computed from the following formula:

$$\text{area of triangle} = \frac{1}{2} \begin{vmatrix} x_1 & y_1 & 1 \\ x_2 & y_2 & 1 \\ x_3 & y_3 & 1 \end{vmatrix} \tag{1}$$

(that is, the area is $\frac{1}{2}$ the absolute value of the determinant).

The triangle T in Figure 4.6a has vertices $(-1, 4)$, $(3, 1)$, and $(2,6)$. Thus the area of T is, by (1),

$$\frac{1}{2} \begin{vmatrix} -1 & 4 & 1 \\ 3 & 1 & 1 \\ 2 & 6 & 1 \end{vmatrix} = \frac{1}{2}|17| = 8.5.$$

The image of T using the linear operator defined by the matrix

$$\mathbf{A} = \begin{bmatrix} 6 & -3 \\ -1 & 1 \end{bmatrix}$$

has vertices $(-18, 5)$, $(15, -2)$, and $(-6, 4)$ (verify). See Figure 4.9. The area of the image of triangle T obtained from the linear operator using matrix **A** is obtained by (1):

$$\frac{1}{2} \begin{vmatrix} -18 & 5 & 1 \\ 15 & -2 & 1 \\ -6 & 4 & 1 \end{vmatrix} = \frac{1}{2}|51| = 25.5.$$

We have $|\mathbf{A}| = 3$ and we see that $3 \times$ area of triangle T = area of the image. This result holds in general. That is, the area of the image = $\|\mathbf{A}\| \times$ the area of the original figure.

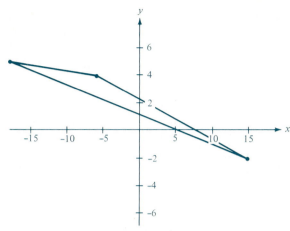

Figure 4.9

FURTHER READING

Rogers, D. F., and J. A. Adams. *Mathematical Elements for Computer Graphics.* New York: McGraw-Hill Book Company, 1976.

4.4 EXERCISES

1. Determine the matrix representation with respect to the natural basis of each of the following linear operators **L**: $R^2 \to R^2$.
   (a) Reflection with respect to the y-axis.
   (b) Reflection with respect to the line $y = x$.
   (c) Rotation of 90° counterclockwise.
   (d) Rotation of 30° clockwise.

2. A linear operator **L**: $R^2 \to R^2$ defined by

$$\mathbf{L}\left(\begin{bmatrix} x \\ y \end{bmatrix}\right) = \begin{bmatrix} rx \\ y \end{bmatrix}$$

is called a **dilation in the x-direction** if $r > 1$ and it is called a **contraction in the x-direction** if $0 < r < 1$. Determine the matrix representation of **L** with respect to the natural basis.

3. A linear operator **L**: $R^2 \to R^2$ defined by

$$\mathbf{L}\left(\begin{bmatrix} x \\ y \end{bmatrix}\right) = \begin{bmatrix} x \\ ry \end{bmatrix}$$

is called a **dilation in the y-direction** if $r > 1$ and it is called a **contraction in the y-direction** if $0 < r < 1$. Determine the matrix representation of **L** with respect to the natural basis.

4. Let T be the triangle with vertices $(5, 0)$, $(0, 3)$, and $(-2, 1)$.
   (a) Find the area of triangle T.
   (b) Find the coordinates of the vertices of the image of T under the linear operator with matrix representation

$$\mathbf{A} = \begin{bmatrix} -2 & 1 \\ 3 & 4 \end{bmatrix}.$$

   (c) Find the area of the triangle whose vertices are obtained in part (b). [*Hint:* Use Equation (1).]

5. Let T be the triangle with vertices $(1, 1)$, $(-3, -3)$, and $(2, -1)$.
   (a) Find the area of the triangle T. [*Hint:* Use Equation (1).]

(b) Find the coordinates of the vertices of the image of T under the linear operator with matrix representation

$$\mathbf{A} = \begin{bmatrix} 4 & -3 \\ -4 & 2 \end{bmatrix}.$$

(c) Find the area of the triangle whose vertices are obtained in part (b).

6. Let Q be the quadrilateral with vertices $(-2, 3)$, $(1, 4)$, $(3, 0)$, and $(-1, -3)$. Find the area of Q.

THEORETICAL EXERCISES

T.1. Prove that a rotation leaves the area of a triangle unchanged.

KEY IDEAS FOR REVIEW

□ **Theorem 4.1.** If $\mathbf{L}: V \rightarrow W$ is a linear transformation, then

$$\mathbf{L}(c_1\mathbf{X}_1 + c_2\mathbf{X}_2 + \cdots + c_k\mathbf{X}_k) = c_1\mathbf{L}(\mathbf{X}_1)$$
$$+ c_2\mathbf{L}(\mathbf{X}_2) + \cdots + c_k\mathbf{L}(\mathbf{X}_k).$$

□ **Theorem 4.3.** Let $\mathbf{L}: V \rightarrow W$ be a linear transformation and let $S = \{\mathbf{X}_1, \mathbf{X}_2, \ldots, \mathbf{X}_n\}$ be a basis for V. If $\mathbf{X}$ is any vector in V, then $\mathbf{L}(\mathbf{X})$ is completely determined by $\{\mathbf{L}(\mathbf{X}_1), \mathbf{L}(\mathbf{X}_2), \ldots, \mathbf{L}(\mathbf{X}_n)\}$.

□ **Theorem 4.4.** If $\mathbf{L}: V \rightarrow W$ is a linear transformation, then ker $\mathbf{L}$ is a subspace of V.

□ **Theorem 4.5.** A linear transformation $\mathbf{L}: V \rightarrow W$ is one-to-one if and only if ker $\mathbf{L} = \{\mathbf{0}_V\}$.

□ **Theorem 4.6.** If $\mathbf{L}: V \rightarrow W$ is a linear transformation, then range $\mathbf{L}$ is a subspace of W.

□ **Theorem 4.7.** If $\mathbf{L}: V \rightarrow W$ is a linear transformation, then

$$\dim (\ker \mathbf{L}) + \dim (\text{range } \mathbf{L}) = \dim V.$$

□ The following statements are equivalent for an $n \times n$ matrix $\mathbf{A}$.
1. $\mathbf{A}$ is nonsingular.
2. $\mathbf{AX} = \mathbf{0}$ has only the trivial solution.
3. $\mathbf{A}$ is row equivalent to $\mathbf{I}_n$.
4. The linear system $\mathbf{AX} = \mathbf{B}$ has a unique solution for every $n \times 1$ matrix $\mathbf{B}$.
5. $|\mathbf{A}| \neq 0$.
6. $\mathbf{A}$ has rank n.
7. The rows (columns) of $\mathbf{A}$ form a linearly independent set of n vectors in R^n.
8. The dimension of the solution space of $\mathbf{AX} = \mathbf{0}$ is zero.
9. The linear transformation $\mathbf{L}: R^n \rightarrow R^n$ defined by $\mathbf{L}(\mathbf{X}) = \mathbf{AX}$, for $\mathbf{X}$ in R^n, is one-to-one and onto.

□ Matrix of a linear transformation $\mathbf{L}: V \rightarrow W$. See Theorem 4.8 on page 211.

SUPPLEMENTARY EXERCISES

1. Is $\mathbf{L}: R^2 \rightarrow R^2$ defined by

$$\mathbf{L}(x, y) = (x - 1, y - x)$$

a linear transformation?

2. Let $\mathbf{X}_0$ be a fixed vector in R^n. Let $\mathbf{L}: R^n \rightarrow R^1$ be defined by $\mathbf{L}(\mathbf{X}) = \mathbf{X} \cdot \mathbf{X}_0$. Show that $\mathbf{L}$ is a linear transformation.

3. Let $\mathbf{L}: P_1 \rightarrow P_1$ be the linear transformation defined by $\mathbf{L}(t - 1) = t + 2$, $\mathbf{L}(t + 1) = 2t + 1$. Find $\mathbf{L}(5t + 1)$. (*Hint:* $\{t - 1, t + 1\}$ is a basis for P_1.)

4. Let $\mathbf{L}: P_2 \rightarrow P_2$ be defined by

$$\mathbf{L}(at^2 + bt + c) = (a - 2c)t^2 + (b - c)t.$$

(a) Find a basis for ker **L**.

(b) Is **L** one-to-one?

5. Let **L**: $R^2 \rightarrow R^3$ be defined by

$$\mathbf{L}(x, y) = (x + y, x - y, x + 2y).$$

(a) Find a basis for range **L**.

(b) Is **L** onto?

6. Find the dimension of the solution space of $\mathbf{AX} = \mathbf{0}$, where

$$\mathbf{A} = \begin{bmatrix} 1 & -2 & -1 & 2 & -1 \\ 4 & -7 & 1 & 4 & 0 \\ 1 & -3 & -6 & 4 & -4 \\ 2 & -3 & 3 & 2 & 1 \end{bmatrix}.$$

7. Let $S = \{t^2 - 1, t + 2, t - 1\}$ be a basis for P_2.

(a) Find the coordinate vector of $2t^2 - 2t + 6$ with respect to S.

(b) If the coordinate vector of $p(t)$ with respect to S

is $\begin{bmatrix} 2 \\ -1 \\ 3 \end{bmatrix}$, find $p(t)$.

8. Let **L**: $P_2 \rightarrow P_2$ be the linear transformation that is defined by $\mathbf{L}(at^2 + bt + c) = (a + 2c)t^2 + (b - c)t + (a - c)$. Let $S = \{t^2, t, 1\}$ and $T = \{t^2 - 1, t, t - 1\}$ be bases for P_2.

(a) Find the matrix of **L** with respect to S and T.

(b) If $p(t) = 2t^2 - 3t + 1$, compute $\mathbf{L}(p(t))$ using the matrix obtained in part (a).

9. Let **L**: $P_1 \rightarrow P_1$ be a linear transformation. Suppose that the matrix of **L** with respect to the basis $S = \{p_1(t), p_2(t)\}$ is

$$\mathbf{A} = \begin{bmatrix} 2 & -3 \\ 1 & 2 \end{bmatrix},$$

where

$$p_1(t) = t - 2 \quad \text{and} \quad p_2(t) = t + 1.$$

(a) Compute $[\mathbf{L}(p_1(t))]_S$ and $[\mathbf{L}(p_2(t))]_S$.

(b) Compute $\mathbf{L}(p_1(t))$ and $\mathbf{L}(p_2(t))$.

(c) Compute $\mathbf{L}(t + 2)$.

10. Let **L**: $P_3 \rightarrow P_3$ be defined by

$$\mathbf{L}(at^3 + bt^2 + ct + d) = 3at^2 + 2bt + c.$$

Find the matrix of **L** with respect to the basis $S = \{t^3, t^2, t, 1\}$ for P_3.

11. Let **L**: $R^3 \rightarrow R^3$ be the linear transformation defined by

$$\mathbf{L}\left(\begin{bmatrix} x \\ y \\ z \end{bmatrix}\right) = \begin{bmatrix} y + z \\ x + z \\ x + y \end{bmatrix}.$$

Let S be the natural basis for R^3 and let

$$T = \left\{ \begin{bmatrix} 0 \\ 1 \\ 1 \end{bmatrix}, \begin{bmatrix} 1 \\ 0 \\ 1 \end{bmatrix}, \begin{bmatrix} 1 \\ 1 \\ 0 \end{bmatrix} \right\}$$

be another basis for R. Find the matrix of **L** with respect to S and T.

12. Let V be the vector space of all $n \times n$ matrices and let **L**: $V \rightarrow R^1$ be defined by $L(\mathbf{A}) = |\mathbf{A}|$, for **A** in V. Is **L** a linear transformation? Justify your answer.

13. Let **L**: $R^1 \rightarrow R^1$ be defined by $\mathbf{L}(\mathbf{X}) = a\mathbf{X} + b$, where a and b are real numbers. Find all values of a and b for which **L** is a linear transformation.

14. Let V be the vector space of all $n \times n$ matrices and let **A** be a fixed $n \times n$ matrix. Define **L**: $V \rightarrow V$ by $\mathbf{L}(\mathbf{B}) = \mathbf{AB} - \mathbf{BA}$ for **B** in V. Is **L** a linear transformation? Justify your answer.

15. Let V be the vector space of all $n \times n$ matrices and let **P** be a fixed nonsingular $n \times n$ matrix. Define **L**: $V \rightarrow V$ by $\mathbf{L}(\mathbf{A}) = \mathbf{P}^{-1}\mathbf{AP}$ for **A** in V. Is **L** a linear transformation? Justify your answer.

THEORETICAL EXERCISES

T.1. Let V be an n-dimensional vector space with basis $S = \{X_1, X_2, \ldots, X_n\}$. Show that $\{[X_1]_S, [X_2]_S, \ldots, [X_n]_S\}$ is the standard basis for R^n.

T.2. Let V be an n-dimensional vector space with basis $S = \{X_1, X_2, \ldots, X_n\}$. If X and Y are vectors in V and c is a scalar, show that

$$[X + Y]_S = [X]_S + [Y]_S$$
$$[cX]_S = c[X]_S.$$

T.3. Let V and W be two vector spaces of dimensions n and m, respectively. If $L_1: V \to W$ and $L_2: V \to W$ are linear transformations, we define $L_1 \boxplus L_2: V \to W$ by $(L_1 \boxplus L_2)(X) = L_1(X) + L_2(X)$, for X in V. Also, if $L: V \to W$ is a linear transformation and c is a scalar, we define $c \boxdot L: V \to W$ by $(c \boxdot L)(X) = cL(X)$, for X in V.
(a) Prove that $L_1 \boxplus L_2$ is a linear transformation.
(b) Prove that $c \boxdot L$ is a linear transformation.
(c) Let $V = R^3$, $W = R^2$, $L_1: V \to W$, and $L_2: V \to W$ be defined by

$$L_1(X) = L_1(x_1, x_2, x_3) = (x_1 + x_2, x_2 + x_3)$$
$$L_2(X) = L_2(x_1, x_2, x_3) = (x_1 + x_3, x_2).$$

Compute $(L_1 \boxplus L_2)(X)$ and $(-2 \boxdot L_1)(X)$.

T.4. Prove that the set U of all linear transformations of an n-dimensional vector space into an m-dimensional vector space W is a vector space under the operations $\boxplus$ and $\boxdot$ defined in Supplementary Exercise T.3.

T.5. Let V_1, V_2, and V_3 be vector spaces of dimensions n, m, and p, respectively. Let $L_1: V_1 \to V_2$ and $L_2: V_2 \to V_3$ be linear transformations. Define $L_2 \circ L_1, V_1 \to V_3$ by

$$(L_2 \circ L_1)(X) = L_2(L_1(X)) \qquad \text{for } X \text{ in } V_1.$$

(a) Prove that $L_2 \circ L_1$ is a linear transformation.
(b) Let $L_1: R^2 \to R^2$ and $L_2: R^2 \to R^2$ be defined by

$$L_1(x_1, x_2) = (x_1, -x_2)$$

and

$$L_2(x_1, x_2) = (x_2, x_1).$$

Compute $L_1 \circ L_2$ and $L_2 \circ L_1$.

T.6. Let $L: R^n \to R^m$ be a linear transformation defined by $L(X) = AX$, X in R^n, where A is an $m \times n$ matrix.
(a) Prove that L is one-to-one if and only if rank $A = n$.
(b) Prove that L is onto if and only if rank $A = m$.

CHAPTER TEST

1. Let $L: R^2 \to R^3$ be a linear transformation for which we know that

$$L\left(\begin{bmatrix} 1 \\ -1 \end{bmatrix}\right) = \begin{bmatrix} 1 \\ 2 \\ -1 \end{bmatrix} \quad \text{and} \quad L\left(\begin{bmatrix} 2 \\ -1 \end{bmatrix}\right) = \begin{bmatrix} 0 \\ 1 \\ 2 \end{bmatrix}.$$

Find $L\left(\begin{bmatrix} 8 \\ -5 \end{bmatrix}\right)$.

2. Let $L: R^3 \to R^3$ be defined by

$$L(x, y, z) = (x + 2y + z, x + y, 2y + z).$$

(a) Find a basis for ker L.
(b) Is L one-to-one?

3. Let $L: R^2 \to R^3$ be defined by

$$L(x, y) = (x + y, x - y, 2x + y).$$

(a) Find a basis for range L.
(b) Is L onto?

4. Find the dimension of the solution space of $AX = 0$, where

$$A = \begin{bmatrix} 1 & 0 & -1 & 2 & 3 \\ 2 & 1 & -2 & 2 & 1 \\ 0 & 1 & -6 & 8 & 8 \\ 1 & -1 & -1 & 4 & 8 \end{bmatrix}.$$

5. Let $L: R^2 \to R^2$ be the linear transformation defined

by $L\left(\begin{bmatrix} x \\ y \end{bmatrix}\right) = \begin{bmatrix} x + 2y \\ x - y \end{bmatrix}$. Let $S = \left\{ \begin{bmatrix} 1 \\ -1 \end{bmatrix}, \begin{bmatrix} 0 \\ 2 \end{bmatrix} \right\}$

and $T = \left\{ \begin{bmatrix} 1 \\ 2 \end{bmatrix}, \begin{bmatrix} -1 \\ 2 \end{bmatrix} \right\}$ be bases for R^2. Find the

matrix of L with respect to S and T.

6. Answer each of the following as true or false.

(a) If $L: P_2 \to P_1$ is the linear transformation defined by

$$L(at^2 + bt + c) = (a - c)t + (b + c),$$

then $t^2 + 2t + 1$ is in ker L.

(b) If $L: R^2 \to R^2$ is the linear transformation defined by $L(x, y) = (x - y, x + y)$, then $(2, 3)$ is in range L.

(c) If A is a 4×7 matrix whose rank is 4, then the homogeneous system $AX = 0$ has a nontrivial solution.

(d) There are linear transformations $L: R^3 \to R^5$ that are onto.

(e) If $L: R^7 \to R^5$ is a linear transformation such that dim (ker L) = 3, then dim (range L) = 2.

5

Eigenvalues
and Eigenvectors

Note. This chapter can now be covered without first covering Chapter 4. The several places in this chapter where Sections 4.1 and 4.3 are used have been clearly marked. These discussions can be omitted without any loss of continuity.

5.1

DIAGONALIZATION

In this chapter every matrix considered is a square matrix. If A is an $n \times n$ matrix, then for X in R^n, AX is also a vector in R^n. A question of considerable importance in a great many applied problems is the determination of vectors X, if there are any, such that X and AX are parallel. Such questions arise in all applications involving vibrations; they arise in aerodynamics, elasticity, nuclear physics, mechanics, chemical engineering, biology, differential equations, and so on. In this section we shall formulate this problem precisely; we also define some pertinent terminology. In the next section we settle the question for symmetric matrices and briefly discuss the situation in the general case.

Definition Let A be an $n \times n$ matrix. The real number λ is called an **eigenvalue** of A if there exists a nonzero vector X in R^n such that

232

$$\mathbf{AX} = \lambda\mathbf{X}. \tag{1}$$

Every nonzero vector $\mathbf{X}$ satisfying (1) is called an **eigenvector of A associated with the eigenvalue λ**. We might mention that the word "eigenvalue" is a hybrid one ("eigen" in German means "proper"). Eigenvalues are also called **proper values**, **characteristic values**, and **latent values**; and eigenvectors are also called **proper vectors**, and so on, accordingly.

Note that $\mathbf{X} = \mathbf{0}$ always satisfies Equation (1), but we insist that an eigenvector $\mathbf{X}$ be a nonzero vector. In many practical applications one encounters matrices with complex entries and vector spaces with scalars that are complex numbers. (See Section A.2.) In such a setting the preceding definition of eigenvalue is modified so that an eigenvalue can be a real *or* a complex number. Such treatments are presented in more advanced books. In this book we require an eigenvalue to be a real number.

Example 1 If $\mathbf{A}$ is the identity matrix $\mathbf{I}_n$, then the only eigenvalue is $\lambda = 1$; every nonzero vector in R^n is an eigenvector of $\mathbf{A}$ associated with the eigenvalue $\lambda = 1$:

$$\mathbf{I}_n\mathbf{X} = 1\mathbf{X}.$$

Example 2 Let

$$\mathbf{A} = \begin{bmatrix} 0 & \frac{1}{2} \\ \frac{1}{2} & 0 \end{bmatrix}.$$

Then

$$\mathbf{A}\begin{bmatrix} 1 \\ 1 \end{bmatrix} = \begin{bmatrix} 0 & \frac{1}{2} \\ \frac{1}{2} & 0 \end{bmatrix}\begin{bmatrix} 1 \\ 1 \end{bmatrix} = \begin{bmatrix} \frac{1}{2} \\ \frac{1}{2} \end{bmatrix} = \frac{1}{2}\begin{bmatrix} 1 \\ 1 \end{bmatrix}$$

so that $\mathbf{X}_1 = \begin{bmatrix} 1 \\ 1 \end{bmatrix}$ is an eigenvector of $\mathbf{A}$ associated with the eigenvalue $\lambda_1 = \frac{1}{2}$. Also,

$$\mathbf{A}\begin{bmatrix} 1 \\ -1 \end{bmatrix} = \begin{bmatrix} 0 & \frac{1}{2} \\ \frac{1}{2} & 0 \end{bmatrix}\begin{bmatrix} 1 \\ -1 \end{bmatrix} = \begin{bmatrix} -\frac{1}{2} \\ \frac{1}{2} \end{bmatrix} = -\frac{1}{2}\begin{bmatrix} 1 \\ -1 \end{bmatrix}$$

so that $\mathbf{X}_2 = \begin{bmatrix} 1 \\ -1 \end{bmatrix}$ is an eigenvector of $\mathbf{A}$ associated with the eigenvalue $\lambda_2 = -\frac{1}{2}$.

Figure 5.1 shows that $\mathbf{X}_1$ and $\mathbf{AX}_1$ are parallel, and $\mathbf{X}_2$ and $\mathbf{AX}_2$ are parallel also. This illustrates the fact that if $\mathbf{X}$ is an eigenvector of $\mathbf{A}$, then $\mathbf{X}$ and $\mathbf{AX}$ are parallel.

In Figure 5.2 we show $\mathbf{X}$ and $\mathbf{AX}$ for the cases $\lambda > 1$, $0 < \lambda < 1$, and $\lambda < 0$.†

An eigenvalue λ of $\mathbf{A}$ can have associated with it many different eigenvectors. In fact, if $\mathbf{X}$ is an eigenvector of $\mathbf{A}$ associated with λ (that is, $\mathbf{AX} = \lambda\mathbf{X}$) and r is any nonzero real number, then

---

† For those who have read Section 4.1: if we define $\mathbf{L}: R^n \rightarrow R^n$ by $\mathbf{L}(\mathbf{X}) = \mathbf{AX}$, then $\mathbf{L}$ is a linear transformation and Figure 5.2 shows that if $\lambda > 1$, then $\mathbf{L}$ is a dilation, and if $0 < \lambda < 1$, then $\mathbf{L}$ is a contraction; if $\lambda < 0$, then $\mathbf{L}$ reverses the direction of $\mathbf{X}$.

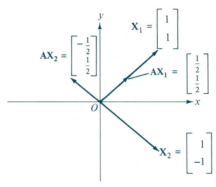

Figure 5.1

$$\mathbf{A}(r\mathbf{X}) = r(\mathbf{A}\mathbf{X}) = r(\lambda\mathbf{X}) = \lambda(r\mathbf{X}).$$

Thus $r\mathbf{X}$ is also an eigenvector of $\mathbf{A}$ associated with λ.

Example 3 Let

$$\mathbf{A} = \begin{bmatrix} 0 & 0 \\ 0 & 1 \end{bmatrix}.$$

Then

$$\mathbf{A}\begin{bmatrix} 1 \\ 0 \end{bmatrix} = \begin{bmatrix} 0 & 0 \\ 0 & 1 \end{bmatrix}\begin{bmatrix} 1 \\ 0 \end{bmatrix} = \begin{bmatrix} 0 \\ 0 \end{bmatrix} = 0\begin{bmatrix} 1 \\ 0 \end{bmatrix}$$

so that $\mathbf{X}_1 = \begin{bmatrix} 1 \\ 0 \end{bmatrix}$ is an eigenvector of $\mathbf{A}$ associated with the eigenvalue $\lambda_1 = 0$. Also,

$\mathbf{X}_2 = \begin{bmatrix} 0 \\ 1 \end{bmatrix}$ is an eigenvector of $\mathbf{A}$ associated with the eigenvalue $\lambda_2 = 1$ (verify).

Example 3 points out the fact that although the zero vector, by definition, cannot be an eigenvector, the number zero can be an eigenvalue.

Figure 5.2

Example 4 Let

$$\mathbf{A} = \begin{bmatrix} 1 & 1 \\ -2 & 4 \end{bmatrix}.$$

We wish to find the eigenvalues of $\mathbf{A}$ and their associated eigenvectors. Thus we wish to find all real numbers λ and all nonzero vectors $\mathbf{X} = \begin{bmatrix} x_1 \\ x_2 \end{bmatrix}$ satisfying (1), that is,

$$\begin{bmatrix} 1 & 1 \\ -2 & 4 \end{bmatrix}\begin{bmatrix} x_1 \\ x_2 \end{bmatrix} = \lambda \begin{bmatrix} x_1 \\ x_2 \end{bmatrix}. \tag{2}$$

Equation (2) becomes

$$\begin{aligned} x_1 + x_2 &= \lambda x_1 \\ -2x_1 + 4x_2 &= \lambda x_2, \end{aligned}$$

or

$$\begin{aligned} (\lambda - 1)x_1 - \quad\quad x_2 &= 0 \\ 2x_1 + (\lambda - 4)x_2 &= 0. \end{aligned} \tag{3}$$

Equation (3) is a homogeneous system of two equations in two unknowns. From Corollary 2.3 of Section 2.2 it follows that the homogeneous system in (3) has a nontrivial solution if and only if the determinant of its coefficient matrix is zero: thus if and only if

$$\begin{vmatrix} \lambda - 1 & -1 \\ 2 & \lambda - 4 \end{vmatrix} = 0.$$

This means that

$$(\lambda - 1)(\lambda - 4) + 2 = 0,$$

or

$$\lambda^2 - 5\lambda + 6 = 0 = (\lambda - 3)(\lambda - 2).$$

Hence

$$\lambda_1 = 2 \quad\quad \text{and} \quad\quad \lambda_2 = 3$$

are the eigenvalues of $\mathbf{A}$. To find all eigenvectors of $\mathbf{A}$ associated with $\lambda_1 = 2$, we form the linear system

$$\mathbf{AX} = 2\mathbf{X},$$

or

$$\begin{bmatrix} 1 & 1 \\ -2 & 4 \end{bmatrix}\begin{bmatrix} x_1 \\ x_2 \end{bmatrix} = 2\begin{bmatrix} x_1 \\ x_2 \end{bmatrix}.$$

This gives

$$x_1 + x_2 = 2x_1$$
$$-2x_1 + 4x_2 = 2x_2$$

or

$$(2 - 1)x_1 - x_2 = 0$$
$$2x_1 + (2 - 4)x_2 = 0$$

or

$$x_1 - x_2 = 0$$
$$2x_1 - 2x_2 = 0.$$

Note that we could have obtained this last homogeneous system by merely substituting $\lambda = 2$ in (3). All solutions to this last system are given by

$$x_1 = x_2$$
$$x_2 = \text{any real number } r.$$

Hence all eigenvectors associated with the eigenvalue $\lambda_1 = 2$ are given by $\begin{bmatrix} r \\ r \end{bmatrix}$, r any nonzero real number. In particular, $\mathbf{X}_1 = \begin{bmatrix} 1 \\ 1 \end{bmatrix}$ is an eigenvector associated with $\lambda_1 = 2$. Similarly, for $\lambda_2 = 3$ we obtain, from (3),

$$(3 - 1)x_1 - x_2 = 0$$
$$2x_1 + (3 - 4)x_2 = 0$$

or

$$2x_1 - x_2 = 0$$
$$2x_1 - x_2 = 0.$$

All solutions to this last homogeneous system are given by

$$x_1 = \tfrac{1}{2}x_2$$
$$x_2 = \text{any real number } r.$$

Hence all eigenvectors associated with the eigenvalue $\lambda_2 = 3$ are given by $\begin{bmatrix} \frac{r}{2} \\ r \end{bmatrix}$, r any nonzero real number. In particular, $\mathbf{X}_2 = \begin{bmatrix} 1 \\ 2 \end{bmatrix}$ is an eigenvector associated with the eigenvalue $\lambda_2 = 3$.

In Examples 1, 2, and 3 we found eigenvalues and eigenvectors by inspection, whereas in Example 4 we proceeded in a more systematic fashion. We use the procedure of Example 4 as our standard method, as follows.

Definition Let $\mathbf{A} = [a_{ij}]$ be an $n \times n$ matrix. The determinant

$$f(\lambda) = |\lambda \mathbf{I}_n - \mathbf{A}| = \begin{vmatrix} \lambda - a_{11} & -a_{12} & \cdots & -a_{1n} \\ -a_{21} & \lambda - a_{22} & \cdots & -a_{2n} \\ \vdots & \vdots & & \vdots \\ -a_{n1} & -a_{n2} & \cdots & \lambda - a_{nn} \end{vmatrix} \qquad (4)$$

is called the **characteristic polynomial of A**. The equation

$$f(\lambda) = |\lambda \mathbf{I}_n - \mathbf{A}| = 0$$

is called the **characteristic equation of A**.

Recall from Chapter 2 that each term in the expansion of the determinant of an $n \times n$ matrix is a product of n elements of the matrix, containing exactly one element from each row and exactly one element from each column. Thus, if we expand $f(\lambda) = |\lambda \mathbf{I}_n - \mathbf{A}|$, we obtain a polynomial of degree n. A polynomial of degree n with real coefficients has n roots, some of which may be imaginary numbers. The expression involving λ^n in the characteristic polynomial of $\mathbf{A}$ comes from the product $(\lambda - a_{11})(\lambda - a_{22}) \cdots (\lambda - a_{nn})$, and so the coefficient of λ^n is 1. We can then write

$$f(\lambda) = |\lambda \mathbf{I}_n - \mathbf{A}| = \lambda^n + a_1 \lambda^{n-1} + a_2 \lambda^{n-2} + \cdots + a_{n-1} \lambda + a_n.$$

If we let $\lambda = 0$ in $|\lambda \mathbf{I}_n - \mathbf{A}|$ as well as in the expression on the right, then we get $|-\mathbf{A}| = a_n$, which shows that the constant term a_n is $(-1)^n |\mathbf{A}|$.

Example 5 Let

$$\mathbf{A} = \begin{bmatrix} 1 & 2 & -1 \\ 1 & 0 & 1 \\ 4 & -4 & 5 \end{bmatrix}.$$

The characteristic polynomial of $\mathbf{A}$ is (verify)

$$f(\lambda) = |\lambda \mathbf{I}_3 - \mathbf{A}| = \begin{vmatrix} \lambda - 1 & -2 & 1 \\ -1 & \lambda - 0 & -1 \\ -4 & 4 & \lambda - 5 \end{vmatrix} = \lambda^3 - 6\lambda^2 + 11\lambda - 6.$$

We now connect the characteristic polynomial of a matrix with its eigenvalues in the following theorem.

Theorem 5.1 *The eigenvalues of* $\mathbf{A}$ *are the real roots of the characteristic polynomial of* $\mathbf{A}$.

Proof
Let λ be an eigenvalue of $\mathbf{A}$ with associated eigenvector $\mathbf{X}$. Then

$$\mathbf{AX} = \lambda \mathbf{X},$$

which can be rewritten as

$$\mathbf{AX} = (\lambda \mathbf{I}_n)\mathbf{X}$$

or

$$(\lambda \mathbf{I}_n - \mathbf{A})\mathbf{X} = \mathbf{0}, \tag{5}$$

a homogeneous system of n equations in n unknowns. This system has a nontrivial solution if and only if the determinant of its coefficient matrix vanishes (Corollary 2.3, Section 2.2), that is, if and only if $|\lambda \mathbf{I}_n - \mathbf{A}| = 0$.

Conversely, if λ is a real root of the characteristic polynomial of $\mathbf{A}$, then $|\lambda \mathbf{I}_n - \mathbf{A}| = 0$, so the homogeneous system (5) has a nontrivial solution $\mathbf{X}$. Hence λ is an eigenvalue of $\mathbf{A}$.

Thus, to find the eigenvalues of a given matrix $\mathbf{A}$, we must find the real roots of its characteristic polynomial $f(\lambda)$. There are many methods for finding approximations to the roots of a polynomial, some of them more effective than others; indeed, many computer codes are available to find the roots of a polynomial. Two results that are sometimes useful in this connection are (1) the product of all the roots of the polynomial $f(\lambda) = \lambda^n + a_1 \lambda^{n-1} + \cdots + a_{n-1}\lambda + a_n$ is $(-1)^n a_n$, and (2) if $a_1, a_2, \ldots, a_n$ are integers, then $f(\lambda)$ cannot have a rational root that is not already an integer. Thus, as possible rational roots of $f(\lambda)$, one need only try the integer factors of a_n. Of course, $f(\lambda)$ might well have irrational roots. However, to minimize the computational effort and as a convenience to the reader, all the characteristic polynomials to be considered in the rest of this chapter have only integer roots, and each of these roots is a factor of the constant term of the characteristic polynomial of $\mathbf{A}$. The corresponding eigenvectors are obtained by substituting for λ in Equation (5) and solving the resulting homogeneous system.

Example 6 Consider the matrix of Example 5. The characteristic polynomial is

$$f(\lambda) = \lambda^3 - 6\lambda^2 + 11\lambda - 6.$$

The possible integer roots of $f(\lambda)$ are ± 1, ± 2, ± 3, and ± 6. By substituting these values in $f(\lambda)$, we find that $f(1) = 0$ so that $\lambda = 1$ is a root of $f(\lambda)$. Hence $(\lambda - 1)$ is a factor of $f(\lambda)$. Dividing $f(\lambda)$ by $(\lambda - 1)$, we obtain

$$f(\lambda) = (\lambda - 1)(\lambda^2 - 5\lambda + 6).$$

Factoring $\lambda^2 - 5\lambda + 6$, we have

$$f(\lambda) = (\lambda - 1)(\lambda - 2)(\lambda - 3).$$

The eigenvalues of $\mathbf{A}$ are then

$$\lambda_1 = 1, \qquad \lambda_2 = 2, \qquad \lambda_3 = 3.$$

To find an eigenvector X_1 associated with $\lambda_1 = 1$, we form the system

$$(1I_3 - A)X = 0,$$

$$\begin{bmatrix} 1-1 & -2 & 1 \\ -1 & 1 & -1 \\ -4 & 4 & 1-5 \end{bmatrix} \begin{bmatrix} x_1 \\ x_2 \\ x_3 \end{bmatrix} = \begin{bmatrix} 0 \\ 0 \\ 0 \end{bmatrix}$$

or

$$\begin{bmatrix} 0 & -2 & 1 \\ -1 & 1 & -1 \\ -4 & 4 & -4 \end{bmatrix} \begin{bmatrix} x_1 \\ x_2 \\ x_3 \end{bmatrix} = \begin{bmatrix} 0 \\ 0 \\ 0 \end{bmatrix}.$$

A solution is $\begin{bmatrix} -\frac{r}{2} \\ \frac{r}{2} \\ r \end{bmatrix}$ for any real number r. Thus $X_1 = \begin{bmatrix} -1 \\ 1 \\ 2 \end{bmatrix}$ is an eigenvector

of A associated with $\lambda_1 = 1$.

To find an eigenvector X_2 associated with $\lambda_2 = 2$, we form the system

$$(2I_3 - A)X = 0,$$

that is,

$$\begin{bmatrix} 2-1 & -2 & 1 \\ -1 & 2 & -1 \\ -4 & 4 & 2-5 \end{bmatrix} \begin{bmatrix} x_1 \\ x_2 \\ x_3 \end{bmatrix} = \begin{bmatrix} 0 \\ 0 \\ 0 \end{bmatrix}$$

or

$$\begin{bmatrix} 1 & -2 & 1 \\ -1 & 2 & -1 \\ -4 & 4 & -3 \end{bmatrix} \begin{bmatrix} x_1 \\ x_2 \\ x_3 \end{bmatrix} = \begin{bmatrix} 0 \\ 0 \\ 0 \end{bmatrix}.$$

A solution is $\begin{bmatrix} -\frac{r}{2} \\ \frac{r}{4} \\ r \end{bmatrix}$ for any real number r. Thus $X_2 = \begin{bmatrix} -2 \\ 1 \\ 4 \end{bmatrix}$ is an eigenvector

of A associated with $\lambda_2 = 2$.

To find an eigenvector X_3 associated with $\lambda_3 = 3$, we form the system

$$(3I_3 - A)X = 0,$$

and find that a solution is $\begin{bmatrix} -\frac{r}{4} \\ \frac{r}{4} \\ r \end{bmatrix}$ for any real number r. Thus $X_3 = \begin{bmatrix} -1 \\ 1 \\ 4 \end{bmatrix}$ is an

eigenvector of A associated with $\lambda_3 = 3$.

Of course, the characteristic polynomial of a given matrix may have imaginary roots and it may even have no real roots. However, for the matrices that we are most interested in, symmetric matrices, all the roots of the characteristic polynomial are real. We shall state this result in Section 5.2.

Example 7 Let $\mathbf{A} = \begin{bmatrix} 0 & 1 \\ -1 & 0 \end{bmatrix}$. Then the characteristic polynomial of $\mathbf{A}$ is

$$f(\lambda) = \lambda^2 + 1,$$

which has no real roots. Thus $\mathbf{A}$ has no eigenvalues.

Similar Matrices

We shall now develop an equivalent formulation for the eigenvalue–eigenvector problem, which will shed further light on the problem.

Definition A matrix $\mathbf{B}$ is said to be **similar** to a matrix $\mathbf{A}$ if there is a nonsingular matrix $\mathbf{P}$ such that

$$\mathbf{B} = \mathbf{P}^{-1}\mathbf{AP}.$$

Example 8 Let

$$\mathbf{A} = \begin{bmatrix} 1 & 1 \\ -2 & 4 \end{bmatrix}$$

be the matrix of Example 4. Let

$$\mathbf{P} = \begin{bmatrix} 1 & 1 \\ 1 & 2 \end{bmatrix}.$$

Then

$$\mathbf{P}^{-1} = \begin{bmatrix} 2 & -1 \\ -1 & 1 \end{bmatrix}$$

and

$$\mathbf{B} = \mathbf{P}^{-1}\mathbf{AP} = \begin{bmatrix} 2 & -1 \\ -1 & 1 \end{bmatrix}\begin{bmatrix} 1 & 1 \\ -2 & 4 \end{bmatrix}\begin{bmatrix} 1 & 1 \\ 1 & 2 \end{bmatrix} = \begin{bmatrix} 2 & 0 \\ 0 & 3 \end{bmatrix}.$$

Thus $\mathbf{B}$ is similar to $\mathbf{A}$.

We shall let the reader (Exercise T.1) prove that the following elementary properties hold for similarity:

1. $\mathbf{A}$ is similar to $\mathbf{A}$.
2. If $\mathbf{B}$ is similar to $\mathbf{A}$, then $\mathbf{A}$ is similar to $\mathbf{B}$.
3. If $\mathbf{A}$ is similar to $\mathbf{B}$ and $\mathbf{B}$ is similar to $\mathbf{C}$, then $\mathbf{A}$ is similar to $\mathbf{C}$.

By property 2 we replace the statements "**A** is similar to **B**" and "**B** is similar to **A**" by "**A** and **B** are similar."

Definition We shall say that the matrix **A** is **diagonalizable** if it is similar to a diagonal matrix. In this case we also say that **A** can be diagonalized.

Example 9 If **A** and **B** are as in Example 8, then **A** is diagonalizable, since it is similar to **B**.

Theorem 5.2 *An n × n matrix **A** is diagonalizable if and only if it has n linearly independent eigenvectors. In this case **A** is similar to a diagonal matrix **D**, with $P^{-1}AP = D$, whose diagonal elements are the eigenvalues of **A**, while **P** is a matrix whose columns are n linearly independent eigenvectors of **A**.*

Proof

Suppose that **A** is similar to **D**. Then

$$P^{-1}AP = D$$

so that

$$AP = PD. \tag{6}$$

Let

$$D = \begin{bmatrix} \lambda_1 & 0 & \cdots & 0 \\ 0 & \lambda_2 & \cdots & 0 \\ \vdots & & & \vdots \\ 0 & \cdots & 0 & \lambda_n \end{bmatrix},$$

and let X_j, $j = 1, 2, \ldots, n$ be the jth column of **P**. Note that the jth column of the matrix **AP** is AX_j and the jth column of **PD** is $\lambda_j X_j$ (see Exercise T.5 in Section 1.2 and verify). Thus from (6) we have

$$AX_j = \lambda_j X_j. \tag{7}$$

Since **P** is a nonsingular matrix, its columns are linearly independent and so are all nonzero. Hence λ_j is an eigenvalue of **A** and X_j is a corresponding eigenvector. Moreover, since **P** is nonsingular, its column vectors are linearly independent (Theorem 2.12, Section 2.2, and Corollary 3.3, Section 3.7).

Conversely, suppose that $\lambda_1, \lambda_2, \ldots, \lambda_n$ are n eigenvalues of **A** and that the corresponding eigenvectors $X_1, X_2, \ldots, X_n$ are linearly independent. Let **P** be the matrix whose jth column is X_j. From Corollary 3.3 of Section 3.7 and Theorem 2.12 of Section 2.2, it follows that **P** is nonsingular. From (7) we obtain (6), which implies that **A** is diagonalizable. This completes the proof.

Observe that in Theorem 5.2 the order of the columns of **P** determines the order of the diagonal entries in **D**.

Example 10 Let **A** be as in Example 4. The eigenvalues are $\lambda_1 = 2$ and $\lambda_2 = 3$. The corresponding eigenvectors $\mathbf{X}_1 = \begin{bmatrix} 1 \\ 1 \end{bmatrix}$ and $\mathbf{X}_2 = \begin{bmatrix} 1 \\ 2 \end{bmatrix}$ are linearly independent. Hence **A** is diagonalizable. Here

$$\mathbf{P} = \begin{bmatrix} 1 & 1 \\ 1 & 2 \end{bmatrix} \quad \text{and} \quad \mathbf{P}^{-1} = \begin{bmatrix} 2 & -1 \\ -1 & 1 \end{bmatrix}.$$

Thus

$$\mathbf{P}^{-1}\mathbf{A}\mathbf{P} = \begin{bmatrix} 2 & -1 \\ -1 & 1 \end{bmatrix}\begin{bmatrix} 1 & 1 \\ -2 & 4 \end{bmatrix}\begin{bmatrix} 1 & 1 \\ 1 & 2 \end{bmatrix} = \begin{bmatrix} 2 & 0 \\ 0 & 3 \end{bmatrix}.$$

On the other hand, if we let $\lambda_1 = 3$ and $\lambda_2 = 2$, then $\mathbf{X}_1 = \begin{bmatrix} 1 \\ 2 \end{bmatrix}$ and $\mathbf{X}_2 = \begin{bmatrix} 1 \\ 1 \end{bmatrix}$. Then

$$\mathbf{P} = \begin{bmatrix} 1 & 1 \\ 2 & 1 \end{bmatrix} \quad \text{and} \quad \mathbf{P}^{-1} = \begin{bmatrix} -1 & 1 \\ 2 & -1 \end{bmatrix}.$$

Hence

$$\mathbf{P}^{-1}\mathbf{A}\mathbf{P} = \begin{bmatrix} -1 & 1 \\ 2 & -1 \end{bmatrix}\begin{bmatrix} 1 & 1 \\ -2 & 4 \end{bmatrix}\begin{bmatrix} 1 & 1 \\ 2 & 1 \end{bmatrix} = \begin{bmatrix} 3 & 0 \\ 0 & 2 \end{bmatrix}.$$

Example 11 Let

$$\mathbf{A} = \begin{bmatrix} 1 & 1 \\ 0 & 1 \end{bmatrix}.$$

The eigenvalues of **A** are $\lambda_1 = 1$ and $\lambda_2 = 1$. Eigenvectors associated with λ_1 and λ_2 are vectors of the form

$$\begin{bmatrix} r \\ 0 \end{bmatrix},$$

where r is any nonzero real number. Since **A** does not have two linearly independent eigenvectors, we conclude that **A** is not diagonalizable.

Corollary 5.1† *Consider the linear transformation* **L**: $R^n \to R^n$ *defined by* $\mathbf{L}(\mathbf{X}) = \mathbf{A}\mathbf{X}$ *for* **X** *in* R^n. *Then* **A** *is diagonalizable with n linearly independent eigenvectors* $\mathbf{X}_1, \mathbf{X}_2, \ldots, \mathbf{X}_n$ *if and only if the matrix of* **L** *with respect to* $S = \{\mathbf{X}_1, \mathbf{X}_2, \ldots, \mathbf{X}_n\}$ *is diagonal.*

Proof

Suppose that **A** is diagonalizable. Then by Theorem 5.2 it has n linearly independent eigenvectors $\mathbf{X}_1, \mathbf{X}_2, \ldots, \mathbf{X}_n$, with corresponding eigenvalues $\lambda_1, \lambda_2, \ldots, \lambda_n$. Since

---

† Uses Section 4.1.

n linearly independent vectors in R^n form a basis (Theorem 3.14 in Section 3.6), we can conclude that $S = \{X_1, X_2, \ldots, X_n\}$ is a basis for R^n. Now

$$L(X_j) = AX_j$$
$$= \lambda_j X_j = 0X_1 + \cdots + 0X_{j-1} + \lambda_j X_j + 0X_{j+1} + \cdots + 0X_n,$$

so the coordinate vector $[L(X_j)]_S$ of $L(X_j)$ with respect to S is

$$\begin{bmatrix} 0 \\ \vdots \\ 0 \\ \lambda_j \\ 0 \\ \vdots \\ 0 \end{bmatrix} \longleftarrow j\text{th row.} \tag{8}$$

Hence the matrix of L with respect to S is (Theorem 4.8 of Section 4.3)

$$\begin{bmatrix} \lambda_1 & 0 & \cdots & 0 \\ 0 & \lambda_2 & \cdots & 0 \\ \vdots & & & \vdots \\ 0 & \cdots & 0 & \lambda_n \end{bmatrix}. \tag{9}$$

Conversely, suppose that there is a basis $S = \{X_1, X_2, \ldots, X_n\}$ for R^n with respect to which the matrix of L is diagonal, say, of the form in (9). Then the coordinate vector of $L(X_j)$ with respect to S is (8), so

$$L(X_j) = 0X_1 + \cdots + 0X_{j-1} + \lambda_j X_j + 0X_{j+1} + \cdots + 0X_n = \lambda_j X_j.$$

Since $L(X_j) = AX_j$, we have

$$AX_j = \lambda_j X_j,$$

which means that $X_1, X_2, \ldots, X_n$ are eigenvectors of A. Since they form a basis for R^n, they are linearly independent, and by Theorem 5.2 we conclude that A is diagonalizable.

The following is a useful theorem because it identifies a large class of matrices that can be diagonalized.

Theorem 5.3   *A matrix A is diagonalizable if all the roots of its characteristic polynomial are real and distinct.*

Proof
Let $\lambda_1, \lambda_2, \ldots, \lambda_n$ be the distinct eigenvalues of A and let $S = \{X_1, X_2, \ldots, X_n\}$ be a set of associated eigenvectors. We wish to show that S is linearly independent.
  Suppose that S is a linearly dependent set of vectors. Then Theorem 3.9 of Section 3.5

implies that some vector $\mathbf{X}_j$ is a linear combination of the preceding vectors in S. We can assume that $S_1 = \{\mathbf{X}_1, \mathbf{X}_2, \ldots, \mathbf{X}_{j-1}\}$ is linearly independent, for otherwise one of the vectors in S_1 is a linear combination of the preceding ones, and we can choose a new set S_2, and so on. We thus have that S_1 is linearly independent and that

$$\mathbf{X}_j = c_1\mathbf{X}_1 + c_2\mathbf{X}_2 + \cdots + c_{j-1}\mathbf{X}_{j-1}, \tag{10}$$

where $c_1, c_2, \ldots, c_{j-1}$ are real numbers. Premultiplying (multiplying on the left) both sides of Equation (10) by $\mathbf{A}$, we obtain

$$\begin{aligned}\mathbf{A}\mathbf{X}_j &= \mathbf{A}(c_1\mathbf{X}_1 + c_2\mathbf{X}_2 + \cdots c_{j-1}\mathbf{X}_{j-1}) \\ &= c_1\mathbf{A}\mathbf{X}_1 + c_2\mathbf{A}\mathbf{X}_2 + \cdots + c_{j-1}\mathbf{A}\mathbf{X}_{j-1}.\end{aligned} \tag{11}$$

Since $\lambda_1, \lambda_2, \ldots, \lambda_j$ are eigenvalues of $\mathbf{A}$ and $\mathbf{X}_1, \mathbf{X}_2, \ldots, \mathbf{X}_j$, its associated eigenvectors, we know that $\mathbf{A}\mathbf{X}_i = \lambda_i\mathbf{X}_i$ for $i = 1, 2, \ldots, j$. Substituting in (11), we have

$$\lambda_j\mathbf{X}_j = c_1\lambda_1\mathbf{X}_1 + c_2\lambda_2\mathbf{X}_2 + \cdots + c_{j-1}\lambda_{j-1}\mathbf{X}_{j-1}. \tag{12}$$

Multiplying (10) by λ_j, we obtain

$$\lambda_j\mathbf{X}_j = \lambda_j c_1\mathbf{X}_1 + \lambda_j c_2\mathbf{X}_2 + \cdots + \lambda_j c_{j-1}\mathbf{X}_{j-1}. \tag{13}$$

Subtracting (13) from (12), we have

$$\begin{aligned}0 &= \lambda_j\mathbf{X}_j - \lambda_j\mathbf{X}_j \\ &= c_1(\lambda_1 - \lambda_j)\mathbf{X}_1 + c_2(\lambda_2 - \lambda_j)\mathbf{X}_2 + \cdots + c_{j-1}(\lambda_{j-1} - \lambda_j)\mathbf{X}_{j-1}.\end{aligned}$$

Since S_1 is linearly independent, we must have

$$c_1(\lambda_1 - \lambda_j) = 0, \qquad c_2(\lambda_2 - \lambda_j) = 0, \ldots, \qquad c_{j-1}(\lambda_{j-1} - \lambda_j) = 0.$$

Now

$$\lambda_1 - \lambda_j \neq 0, \qquad \lambda_2 - \lambda_j \neq 0, \ldots, \qquad \lambda_{j-1} - \lambda_j \neq 0$$

(because the λ's are distinct), which implies that

$$c_1 = c_2 = \cdots = c_{j-1} = 0.$$

From (10) we conclude that $\mathbf{X}_j = \mathbf{0}$, which is impossible if $\mathbf{X}_j$ is an eigenvector. Hence S is linearly independent, and from Theorem 5.2 it follows that $\mathbf{A}$ is diagonalizable.

If all the roots of the characteristic polynomial of $\mathbf{A}$ are real and not all distinct, then $\mathbf{A}$ may or may not be diagonalizable. The characteristic polynomial of $\mathbf{A}$ can be written as the product of n factors, each of the form $\lambda - \lambda_j$, where λ_j is a root of the characteristic polynomial. Now the eigenvalues of $\mathbf{A}$ are the real roots of the characteristic polynomial of $\mathbf{A}$. Thus the characteristic polynomial can be written as

$$(\lambda - \lambda_1)^{k_1}(\lambda - \lambda_2)^{k_2} \cdots (\lambda - \lambda_r)^{k_r},$$

where $\lambda_1, \lambda_2, \ldots, \lambda_r$ are the distinct eigenvalues of $\mathbf{A}$, and $k_1, k_2, \ldots, k_r$ are integers whose sum is n. The integer k_i is called the **multiplicity** of λ_i. Thus in Example 11, $\lambda = 1$

is an eigenvalue of $\mathbf{A} = \begin{bmatrix} 1 & 1 \\ 0 & 1 \end{bmatrix}$ of multiplicity 2. It can be shown that if the roots of the characteristic polynomial of $\mathbf{A}$ are all real, then $\mathbf{A}$ can be diagonalized if and only if for each eigenvalue of multiplicity k we can find k linearly independent eigenvectors. This means that the solution space of the linear system $(\lambda \mathbf{I}_n - \mathbf{A})\mathbf{X} = \mathbf{0}$ has dimension k. It can also be shown that if λ is an eigenvalue of $\mathbf{A}$ of multiplicity k, then we can never find more than k linearly independent eigenvectors associated with λ. We consider the following examples.

Example 12 Let

$$\mathbf{A} = \begin{bmatrix} 0 & 0 & 1 \\ 0 & 1 & 2 \\ 0 & 0 & 1 \end{bmatrix}.$$

The characteristic polynomial of $\mathbf{A}$ is $f(\lambda) = \lambda(\lambda - 1)^2$, so the eigenvalues of $\mathbf{A}$ are $\lambda_1 = 0$, $\lambda_2 = 1$, and $\lambda_3 = 1$. Thus $\lambda_2 = 1$ is an eigenvalue of multiplicity 2. We now consider the eigenvectors associated with the eigenvalues $\lambda_2 = \lambda_3 = 1$. They are obtained by solving the linear system $(1\mathbf{I}_3 - \mathbf{A})\mathbf{X} = \mathbf{0}$:

$$\begin{bmatrix} 1 & 0 & -1 \\ 0 & 0 & -2 \\ 0 & 0 & 0 \end{bmatrix} \begin{bmatrix} x_1 \\ x_2 \\ x_3 \end{bmatrix} = \begin{bmatrix} 0 \\ 0 \\ 0 \end{bmatrix}.$$

A solution is any vector of the form $\begin{bmatrix} 0 \\ r \\ 0 \end{bmatrix}$, where r is any real number, so the dimension of the solution space of $(1\mathbf{I}_3 - \mathbf{A})\mathbf{X} = \mathbf{0}$ is 1. There do not exist two linearly independent eigenvectors associated with $\lambda_2 = 1$. Thus $\mathbf{A}$ cannot be diagonalized.

Example 13 Let

$$\mathbf{A} = \begin{bmatrix} 0 & 0 & 0 \\ 0 & 1 & 0 \\ 1 & 0 & 1 \end{bmatrix}.$$

The characteristic polynomial of $\mathbf{A}$ is $f(\lambda) = \lambda(\lambda - 1)^2$, so the eigenvalues of $\mathbf{A}$ are $\lambda_1 = 0$, $\lambda_2 = 1$, $\lambda_3 = 1$; $\lambda_2 = 1$ is again an eigenvalue of multiplicity 2. Now we again consider the solution space of $(1\mathbf{I}_3 - \mathbf{A})\mathbf{X} = \mathbf{0}$, that is, of

$$\begin{bmatrix} 1 & 0 & 0 \\ 0 & 0 & 0 \\ -1 & 0 & 0 \end{bmatrix} \begin{bmatrix} x_1 \\ x_2 \\ x_3 \end{bmatrix} = \begin{bmatrix} 0 \\ 0 \\ 0 \end{bmatrix}.$$

A solution is any vector of the form $\begin{bmatrix} 0 \\ r \\ s \end{bmatrix}$ for any real numbers r and s. Thus we can take as eigenvectors X_2 and X_3 the vectors $X_2 = \begin{bmatrix} 0 \\ 1 \\ 0 \end{bmatrix}$ and $X_3 = \begin{bmatrix} 0 \\ 0 \\ 1 \end{bmatrix}$.

Now we look for an eigenvector associated with $\lambda_1 = 0$. We have to solve $(0I_3 - A)X = 0$, or

$$\begin{bmatrix} 0 & 0 & 0 \\ 0 & -1 & 0 \\ -1 & 0 & -1 \end{bmatrix} \begin{bmatrix} x_1 \\ x_2 \\ x_3 \end{bmatrix} = \begin{bmatrix} 0 \\ 0 \\ 0 \end{bmatrix}.$$

A solution is any vector of the form $\begin{bmatrix} t \\ 0 \\ -t \end{bmatrix}$ for any real number t. Thus

$X_1 = \begin{bmatrix} 1 \\ 0 \\ -1 \end{bmatrix}$ is an eigenvector associated with $\lambda_1 = 0$. Since X_1, X_2, and X_3

are linearly independent, A can be diagonalized.

Thus an $n \times n$ matrix may fail to be diagonalizable either because not all the roots of its characteristic polynomial are real numbers, or because it does not have n linearly independent eigenvectors.

Eigenvalues and eigenvectors satisfy many important properties. For example, if A is an upper (lower) triangular matrix, then the eigenvalues of A are the elements on the main diagonal of A. Also, let λ be a fixed eigenvalue of A. The set S consisting of all eigenvectors of A associated with λ as well as the zero vector is a subspace of R^n (Exercise T.2) called the **eigenspace associated with λ**. Other properties are developed in the exercises of this section.

The procedure for diagonalizing a matrix A is as follows.

STEP 1. Form the characteristic polynomial $f(\lambda) = |\lambda I_n - A|$ of A.

STEP 2. Find the roots of the characteristic polynomial of A. If the roots are not all real, then A cannot be diagonalized.

STEP 3. For each eigenvalue λ_i of A of multiplicity k_i, find a basis for the solution space of $(\lambda_i I_n - A)X = 0$ (the eigenspace of λ_i). If the dimension of the eigenspace is less than k_i, then A is not diagonalizable. We thus determine n linearly independent eigenvectors of A.

STEP 4. Let P be the matrix whose columns are the n linearly independent eigenvectors determined in Step 3. Then $P^{-1}AP = D$, a diagonal matrix whose diagonal elements are the eigenvalues of A that correspond to the columns of P.

It must be pointed out that the method for finding the eigenvalues of a matrix by obtaining the roots of the characteristic polynomial is not a practical one for $n > 4$, owing to the need for evaluating a determinant. An efficient numerical method for finding eigenvalues will be presented in Section 8.3.

APPLICATION (POPULATION GROWTH).   Consider a population of animals that can live to a maximum age of n years (or any other time unit). Suppose that the number of males in the population is always a fixed percentage of the female population. Thus, in studying the growth of the entire population, we can ignore the male population and concentrate our attention on the female population. We divide the female population into $n + 1$ age groups as follows. Let

$$x_i^{(k)} = \text{number of females of age } i \text{ who are alive at time } k, \quad 0 \le k \le n;$$
$$f_i = \text{fraction of females of age } i \text{ who will be alive a year later;}$$
$$b_i = \text{average number of females born to a female of age } i.$$

Let

$$\mathbf{X}^{(k)} = \begin{bmatrix} x_0^{(k)} \\ x_1^{(k)} \\ \vdots \\ x_n^{(k)} \end{bmatrix} \qquad (0 \le k \le n)$$

denote the age distribution vector at time k.

The number of females in the first age group (age zero) at time $k + 1$ is merely the total number of females born from time k to time $k + 1$. There are $x_0^{(k)}$ females in the first age group at time k and each of these females, on the average, produces b_0 female offspring, so the first age group produces a total of $b_0 x_0^{(k)}$ females. Similarly, the $x_1^{(k)}$ females in the second age group (age 1) produces a total of $b_1 x_1^{(k)}$ females. Thus

$$x_0^{(k+1)} = b_0 x_0^{(k)} + b_1 x_1^{(k)} + \cdots + b_n x_n^{(k)}. \tag{14}$$

The number $x_1^{(k+1)}$ of females in the second age group at time $k + 1$ is the number of females from the first age group at time k who are alive a year later. Thus

$$x_1^{(k+1)} = \left(\begin{array}{c} \text{fraction of females in} \\ \text{first age group who are} \\ \text{alive a year later} \end{array} \right) \times \left(\begin{array}{c} \text{number of females in} \\ \text{first age group} \end{array} \right)$$

or

$$x_1^{(k+1)} = f_0 x_0^{(k)},$$

and in general

$$x_j^{(k+1)} = f_{j-1} x_{j-1}^{(k)} \qquad (1 \le j \le n). \tag{15}$$

We can write (14) and (15) using matrix notation as

$$\mathbf{X}^{(k+1)} = \mathbf{A}\mathbf{X}^{(k)} \qquad (1 \le k \le n), \tag{16}$$

where

$$
\mathbf{A} = \begin{bmatrix}
b_0 & b_1 & b_2 & \cdots & b_{n-1} & b_n \\
f_0 & 0 & 0 & \cdots & 0 & 0 \\
0 & f_1 & 0 & \cdots & 0 & 0 \\
\vdots & \vdots & \vdots & & \vdots & \vdots \\
0 & 0 & 0 & \cdots & f_{n-1} & 0
\end{bmatrix}.
$$

We can use Equation (16) to try to determine a distribution of the population by age groups at time $k + 1$ so that the number of females in each age group at time $k + 1$ will be a fixed multiple of the number in the corresponding age group at time k. That is, if λ is the multiplier, we want

$$
\mathbf{X}^{(k+1)} = \lambda \mathbf{X}^{(k)}
$$

or

$$
\mathbf{A}\mathbf{X}^{(k)} = \lambda \mathbf{X}^{(k)}.
$$

Thus λ is an eigenvalue of $\mathbf{A}$ and $\mathbf{X}^{(k)}$ is a corresponding eigenvector. If $\lambda = 1$, the number of females in each age group will be the same year after year. If we can find an eigenvector $\mathbf{X}^{(k)}$ corresponding to the eigenvalue $\lambda = 1$, we say that we have a **stable age distribution**.

Example 14
(Population Growth)

Consider a beetle that can live to a maximum age of 2 years and whose matrix is

$$
\mathbf{A} = \begin{bmatrix}
0 & 0 & 6 \\
\frac{1}{2} & 0 & 0 \\
0 & \frac{1}{3} & 0
\end{bmatrix}.
$$

We find that $\lambda = 1$ is an eigenvalue of $\mathbf{A}$ with corresponding eigenvector

$$
\begin{bmatrix}
6 \\
3 \\
1
\end{bmatrix}.
$$

Thus, if the numbers of females in the three age groups are proportional to $6 : 3 : 1$, we have a stable age distribution. That is, if we have 600 females in the first age group, 300 in the second, and 100 in the third, then year after year the number of females in each age group will remain the same.

Population growth problems of this type have applications to animal harvesting.
The theoretical exercises in this section contain many useful properties of eigenvalues. The reader is encouraged to develop a list of facts about eigenvalues, eigenvectors, and diagonalizable matrices.

APPLICATIONS.   At this point the reader may also turn to the following applications: Section 7.8, The Fibonacci Sequence; and Section 7.9, Differential Equations (calculus required for the latter).

5.1 EXERCISES

In Exercises 1 through 3, find the characteristic polynomial of each matrix.

1. $\begin{bmatrix} 1 & 2 & 1 \\ 0 & 1 & 2 \\ -1 & 3 & 2 \end{bmatrix}$.   **2.** $\begin{bmatrix} 2 & 1 \\ -1 & 3 \end{bmatrix}$.

3. $\begin{bmatrix} 4 & -1 & 3 \\ 0 & 2 & 1 \\ 0 & 0 & 3 \end{bmatrix}$.

In Exercises 4 through 11, find the characteristic polynomial, eigenvalues, and eigenvectors of each matrix.

4. $\begin{bmatrix} 0 & 1 & 2 \\ 0 & 0 & 3 \\ 0 & 0 & 0 \end{bmatrix}$.   **5.** $\begin{bmatrix} 1 & 0 & 0 \\ -1 & 3 & 0 \\ 3 & 2 & -2 \end{bmatrix}$.

6. $\begin{bmatrix} 1 & 1 \\ 1 & 1 \end{bmatrix}$.   **7.** $\begin{bmatrix} 1 & -1 \\ 2 & 4 \end{bmatrix}$.

8. $\begin{bmatrix} 2 & -2 & 3 \\ 0 & 3 & -2 \\ 0 & -1 & 2 \end{bmatrix}$.   **9.** $\begin{bmatrix} 2 & 2 & 3 \\ 1 & 2 & 1 \\ 2 & -2 & 1 \end{bmatrix}$.

10. $\begin{bmatrix} 2 & 0 & 0 \\ 3 & -1 & 0 \\ 0 & 4 & 3 \end{bmatrix}$.   **11.** $\begin{bmatrix} 1 & 2 & 3 & 4 \\ 0 & -1 & 3 & 2 \\ 0 & 0 & 3 & 3 \\ 0 & 0 & 0 & 2 \end{bmatrix}$.

In Exercises 12 through 16, find which of the matrices are diagonalizable.

12. $\begin{bmatrix} 1 & 4 \\ 1 & -2 \end{bmatrix}$.   **13.** $\begin{bmatrix} 1 & 0 \\ -2 & 1 \end{bmatrix}$.

14. $\begin{bmatrix} 1 & 1 & -2 \\ 4 & 0 & 4 \\ 1 & -1 & 4 \end{bmatrix}$.   **15.** $\begin{bmatrix} 1 & 2 & 3 \\ 0 & -1 & 2 \\ 0 & 0 & 2 \end{bmatrix}$.

16. $\begin{bmatrix} 3 & 1 & 0 \\ 0 & 3 & 1 \\ 0 & 0 & 3 \end{bmatrix}$.

In Exercises 17 through 21, find for each matrix $\mathbf{A}$, if possible, a nonsingular matrix $\mathbf{P}$ such that $\mathbf{P}^{-1}\mathbf{AP}$ is diagonal.

17. $\begin{bmatrix} 4 & 2 & 3 \\ 2 & 1 & 2 \\ -1 & -2 & 0 \end{bmatrix}$.   **18.** $\begin{bmatrix} 1 & 1 & 2 \\ 0 & 1 & 0 \\ 0 & 1 & 3 \end{bmatrix}$.

19. $\begin{bmatrix} 1 & 2 & 3 \\ 0 & 1 & 0 \\ 2 & 1 & 2 \end{bmatrix}$.   **20.** $\begin{bmatrix} 0 & -1 \\ 2 & 3 \end{bmatrix}$.

21. $\begin{bmatrix} 3 & -2 & 1 \\ 0 & 2 & 0 \\ 0 & 0 & 0 \end{bmatrix}$.

In Exercises 22 and 23, find bases for the eigenspaces associated with each eigenvalue.

22. $\begin{bmatrix} 2 & 3 & 0 \\ 0 & 1 & 0 \\ 0 & 0 & 2 \end{bmatrix}$.   **23.** $\begin{bmatrix} 2 & 2 & 3 & 4 \\ 0 & 2 & 3 & 2 \\ 0 & 0 & 1 & 1 \\ 0 & 0 & 0 & 1 \end{bmatrix}$.

24. Let $\mathbf{D} = \begin{bmatrix} 2 & 0 \\ 0 & -2 \end{bmatrix}$. Compute $\mathbf{D}^9$.

25. Let $\mathbf{A} = \begin{bmatrix} 3 & -5 \\ 1 & -3 \end{bmatrix}$. Compute $\mathbf{A}^9$. (*Hint:* Find a matrix $\mathbf{P}$ such that $\mathbf{P}^{-1}\mathbf{AP}$ is a diagonal matrix $\mathbf{D}$ and show that $\mathbf{A}^9 = \mathbf{PD}^9\mathbf{P}^{-1}$.)

26. Consider a living organism that can live to a maximum age of 2 years and whose matrix is

$$\mathbf{A} = \begin{bmatrix} 0 & 0 & 8 \\ \frac{1}{4} & 0 & 0 \\ 0 & \frac{1}{2} & 0 \end{bmatrix}.$$

Find a stable age distribution.

THEORETICAL EXERCISES

T.1. Prove:
  (a) **A** is similar to **A**.
  (b) If **B** is similar to **A**, then **A** is similar to **B**.
  (c) If **A** is similar to **B** and **B** is similar to **C**, then **A** is similar to **C**.

T.2. Let λ be a fixed eigenvalue of **A**. Prove that the set S consisting of all eigenvectors of **A** associated with λ, as well as the zero vector, is a subspace of R^n. This subspace is called the **eigenspace associated with λ**.

T.3. Prove that if **A** and **B** are similar matrices, they have the same characteristic polynomials and hence the same eigenvalues.

T.4. Prove that if **A** is an upper (lower) triangular matrix, then the eigenvalues of **A** are the elements on the main diagonal of **A**.

T.5. Prove that **A** and $\mathbf{A}^T$ have the same eigenvalues. What, if anything, can we say about the associated eigenvectors of **A** and $\mathbf{A}^T$?

T.6. If λ is an eigenvalue of **A** with associated eigenvector **X**, prove that λ^k is an eigenvalue of $\mathbf{A}^k = \mathbf{A} \cdot \mathbf{A} \cdots \mathbf{A}$ (k factors) with associated eigenvector **X**, where k is a positive integer.

T.7. An $n \times n$ matrix **A** is called **nilpotent** if $\mathbf{A}^k = \mathbf{0}$ for some positive integer k. Prove that if **A** is nilpotent, then the only eigenvalue of **A** is 0. (*Hint:* Use Exercise T.6.)

T.8. Let **A** be an $n \times n$ matrix.
  (a) Show that $|\mathbf{A}|$ is the product of all the roots of the characteristic polynomial of **A**.
  (b) Show that **A** is singular if and only if 0 is an eigenvalue of **A**.

T.9. Let λ be an eigenvalue of the nonsingular matrix **A** with associated eigenvector **X**. Show that $1/\lambda$ is an eigenvalue of $\mathbf{A}^{-1}$ with associated eigenvector **X**.

T.10. Let

$$\mathbf{A} = \begin{bmatrix} a & b \\ c & d \end{bmatrix}.$$

Find necessary and sufficient conditions for **A** to be diagonalizable.

T.11. If **A** and **B** are nonsingular, show that **AB** and **BA** are similar.

T.12. Prove that if **A** is diagonalizable, then (a) $\mathbf{A}^T$ is diagonalizable and (b) $\mathbf{A}^k$ is diagonalizable, where k is a positive integer.

5.2

DIAGONALIZATION OF SYMMETRIC MATRICES

In this section we consider the diagonalization of symmetric matrices ($\mathbf{A} = \mathbf{A}^T$). We restrict our attention to this case because it is easier to handle than that of general matrices and also because symmetric matrices arise in many applied problems.

As an example of such a problem, consider the task of identifying the conic represented by the equation

$$2x^2 + 2xy + 2y^2 = 9,$$

which can be written in matrix form as

$$[x \quad y] \begin{bmatrix} 2 & 1 \\ 1 & 2 \end{bmatrix} \begin{bmatrix} x \\ y \end{bmatrix} = 9.$$

Observe that the matrix used here is a symmetric matrix. This problem is handled in Section 7.2. We shall merely remark here that the solution calls for the determination of the eigenvalues and eigenvectors of the matrix

$$\begin{bmatrix} 2 & 1 \\ 1 & 2 \end{bmatrix}.$$

The x- and y-axes are then rotated to a new set of axes, which lie along the eigenvectors of the matrix. In the new set of axes, the given conic can be readily identified.

In Example 7 of Section 5.1 we saw a matrix with the property that the roots of its characteristic polynomial were imaginary numbers. This cannot happen for a symmetric matrix, as the following theorem asserts. We omit the proof.

Theorem 5.4 *All the roots of the characteristic polynomial of a symmetric matrix are real numbers.*

Corollary 5.2 *If $\mathbf{A}$ is a symmetric matrix all of whose eigenvalues are distinct, then $\mathbf{A}$ is diagonalizable.*

Proof

Since $\mathbf{A}$ is symmetric, all its eigenvalues are real. From Theorem 5.3 it now follows that $\mathbf{A}$ can be diagonalized.

Theorem 5.5 *If $\mathbf{A}$ is a symmetric matrix, then eigenvectors that belong to distinct eigenvalues of $\mathbf{A}$ are orthogonal.*

Proof

First, we shall let the reader verify (Exercise T.1) the property that if $\mathbf{X}$ and $\mathbf{Y}$ are vectors in R^n, then

$$(\mathbf{AX}) \cdot \mathbf{Y} = \mathbf{X} \cdot (\mathbf{A}^T \mathbf{Y}). \tag{1}$$

Now let $\mathbf{X}_1$ and $\mathbf{X}_2$ be eigenvectors of $\mathbf{A}$ associated with the distinct eigenvalues λ_1 and λ_2 of $\mathbf{A}$. We then have

$$\mathbf{AX}_1 = \lambda_1 \mathbf{X}_1 \qquad \text{and} \qquad \mathbf{AX}_2 = \lambda_2 \mathbf{X}_2.$$

Now

$$\begin{aligned} \lambda_1(\mathbf{X}_1 \cdot \mathbf{X}_2) = (\lambda_1 \mathbf{X}_1) \cdot \mathbf{X}_2 &= (\mathbf{AX}_1) \cdot \mathbf{X}_2 \\ &= \mathbf{X}_1 \cdot (\mathbf{A}^T \mathbf{X}_2) = \mathbf{X}_1 \cdot (\mathbf{AX}_2) \\ &= \mathbf{X}_1 \cdot (\lambda_2 \mathbf{X}_2) = \lambda_2(\mathbf{X}_1 \cdot \mathbf{X}_2), \end{aligned} \tag{2}$$

where we have used the fact that $\mathbf{A} = \mathbf{A}^T$. Thus

$$\lambda_1(\mathbf{X}_1 \cdot \mathbf{X}_2) = \lambda_2(\mathbf{X}_1 \cdot \mathbf{X}_2)$$

and subtracting, we obtain

$$\begin{aligned} 0 &= \lambda_1(\mathbf{X}_1 \cdot \mathbf{X}_2) - \lambda_2(\mathbf{X}_1 \cdot \mathbf{X}_2) \\ &= (\lambda_1 - \lambda_2)(\mathbf{X}_1 \cdot \mathbf{X}_2). \end{aligned} \tag{3}$$

Since $\lambda_1 \neq \lambda_2$, we conclude that $\mathbf{X}_1 \cdot \mathbf{X}_2 = 0$.

Example 1 Let

$$\mathbf{A} = \begin{bmatrix} 0 & 0 & -2 \\ 0 & -2 & 0 \\ -2 & 0 & 3 \end{bmatrix}.$$

We find that the characteristic polynomial of $\mathbf{A}$ is (verify)

$$f(\lambda) = (\lambda + 2)(\lambda - 4)(\lambda + 1),$$

so the eigenvalues of $\mathbf{A}$ are

$$\lambda_1 = -2, \qquad \lambda_2 = 4, \qquad \text{and} \qquad \lambda_3 = -1.$$

Then we find the associated eigenvectors by solving the linear system $(\lambda \mathbf{I}_3 - \mathbf{A})\mathbf{X} = \mathbf{0}$; and obtain the respective eigenvectors

$$\mathbf{X}_1 = \begin{bmatrix} 0 \\ 1 \\ 0 \end{bmatrix}, \qquad \mathbf{X}_2 = \begin{bmatrix} -1 \\ 0 \\ 2 \end{bmatrix}, \qquad \text{and} \qquad \mathbf{X}_3 = \begin{bmatrix} 2 \\ 0 \\ 1 \end{bmatrix}.$$

It is clear that $\{\mathbf{X}_1, \mathbf{X}_2, \mathbf{X}_3\}$ is an orthogonal set of vectors in R^3 (and is thus linearly independent by Theorem 3.18, Section 3.8). Thus $\mathbf{A}$ is diagonalizable and is similar to

$$\mathbf{D} = \begin{bmatrix} -2 & 0 & 0 \\ 0 & 4 & 0 \\ 0 & 0 & -1 \end{bmatrix}.$$

We recall that if $\mathbf{A}$ can be diagonalized, then there exists a nonsingular matrix $\mathbf{P}$ such that $\mathbf{P}^{-1}\mathbf{A}\mathbf{P}$ is diagonal. Moreover, the columns of $\mathbf{P}$ are eigenvectors of $\mathbf{A}$. Now, if the eigenvectors of $\mathbf{A}$ form an orthogonal set S, as happens when $\mathbf{A}$ is symmetric and the eigenvalues of $\mathbf{A}$ are distinct, then since any scalar multiple of an eigenvector of $\mathbf{A}$ is also an eigenvector of $\mathbf{A}$, we can normalize S to obtain an orthonormal set

$$T = \{\mathbf{X}_1, \mathbf{X}_2, \ldots, \mathbf{X}_n\}$$

of eigenvectors of $\mathbf{A}$. The jth column of P is the eigenvector $\mathbf{X}_j$ associated with λ_j, and we now examine what type of matrix $\mathbf{P}$ must be. We can write $\mathbf{P}$ as

$$\mathbf{P} = [\mathbf{X}_1\ \mathbf{X}_2 \ldots\ \mathbf{X}_n].$$

Then

$$\mathbf{P}^T = \begin{bmatrix} \mathbf{X}_1^T \\ \mathbf{X}_2^T \\ \vdots \\ \mathbf{X}_n^T \end{bmatrix},$$

where $\mathbf{X}_i^T$, $1 \leq i \leq n$, is the transpose of the $n \times 1$ matrix (or vector) $\mathbf{X}_i$. We find that the i, jth entry in $\mathbf{P}^T\mathbf{P}$ is $\mathbf{X}_i \cdot \mathbf{X}_j$ (verify). Since

$$\mathbf{X}_i \cdot \mathbf{X}_j = \begin{cases} 1 & \text{if } i = j \\ 0 & \text{if } i \neq j, \end{cases}$$

then $\mathbf{P}^T\mathbf{P} = \mathbf{I}_n$. Thus $\mathbf{P}^T = \mathbf{P}^{-1}$. Such matrices are important enough to have a special name.

Definition A nonsingular matrix $\mathbf{A}$ is called **orthogonal** if

$$\mathbf{A}^{-1} = \mathbf{A}^T.$$

Of course, we can also say that a nonsingular matrix $\mathbf{A}$ is orthogonal if $\mathbf{A}^T\mathbf{A} = \mathbf{I}_n$.

Example 2 Let

$$\mathbf{A} = \begin{bmatrix} \frac{2}{3} & -\frac{2}{3} & \frac{1}{3} \\ \frac{2}{3} & \frac{1}{3} & -\frac{2}{3} \\ \frac{1}{3} & \frac{2}{3} & \frac{2}{3} \end{bmatrix}.$$

It is easy to check that $\mathbf{A}^T\mathbf{A} = \mathbf{I}_n$. Hence $\mathbf{A}$ is an orthogonal matrix.

Example 3 Let $\mathbf{A}$ be the matrix of Example 1. We already know that the set of eigenvectors

$$\left\{ \begin{bmatrix} 0 \\ 1 \\ 0 \end{bmatrix}, \begin{bmatrix} -1 \\ 0 \\ 2 \end{bmatrix}, \begin{bmatrix} 2 \\ 0 \\ 1 \end{bmatrix} \right\}$$

is orthogonal. If we normalize these vectors, we find that

$$T = \left\{ \begin{bmatrix} 0 \\ 1 \\ 0 \end{bmatrix}, \begin{bmatrix} -\dfrac{1}{\sqrt{5}} \\ 0 \\ \dfrac{2}{\sqrt{5}} \end{bmatrix}, \begin{bmatrix} \dfrac{2}{\sqrt{5}} \\ 0 \\ \dfrac{1}{\sqrt{5}} \end{bmatrix} \right\}$$

is an orthonormal set of vectors. The matrix $\mathbf{P}$ such that $\mathbf{P}^{-1}\mathbf{A}\mathbf{P}$ is diagonal is the matrix whose columns are the vectors in T. Thus

$$\mathbf{P} = \begin{bmatrix} 0 & -\dfrac{1}{\sqrt{5}} & \dfrac{2}{\sqrt{5}} \\ 1 & 0 & 0 \\ 0 & \dfrac{2}{\sqrt{5}} & \dfrac{1}{\sqrt{5}} \end{bmatrix}.$$

We leave it to the reader to verify (Exercise 4) that **P** is an orthogonal matrix and that

$$\mathbf{P}^{-1}\mathbf{AP} = \mathbf{P}^{T}\mathbf{AP} = \begin{bmatrix} -2 & 0 & 0 \\ 0 & 4 & 0 \\ 0 & 0 & -1 \end{bmatrix}.$$

The following theorem is not difficult to show, and we leave its proof to the reader (Exercise T.2).

Theorem 5.6 *The $n \times n$ matrix **A** is orthogonal if and only if the columns (and rows) of **A** form an orthonormal set of vectors in R^n.*

If **A** is an orthogonal matrix, then it is easy to show that $|\mathbf{A}| = \pm 1$ (Exercise T.3).

REMARK (For students who have read Sections 4.1 and 4.3.) We now look at the geometrical implications of orthogonal matrices. Let **A** be an orthogonal $n \times n$ matrix and consider the linear transformation $\mathbf{L}: R^n \rightarrow R^n$ defined by $\mathbf{L(X)} = \mathbf{AX}$, for **X** in R^n. We now compute $\mathbf{L(X)} \cdot \mathbf{L(Y)}$ for any vectors **X** and **Y** in R^n. We have, using Equation (1) and the fact that $\mathbf{A}^{T}\mathbf{A} = \mathbf{I}_n$,

$$\mathbf{L(X)} \cdot \mathbf{L(Y)} = (\mathbf{AX}) \cdot (\mathbf{AY}) = \mathbf{X} \cdot (\mathbf{A}^{T}\mathbf{AY}) = \mathbf{X} \cdot (\mathbf{I}_n\mathbf{Y}) = \mathbf{X} \cdot \mathbf{Y}.$$

This means that **L** preserves the inner product of two vectors, and consequently **L** preserves length. It is, of course, clear that if θ is the angle between vectors **X** and **Y** in R^n, then the angle between $\mathbf{L(X)}$ and $\mathbf{L(Y)}$ is also θ (Exercise T.4). Conversely, let $\mathbf{L}: R^n \rightarrow R^n$ be a linear transformation that preserves length. Then it can be shown (Exercise T.10) that

$$\mathbf{L(X)} \cdot \mathbf{L(Y)} = \mathbf{X} \cdot \mathbf{Y},$$

for any **X** and **Y** in R^n. Let **A** be the matrix of **L** with respect to the natural basis for R^n. Then

$$\mathbf{L(X)} = \mathbf{AX}.$$

If **X** and **Y** are any vectors in R^n, we have, using (1),

$$\mathbf{X} \cdot \mathbf{Y} = \mathbf{L(X)} \cdot \mathbf{L(Y)} = (\mathbf{AX}) \cdot (\mathbf{AY}) = \mathbf{X} \cdot (\mathbf{A}^{T}\mathbf{AY}).$$

Since this holds for all **X** in R^n, we conclude by Exercise T.8 of Section 3.2 that

$$\mathbf{A}^{T}\mathbf{AY} = \mathbf{Y}$$

for any **Y** in R^n. It then follows that $\mathbf{A}^{T}\mathbf{A} = \mathbf{I}_n$ (Exercise T.9) so that **A** is an orthogonal matrix. Other properties of orthogonal matrices are examined in the theoretical exercises. This concludes the remark.

We now turn to the general situation for a symmetric matrix; even if **A** has eigenvalues whose multiplicities are greater than one, it turns out that we can still diagonalize **A**. We omit the proof of the following theorem.

Theorem 5.7 *If* **A** *is a symmetric n × n matrix, then there exists an orthogonal matrix* **P** *such that* $\mathbf{P}^{-1}\mathbf{AP} = \mathbf{D}$, *a diagonal matrix. The eigenvalues of* **A** *lie on the main diagonal of* **D**.

It can be shown that if **A** has an eigenvalue λ of multiplicity k, then the solution space of the linear system $(\lambda\mathbf{I}_n - \mathbf{A})\mathbf{X} = \mathbf{0}$ (the eigenspace of λ) has dimension k. This means that there exist k linearly independent eigenvectors of **A** associated with the eigenvalue λ. By the Gram–Schmidt process, we can choose an orthonormal basis for this solution space. Thus we obtain a set of k orthonormal eigenvectors associated with the eigenvalue λ. Since eigenvectors associated with distinct eigenvalues are orthogonal, if we form the set of all eigenvectors we get an orthonormal set. Hence the matrix **P** whose columns are the eigenvectors is orthogonal.

The procedure for diagonalizing a symmetric matrix **A** by an orthogonal matrix **P** is as follows.

STEP 1. Form the characteristic polynomial $f(\lambda) = |\lambda\mathbf{I}_n - \mathbf{A}|$.

STEP 2. Find the roots of the characteristic polynomial of **A**. These will all be real.

STEP 3. For each eigenvalue λ_i of **A** of multiplicity k_i, find a basis of k_i eigenvectors for the solution space of $(\lambda_i\mathbf{I}_n - \mathbf{A})\mathbf{X} = \mathbf{0}$ (the eigenspace of λ_i).

STEP 4. For each eigenspace, transform the basis obtained in Step 3 to an orthonormal basis by the Gram–Schmidt process. The totality of all these orthonormal bases determines an orthonormal set of n linearly independent eigenvectors of **A**.

STEP 5. Let **P** be the matrix whose columns are the n linearly independent eigenvectors determined in Step 4. Then **P** is an orthogonal matrix and $\mathbf{P}^{-1}\mathbf{AP} = \mathbf{D}$, a diagonal matrix whose diagonal elements are the eigenvalues of **A** that correspond to the columns of **P**.

Example 4 Let

$$\mathbf{A} = \begin{bmatrix} 0 & 2 & 2 \\ 2 & 0 & 2 \\ 2 & 2 & 0 \end{bmatrix}.$$

The characteristic polynomial of **A** is

$$f(\lambda) = (\lambda + 2)^2(\lambda - 4),$$

so the eigenvalues are

$$\lambda_1 = -2, \qquad \lambda_2 = -2, \qquad \text{and} \qquad \lambda_3 = 4.$$

That is, -2 is an eigenvalue whose multiplicity is 2. To find the eigenvectors associated with λ_1 and λ_2, we solve the homogeneous linear system $(-2I_3 - A)X = 0$:

$$\begin{bmatrix} -2 & -2 & -2 \\ -2 & -2 & -2 \\ -2 & -2 & -2 \end{bmatrix} \begin{bmatrix} x_1 \\ x_2 \\ x_3 \end{bmatrix} = \begin{bmatrix} 0 \\ 0 \\ 0 \end{bmatrix}. \tag{4}$$

A basis for the solution space of (4) consists of the eigenvectors

$$X_1 = \begin{bmatrix} -1 \\ 1 \\ 0 \end{bmatrix} \qquad \text{and} \qquad X_2 = \begin{bmatrix} -1 \\ 0 \\ 1 \end{bmatrix}.$$

Now X_1 and X_2 are not orthogonal, since $X_1 \cdot X_2 \neq 0$. We can use the Gram–Schmidt process to obtain an orthonormal basis for the solution space of (4) (the eigenspace of $\lambda_1 = -2$) as follows. Let

$$Y_1 = X_1 = \begin{bmatrix} -1 \\ 1 \\ 0 \end{bmatrix}$$

and

$$Y_2 = X_2 - \left(\frac{X_2 \cdot Y_1}{Y_1 \cdot Y_1} \right) Y_1 = \begin{bmatrix} -\frac{1}{2} \\ -\frac{1}{2} \\ 1 \end{bmatrix}.$$

Let

$$Y_2^* = 2Y_2 = \begin{bmatrix} -1 \\ -1 \\ 2 \end{bmatrix}.$$

The set $\{Y_1, Y_2^*\}$ is an orthogonal set of vectors. Normalizing these eigenvectors, we obtain

$$Z_1 = \frac{Y_1}{|Y_1|} = \frac{1}{\sqrt{2}} \begin{bmatrix} -1 \\ 1 \\ 0 \end{bmatrix} \qquad \text{and} \qquad Z_2 = \frac{Y_2^*}{|Y_2^*|} = \frac{1}{\sqrt{6}} \begin{bmatrix} -1 \\ -1 \\ 2 \end{bmatrix}.$$

The set $\{Z_1, Z_2\}$ is an orthonormal basis of eigenvectors of A for the solution space of (4). Now we find a basis for the solution space of $(4I_3 - A)X = 0$,

$$\begin{bmatrix} 4 & -2 & -2 \\ -2 & 4 & -2 \\ -2 & -2 & 4 \end{bmatrix} \begin{bmatrix} x_1 \\ x_2 \\ x_3 \end{bmatrix} = \begin{bmatrix} 0 \\ 0 \\ 0 \end{bmatrix}, \tag{5}$$

to consist of

$$\mathbf{X}_3 = \begin{bmatrix} 1 \\ 1 \\ 1 \end{bmatrix}.$$

Normalizing this vector, we have the eigenvector

$$\mathbf{Z}_3 = \frac{1}{\sqrt{3}} \begin{bmatrix} 1 \\ 1 \\ 1 \end{bmatrix}$$

as an orthonormal basis for the solution space of (5). Since eigenvectors associated with distinct eigenvalues are orthogonal, we observe that $\mathbf{Z}_3$ is orthogonal to both $\mathbf{Z}_1$ and $\mathbf{Z}_2$. Thus the set $\{\mathbf{Z}_1, \mathbf{Z}_2, \mathbf{Z}_3\}$ is an orthonormal basis of R^3 consisting of eigenvectors of $\mathbf{A}$. The matrix $\mathbf{P}$ is the matrix whose jth column is $\mathbf{Z}_j$:

$$\mathbf{P} = \begin{bmatrix} -\dfrac{1}{\sqrt{2}} & -\dfrac{1}{\sqrt{6}} & \dfrac{1}{\sqrt{3}} \\ \dfrac{1}{\sqrt{2}} & -\dfrac{1}{\sqrt{6}} & \dfrac{1}{\sqrt{3}} \\ 0 & \dfrac{2}{\sqrt{6}} & \dfrac{1}{\sqrt{3}} \end{bmatrix}.$$

We leave it to the reader to verify that

$$\mathbf{P}^{-1}\mathbf{A}\mathbf{P} = \mathbf{P}^T\mathbf{A}\mathbf{P} = \begin{bmatrix} -2 & 0 & 0 \\ 0 & -2 & 0 \\ 0 & 0 & 4 \end{bmatrix}.$$

Example 5 Let

$$\mathbf{A} = \begin{bmatrix} 1 & 2 & 0 & 0 \\ 2 & 1 & 0 & 0 \\ 0 & 0 & 1 & 2 \\ 0 & 0 & 2 & 1 \end{bmatrix}.$$

The characteristic polynomial of $\mathbf{A}$ is

$$f(\lambda) = (\lambda + 1)^2(\lambda - 3)^2,$$

so the eigenvalues of $\mathbf{A}$ are

$$\lambda_1 = -1, \qquad \lambda_2 = -1, \qquad \lambda_3 = 3, \qquad \text{and} \qquad \lambda_4 = 3.$$

We find (verify) that a basis for the solution space of

$$(-1\mathbf{I}_3 - \mathbf{A})\mathbf{X} = \mathbf{0} \tag{6}$$

consists of the eigenvectors

$$\mathbf{X}_1 = \begin{bmatrix} 1 \\ -1 \\ 0 \\ 0 \end{bmatrix} \quad \text{and} \quad \mathbf{X}_2 = \begin{bmatrix} 0 \\ 0 \\ 1 \\ -1 \end{bmatrix},$$

which are orthogonal. Normalizing these eigenvectors, we obtain

$$\mathbf{Z}_1 = \begin{bmatrix} \dfrac{1}{\sqrt{2}} \\ -\dfrac{1}{\sqrt{2}} \\ 0 \\ 0 \end{bmatrix} \quad \text{and} \quad \mathbf{Z}_2 = \begin{bmatrix} 0 \\ 0 \\ \dfrac{1}{\sqrt{2}} \\ -\dfrac{1}{\sqrt{2}} \end{bmatrix}$$

as an orthonormal basis of eigenvectors for the solution space of (6). We also find (verify) that a basis for the solution space of

$$(3\mathbf{I}_3 - \mathbf{A})\mathbf{X} = \mathbf{0} \tag{7}$$

consists of the eigenvectors

$$\mathbf{X}_3 = \begin{bmatrix} 1 \\ 1 \\ 0 \\ 0 \end{bmatrix} \quad \text{and} \quad \mathbf{X}_4 = \begin{bmatrix} 0 \\ 0 \\ 1 \\ 1 \end{bmatrix},$$

which are orthogonal. Normalizing these eigenvectors, we obtain

$$\mathbf{Z}_3 = \begin{bmatrix} \dfrac{1}{\sqrt{2}} \\ \dfrac{1}{\sqrt{2}} \\ 0 \\ 0 \end{bmatrix} \quad \text{and} \quad \mathbf{Z}_4 = \begin{bmatrix} 0 \\ 0 \\ \dfrac{1}{\sqrt{2}} \\ \dfrac{1}{\sqrt{2}} \end{bmatrix}$$

as an orthonormal basis of eigenvectors for the solution space of (7). Since eigenvectors associated with distinct eigenvalues are orthogonal, we conclude that

$$\{\mathbf{Z}_1, \mathbf{Z}_2, \mathbf{Z}_3, \mathbf{Z}_4\}$$

is an orthonormal basis of R^4 consisting of eigenvectors of $\mathbf{A}$. The matrix $\mathbf{P}$ is the matrix whose jth column is $\mathbf{Z}_j$:

$$
\mathbf{P} = \begin{bmatrix}
\dfrac{1}{\sqrt{2}} & 0 & \dfrac{1}{\sqrt{2}} & 0 \\[2ex]
-\dfrac{1}{\sqrt{2}} & 0 & \dfrac{1}{\sqrt{2}} & 0 \\[2ex]
0 & \dfrac{1}{\sqrt{2}} & 0 & \dfrac{1}{\sqrt{2}} \\[2ex]
0 & -\dfrac{1}{\sqrt{2}} & 0 & \dfrac{1}{\sqrt{2}}
\end{bmatrix}.
$$

Suppose now that $\mathbf{A}$ is an $n \times n$ matrix for which we can find an orthogonal matrix $\mathbf{P}$ such that $\mathbf{P}^{-1}\mathbf{A}\mathbf{P}$ is a diagonal matrix $\mathbf{D}$. Thus $\mathbf{P}^{-1}\mathbf{A}\mathbf{P} = \mathbf{D}$, or $\mathbf{A} = \mathbf{P}\mathbf{D}\mathbf{P}^{-1}$. Since $\mathbf{P}^{-1} = \mathbf{P}^{T}$, we can write $\mathbf{A} = \mathbf{P}\mathbf{D}\mathbf{P}^{T}$. Then $\mathbf{A}^{T} = (\mathbf{P}\mathbf{D}\mathbf{P}^{T})^{T} = (\mathbf{P}^{T})^{T}\mathbf{D}^{T}\mathbf{P}^{T} = \mathbf{P}\mathbf{D}\mathbf{P}^{T} = \mathbf{A}$ ($\mathbf{D} = \mathbf{D}^{T}$, since $\mathbf{D}$ is a diagonal matrix). Thus $\mathbf{A}$ is symmetric.

Some remarks about nonsymmetric matrices are in order at this point. Theorem 5.3 assures us that $\mathbf{A}$ is diagonalizable if all the roots of its characteristic polynomial are real and distinct. We also studied examples, in Section 5.1, of nonsymmetric matrices with repeated eigenvalues that were diagonalizable and others that were not diagonalizable. There are some striking differences between the symmetric and nonsymmetric cases, which we now summarize. If $\mathbf{A}$ is nonsymmetric, then the roots of its characteristic polynomial need not all be real numbers; if an eigenvalue λ has multiplicity k, then the solution space of $(\lambda\mathbf{I}_n - \mathbf{A})\mathbf{X} = \mathbf{0}$ may have dimension $< k$; if the roots of the characteristic polynomial of $\mathbf{A}$ are all real, it is still possible for $\mathbf{A}$ not to have n linearly independent eigenvectors (which means that $\mathbf{A}$ cannot be diagonalized); eigenvectors associated with distinct eigenvalues need not be orthogonal. Thus, in Example 13 of Section 5.1, the eigenvectors $\mathbf{X}_1$ and $\mathbf{X}_3$ associated with the eigenvalues $\lambda_1 = 0$ and $\lambda_3 = 1$, respectively, are not orthogonal. If a matrix $\mathbf{A}$ cannot be diagonalized, then we can often find a matrix $\mathbf{B}$ similar to $\mathbf{A}$ which is "nearly diagonal." The matrix $\mathbf{B}$ is said to be in **Jordan canonical form**; its treatment lies beyond the scope of this book but is studied in advanced books on linear algebra.[†]

It should also be noted that in many applications, we need only find a diagonal matrix $\mathbf{D}$ that is similar to the given matrix $\mathbf{A}$; that is, we do not need the orthogonal matrix $\mathbf{P}$ such that $\mathbf{P}^{-1}\mathbf{A}\mathbf{P} = \mathbf{D}$. Many of the matrices to be diagonalized in applied problems are either symmetric, or all the roots of their characteristic polynomial are real. Of course, the methods for finding eigenvalues that have been presented in this chapter are not recommended for matrices with more than four rows because of the need to evaluate determinants. In Section 8.3 we shall consider an efficient numerical method for diagonalizing a symmetric matrix.

APPLICATIONS.   At this point the reader may turn to Section 7.2, Quadratic Forms.

[†] See, for example, K. Hoffman and R. Kunze, *Linear Algebra,* 2nd ed. (Englewood Cliffs, N.J.: Prentice-Hall, Inc., 1971).

5.2 EXERCISES

1. Verify that

$$P = \begin{bmatrix} \frac{2}{3} & -\frac{2}{3} & \frac{1}{3} \\ \frac{2}{3} & \frac{1}{3} & -\frac{2}{3} \\ \frac{1}{3} & \frac{2}{3} & \frac{2}{3} \end{bmatrix}$$

is an orthogonal matrix.

2. Find the inverse of each of the following orthogonal matrices.

(a) $\mathbf{A} = \begin{bmatrix} 1 & 0 & 0 \\ 0 & \cos\theta & \sin\theta \\ 0 & -\sin\theta & \cos\theta \end{bmatrix}$.

(b) $\mathbf{B} = \begin{bmatrix} 1 & 0 & 0 \\ 0 & \dfrac{1}{\sqrt{2}} & \dfrac{1}{\sqrt{2}} \\ 0 & -\dfrac{1}{\sqrt{2}} & -\dfrac{1}{\sqrt{2}} \end{bmatrix}$.

3. Verify Theorem 5.6 for the matrices in Exercise 2.

4. Verify that the matrix $\mathbf{P}$ in Example 3 is an orthogonal matrix.

5. For the orthogonal matrix

$$\mathbf{A} = \begin{bmatrix} \dfrac{1}{\sqrt{2}} & -\dfrac{1}{\sqrt{2}} \\ -\dfrac{1}{\sqrt{2}} & -\dfrac{1}{\sqrt{2}} \end{bmatrix},$$

verify that $(\mathbf{AX}) \cdot (\mathbf{AY}) = \mathbf{X} \cdot \mathbf{Y}$ for any $\mathbf{X}$ and $\mathbf{Y}$ in R^2.

6. (**Uses Sections 4.1 and 4.3**). Let $L: R^2 \to R^2$ be the linear transformation performing a counterclockwise rotation through $45°$; and let $\mathbf{A}$ be the matrix of $\mathbf{L}$ with respect to the natural basis for R^2. Show that $\mathbf{A}$ is orthogonal.

In Exercises 7 through 12, diagonalize each given matrix $\mathbf{A}$ and find an orthogonal matrix $\mathbf{P}$ such that $\mathbf{P}^{-1}\mathbf{AP}$ is diagonal.

7. $\begin{bmatrix} 2 & 2 \\ 2 & 2 \end{bmatrix}$. **8.** $\begin{bmatrix} 0 & 0 & 1 \\ 0 & 0 & 0 \\ 1 & 0 & 0 \end{bmatrix}$.

9. $\begin{bmatrix} 0 & 0 & 0 \\ 0 & 2 & 2 \\ 0 & 2 & 2 \end{bmatrix}$. **10.** $\begin{bmatrix} 0 & 0 & 0 & 0 \\ 0 & 0 & 0 & 0 \\ 0 & 0 & 0 & 1 \\ 0 & 0 & 1 & 0 \end{bmatrix}$.

11. $\begin{bmatrix} 0 & -1 & -1 \\ -1 & 0 & -1 \\ -1 & -1 & 0 \end{bmatrix}$. **12.** $\begin{bmatrix} -1 & 2 & 2 \\ 2 & -1 & 2 \\ 2 & 2 & -1 \end{bmatrix}$.

In Exercises 13 through 20, diagonalize each given matrix by an orthogonal matrix.

13. $\begin{bmatrix} 2 & 1 \\ 1 & 2 \end{bmatrix}$. **14.** $\begin{bmatrix} 2 & 2 & 0 & 0 \\ 2 & 2 & 0 & 0 \\ 0 & 0 & 2 & 2 \\ 0 & 0 & 2 & 2 \end{bmatrix}$.

15. $\begin{bmatrix} 1 & 1 & 0 \\ 1 & 1 & 0 \\ 0 & 0 & 1 \end{bmatrix}$. **16.** $\begin{bmatrix} 1 & 0 & 0 \\ 0 & 3 & -2 \\ 0 & -2 & 3 \end{bmatrix}$.

17. $\begin{bmatrix} 1 & 0 & 0 \\ 0 & 1 & 1 \\ 0 & 1 & 1 \end{bmatrix}$. **18.** $\begin{bmatrix} 0 & 0 & 0 & 1 \\ 0 & 0 & 0 & 0 \\ 0 & 0 & 0 & 0 \\ 1 & 0 & 0 & 0 \end{bmatrix}$.

19. $\begin{bmatrix} 2 & 1 & 1 \\ 1 & 2 & 1 \\ 1 & 1 & 2 \end{bmatrix}$. **20.** $\begin{bmatrix} -3 & 0 & -1 \\ 0 & -2 & 0 \\ -1 & 0 & -3 \end{bmatrix}$.

THEORETICAL EXERCISES

T.1. Show that if $\mathbf{X}$ and $\mathbf{Y}$ are vectors in R^n, then $(\mathbf{AX}) \cdot \mathbf{Y} = \mathbf{X} \cdot (\mathbf{A}^T\mathbf{Y})$.

T.2. Prove Theorem 5.6.

T.3. Show that if $\mathbf{A}$ is an orthogonal matrix, then $|\mathbf{A}| = \pm 1$.

T.4. (**Uses Section 4.1**). Let $\mathbf{A}$ be an $n \times n$ orthogonal matrix and let $L: R^n \to R^n$ be the linear transformation defined by $L(X) = AX$, for X in R^n. Let θ be the angle between vectors X and Y in R^n. Prove that if A is orthogonal, then the angle between $L(X)$ and $L(Y)$ is also θ.

T.5. Prove Theorem 5.7 for the 2×2 case by studying the possible roots of the characteristic polynomial of A.

T.6. Show that if A and B are orthogonal matrices, then AB is an orthogonal matrix.

T.7. Show that if A is an orthogonal matrix, then A^{-1} is orthogonal.

T.8. (a) Verify that the matrix

$$\begin{bmatrix} \cos \theta & \sin \theta \\ -\sin \theta & \cos \theta \end{bmatrix}$$

is orthogonal.

(b) Prove that if A is an orthogonal 2×2 matrix, then there exists a real number θ such that either

$$A = \begin{bmatrix} \cos \theta & \sin \theta \\ -\sin \theta & \cos \theta \end{bmatrix}$$

or

$$A = \begin{bmatrix} \cos \theta & \sin \theta \\ \sin \theta & -\cos \theta \end{bmatrix}.$$

T.9. Prove that if $A^T A Y = Y$ for any Y in R^n, then $A^T A = I_n$.

T.10. (Uses Section 4.1). Prove that if a linear transformation $L: R^n \to R^n$ preserves length ($\|L(X)\| = \|X\|$ for all X in R^n), then it also preserves the inner product $[L(X) \cdot L(Y) = X \cdot Y$ for all X and Y in $R^n]$.

KEY IDEAS FOR REVIEW

☐ **Theorem 5.1.** The eigenvalues of A are the real roots of the characteristic polynomial of A.

☐ **Theorem 5.2.** An $n \times n$ matrix A is diagonalizable if and only if it has n linearly independent eigenvectors. In this case A is similar to a diagonal matrix D, with $D = P^{-1}AP$, whose diagonal elements are the eigenvalues of A, and P is a matrix whose columns are n linearly independent eigenvectors of A.

☐ **Theorem 5.4.** All the roots of the characteristic polynomial of a symmetric matrix are real numbers.

☐ **Theorem 5.5.** If A is a symmetric matrix, then eigenvectors that belong to distinct eigenvalues of A are orthogonal.

☐ **Theorem 5.6.** The $n \times n$ matrix A is orthogonal if and only if the columns of A form an orthonormal set of vectors in R^n.

☐ **Theorem 5.7.** If A is a symmetric $n \times n$ matrix, then there exists an orthogonal matrix $P(P^{-1} = P^T)$ such that $P^{-1}AP = D$, a diagonal matrix. The eigenvalues of A lie on the main diagonal of D.

SUPPLEMENTARY EXERCISES

1. Find the characteristic polynomial, eigenvalues, and eigenvectors of the matrix

$$\begin{bmatrix} -2 & 0 & 0 \\ 3 & 2 & 3 \\ 4 & -1 & 6 \end{bmatrix}.$$

In Exercises 2 and 3, determine whether the given matrix is diagonalizable.

2. $A = \begin{bmatrix} -1 & 0 & 0 \\ 3 & 2 & 0 \\ 4 & -1 & 2 \end{bmatrix}$.   3. $A = \begin{bmatrix} 2 & 2 & 0 \\ 5 & -1 & 3 \\ 0 & 0 & 0 \end{bmatrix}.$

In Exercises 4 and 5, find, if possible, a nonsingular matrix P and a diagonal matrix D so that A is similar to D.

4. $\mathbf{A} = \begin{bmatrix} 1 & 0 & 0 \\ 2 & -1 & -6 \\ 2 & 0 & -2 \end{bmatrix}$.   **5.** $\mathbf{A} = \begin{bmatrix} 1 & 0 & 1 \\ 0 & 1 & 0 \\ -1 & 0 & 1 \end{bmatrix}$.

6. If possible, find a diagonal matrix $\mathbf{D}$ so that

$$\mathbf{A} = \begin{bmatrix} 0 & 0 & 0 \\ 0 & 1 & 0 \\ 2 & 2 & 1 \end{bmatrix}$$

is similar to $\mathbf{D}$.

7. Find bases for the eigenspaces associated with each eigenvalue of the matrix

$$\mathbf{A} = \begin{bmatrix} 0 & 0 & 1 \\ 0 & 2 & 0 \\ 0 & 0 & 2 \end{bmatrix}.$$

8. Is the matrix

$$\begin{bmatrix} \dfrac{2}{3} & \dfrac{1}{\sqrt{5}} & 1 \\[2mm] \dfrac{2}{3} & 0 & 0 \\[2mm] \dfrac{1}{3} & -\dfrac{2}{\sqrt{5}} & 0 \end{bmatrix}$$

orthogonal?

In Exercises 9 and 10, find an orthogonal matrix $\mathbf{P}$ and a diagonal matrix $\mathbf{D}$ so that $\mathbf{P}^{-1}\mathbf{AP} = \mathbf{D}$.

9. $\mathbf{A} = \begin{bmatrix} 1 & 1 & 1 \\ 1 & 1 & 1 \\ 1 & 1 & 1 \end{bmatrix}$.   **10.** $\mathbf{A} = \begin{bmatrix} -3 & 0 & -4 \\ 0 & 5 & 0 \\ -4 & 0 & 3 \end{bmatrix}$.

11. If $\mathbf{A}^2 = \mathbf{A}$, what are possible values for the eigenvalues of $\mathbf{A}$? Justify your answer.

THEORETICAL EXERCISES

T.1. Prove that if a matrix $\mathbf{A}$ is similar to a diagonal matrix $\mathbf{D}$, then $\mathrm{Tr}(\mathbf{A}) = \mathrm{Tr}(\mathbf{D})$, where $\mathrm{Tr}(\mathbf{A})$ is the trace of $\mathbf{A}$. [*Hint:* See Supplementary Exercise T.1 in Chapter 1, where part (c) establishes $\mathrm{Tr}\,(\mathbf{AB}) = \mathrm{Tr}(\mathbf{BA})$.]

T.2. Prove that $\mathrm{Tr}(\mathbf{A})$ is the sum of the eigenvalues of $\mathbf{A}$ (see Supplementary Exercise T.1).

T.3 Let $\mathbf{A}$ and $\mathbf{B}$ be similar matrices. Prove:
(a) $\mathbf{A}^T$ and $\mathbf{B}^T$ are similar.

(b) rank $\mathbf{A}$ = rank $\mathbf{B}$.
(c) $\mathbf{A}$ is nonsingular if and only if $\mathbf{B}$ is nonsingular.
(d) If $\mathbf{A}$ and $\mathbf{B}$ are nonsingular, then $\mathbf{A}^{-1}$ and $\mathbf{B}^{-1}$ are similar.
(e) $\mathrm{Tr}(\mathbf{A}) = \mathrm{Tr}(\mathbf{B})$.

T.4. Prove that if $\mathbf{A}$ is an orthogonal matrix, then $\mathbf{A}^T$ is also orthogonal.

T.5. Let $\mathbf{A}$ be an orthogonal matrix. Prove that $c\mathbf{A}$ is orthogonal if and only if $c = \pm 1$.

CHAPTER TEST

1. If possible, find a nonsingular matrix $\mathbf{P}$ and a diagonal matrix $\mathbf{D}$ so that $\mathbf{A}$ is similar to $\mathbf{D}$, where

$$\mathbf{A} = \begin{bmatrix} 1 & 0 & 0 \\ 5 & 2 & 0 \\ 4 & 3 & 2 \end{bmatrix}.$$

2. Verify that the following matrix is orthogonal.

$$\begin{bmatrix} -\dfrac{1}{\sqrt{2}} & \dfrac{1}{\sqrt{3}} & \dfrac{1}{\sqrt{6}} \\[2mm] 0 & \dfrac{1}{\sqrt{3}} & -\dfrac{2}{\sqrt{6}} \\[2mm] \dfrac{1}{\sqrt{2}} & \dfrac{1}{\sqrt{3}} & \dfrac{1}{\sqrt{6}} \end{bmatrix}.$$

3. Find a nonsingular matrix $\mathbf{P}$ and a diagonal matrix $\mathbf{D}$ such that $\mathbf{A}$ is similar to $\mathbf{D}$, where

$$\mathbf{A} = \begin{bmatrix} -1 & -4 & -8 \\ -4 & -7 & 4 \\ -8 & 4 & -1 \end{bmatrix}.$$

4. Answer each of the following as true or false.
   (a) If $\mathbf{A}$ is an $n \times n$ orthogonal matrix, then rank $\mathbf{A} < n$.

(b) If $\mathbf{A}$ is diagonalizable, then each of its eigenvalues has multiplicity one.
(c) If none of the eigenvalues of $\mathbf{A}$ are zero, then $|\mathbf{A}| \neq 0$.
(d) If $\mathbf{A}$ and $\mathbf{B}$ are similar, then $|\mathbf{A}| = |\mathbf{B}|$.
(e) If $\mathbf{X}$ and $\mathbf{Y}$ are eigenvectors of $\mathbf{A}$ associated with the distinct eigenvalues λ_1 and λ_2, respectively, then $\mathbf{X} + \mathbf{Y}$ is an eigenvector of $\mathbf{A}$ associated with the eigenvalue $\lambda_1 + \lambda_2$.

Cumulative Review
of Part I

Answer each of the following as true (T) or false (F). Justify your answer.

1. If an $n \times n$ matrix $\mathbf{A}$ is singular, then $\mathbf{A}$ has either a row or a column of zeros.
2. A diagonal matrix is nonsingular if and only if none of the entries on its main diagonal are zero.
3. Two vectors in R^3 always span a two-dimensional subspace.
4. If $|\mathbf{AB}| = 12$ and $|\mathbf{A}| = 4$, then $|\mathbf{B}| = 48$.
5. Let $\mathbf{L}: R^6 \to R^{10}$ be a linear transformation defined by $\mathbf{L(X)} = \mathbf{AX}$ for $\mathbf{X}$ in R^6. If dim (range $\mathbf{L}$) = 3, then dim ker $\mathbf{L} = 7$.
6. If $\mathbf{AB}$ is singular, then $\mathbf{A}$ is singular or $\mathbf{B}$ is singular.
7. Let $\mathbf{L}: R^n \to R^n$ be a linear transformation defined by $\mathbf{L(X)} = \mathbf{AX}$. Then $\mathbf{L}$ is onto if and only if $|\mathbf{A}| \neq 0$.
8. The columns of a 5×8 matrix whose rank is 5 form a linearly dependent set.
9. If $\mathbf{A}$ is an $m \times n$ matrix with $m < n$, then the linear system $\mathbf{AX} = \mathbf{B}$ has a solution for every $m \times 1$ matrix $\mathbf{B}$.
10. Let $\mathbf{L}: R^n \to R^n$ be the linear transformation defined by $\mathbf{L(X)} = \mathbf{AX}$, for $\mathbf{X}$ in R^n. Then $\mathbf{L}$ is onto if and only if $\mathbf{A}$ is nonsingular.
11. If $\mathbf{A}$ is an $n \times n$ matrix, then rank $\mathbf{A} < n$ if and only if some eigenvalue of $\mathbf{A}$ is zero.
12. Let $\mathbf{A}$ be an $n \times n$ matrix. If $\mathbf{AX} = \mathbf{AY}$, then $\mathbf{X} = \mathbf{Y}$.
13. Span $\{(1, 1, 0), (0, 1, -1), (1, 0, 1)\} = R^3$.

14. If no row in an $n \times n$ matrix is a multiple of another row of $\mathbf{A}$, then $|\mathbf{A}| \neq 0$.
15. If $\mathbf{L}: V \to W$ is a linear transformation, then ker $\mathbf{L} = V$ if and only if range $\mathbf{L} = \{\mathbf{0}_W\}$.
16. The inverse of a nonsingular diagonal matrix is a diagonal matrix.
17. If $\mathbf{A}$ and $\mathbf{B}$ are $n \times n$ matrices, then $(\mathbf{A} + \mathbf{B})^2 = \mathbf{A}^2 + 2\mathbf{AB} + \mathbf{B}^2$.
18. If $\mathbf{X}$ and $\mathbf{Y}$ are vectors in R^n, then $\|\mathbf{X} - \mathbf{Y}\|^2 = \|\mathbf{X}\|^2 - \|\mathbf{Y}\|^2$.
19. There are real vector spaces containing exactly seven vectors.
20. If $\mathbf{X}_1, \mathbf{X}_2, \ldots, \mathbf{X}_k$ are orthogonal to $\mathbf{Y}$, then every vector in span $\{\mathbf{X}_1, \mathbf{X}_2, \ldots, \mathbf{X}_k\}$ is orthogonal to $\mathbf{Y}$.
21. $|\mathbf{A}| = 0$ if and only if some eigenvalue of $\mathbf{A}$ is zero.
22. If λ is an eigenvalue of $\mathbf{A}$ of multiplicity k, then the dimension of the eigenspace associated with λ is k.
23. If $\mathbf{A}$ and $\mathbf{B}$ are $n \times n$ matrices, then $(\mathbf{A}^T\mathbf{B}^T)^T = \mathbf{BA}$.
24. Let $\mathbf{A}$ be an $n \times n$ singular matrix. If the linear system $\mathbf{AX} = \mathbf{B}$ has a solution for $\mathbf{B} \neq \mathbf{0}$, then it has infinitely many solutions.
25. Let $\mathbf{L}: R^n \to R^n$ be a linear transformation defined by $\mathbf{L(X)} = \mathbf{AX}$, for $\mathbf{X}$ in R^n. Then $\mathbf{A}$ is singular if and only if ker $\mathbf{L} = \{\mathbf{0}_V\}$.

26. The product of two diagonal matrices is always a diagonal matrix.

27. If $\mathbf{A}$ is an $n \times n$ matrix that is row equivalent to $\mathbf{I}_n$, then $\mathbf{A}$ is singular.

28. If

$$\mathbf{A} = \begin{bmatrix} a - 3 & 2 & 1 \\ -1 & a & 1 \\ 0 & 0 & 1 \end{bmatrix}$$

then the only value of a for which the linear system $\mathbf{AX} = \mathbf{0}$ has a nontrivial solution is $a = 2$.

29. If $\mathbf{A}$ is a 3×3 matrix and $|\mathbf{A}| = 3$, then $|\frac{1}{2}\mathbf{A}^{-1}| = \frac{8}{3}$.

30. The linear system $\mathbf{AX} = \mathbf{B}$ has a solution if and only if $\mathbf{B}$ is in the column space of $\mathbf{A}$.

31. Let $\mathbf{L}: R^n \to R^n$ be a linear transformation defined by $\mathbf{L}(\mathbf{X}) = \mathbf{AX}$, for $\mathbf{X}$ in R^n. Then dim (range $\mathbf{L}$) $= n$ if and only if rank $\mathbf{A} = n$.

32. If W is a subspace of a finite-dimensional vector space V such that dim $W = $ dim V, then $W = V$.

33. If $\mathbf{A}$ is similar to $\mathbf{B}$, then rank $\mathbf{A} = $ rank $\mathbf{B}$.

34. The set of all vectors of the form $(a, b, -a)$ is a subspace of R^3.

35. If the columns of an $n \times n$ matrix span R^n, then the rows are linearly independent.

36. If $\mathbf{A}$ is an $n \times n$ matrix, then rank $\mathbf{A} = n$.

37. Let λ_1 and λ_2 be eigenvalues of $\mathbf{A}$ with associated eigenvectors $\mathbf{X}_1$ and $\mathbf{X}_2$. If $\lambda_1 = \lambda_2$, then $\mathbf{X}_1$ and $\mathbf{X}_2$ are linearly independent.

38. Every orthogonal set of n vectors in R^n is a basis for R^n.

39. The set of all solutions to the linear system $\mathbf{AX} = \mathbf{B}$, where $\mathbf{A}$ is $m \times n$ and $\mathbf{B} \neq \mathbf{0}$, is a subspace of R^n.

40. If $\mathbf{X}$ is a vector in R^n such that $\mathbf{X} \cdot \mathbf{Y} = 0$ for all $\mathbf{Y}$ in R^n, then $\mathbf{X} = \mathbf{0}$.

41. Every set of linearly independent vectors in R^3 contains three vectors.

42. If $|\mathbf{A}| = 0$, then the linear system $\mathbf{AX} = \mathbf{B}, \mathbf{B} \neq \mathbf{0}$, has no solution.

43. Let V be an n-dimensional vector space. If a set of m vectors spans V, then $m = n$.

44. Every set of five orthonormal vectors is a basis for R^5.

45. It is possible for a vector space to have more than one orthonormal basis.

46. If the set of vectors S spans a vector space V, then every subset of S also spans V.

47. If an $n \times n$ matrix $\mathbf{A}$ is diagonalizable, then A^3 is diagonalizable.

48. If $\mathbf{X}$ and $\mathbf{Y}$ are eigenvectors of $\mathbf{A}$ associated with the eigenvalue λ, then $\mathbf{X} + \mathbf{Y}$ is also an eigenvector of $\mathbf{A}$ associated with λ.

49. If $\mathbf{X}_0$ is a nontrivial solution to the homogeneous system $\mathbf{AX} = \mathbf{0}$, where $\mathbf{A}$ is $n \times n$, then $\mathbf{X}_0$ is an eigenvector of $\mathbf{A}$ associated with the eigenvalue 0.

50. If $\mathbf{A}$ is similar to $\mathbf{B}$, then $\mathbf{A}^n$ is similar to $\mathbf{B}^n$ for each positive integer n.

APPLICATIONS

6

Linear Programming

Prerequisite. Section 1.4, Solutions of Equations.

In this chapter we provide an introduction to the basic ideas and techniques of linear programming. Linear programming is a new area of applied mathematics, developed in the late 1940s, to solve a number of problems for the federal government. Since then, it has been applied to amazingly numerous problems in many areas. As a vital tool in management science and operations research, it has resulted in enormous savings of money. In the first section we give some examples of linear programming problems, formulate their mathematical models, and describe a geometric solution method. In the second section we present an algebraic method for solving linear programming problems. In the third section, dealing with duality, we discuss several applied interpretations of two related linear programming problems.

6.1

THE LINEAR PROGRAMMING PROBLEM; GEOMETRIC SOLUTION

In many problems in business and industry we are interested in making decisions that will maximize or minimize some quantity. For example, a plant manager may want to determine the most economical way of shipping goods from the factory to the markets, a hospital may want to design a diet satisfying certain nutritional requirements at a minimum cost, an investor may want to select investments that will maximize profits, or a manufacturer may wish to blend ingredients, subject to given specifications, to maximize profit. In this section we give several examples of linear programming problems and show how mathematical models for them can be formulated. Their geometric solution is also considered in this section.

Example 1
(A Production
Problem) A small manufacturer of photographic products prepares two types of film developers each day: Fine and Extra Fine, using as the raw material solutions A and B. Suppose that each quart of Fine contains 2 ounces of solution A and 1 ounce of solution B, while each quart of Extra Fine contains 1 ounce of solution A and 2 ounces of solution B. Suppose also that the profit on each quart of Fine is 8 cents and that it is 10 cents on each quart of Extra Fine. If the firm has 50 ounces of solution A and 70 ounces of solution B available each day, how many quarts of Fine and how many quarts of Extra Fine should be made each day to maximize the profit (assuming that the shop can sell all that is made)?

Mathematical Formulation

Let x be the number of quarts of Fine to be made and let y be the number of quarts of Extra Fine to be made. Since each quart of Fine contains 2 ounces of solution A and each quart of Extra Fine contains 1 ounce of solution A, the total amount of solution A required is

$$2x + y.$$

Similarly, since each quart of Fine contains one ounce of solution B and each quart of Extra Fine contains 2 ounces of solution B, the total amount of solution B required is

$$x + 2y.$$

Since we only have 50 ounces of solution A and 70 ounces of solution B available, we must have

$$2x + y \leqslant 50$$
$$x + 2y \leqslant 70.$$

Of course, x and y cannot be negative, so we must also have

$$x \geqslant 0 \quad \text{and} \quad y \geqslant 0.$$

Since the profit on each quart of Fine is 8 cents and the profit on each quart of Extra Fine is 10 cents, the total profit (in cents) is

$$z = 8x + 10y.$$

Our problem can be stated in mathematical form as: Find values of x and y that will maximize

$$z = 8x + 10y$$

subject to the following restrictions that must be satisfied by x and y:

$$2x + y \leq 50$$
$$x + 2y \leq 70$$
$$x \geq 0$$
$$y \geq 0.$$

Example 2
(Pollution) A manufacturer of a certain chemical product has two plants where the product is made. Plant X can make at most 30 tons per week and plant Y can make at most 40 tons per week. The manufacturer wants to make a total of at least 50 tons per week. The amount of particulate matter found weekly in the atmosphere over a nearby town is measured and found to be 20 pounds for each ton of the product made by plant X and 30 pounds for each ton of the product made at plant Y. How many tons should be made weekly at each plant to minimize the total amount of particulate matter in the atmosphere?

Mathematical Formulation

Let x and y denote the number of tons of the product made at plants X and Y, respectively, per week. Then the total amount manufactured per week is

$$x + y.$$

Since we want to make at least 50 tons per week, we must have

$$x + y \geq 50.$$

Since plant X can make at most 30 tons, we must have

$$x \leq 30.$$

Similarly, since plant Y can make at most 40 tons, we must have

$$y \leq 40.$$

Of course, x and y cannot be negative, so we require

$$x \geq 0$$
$$y \geq 0.$$

The total amount of particulate matter (in pounds) is

$$z = 20x + 30y,$$

which we want to minimize. Thus a mathematical statement of our problem is: Find values of x and y that will minimize

$$z = 20x + 30y$$

subject to the following restrictions that must be satisfied by x and y:

$$x + y \geq 50$$
$$x \leq 30$$
$$y \leq 40$$
$$x \geq 0$$
$$y \geq 0.$$

Example 3
(The Diet
Problem)
A nutritionist is planning a menu that includes foods A and B as its main staples. Suppose that each ounce of food A contains 2 units of protein, 1 unit of iron, and 1 unit of thiamine; each ounce of food B contains 1 unit of protein, 1 unit of iron, and 3 units of thiamine. Suppose that each ounce of A costs 30 cents, while each ounce of B costs 40 cents. The nutritionist wants the meal to provide at least 12 units of protein, at least 9 units of iron, and at least 15 units of thiamine. How many ounces of each of the foods should be used to minimize the cost of the meal?

Mathematical Formulation

Let x denote the number of ounces of food A to be used and let y be the number of ounces of food B. The number of units of protein supplied by the meal is

$$2x + y,$$

so we must have

$$2x + y \geqslant 12.$$

The number of units of iron supplied by the meal is

$$x + y,$$

so we must have

$$x + y \geqslant 9.$$

Since the number of units of thiamine supplied by the meal is

$$x + 3y,$$

we must have

$$x + 3y \geqslant 15.$$

Of course, we also require

$$x \geqslant 0, \qquad y \geqslant 0.$$

The cost of the meal (in cents) is

$$z = 30x + 40y,$$

which we want to minimize. Thus a mathematical formulation of our problem is: Find values of x and y that will minimize

$$z = 30x + 40y$$

subject to the restrictions

$$
\begin{aligned}
2x + y &\geqslant 12 \\
x + y &\geqslant 9 \\
x + 3y &\geqslant 15 \\
x &\geqslant 0 \\
y &\geqslant 0.
\end{aligned}
$$

These examples are typical of linear programming problems, the general form being: Find values of $x_1, x_2, \ldots, x_n$ that will minimize or maximize

$$z = c_1x_1 + c_2x_2 + \cdots + c_nx_n \tag{1}$$

subject to

$$\left.\begin{array}{l}
a_{11}x_1 + a_{12}x_2 + \cdots + a_{1n}x_n \ (\leqslant)(\geqslant)(=) \ b_1 \\
a_{21}x_1 + a_{22}x_2 + \cdots + a_{2n}x_n \ (\leqslant)(\geqslant)(=) \ b_2 \\
\vdots \qquad \vdots \qquad\qquad \vdots \qquad\qquad \vdots \\
a_{m1}x_1 + a_{m2}x_2 + \cdots + a_{mn}x_n \ (\leqslant)(\geqslant)(=) \ b_m
\end{array}\right\} \tag{2}$$

$$x_j \geqslant 0 \qquad (1 \leqslant j \leqslant n), \tag{3}$$

where in (2) one and only one of the symbols $\leqslant$, $=$, $\geqslant$ occurs in each statement. The linear function in (1) is called the **objective function**. The equalities or inequalities in (2) and (3) are called **constraints**. The term **linear** in linear programming means that the objective function (1) and each of the constraints in (2) are linear functions of the variables $x_1, x_2, \ldots, x_n$. The word "programming," *not* to be confused with its use in computer programming, refers to the applications in planning or allocation problems.

Geometric Solution

We now develop a geometric method for solving linear programming problems in two variables. This will enable us to solve the problems posed earlier. Since linear programming deals with systems of linear inequalities, we first consider these from a geometric point of view.

Consider Example 1. Find the set of points that maximize

$$z = 8x + 10y \tag{4}$$

subject to the constraints

$$\begin{aligned}
2x + y &\leqslant 50 \\
x + 2y &\leqslant 70 \\
x &\geqslant 0 \\
y &\geqslant 0.
\end{aligned} \tag{5}$$

The set of points satisfying the system of four inequalities (5) consists of those points satisfying each of the inequalities. In Figure 6.1(a) we show the set of points satisfying the inequality $x \geqslant 0$, in Figure 6.1(b) the set of points satisfying the inequality $y \geqslant 0$. The set of points satisfying both inequalities $x \geqslant 0$ and $y \geqslant 0$ is the intersection of the regions in Figure 6.1(a) and (b). This set of points, the set of all points in the first quadrant, is shown in Figure 6.1(c).

We must now consider the set of points satisfying the inequality

$$2x + y \leqslant 50. \tag{6}$$

(a) Set of points satisfying
$x \geqslant 0$.

(b) Set of points satisfying
$y \geqslant 0$.

(c) Set of points satisfying
$x \geqslant 0$ and $y \geqslant 0$.

Figure 6.1

First, consider the set of points satisfying the strict inequality

$$2x + y < 50. \tag{7}$$

Now the straight line

$$2x + y = 50 \tag{8}$$

divides the set of points not on the line into two regions (Figure 6.2); one region contains all points satisfying (7) and the other region contains all points not satisfying (7). The line itself, drawn in dashed form, belongs to neither region.

To determine the region determined by inequality (7), we choose any point not on the line as a test point and see on which side of the line it lies. The point (10, 20) does not lie on the line (8) and can thus serve as a test point. Since its coordinates satisfy inequality (7), the correct region determined by (7) is shown in Figure 6.3(a). Another possible test point is (20, 20), since it does not lie on the line. Its coordinates do not satisfy inequality (7), so the test point does not lie in the correct region. Thus the correct region would again be as shown in Figure 6.3(a).

The set of points satisfying inequality (6) consists of the set of points satisfying (7) as well as the points on the line (8). Thus the region, shown in Figure 6.3(b), includes the straight line, which is drawn solidly.

Figure 6.2

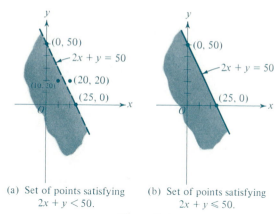

(a) Set of points satisfying   (b) Set of points satisfying
        $2x + y < 50$. $2x + y \leq 50$.

Figure 6.3

Next, the set of points satisfying the inequality

$$x + 2y \leq 70 \tag{9}$$

is shown in Figure 6.4.

The set of points satisfying inequalities (6) and (9) is the intersection of the regions in Figures 6.3(b) and 6.4; it is the shaded region shown in Figure 6.5.

Finally, the set of points satisfying the four inequalities (5) is the intersection of the regions in Figures 6.1(c) and 6.5. That is, it is the set of points in Figure 6.5 that lie in the first quadrant. This set of points is shown in Figure 6.6.

Thus the set of points satisfying a system of inequalities is the intersection of the set of points satisfying each of the inequalities.

A point satisfying the constraints of a linear programming problem is called a **feasible solution**; the set of all such points is the **feasible region**. It should be noted that not every linear programming problem has a feasible region, as the next example shows.

Example 4 Find the set of points that maximize

$$z = 8x + 10y \tag{10}$$

Figure 6.4

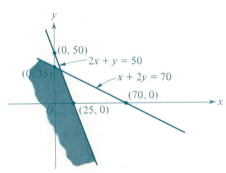

Figure 6.5

The set of points satisfying $2x + y \leqslant 50$ and $x + 2y \leqslant 70$.

subject to the constraints

$$
\begin{aligned}
2x + 3y &\leqslant 6 \\
x + 2y &\geqslant 6 \\
x &\geqslant 0 \\
y &\geqslant 0.
\end{aligned}
\tag{11}
$$

   In Figure 6.7 we have labeled as Region I the set of points satisfying the first, third, and fourth inequalities; that is, Region I is the set of points in the first quadrant satisfying the first inequality in (11). Similarly, Region II is the set of points satisfying the second, third, and fourth inequalities. It is clear that there are no points satisfying all four inequalities.

   To solve our given linear programming problem, we must find a feasible solution that makes the objective function (1) as large as possible. Such a solution is called an **optimal solution**. Since there are infinitely many feasible points, it appears, at first glance, quite difficult to tell whether a feasible solution is optimal or not. In Figure 6.8 we have reproduced Figure 6.6 by showing the feasible region for our linear programming problem along with a number of indicated feasible points: O, A, B, C, D, and E. Points F and G are not in the feasible region.

   In Table 6.1 we have tabulated the value of the objective function (in cents) for each of the points O, A, B, C, D, and E. Among the feasible solutions O, A, B, C, D, and E we

Figure 6.6

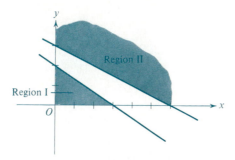

Figure 6.7

see that the value of the objective function is largest for the point B (10, 30). However, we do not know if there is another feasible solution at which the value of the objective function will be larger than it is at B. Since there are infinitely many feasible solutions in the region, we cannot examine the value of the objective function at every one of these. However, we shall be able to find optimal solutions without examining all feasible solutions, but we first consider some auxiliary notions.

Table 6.1

Point	Value of $z = 8x + 10y$ (cents)
$O(0, 0)$	0
$A(0, 35)$	350
$B(10, 30)$	380
$C(25, 0)$	200
$D(5, 10)$	140
$E(20, 5)$	210

Figure 6.8

Figure 6.9

Definition The **line segment** joining the distinct points $\mathbf{X}_1$ and $\mathbf{X}_2$ of R^n is the set of all points in R^n of the form $\lambda\mathbf{X}_1 + (1 - \lambda)\mathbf{X}_2$, $0 \leq \lambda \leq 1$. Observe that if $\lambda = 0$, we get $\mathbf{X}_2$, and if $\lambda = 1$, we get $\mathbf{X}_1$.

Definition A nonempty set S of points in R^n is called **convex** if the line segment joining any two points in S also lies entirely in S.

Example 5 The shaded sets in Figure 6.9(a), (b), (c), and (d) are convex sets in R^2. The shaded sets in Figure 6.10(a), (b), and (c) are *not* convex sets in R^2, since the line joining the indicated points does not lie entirely in the set.

Example 6 It is not too difficult to show that the feasible region of a linear programming problem is a convex set. The proof of this result for a broad class of linear programming problems is outlined in Exercise T.1 of Section 6.2.

The reader may also check that the feasible regions for the linear programming problems in Examples 2 and 3 are convex.

We shall now limit our further discussion of convex sets to such sets in R^2 although the ideas presented here can be generalized to convex sets in R^n.

Convex sets are either **bounded** or **unbounded**. A bounded convex set is one that can be enclosed by a sufficiently large rectangle; an unbounded convex set is one that cannot be so enclosed. The convex sets in Figure 6.9(a) and (b) are bounded; the convex sets in Figure 6.9(c) and (d) are unbounded.

Definition An **extreme point** in a convex set S is a point *in* S that is not an interior point of any line segment in S.

Figure 6.10

(a)   (b)   (c)

Figure 6.11

Example 7 In Figure 6.11 we have marked the extreme points of the convex sets in Figure 6.9(a), (b), and (c). The convex set in Figure 6.9(d) has no extreme points.

The basic result connecting convex sets, extreme points, and linear programming problems is the following theorem, whose proof we omit.

Theorem 6.1 *Let S be the feasible region of a linear programming problem.*

   (a) If S is bounded, then the objective function

$$z = ax + by$$

assumes both a maximum and a minimum value on S; these values occur at extreme points of S.

   (b) If S is unbounded, then there may or may not be a maximum or minimum value on S. If a maximum or minimum value does exist on S, it occurs at an extreme point.

Thus, when the feasible region S of a linear programming problem is bounded, a method for solving the problem consists in finding the extreme points of S and evaluating the objective function $z = ax + by$ at each extreme point. An optimal solution is an extreme point at which the value of z is a maximum or a minimum.

The procedure for solving a linear programming problem in two variables geometrically is as follows.

STEP 1. Sketch the feasible region S.

STEP 2. Determine all the extreme points of S.

STEP 3. Evaluate the objective function at each extreme point.

STEP 4. Choose an extreme point at which the objective function is largest (smallest) for a maximization (minimization) problem.

Example 8 Consider Example 1 again. The feasible region shown in Figure 6.8 is bounded and its extreme points are $O(0, 0)$, $A(0, 35)$, $B(10, 30)$, and $C(25, 0)$. From Table 6.1 we see that the value of z is a maximum at the extreme point $B(10, 30)$. Thus the optimal solution is

$$x = 10, \qquad y = 30.$$

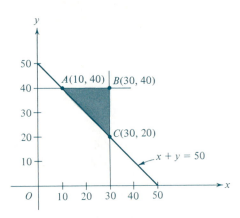

Figure 6.12

This means that the manufacturer should make 10 quarts of Fine and 30 quarts of Extra Fine to maximize its profit. If the firm follows this course of action, its maximum profit will be $3.80 each day.

Referring back to Figure 6.8, we note that the points (70, 0) and (0, 50) are points of intersection of boundary lines. However, they are not extreme points in the feasible region, since they are *not* feasible solutions; that is, they do not lie in the feasible region.

Example 9 Solve Example 2 geometrically.

Solution
The feasible region S of this linear programming problem is shown in Figure 6.12 (verify). Since S is bounded, we can find the minimum value of z by evaluating z at each of the extreme points. In Table 6.2 we have tabulated the value of the objective function for each of the points $A(10, 40)$, $B(30, 40)$, and $C(30, 20)$. The value of z is a minimum at the extreme point $C(30, 20)$. Thus the optimal solution is

$$x = 30, \qquad y = 20,$$

which means that the manufacturer should make 30 tons of the product at plant X and 20 tons at plant Y. If this course of action is followed, the total amount of particulate matter over the town will be 1200 pounds weekly.

Table 6.2

Point	Value of $z = 20x + 30y$ (pounds)
A(10, 40)	1400
B(30, 40)	1800
C(30, 20)	1200

The feasible region of Example 3 is not bounded (verify). We shall not deal further with such problems in this book.

In general, a linear programming problem may have

1. No feasible solution; that is, there are no points satisfying all the constraints in Equations (2) and (3).
2. A unique optimal solution.
3. More than one optimal solution (see Exercise 4).
4. No maximum (or minimum) value in the feasible region; that is, it may be possible to choose a point in the feasible region to make the objective function as large (or as small) as we please.

Slack Variables

We shall now focus our attention on a special class of linear programming problems and will show that every linear programming problem can be transformed into a problem of the special type.

Definition We shall refer to the following linear programming problem as a **standard linear programming problem**. Find values of $x_1, x_2, \ldots, x_n$ that will maximize

$$z = c_1 x_1 + c_2 x_2 + \cdots + c_n x_n \tag{12}$$

subject to the constraints

$$\left.\begin{array}{l} a_{11}x_1 + a_{12}x_2 + \cdots + a_{1n}x_n \leqslant b_1 \\ a_{21}x_1 + a_{22}x_2 + \cdots + a_{2n}x_n \leqslant b_2 \\ \quad\vdots \qquad\quad \vdots \qquad\qquad\quad \vdots \qquad\;\; \vdots \\ a_{m1}x_1 + a_{m2}x_2 + \cdots + a_{mn}x_n \leqslant b_m. \end{array}\right\} \tag{13}$$

$$x_j \geqslant 0 \qquad (1 \leqslant j \leqslant n). \tag{14}$$

Example 10 Example 1 is a standard linear programming problem.

Example 11 The linear programming problem:

$$\text{Minimize} \qquad z = 3x - 4y$$

subject to

$$\begin{array}{r} 2x - 3y \leqslant 6 \\ x + y \leqslant 8 \\ x \geqslant 0 \\ y \geqslant 0 \end{array}$$

is *not* a standard linear programming problem, since the objective function is to be minimized and not maximized.

Example 12   The linear programming problem:

$$\text{Maximize} \qquad z = 12x - 15y$$

subject to

$$
\begin{aligned}
3x - y &\geqslant 4 \\
2x + 3y &\leqslant 6 \\
x &\geqslant 0 \\
y &\geqslant 0
\end{aligned}
$$

is *not* a standard linear programming problem, since one of the inequalities is of the form $\geqslant$; in a standard linear programming problem every inequality of (13) must be of the form $\leqslant$.

Example 13   The linear programming problem:

$$\text{Maximize} \qquad z = 8x + 10y$$

subject to

$$
\begin{aligned}
3x + y &= 4 \\
2x - 3y &\leqslant 5 \\
x &\geqslant 0 \\
y &\geqslant 0
\end{aligned}
$$

is *not* a standard linear programming problem, since the first constraint is an equation and not an inequality of the form $\leqslant$.

Every linear programming problem can be transformed into a standard linear programming problem.

Minimization Problem as a Maximization Problem

Every maximization problem can be viewed as a minimization problem, and conversely. This follows from the observation that

$$
\begin{aligned}
\text{minimum of } & c_1 x_1 + c_2 x_2 + \cdots + c_n x_n \\
& = -\text{maximum of } -(c_1 x_1 + c_2 x_2 + \cdots + c_n x_n).
\end{aligned} \qquad (15)
$$

Reversing an Inequality

Consider the inequality

$$d_1 x_1 + d_2 x_2 + \cdots + d_n x_n \geqslant -b.$$

Multiplying both sides of this inequality by -1 reverses the inequality, yielding

$$-d_1 x_1 - d_2 x_2 - \cdots - d_n x_n \leqslant b.$$

Example 14 Consider the linear programming problem:

$$\text{Minimize} \qquad w = 5x - 2y$$

subject to

$$
\begin{aligned}
2x - 3y &\geq -5 \\
3x + 2y &\leq 12 \\
x &\geq 0 \\
y &\geq 0.
\end{aligned}
\tag{16}
$$

Using (15) and multiplying the first inequality in (16) by (-1), we then obtain the following standard linear programming problem:

$$\text{Maximize} \qquad z = -(5x - 2y) = -5x + 2y$$

subject to

$$
\begin{aligned}
-2x + 3y &\leq 5 \\
3x + 2y &\leq 12 \\
x &\geq 0 \\
y &\geq 0.
\end{aligned}
$$

Once this new maximization problem has been solved, we take the negative of the maximum value of z to obtain the minimum value of w.

It is not too difficult to change a number of equalities into inequalities of the form $\leq$. Thus Example 13 can be transformed into a standard linear programming problem. In this book we shall not encounter linear programming problems of the type in Example 13, so we shall not continue to examine such problems.

It is not easy to handle systems of linear inequalities algebraically. However, as we have seen in Chapter 1, it is not difficult at all to deal with systems of linear equations. Accordingly, we shall change our given standard linear programming problem into a problem where we must find nonnegative variables maximizing a linear objective function and satisfying a system of linear equations. Every solution to the given problem yields a solution to the new problem, and conversely, every solution to the new problem yields a solution to the given problem.

Consider the constraint

$$d_1 x_1 + d_2 x_2 + \cdots + d_n x_n \leq b. \tag{17}$$

Since the left side of (17) is not larger than the right side, we can make (17) into an equation by adding the unknown nonnegative quantity u to its left side, to obtain

$$d_1 x_1 + d_2 x_2 + \cdots + d_n x_n + u = b. \tag{18}$$

The quantity u in (18) is called a **slack variable**, since it takes up the slack between the two sides of the inequality.

We now change each of the constraints in (2) (which we assume represents a standard linear programming problem with only the $\leq$-type inequalities) into an equation by introducing a nonnegative slack variable. Thus the ith inequality

$$a_{i1}x_1 + a_{i2}x_2 + \cdots + a_{in}x_n \leqslant b_i \qquad (1 \leqslant i \leqslant m) \qquad (19)$$

is converted into the equation

$$a_{i1}x_1 + a_{i2}x_2 + \cdots + a_{in}x_n + x_{n+i} = b_i \qquad (1 \leqslant i \leqslant m),$$

by introducing the nonnegative slack variable x_{n+i}. Our new problem can now be stated as follows.

New Problem

Find values of $x_1, x_2, \ldots, x_n, x_{n+1}, \ldots, x_{n+m}$ that will maximize

$$z = c_1x_1 + c_2x_2 + \cdots + c_nx_n \qquad (20)$$

subject to

$$\left.\begin{array}{l} a_{11}x_1 + a_{12}x_2 + \cdots + a_{1n}x_n + x_{n+1} \qquad\qquad\qquad = b_1 \\ a_{21}x_1 + a_{22}x_2 + \cdots + a_{2n}x_n \qquad + x_{n+2} \qquad\qquad = b_2 \\ \quad\vdots \qquad\quad \vdots \qquad\qquad\quad \vdots \qquad\qquad\qquad \ddots \qquad\quad \vdots \\ a_{m1}x_1 + a_{m2}x_2 + \cdots + a_{mn}x_n \qquad\qquad\qquad + x_{n+m} = b_m \end{array}\right\} \qquad (21)$$

$$x_1 \geqslant 0, \ldots, x_n \geqslant 0, x_{n+1} \geqslant 0, x_{n+2} \geqslant 0, \ldots, x_{n+m} \geqslant 0. \qquad (22)$$

Thus the new problem has m equations in $m + n$ unknowns. Solving the original problem is equivalent to solving the new problem in the following sense. If $x_1, x_2, \ldots, x_n$ is a feasible solution to the given problem as defined by (1), (2), and (3), then

$$x_1 \geqslant 0, x_2 \geqslant 0, \ldots, x_n \geqslant 0.$$

Also, $x_1, x_2, \ldots, x_n$ satisfy each of the constraints in (2). Let x_{n+i}, $1 \leqslant i \leqslant m$, be defined by

$$x_{n+i} = b_i - a_{i1}x_1 - a_{i2}x_2 - \cdots - a_{in}x_n.$$

That is, x_{n+i} is the difference between the right side of inequality (19) and its left side. Then

$$x_{n+1} \geqslant 0, x_{n+2} \geqslant 0, \ldots, x_{n+m} \geqslant 0$$

so that $x_1, x_2, \ldots, x_n, x_{n+1}, \ldots, x_{n+m}$ satisfy (21) and (22).

Conversely, suppose that $x_1, x_2, \ldots, x_n, x_{n+1}, \ldots, x_{n+m}$ satisfy (21) and (22). It is then clear that $x_1, x_2, \ldots, x_n$ satisfy (2) and (3).

Example 15 Consider the problem of Example 1. Introducing the slack variables u and v, we formulate our new problem as: Find values of x, y, u, and v that will maximize

$$z = 8x + 10y$$

subject to

$$2x + y + u = 50$$
$$x + 2y + v = 70$$
$$x \geqslant 0, \quad y \geqslant 0, \quad u \geqslant 0, \quad v \geqslant 0.$$

The slack variable u is the difference between the total amount of solution A available, 50 ounces, and the amount $2x + y$ of solution A actually used. The slack variable v is the difference between the total amount of solution B available, 70 ounces, and the amount $x + 2y$ of solution B actually used.

Consider the feasible solution to the given problem

$$x = 5, \qquad y = 10,$$

which represents point D in Figure 6.8. We then obtain the slack variables

$$u = 50 - 2(5) - 10$$
$$= 50 - 10 - 10 = 30$$

and

$$v = 70 - 5 - 2(10)$$
$$= 70 - 5 - 20 = 45$$

so that

$$x = 5, \quad y = 10, \quad u = 30, \quad v = 45$$

is a feasible solution to the new problem. Of course, the solution $x = 5, y = 10$ is not an optimal solution, since $z = 8(5) + 10(10) = 140$, and we recall that the maximum value of z is attained for

$$x = 10, \quad y = 30.$$

In this case, the corresponding optimal solution to the new problem is

$$x = 10, \quad y = 30, \quad u = 0, \quad v = 0.$$

6.1 EXERCISES

In Exercises 1 through 9, formulate mathematically each linear programming problem.

1. A steel producer makes two types of steel: regular and special. A ton of regular steel requires 2 hours in the open-hearth furnace and 5 hours in the soaking pit; a ton of special steel requires 2 hours in the open-hearth furnace and 3 hours in the soaking pit. The open-hearth furnace is available 8 hours per day and the soaking pit is available 15 hours per day. The profit on a ton of regular steel is $120 and it is $100 on a ton of special steel. Determine how many tons of each type of steel should be made to maximize the profit.

2. A trust fund is planning to invest up to $6000 in two types of bonds: A and B. Bond A is safer than bond B and carries a dividend of 8 percent, and bond B carries a dividend of 10 percent. Suppose that the fund's rules state that no more than $4000 may be invested in bond B, while at least $1500 must be invested in bond A. How much should be invested in each type of bond to maximize the fund's return?

3. Solve Exercise 2 if the fund has the following additional rule: "The amount invested in bond B cannot exceed one half the amount invested in bond A."

4. A trash-removal company carries industrial waste in sealed containers in its fleet of trucks. Suppose that each container from the Smith Corporation weighs 6 pounds and is 3 cubic feet in volume, while each container from the Johnson Corporation weighs 12 pounds and is 1 cubic foot in volume. The company charges the Smith Corporation 30 cents for each container carried on a trip, and 60 cents for each container from the Johnson Corporation. If a truck cannot carry more than 18,000 pounds or more than 1800 cubic feet in volume, how many containers from each customer should the company carry in a truck on each trip to maximize the revenue per truckload?

5. A television producer designs a program based on a comedian and time for commercials. The advertiser insists on at least 2 minutes of advertising time, the station insists on no more than 4 minutes of advertising time, and the comedian insists on at least 24 minutes of the comedy program. Also, the total time allotted for the advertising and comedy portions of the program cannot exceed 30 minutes. If it has been determined that each minute of advertising (very creative) attracts 40,000 viewers and each minute of the comedy program attracts 45,000 viewers, how should the time be divided between advertising and programming to maximize the number of viewer-minutes?

6. A small generator burns two types of fuel: low sulfur (L) and high sulfur (H) to produce electricity. For each hour of use, each gallon of L emits 3 units of sulfur dioxide, generates 4 kilowatts, and costs 60 cents, while each gallon of H emits 5 units of sulfur dioxide, generates 4 kilowatts, and costs 50 cents. The environmental protection agency insists that the maximum amount of sulfur dioxide that may be emitted per hour is 15 units. Suppose that at least 16 kilowatts must be generated per hour. How many gallons of L and how many gallons of H should be used hourly to minimize the cost of the fuel used?

7. The Protein Diet Club serves a luncheon consisting of two dishes, A and B. Suppose that each unit of A has 1 gram of fat, 1 gram of carbohydrate, and 4 grams of protein, whereas each unit of B has 2 grams of fat, 1 gram of carbohydrate, and 6 grams of protein. If the dietician planning the luncheon wants to provide no more than 10 grams of fat or more than 7 grams of carbohydrate, how many units of A and how many units of B should be served to maximize the amount of protein consumed?

8. In designing a new transportation system, a company is considering two types of buses, types A and B. A type A bus can carry 40 passengers and requires 2 mechanics for servicing, a type B bus can carry 60 passengers and requires 3 mechanics for servicing. Suppose that the company must transport at least 300 people daily and that insurance rules for the size of the garage allow no more than 12 mechanics on the payroll. If each type A bus costs $20,000 and each type B bus costs $25,000, how many buses of each type should be bought to minimize the cost?

9. An animal feed producer mixes two types of grain: A and B. Each unit of grain A contains 2 grams of fat, 1 gram of protein, and 80 calories. Each unit of grain B contains 3 grams of fat, 3 grams of protein, and 60 calories. Suppose that the producer wants each unit of the final product to yield at least 18 grams of fat, at least 12 grams of protein, and at least 480 calories. If each unit of A costs 10 cents and each unit of B costs 12 cents, how many units of each type of grain should the producer use to minimize the cost?

In Exercises 10 through 13, sketch the set of points satisfying the given system of inequalities.

10. $\quad x \leqslant 4$

$\quad x \geqslant 2$

$\quad y \leqslant 4$

$\quad y \geqslant 1$

$\quad x + y \leqslant 6.$

11. $2x - y \leqslant 6$

$\quad 2x + y \leqslant 10$

$\quad\quad x \geqslant 0$

$\quad\quad y \geqslant 0.$

12. $\quad x + \ y \leqslant \ 3$

$\quad 5x + 4y \geqslant 20$

$\quad\quad\quad x \geqslant \ 0$

$\quad\quad\quad y \geqslant \ 0.$

13. $x + \ y \geqslant 4$

$\quad x + 4y \geqslant 8$

$\quad\quad x \geqslant 0$

$\quad\quad y \geqslant 0.$

In Exercises 14 and 15, solve the given linear programming problem geometrically.

14. Maximize $z = 3x + 2y$
subject to

$$2x - 3y \leq 6$$
$$x + y \leq 4$$
$$x \geq 0$$
$$y \geq 0.$$

15. Minimize $z = 3x - y$
subject to

$$-3x + 2y \leq 6$$
$$5x + 4y \geq 20$$
$$8x + 3y \leq 24$$
$$x \geq 0$$
$$y \geq 0.$$

16. Solve the problem in Exercise 1 geometrically.

17. Solve the problem in Exercise 2 geometrically.

18. Solve the problem in Exercise 3 geometrically.

19. Solve the problem in Exercise 4 geometrically.

20. Solve the problem in Exercise 5 geometrically.

21. Solve the problem in Exercise 6 geometrically.

22. Solve the problem in Exercise 7 geometrically.

23. Solve the problem in Exercise 8 geometrically.

24. Which of the following are standard linear programming problems?
(a) Maximize $z = 2x - 3y$
subject to

$$2x - 3y \leq 4$$
$$3x + 2y \geq 5.$$

(b) Maximize $z = 2x + 3y$
subject to

$$2x + 3y \leq 4$$
$$3x + 2y \leq 5$$
$$x \geq 0$$
$$y \geq 0.$$

(c) Minimize $z = 2x_1 - 3x_2 + x_3$
subject to

$$2x_1 + 3x_2 + 2x_3 \leq 6$$
$$3x_1 \quad\quad - 2x_3 \leq 4$$
$$x_1 \leq 0$$
$$x_2 \geq 0$$
$$x_3 \geq 0.$$

(d) Maximize $z = 2x + 2y$
subject to

$$2x + 3y \leq 4$$
$$3x \leq 5$$
$$x \geq 0.$$

25. Which of the following are standard linear programming problems?
(a) Maximize $z = 3x_1 + 2x_2 + x_3$
subject to

$$2x_1 + 3x_2 + x_3 \leq 4$$
$$3x_1 - 2x_2 \quad\quad \leq 5$$
$$x_1 \geq 0$$
$$x_2 \geq 0$$
$$x_3 \geq 0.$$

(b) Minimize $z = 2x + 3y$
subject to

$$2x + 4y \leq 2$$
$$3x + 2y \leq 4$$
$$x \geq 0$$
$$y \geq 0.$$

(c) Maximize $z = 3x_1 + 4x_2 + x_3$
subject to

$$2x_1 + 4x_2 + x_3 \leq 2$$
$$3x_1 - 2x_2 + x_3 \leq 4$$
$$2x_1 \quad\quad + x_3 \leq 8$$
$$x_1 \geq 0$$
$$x_2 \geq 0.$$

(d) Maximize $z = 2x_1 + 3x_2 + x_3$
subject to

$$2x_1 + 3x_2 + 5x_3 \leq 8$$
$$3x_1 - 2x_2 + 2x_3 = 4$$
$$2x_1 + x_2 + 3x_3 \leq 6$$
$$x_1 \geq 0$$
$$x_2 \geq 0$$
$$x_3 \geq 0.$$

In Exercises 26 and 27, formulate each problem as a standard linear programming problem.

26. Minimize $z = -2x_1 + 3x_2 + 2x_3$
subject to

$$2x_1 + x_2 + 2x_3 \leq 12$$
$$x_1 + x_2 - 3x_3 \leq 8$$
$$x_1 \geq 0$$
$$x_2 \geq 0$$
$$x_3 \geq 0.$$

27. Maximize $z = 3x_1 - x_2 + 6x_3$
subject to

$$2x_1 + 4x_2 + x_3 \leq 4$$
$$-3x_1 + 2x_2 - 3x_3 \geq -4$$
$$2x_1 + x_2 - x_3 \leq 8$$
$$x_1 \geq 0$$
$$x_2 \geq 0$$
$$x_3 \geq 0.$$

In Exercises 28 and 29, formulate the given linear programming problem as a new problem with slack variables.

28. Maximize $z = 2x + 8y$
subject to

$$2x + 3y \leq 18$$
$$3x - 2y \leq 6$$
$$x \geq 0$$
$$y \geq 0.$$

29. Maximize $z = 2x_1 + 3x_2 + 7x_3$
subject to

$$3x_1 + x_2 - 4x_3 \leq 3$$
$$x_1 - 2x_2 + 6x_3 \leq 21$$
$$x_1 - x_2 - x_3 \leq 9$$
$$x_1 \geq 0$$
$$x_2 \geq 0$$
$$x_3 \geq 0.$$

6.2

THE SIMPLEX METHOD

The simplex method for solving linear programming problems was developed by George B. Dantzig[†] in connection with his work on planning problems for the federal government. In this section we present the essential features of the method, illustrating them with examples. A number of proofs will be omitted, and for further details the interested reader may consult the references given at the end of this chapter. It is convenient to introduce matrix terminology into our further discussion of linear programming.

---

† George B. Dantzig (1914–   ) was born in Portland, Oregon. He received a bachelor's degree from the University of Maryland, a master's degree from the University of Michigan, and a Ph.D. degree from the University of California at Berkeley. He is presently Professor of Operations Research and Computer Science at Stanford University. His simplex method for solving linear programming problems, developed in 1947, was a major breakthrough in the newly developed area of operations research. The rapid growth of computing power coupled with its decreasing cost has produced a great many computer implementations of the simplex method and has resulted in savings of billions of dollars for industry and government. Professor Dantzig has held important positions in government, industry, and universities, and is the recipient of numerous honors and awards.

Matrix Notation

We again restrict our attention to the standard linear programming problem: Maximize

$$c_1x_1 + c_2x_2 + \cdots + c_nx_n \tag{1}$$

subject to

$$\left.\begin{array}{l} a_{11}x_1 + a_{12}x_2 + \cdots + a_{1n}x_n \leqslant b_1 \\ a_{21}x_1 + a_{22}x_2 + \cdots + a_{2n}x_n \leqslant b_2 \\ \vdots \qquad\qquad \vdots \qquad\qquad\quad \vdots \quad\;\; \vdots \\ a_{m1}x_1 + a_{m2}x_2 + \cdots + a_{mn}x_n \leqslant b_m \end{array}\right\} \tag{2}$$

$$x_j \geqslant 0 \qquad (1 \leqslant j \leqslant n). \tag{3}$$

If we let

$$\mathbf{A} = \begin{bmatrix} a_{11} & a_{12} & \cdots & a_{1n} \\ a_{21} & a_{22} & \cdots & a_{2n} \\ \vdots & \vdots & & \vdots \\ a_{m1} & a_{m2} & \cdots & a_{mn} \end{bmatrix}, \qquad \mathbf{B} = \begin{bmatrix} b_1 \\ b_2 \\ \vdots \\ b_m \end{bmatrix}, \qquad \mathbf{X} = \begin{bmatrix} x_1 \\ x_2 \\ \vdots \\ x_n \end{bmatrix},$$

and

$$\mathbf{C} = [c_1 \quad c_2 \quad \cdots \quad c_n],$$

then the given problem can be stated as follows: Find a vector $\mathbf{X}$ in R^n that will maximize the objective function

$$z = \mathbf{CX} \tag{4}$$

subject to

$$\mathbf{AX} \leqslant \mathbf{B} \tag{5}$$
$$\mathbf{X} \geqslant \mathbf{0}, \tag{6}$$

where $\mathbf{X} \geqslant \mathbf{0}$ means that each entry of $\mathbf{X}$ is nonnegative and $\mathbf{AX} \leqslant \mathbf{B}$ means that each entry of $\mathbf{AX}$ is less than or equal to the corresponding entry in $\mathbf{B}$.

A vector $\mathbf{X}$ in R^n satisfying (5) and (6) is called a **feasible solution** to the given problem, and a feasible solution maximizing the objective function (4) is called an **optimal solution**.

Example 1 We can write the problem in Example 1 of Section 6.1 in matrix form as follows: Find a vector $\mathbf{X}$ in R^2 that will maximize

$$z = [8 \quad 10]\begin{bmatrix} x \\ y \end{bmatrix}$$

subject to

$$\begin{bmatrix} 2 & 1 \\ 1 & 2 \end{bmatrix} \begin{bmatrix} x \\ y \end{bmatrix} \le \begin{bmatrix} 50 \\ 70 \end{bmatrix}$$

$$\begin{bmatrix} x \\ y \end{bmatrix} \ge \begin{bmatrix} 0 \\ 0 \end{bmatrix}.$$

Feasible solutions include the vectors

$$\begin{bmatrix} 0 \\ 0 \end{bmatrix}, \quad \begin{bmatrix} 0 \\ 35 \end{bmatrix}, \quad \begin{bmatrix} 10 \\ 30 \end{bmatrix}, \quad \begin{bmatrix} 25 \\ 0 \end{bmatrix}, \quad \begin{bmatrix} 5 \\ 10 \end{bmatrix}, \quad \text{and} \quad \begin{bmatrix} 20 \\ 5 \end{bmatrix}.$$

An optimal solution is the vector

$$\begin{bmatrix} 10 \\ 30 \end{bmatrix}.$$

The new problem with slack variables can also be written in matrix form as follows: Find a vector $\mathbf{X}$ that will maximize

$$z = \mathbf{CX} \tag{7}$$

subject to

$$\mathbf{AX} = \mathbf{B} \tag{8}$$
$$\mathbf{X} \ge \mathbf{0}, \tag{9}$$

where now

$$\mathbf{A} = \begin{bmatrix} a_{11} & a_{12} & \cdots & a_{1n} & 1 & 0 & \cdots & 0 \\ a_{21} & a_{22} & \cdots & a_{2n} & 0 & 1 & \cdots & 0 \\ \vdots & \vdots & & \vdots & \vdots & \vdots & & \vdots \\ a_{m1} & a_{m2} & \cdots & a_{mn} & 0 & 0 & \cdots & 1 \end{bmatrix}, \quad \mathbf{B} = \begin{bmatrix} b_1 \\ b_2 \\ \vdots \\ b_m \end{bmatrix},$$

$$\mathbf{X} = \begin{bmatrix} x_1 \\ x_2 \\ \vdots \\ x_n \\ x_{n+1} \\ \vdots \\ x_{n+m} \end{bmatrix}, \quad \text{and} \quad \mathbf{C} = [c_1 \quad c_2 \quad \cdots \quad c_n \quad 0 \quad \cdots \quad 0].$$

A vector $\mathbf{X}$ satisfying (8) and (9) is called a **feasible solution** to the new problem, and a feasible solution maximizing the objective function (7) is called an **optimal solution**. Throughout this chapter, we now make the additional assumption that in all standard linear programming problems

$$b_1 \ge 0, \quad b_2 \ge 0, \ldots, b_m \ge 0.$$

We shall use Example 1 of Section 6.1 as our principal illustrative example in this section.

Illustrative Problem

Find values of x and y that will maximize

$$z = 8x + 10y \tag{10}$$

subject to

$$2x + y \leqslant 50 \tag{11}$$
$$x + 2y \leqslant 70$$
$$x \geqslant 0, \quad y \geqslant 0. \tag{12}$$

The new problem with slack variables u and v is: Find values of x, y, u, and v that will maximize

$$z = 8x + 10y \tag{13}$$

subject to

$$2x + y + u \phantom{{}+ v} = 50 \tag{14}$$
$$x + 2y + v = 70$$
$$x \geqslant 0, \quad y \geqslant 0, \quad u \geqslant 0, \quad v \geqslant 0. \tag{15}$$

Definition   The vector $\mathbf{X}$ in R^{n+m} is called a **basic solution** to the new problem if it is obtained by setting n of the variables in (8) equal to zero and solving for the remaining m variables. The m variables that we solve for are called **basic variables**, and the n variables set equal to zero are called **nonbasic variables**. The vector $\mathbf{X}$ is called a **basic feasible solution** if it is a basic solution that also satisfies (9).

Basic feasible solutions are important because the following theorem can be established.

Theorem 6.2   *If a linear programming problem has an optimal solution, then it has a basic feasible solution that is optimal.*

Thus to solve a linear programming problem we need only search for basic feasible solutions. In our illustrative example we can select two of the four variables x, y, u, and v as nonbasic variables by setting them equal to zero and solving for the remaining two variables; that is, we solve for the basic variables. Thus, if

$$x = y = 0,$$

then

$$u = 50, \qquad v = 70.$$

The vector

$$\mathbf{X}_1 = \begin{bmatrix} 0 \\ 0 \\ 50 \\ 70 \end{bmatrix}$$

is a basic solution, which gives rise to the feasible solution

$$\begin{bmatrix} 0 \\ 0 \end{bmatrix}$$

to the original problem specified by (10), (11), and (12). The variables x and y are nonbasic, and the variables u and v are basic. The convex region of solutions to the original problem has been sketched in Figure 6.8. Thus the vector $\mathbf{X}_1$ corresponds to the extreme point O.

If we let the variables x and u be nonbasic ($x = u = 0$), then $y = 50$ and $v = -30$. The vector

$$\mathbf{X}_2 = \begin{bmatrix} 0 \\ 50 \\ 0 \\ -30 \end{bmatrix}$$

is a basic solution that is not feasible, since v is negative. It corresponds to the point F in Figure 6.8, which is not a feasible solution to the original problem.

In Table 6.3 we have tabulated all the possible choices for basic solutions. The nonbasic variables are shaded and the corresponding point from Figure 6.8 is indicated in the table. It can be seen in this example, and proved in general, that every basic feasible solution determines an extreme point, and conversely, each extreme point determines a basic feasible solution.

One method of solving our linear programming problem would be to obtain all the basic solutions, discard those which are not feasible, and evaluate the objective function at each basic feasible solution, selecting that one, or ones, for which we get a maximum value of the objective function.

Table 6.3

x	y	u	v	Type of Solution	Corresponding Point in Figure 6.8
0	0	50	70	Basic feasible solution	O
0	50	0	−30	Not a basic feasible solution	F
0	35	15	0	Basic feasible solution	A
25	0	0	45	Basic feasible solution	C
70	0	−90	0	Not a basic feasible solution	G
10	30	0	0	Basic feasible solution	B

The number of possible basic solutions is

$$\binom{n+m}{n} = \frac{(n+m)!}{m!n!}.$$

That is, it is the number of ways selecting n objects out of $m+n$ given objects.

The simplex method is a procedure that enables us to go from a given extreme point (basic feasible solution) to an adjacent extreme point in such a way that the value of the objective function increases as we move from extreme point to extreme point until we either obtain an optimal solution or find that the given problem has no finite optimal solution. The simplex method thus consists of two steps: (1) a way of checking whether a given basic feasible solution is an optimal solution, and (2) a way of obtaining another basic feasible solution with a larger value of the objective function. In most practical problems the simplex method does not consider every basic feasible solution; rather, it works with only a small number of these. However, examples have been given where a large number of basic feasible solutions have been examined by the simplex method. We shall now turn to a detailed discussion of this powerful method, using our illustrative example as a guide.

Selecting an Initial Basic Feasible Solution

We can take all the original (nonslack) variables as our nonbasic variables and set them equal to zero. We then solve for the slack variables, our basic variables. In our example, we set

$$x = y = 0$$

and solve for u and v:

$$u = 50, \qquad v = 70.$$

Thus the initial basic feasible solution is the vector

$$\begin{bmatrix} 0 \\ 0 \\ 50 \\ 70 \end{bmatrix},$$

which yields the origin as an extreme point.

It is convenient to develop a tabular method for displaying the given problem and the initial basic feasible solution. First, we write (13) as the equation

$$-8x - 10y + z = 0, \tag{13'}$$

with z being considered as another variable. We now form the **initial tableau** (Tableau 1). The variables x, y, u, v, and z are written in the top row as labels on the corresponding

Tableau 1

	x	y	u	v	z	
u	2	1	1	0	0	50
v	1	2	0	1	0	70
	-8	-10	0	0	1	0

columns. Constraints (14) are entered in the top two rows followed by Equation (13') in the bottom row. The bottom row of the tableau is called the **objective row.** Along the left-hand side of the tableau we indicate which variable is a basic variable in the corresponding equation. Thus u is a basic variable in the first equation and v is a basic variable in the second equation. A basic variable can also be described as a variable, other than z, which is present in exactly one equation, and there it appears with a coefficient of $+1$. In the tableau, the value of the basic variable is explicitly given in the rightmost column. In general, for the problem given by (7), (8), and (9), the initial tableau is Tableau 2.

Tableau 2

	x_1	x_2	$\cdots$	x_n	x_{n+1}	x_{n+2}	$\cdots$	x_{n+m}	z	
x_{n+1}	a_{11}	a_{12}	$\cdots$	a_{1n}	1	0	$\cdots$	0	0	b_1
x_{n+2}	a_{21}	a_{22}	$\cdots$	a_{2n}	0	1	$\cdots$	0	0	b_2
$\vdots$	$\vdots$						$\vdots$		0	$\vdots$
x_{n+m}	a_{m1}	a_{m2}	$\cdots$	a_{mn}	0	0	$\cdots$	1	0	b_m
	$-c_1$	$-c_2$		$-c_n$	0	0	$\cdots$	0	1	0

The initial tableau (Tableau 1) shows the values of the basic variables u and v, and therefore the nonbasic variables have values

$$x = 0, \qquad y = 0.$$

The value of the objective function for this initial basic feasible solution is

$$c_1 x + c_2 y + 0(u) + 0(v) = 8(0) + 10(0) + 0(u) + 0(v) = 0,$$

which is the entry in the objective row and rightmost column. It is clear that this solution is not optimal, for using the bottom row of the initial tableau, we can write

$$z = 0 + 8x + 10y - 0u - 0v. \tag{16}$$

Now the value of z can be increased by increasing either x or y, since both of these variables appear in (16) with positive coefficients. Since (16) contains terms with positive coefficients if and only if the objective row of our initial tableau has negative entries under the columns labeled with variables, we see that we can increase z by increasing any variable with a negative entry in the objective row. Thus we obtain the following optimality criterion for determining whether the feasible solution indicated in a tableau is an optimal solution yielding a maximum value for the objective function z.

Optimality Criterion.   If the objective row of a tableau has no negative entries in the columns labeled with variables, then the indicated solution is optimal and we can stop our computation.

Selecting the Entering Variable

If the objective row of a tableau has negative entries in the columns labeled with variables, then the indicated solution is not optimal and we must continue our search for an optimal solution.

The simplex method moves from one extreme point to an adjacent extreme point in such a way that the value of the objective function increases. This is done by increasing *one* variable at a time. The largest increases in z per unit increase in a variable occurs for the variable with the most negative entry in the objective row. In Tableau 1, the most negative entry in the objective row is -10, and since it occurs in the y-column, this is the variable to be increased. The variable to be increased is called the **entering variable**, because in the next iteration it will become a basic variable, thereby *entering* the set of basic variables. If there are several candidates for entering variables, choose one. An increase in y must be accompanied by a decrease in some of the other variables. This can be seen if we solve the equations in (14) for u and v:

$$u = 50 - 2x - y$$
$$v = 70 - x - 2y.$$

Since we only increase y, we keep $x = 0$, and obtain

$$u = 50 - y$$
$$v = 70 - 2y \tag{17}$$

so that as y increases, both u and v decrease. Equations (17) also show by how much we can increase y. That is, since u and v must be nonnegative, we must have

$$y \leqslant \tfrac{50}{1} = 50$$
$$y \leqslant \tfrac{70}{2} = 35.$$

We now see that the allowable increase in y can be no larger than the smaller of the ratios $\tfrac{50}{1}$ and $\tfrac{70}{2}$. Taking y as 35, we obtain the new basic feasible solution,

$$x = 0, \quad y = 35, \quad u = 15, \quad v = 0.$$

The basic variables are y and u; the variables x and v are nonbasic. The objective function for this solution now has the value

$$z = 8(0) + 10(35) + 0(15) + 0(0) = 350,$$

which is much better than the earlier value of 0. This solution yields the extreme point A in Figure 6.8, which is adjacent to O.

Choosing the Departing Variable

Since the variable $v = 0$, it is not basic and is called the **departing variable**, for it has *departed* from the set of basic variables. The column of the entering variable is called the **pivotal column**; the row that is labeled by the departing variable is called the **pivotal row**.

Let us look more closely at the selection of the departing variable. The choice of this variable was closely related to the determination of how far we could increase the entering variable (y in our example). To obtain this number, we formed the ratios (called θ**-ratios**) of the entries above the objective row in the rightmost column of the tableau by the corresponding entries in the pivotal column. The smallest of these ratios tells how far the entering variable can be increased. The basic variable labeling the row for which this smallest ratio occurs (the pivotal row) is then the departing variable. In our example, the ratios, formed by using the rightmost column and the y-column, are $\frac{50}{1}$ and $\frac{70}{2}$. The smaller of these ratios, 35, occurs for the second row, which means that the second row is the pivotal row and the basic variable, v, labeling it becomes the departing variable and is no longer basic. If the smallest of the ratios is not selected, then one of the variables in the new solution becomes negative and the new solution is no longer feasible (Exercise T.2). What happens if there are entries in the pivotal column that are either zero or negative? If any entry in the pivotal column is negative, then the corresponding ratio is also negative; in this case the equation associated with the negative entry imposes no restriction on how far the entering variable can be increased. Suppose, for example, that the y-column in our initial tableau is

$$\begin{bmatrix} -3 \\ 2 \end{bmatrix} \quad \text{instead of} \quad \begin{bmatrix} 1 \\ 2 \end{bmatrix}.$$

Then instead of (17) we have

$$u = 50 + 3y$$
$$v = 70 - 2y,$$

and since u must be nonnegative, we have

$$y \geqslant -\tfrac{50}{3},$$

which puts no limitation at all on how far y can be increased. If an entry in the pivotal column is zero, then the corresponding ratio cannot be formed (we cannot divide by zero), and again the associated equation puts no limitation on how far the entering variable can be increased. Thus, in forming the ratios, we only use the positive entries above the objective row in the pivotal column.

If all the entries above the objective row in the pivotal column are either zero or negative, then the entering variable can be made as large as we please. This means that the problem has no finite optimal solution.

Obtaining a New Tableau

We must now obtain a new tableau indicating the new basic variables and the new basic feasible solution. Solving the second equation in (14) for y, we obtain

$$y = 35 - \tfrac{1}{2}x - \tfrac{1}{2}v, \tag{18}$$

and substituting this expression for y in the first equation in (14), we have

$$\tfrac{3}{2}x + u - \tfrac{1}{2}v = 15. \tag{19}$$

The second equation in (14) can be written, upon dividing by 2 (the coefficient of y), as

$$\tfrac{1}{2}x + y + \tfrac{1}{2}v = 35. \tag{20}$$

Substituting (18) for y in (13′), we have

$$-3x + 5v + z = 350. \tag{21}$$

Since $x = 0$, $v = 0$, we obtain the value of z for the current basic feasible solution as

$$z = 350.$$

Equations (19), (20), and (21) yield our new tableau (Tableau 3). Observe in the tableau that we have labeled the basic variables in each row.

Tableau 3

	x	y	u	v	z	
u	$\tfrac{3}{2}$	0	1	$-\tfrac{1}{2}$	0	15
y	$\tfrac{1}{2}$	1	0	$\tfrac{1}{2}$	0	35
	-3	0	0	5	1	350

Comparing Tableau 1 with Tableau 3, we observe that we can transform the former to the latter by elementary row operations as follows.

STEP 1. Locate and circle the entry in the pivotal row and pivotal column. This entry is called the **pivot**. Mark the pivotal column by placing an arrow ↓ above the entering variable and mark the pivotal row by placing an arrow ← to the left of the departing variable.

STEP 2. If the pivot is k, multiply the pivotal row by $1/k$, making the entry that was the pivot a 1.

STEP 3. Add the appropriate multiples of the pivotal row to all other rows (including the objective row) so that all elements in the pivotal column except for the 1 where the pivot was located become zero.

STEP 4. In the new tableau replace the label on the pivotal row by the entering variable.

These four steps form a process called **pivotal elimination**. It is one of the iterations of the procedure described in Section 1.4 for transforming a matrix to reduced row echelon form.

We now repeat Tableau 1, with the arrows placed next to the entering and departing variables and with the pivot circled.

Tableau 1

	x	y	u	v	z	
u	2	1	1	0	0	50
← v	1	②	0	1	0	70
	−8	−10	0	0	1	0

Performing pivotal elimination on Tableau 1 yields Tableau 3. We now repeat the entire procedure, using Tableau 3 as our initial tableau. Since the most negative entry in the objective row of Tableau 3, -3, occurs in the first column, x is the entering variable and the first column is the pivotal column. To determine the departing variable, we form the ratios of the entries in the rightmost column (except for the objective row) by the corresponding entries of the pivotal column for those entries in the pivotal column which are positive and select the smallest of these ratios. Since both entries in the pivotal column are positive, the ratios are $15/\frac{3}{2}$ and $35/\frac{1}{2}$. The smaller of these, $15/\frac{3}{2} = 10$, occurs for the first row, so the departing variable is u and the pivotal row is the first row. Tableau 3, with the pivotal column, pivotal row, and circled pivot, is shown again here.

Tableau 3

	x	y	u	v	z	
← u	③⁄₂	0	1	$-\frac{1}{2}$	0	15
y	$\frac{1}{2}$	1	0	$\frac{1}{2}$	0	35
	−3	0	0	5	1	350

Performing pivotal elimination on Tableau 3 yields Tableau 4.

Tableau 4

	x	y	u	v	z	
x	1	0	$\frac{2}{3}$	$-\frac{1}{3}$	0	10
y	0	1	$-\frac{1}{3}$	$\frac{2}{3}$	0	30
	0	0	2	4	1	380

Since the objective row of Tableau 4 has no negative entries in the columns labeled with variables, we conclude, by the optimality criterion, that we are finished and that the indicated solution is optimal. Thus the optimal solution is

$$x = 10, \quad y = 30, \quad u = 0, \quad v = 0,$$

which corresponds to the extreme point $B(10, 30)$. Thus the simplex method started from the extreme point $O(0, 0)$ moved to the adjacent extreme point $A(0, 35)$ and then to the extreme point $B(10, 30)$, which is adjacent to A. The value of the objective function increased from 0 to 350 to 380, respectively.

We now summarize the simplex method.

The procedure for carrying out the simplex method is as follows.

STEP 1. Set up the initial tableau.

STEP 2. Apply the optimality test. If the objective row has no negative entries in the columns labeled with variables, then the indicated solution is optimal; we stop our computations.

STEP 3. Choose a pivotal column by determining the column with the most negative entry in the objective row. If there are several candidates for a pivotal column, choose any one.

STEP 4. Choose a pivotal row. Form the ratios of the entries above the objective row in the rightmost column by the corresponding entries of the pivotal column for those entries in the pivotal column which are positive. The pivotal row is the row for which the smallest of these ratios occurs. If there is a tie, so that the smallest ratio occurs at more than one row, choose any one of the qualifying rows. If none of the entries in the pivotal column above the objective row is positive, the problem has no finite optimum. We stop our computation.

STEP 5. Perform pivotal elimination to construct a new tableau and return to Step 2.

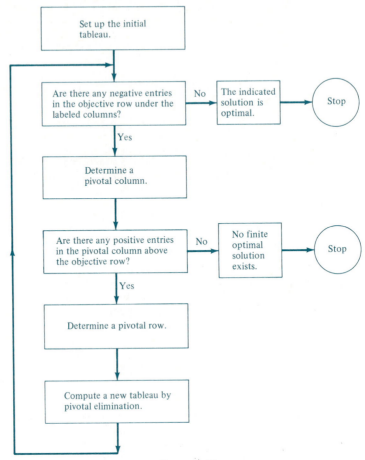

Figure 6.13

Flow chart for the simplex method.

Figure 6.13 gives a flow chart for the simplex method.

We have restricted *our* discussion of the simplex method to standard linear programming problems in which all the right-hand entries are nonnegative. It should be noted that the method applies to the general linear programming problem. For additional details we refer the reader to the references at the end of this chapter.

Example 2 Maximize

$$z = 4x_1 + 8x_2 + 5x_3$$

subject to

$$x_1 + 2x_2 + 3x_3 \leq 18$$
$$x_1 + 4x_2 + x_3 \leq 6$$
$$2x_1 + 6x_2 + 4x_3 \leq 15$$
$$x_1 \geq 0, \quad x_2 \geq 0, \quad x_3 \geq 0.$$

The new problem with slack variables is

$$\text{Maximize} \qquad z = 4x_1 + 8x_2 + 5x_3$$

subject to

$$
\begin{aligned}
x_1 + 2x_2 + 3x_3 + x_4 \qquad\qquad\qquad &= 18 \\
x_1 + 4x_2 + x_3 \qquad + x_5 \qquad\qquad &= 6 \\
2x_1 + 6x_2 + 4x_3 \qquad\qquad\qquad + x_6 &= 15 \\
x_1 \geqslant 0, \quad x_2 \geqslant 0, \quad x_3 \geqslant 0, \quad x_4 \geqslant 0, \quad x_5 \geqslant 0, \quad x_6 \geqslant 0.
\end{aligned}
$$

The initial tableau and the succeeding tableaus are

	x_1	x_2	x_3	x_4	x_5	x_6	z	
x_4	1	2	3	1	0	0	0	18
$\leftarrow x_5$	1	④	1	0	1	0	0	6
x_6	2	6	4	0	0	1	0	15
	-4	-8	-5	0	0	0	1	0

	x_1	x_2	x_3	x_4	x_5	x_6	z	
x_4	$\frac{1}{2}$	0	$\frac{5}{2}$	1	$-\frac{1}{2}$	0	0	15
x_2	$\frac{1}{4}$	1	$\frac{1}{4}$	0	$\frac{1}{4}$	0	0	$\frac{3}{2}$
$\leftarrow x_6$	$\frac{1}{2}$	0	$⑤\!\!/②$	0	$-\frac{3}{2}$	1	0	6
	-2	0	-3	0	2	0	1	12

	x_1	x_2	x_3	x_4	x_5	x_6	z	
x_4	0	0	0	1	1	-1	0	9
$\leftarrow x_2$	①	1	0	0	$\frac{2}{5}$	$-\frac{1}{10}$	0	$\frac{9}{10}$
x_3	$\frac{1}{5}$	0	1	0	$-\frac{3}{5}$	$\frac{2}{5}$	0	$\frac{12}{5}$
	$-\frac{7}{5}$	0	0	0	$\frac{1}{5}$	$\frac{6}{5}$	1	$\frac{96}{5}$

	x_1	x_2	x_3	x_4	x_5	x_6	z	
x_4	0	0	0	1	1	-1	0	9
x_1	1	5	0	0	2	$-\frac{1}{2}$	0	$\frac{9}{2}$
x_3	0	-1	1	0	-1	$\frac{1}{2}$	0	$\frac{3}{2}$
	0	7	0	0	3	$\frac{1}{2}$	1	$\frac{51}{2}$

Hence an optimal solution is

$$x_1 = \tfrac{9}{2}, \qquad x_2 = 0, \quad x_3 = \tfrac{3}{2}.$$

The slack variables are

$$x_4 = 9, \qquad x_5 = 0, \qquad x_6 = 0,$$

and the optimal value of z is $\frac{51}{2}$.

A number of difficulties can arise in using the simplex method. We shall briefly describe one of these, and we again refer the reader to the references at the end of this chapter for an extended discussion of computational considerations.

Degeneracy

Suppose that a basic variable becomes zero in one of the tableaus in the simplex method. Then one of the ratios used to determine the next pivotal row may be zero, in which case the pivotal row is the one labeled with the basic variable that is zero (in this case, zero is the smallest of the ratios). Recall that the entering variable is increased from zero to the smallest ratio. Since the smallest ratio is zero, the entering variable remains at the value zero. In the new tableau all the old basic variables have the same values that they had in the old tableau; the value of the objective function has not been increased. The new tableau looks like the old one, except that a basic variable with value zero has become nonbasic, and its place has been taken by a nonbasic variable also entering with value zero. A basic feasible solution in which one or more of the basic variables are zero is called a **degenerate basic feasible solution**. It can be shown that when a tie occurs for the smallest ratio in determining the pivotal row, a degenerate solution will arise and there are several optimal solutions. When no degenerate solution occurs, the value of the objective function improves as we move from one basic feasible solution to another. The procedure stops after a finite number of steps, since the number of basic feasible solutions is finite. However, if we have a degenerate basic feasible solution, we may return to a basic feasible solution that had already been found at an earlier iteration and thus enter an infinite cycle. Fortunately, cycling is encountered only occasionally in practical linear programming problems and several examples have been carefully constructed to show that cycling can occur. In actual practice, when degeneracy occurs, it is handled by merely ignoring it. The value of the objective function may remain constant for a few iterations and will then start to increase. Moreover, a technique called **perturbation** has been developed for handling degeneracy. This technique calls for making slight changes in the rightmost column of the tableau so that the troublesome ties in the ratios will no longer occur.

There are many computer programs implementing the simplex method and other algorithms in the area of mathematical programming (which includes integer programming and nonlinear programming). Some of these programs can do extensive data manipulation to prepare the input for the problem, solve the problem, and then prepare elaborate reports that can be used by management in decision making. Moreover, some programs can handle large problems having as many as 8200 inequalities and 100,000 unknowns.

In 1984 N. Karmarkar developed a new method for solving linear programming problems that is claimed to be much faster than the simplex method for very large problems. Much of the current work in linear programming is concentrating on a better understanding of Karmarkar's method and of its applications to practical problems. A description of the method may be found in Strang's book mentioned in Further Reading at the end of this chapter.

6.2 EXERCISES

In Exercises 1 through 4, write the initial simplex tableau for each given linear programming problem.

1. Maximize $z = 3x + 7y$
  subject to
$$3x - 2y \leqslant 7$$
$$2x + 5y \leqslant 6$$
$$2x + 3y \leqslant 8$$
$$x \geqslant 0, \quad y \geqslant 0.$$

2. Maximize $z = 2x_1 + 3x_2 - 4x_3$
  subject to
$$3x_1 - 2x_2 + x_3 \leqslant 4$$
$$2x_1 + 4x_2 + 5x_3 \leqslant 6$$
$$x_1 \geqslant 0, \quad x_2 \geqslant 0, \quad x_3 \geqslant 0.$$

3. Maximize $z = 2x_1 + 2x_2 + 3x_3 + x_4$
  subject to
$$3x_1 - 2x_2 + x_3 + x_4 \leqslant 6$$
$$x_1 + x_2 + x_3 + x_4 \leqslant 8$$
$$2x_1 - 3x_2 - x_3 + 2x_4 \leqslant 10$$
$$x_1 \geqslant 0, \quad x_2 \geqslant 0, \quad x_3 \geqslant 0, \quad x_4 \geqslant 0.$$

4. Maximize $z = 2x_1 - 3x_2 + x_3$
  subject to
$$x_1 - 2x_2 + 4x_3 \leqslant 5$$
$$2x_1 + 2x_2 + 4x_3 \leqslant 5$$
$$3x_1 + x_2 - x_3 \leqslant 7$$
$$x_1 \geqslant 0, \quad x_2 \geqslant 0, \quad x_3 \geqslant 0.$$

In Exercises 5 through 11, solve each given linear programming by the simplex method. Some of these problems may have no finite optimal solution.

5. Maximize $z = 2x + 3y$
  subject to
$$3x + 5y \leqslant 6$$
$$2x + 3y \leqslant 7$$
$$x \geqslant 0, \quad y \geqslant 0.$$

6. Maximize $z = 2x + 5y$
  subject to
$$3x + 7y \leqslant 6$$
$$2x + 6y \leqslant 7$$
$$3x + 2y \leqslant 5$$
$$x \geqslant 0, \quad y \geqslant 0.$$

7. Maximize $z = 2x + 5y$
  subject to
$$2x - 3y \leqslant 4$$
$$x - 2y \leqslant 6$$
$$x \geqslant 0, \quad y \geqslant 0.$$

8. Maximize $z = 3x_1 + 2x_2 + 4x_3$
  subject to
$$x_1 - x_2 - x_3 \leqslant 6$$
$$-2x_1 + x_2 - 2x_3 \leqslant 7$$
$$3x_1 + x_2 - 4x_3 \leqslant 8$$
$$x_1 \geqslant 0, \quad x_2 \geqslant 0, \quad x_3 \geqslant 0.$$

9. Maximize $z = 2x_1 - 4x_2 + 5x_3$
subject to

$$3x_1 + 2x_2 + x_3 \leqslant 6$$
$$3x_1 - 6x_2 + 7x_3 \leqslant 9$$
$$x_1 \geqslant 0, \quad x_2 \geqslant 0, \quad x_3 \geqslant 0.$$

10. Maximize $z = 2x_1 + 4x_2 - 3x_3$
subject to

$$5x_1 + 2x_2 + x_3 \leqslant 5$$
$$3x_1 - 2x_2 + 3x_3 \leqslant 10$$
$$4x_1 + 5x_2 - x_3 \leqslant 20$$
$$x_1 \geqslant 0, \quad x_2 \geqslant 0, \quad x_3 \geqslant 0.$$

11. Maximize $z = x_1 + 2x_2 - x_3 + 5x_4$
subject to

$$2x_1 + 3x_2 + x_3 - x_4 \leqslant 8$$
$$3x_1 + x_2 - 4x_3 + 5x_4 \leqslant 9$$
$$x_1 \geqslant 0, \quad x_2 \geqslant 0, \quad x_3 \geqslant 0, \quad x_4 \geqslant 0.$$

12. Solve Exercise 1 in Section 6.1 by the simplex
method.

13. Solve Exercise 4 in Section 6.1 by the simplex
method.

14. Solve Exercise 7 in Section 6.1 by the simplex
method.

15. A power plant burns coal, oil, and gas to generate
electricity. Suppose that each ton of coal generates
600 kilowatt hours, emits 20 units of sulfur dioxide
and 15 units of particulate matter, and cost $200;
each ton of oil generates 550 kilowatt hours, emits
18 units of sulfur dioxide and 12 units of particulate
matter, and costs $220; each ton of gas generates
500 kilowatt hours, emits 15 units of sulfur dioxide
and 10 units of particulate matter, and costs $250.
The environmental protection agency restricts the
daily emission of sulfur dioxide to no more than 60
units and no more than 75 units of particulate matter. If the power plant wants to spend no more than
$2000 per day on fuel, how much fuel of each type
should be bought to maximize the amount of energy
generated?

THEORETICAL EXERCISES

T.1. Consider the standard linear programming problem: Maximize $z = \mathbf{CX}$
subject to

$$\mathbf{AX} \leqslant \mathbf{B}$$
$$\mathbf{X} \geqslant \mathbf{0},$$

where $\mathbf{A}$ is an $m \times n$ matrix. Show that the feasi-

ble region (the set of all feasible solutions) of this
problem is a convex set.
[*Hint:* If $\mathbf{X}$ and $\mathbf{Y}$ are in R^n, the line segment joining them is the set of points $r\mathbf{X} + (1 - r)\,\mathbf{Y}$,
where $0 \leqslant r \leqslant 1$.]

T.2. Suppose that in selecting the departing variable,
the minimum θ-ratio is not chosen. Show that the
resulting solution is not feasible.

_____6.3_____

DUALITY

In this section we show how to associate a minimization problem with each linear programming problem in standard form. There are some interesting interpretations of the
associated problem which we also discuss.

Consider the following pair of linear programming problems:

$$\text{Maximize} \quad z = \mathbf{CX}$$

subject to

$$\mathbf{AX} \leqslant \mathbf{B}$$
$$\mathbf{X} \geqslant \mathbf{0} \tag{1}$$

and

$$\text{Minimize} \quad z' = \mathbf{B}^T\mathbf{Y}$$

subject to

$$\mathbf{A}^T\mathbf{Y} \geqslant \mathbf{C}^T$$
$$\mathbf{Y} \geqslant \mathbf{0}, \tag{2}$$

where $\mathbf{A}$ is $m \times n$, $\mathbf{B}$ is $m \times 1$, $\mathbf{C}$ is $1 \times n$, $\mathbf{X}$ is $n \times 1$, and $\mathbf{Y}$ is $m \times 1$.

These problems are called **dual problems**. The problem given by (1) is called the **primal problem**; the problem given by (2) is called the **dual problem**.

Example 1 If the primal problem is

$$\text{Maximize} \quad z = [3 \quad 4]\begin{bmatrix} x_1 \\ x_2 \end{bmatrix}$$

subject to

$$\begin{bmatrix} 2 & 3 \\ 3 & -1 \\ 5 & 4 \end{bmatrix}\begin{bmatrix} x_1 \\ x_2 \end{bmatrix} \leqslant \begin{bmatrix} 3 \\ 4 \\ 2 \end{bmatrix}$$

$$x_1 \geqslant 0, \quad x_2 \geqslant 0,$$

then the dual problem is

$$\text{Minimize} \quad z' = [3 \quad 4 \quad 2]\begin{bmatrix} y_1 \\ y_2 \\ y_3 \end{bmatrix}$$

subject to

$$\begin{bmatrix} 2 & 3 & 5 \\ 3 & -1 & 4 \end{bmatrix}\begin{bmatrix} y_1 \\ y_2 \\ y_3 \end{bmatrix} \geqslant \begin{bmatrix} 3 \\ 4 \end{bmatrix}$$

$$y_1 \geqslant 0, \quad y_2 \geqslant 0, \quad y_3 \geqslant 0.$$

Observe that in formulating the dual problem, the coefficients of the ith constraint of the primal problem become the coefficients of the variable y_i in the constraints of the dual problem. Conversely, the coefficients of x_j in the primal problem become the coefficients of the jth constraint in the dual problem. Moreover, the coefficients of the objective function of the primal problem become the right-hand sides of the constraints of the dual problem, and conversely.

Since problem (2) can be rewritten as a standard linear programming problem, we can ask for the dual of the dual problem. The answer is provided by the following theorem.

Theorem 6.3 *Given a primal problem, as in (1), the dual of its dual problem is the primal problem.*

Proof

Consider the dual problem (2), which we rewrite in standard form as

$$\text{Maximize} \quad z' = -\mathbf{B}^T\mathbf{Y}$$

subject to

$$-\mathbf{A}^T\mathbf{Y} \leqslant -\mathbf{C}^T \tag{3}$$
$$\mathbf{Y} \geqslant \mathbf{0}.$$

Now the dual of (3) is

$$\text{Minimize} \quad z'' = (-\mathbf{C}^T)^T\mathbf{W}$$

subject to

$$(-\mathbf{A}^T)^T\mathbf{W} \geqslant (-\mathbf{B}^T)^T$$
$$\mathbf{W} \geqslant \mathbf{0}$$

or

$$\text{Maximize} \quad z'' = \mathbf{CW}$$

subject to

$$\mathbf{AW} \leqslant \mathbf{B} \tag{4}$$
$$\mathbf{W} \geqslant \mathbf{0}.$$

Writing $\mathbf{W} = \mathbf{X}$, we see that problem (4) is the primal problem.

Example 2 Find the dual problem of the linear programming problem

$$\text{Minimize} \quad z' = [2 \quad 3]\begin{bmatrix} y_1 \\ y_2 \end{bmatrix}$$

subject to

$$\begin{bmatrix} 3 & 4 \\ 1 & 2 \\ 5 & 3 \end{bmatrix}\begin{bmatrix} y_1 \\ y_2 \end{bmatrix} \geqslant \begin{bmatrix} 5 \\ 2 \\ 7 \end{bmatrix}$$
$$y_1 \geqslant 0, \quad y_2 \geqslant 0.$$

Solution

The given problem is the dual of the primal problem that we seek to formulate. This primal problem is obtained by taking the dual of the given problem, obtaining

$$\text{Maximize} \quad z = 5x_1 + 2x_2 + 7x_3$$

subject to

$$3x_1 + x_2 + 5x_3 \leqslant 2$$
$$4x_1 + 2x_2 + 3x_3 \leqslant 3$$
$$x_1 \geqslant 0, \quad x_2 \geqslant 0, \quad x_3 \geqslant 0.$$

The following theorem, whose proof we omit, gives the relations between the optimal solutions to the primal and dual problems.

Theorem 6.4 *If either the primal problem or dual problem has an optimal solution with finite objective*
(Duality *value, then the other problem also has an optimal solution. Moreover, the objective*
Theorem) *values of the two problems are equal.*

It can also be shown that when the primal problem is solved by the simplex method, the final tableau contains the optimal solution to the dual problem in the objective row under the columns of the slack variables. That is, y_1, the first dual variable, is found in the objective row under the first slack variable, y_2 is found under the second slack variable, and so on. We can use these facts to solve the diet problem, Example 3 in Section 6.1, as follows.

Example 3 Consider the diet problem of Example 3, Section 6.1 (using z' instead of z and y_1, y_2 instead of x, y):

$$\text{Minimize:} \quad z' = 30y_1 + 40y_2$$

subject to

$$2y_1 + y_2 \geqslant 12$$
$$y_1 + y_2 \geqslant 9$$
$$y_1 + 3y_2 \geqslant 15 \tag{5}$$
$$y_1 \geqslant 0, \quad y_2 \geqslant 0.$$

The dual of this problem is

$$\text{Maximize} \quad z = 12x_1 + 9x_2 + 15x_3$$

subject to

$$2x_1 + x_2 + x_3 \leqslant 30$$
$$x_1 + x_2 + 3x_3 \leqslant 40 \tag{6}$$
$$x_1 \geqslant 0, \quad x_2 \geqslant 0, \quad x_3 \geqslant 0.$$

This is a standard linear programming problem. Introducing the slack variables x_4 and x_5, we obtain

$$\text{Maximize} \quad z = 12x_1 + 9x_2 + 15x_3$$

subject to

$$2x_1 + x_2 + x_3 + x_4 \qquad = 30$$
$$x_1 + x_2 + 3x_3 \qquad + x_5 = 40$$
$$x_1 \geqslant 0, \quad x_2 \geqslant 0, \quad x_3 \geqslant 0, \quad x_4 \geqslant 0, \quad x_5 \geqslant 0.$$

We now apply the simplex method, obtaining the following tableaux.

	x_1	x_2	x_3	x_4	x_5	z	
x_4	2	1	1	1	0	0	30
$\leftarrow x_5$	1	1	③	0	1	0	40
	-12	-9	-15	0	0	1	0

	x_1	x_2	x_3	x_4	x_5	z	
$\leftarrow x_4$	⑤⁄₃	$\frac{2}{3}$	0	1	$-\frac{1}{3}$	0	$\frac{50}{3}$
x_3	$\frac{1}{3}$	$\frac{1}{3}$	1	0	$\frac{1}{3}$	0	$\frac{40}{3}$
	-7	-4	0	0	5	1	200

	x_1	x_2	x_3	x_4	x_5	z	
$\leftarrow x_1$	1	②⁄₅	0	$\frac{3}{5}$	$-\frac{1}{5}$	0	10
x_3	0	$\frac{1}{5}$	1	$-\frac{1}{5}$	$\frac{2}{5}$	0	10
	0	$-\frac{6}{5}$	0	$\frac{21}{5}$	$\frac{18}{5}$	1	270

	x_1	x_2	x_3	x_4	x_5	z	
x_2	$\frac{5}{2}$	1	0	$\frac{3}{2}$	$-\frac{1}{2}$	0	25
x_3	$-\frac{1}{2}$	0	1	$-\frac{1}{2}$	$\frac{1}{2}$	0	5
	3	0	0	6	3	1	300

The optimal solution to problem (6) is

$$x_1 = 0, \qquad x_2 = 25, \qquad x_3 = 5,$$

and the value of z is 300.

The optimal solution to the given problem (5), the dual of (6), is found in the objective row under the x_4 and x_5 columns:

$$y_1 = 6, \qquad y_2 = 3.$$

The value of $z' = 30(6) + 40(3) = 300$, as we expected from Theorem 6.4 (the Duality Theorem).

Economic Interpretation of the Dual Problem

We now give two economic interpretations of the dual linear programming problem. The primal problem (1) can be interpreted as follows. We have m resources, inputs or raw materials, and n activities, each of which produces a product. Let

$$b_i = \text{amount of resource } i \text{ that is available } (1 \leq i \leq m);$$
$$a_{ij} = \text{amount of resource } i \text{ that is used by 1 unit of activity } j;$$
$$c_j = \text{contribution to the total profit } z \text{ from 1 unit of activity } j$$
$$(1 \leq j \leq n);$$
$$x_j = \text{level of activity } j.$$

We illustrate these interpretations with Example 1 of Section 6.1, the photo shop problem, which we restate for the convenience of the reader (we use x_1 and x_2 instead of x and y as in Section 6.1).

$$\text{Maximize} \qquad z = 8x_1 + 10x_2$$

subject to

$$2x_1 + x_2 \leq 50$$
$$x_1 + 2x_2 \leq 70$$
$$x_1 \geq 0, \quad x_2 \geq 0.$$

Here $b_1 = 50$ and $b_2 = 70$ are the amounts of solutions A and B, respectively, available each day; activity 1 is the preparation of 1 quart of Fine developer, activity 2 is the preparation of 1 quart of Extra Fine developer; $a_{12} = 1$ is the amount of solution A used in making one quart of Extra Fine and so on; $c_1 = 8$ cents is the profit on 1 quart of Fine, $c_2 = 10$ cents is the profit on one quart of Extra Fine; x_1 and x_2 are the number of quarts of Fine and Extra Fine, respectively, to be made. The dual of the photo shop problem is

$$\text{Minimize} \qquad z' = 50y_1 + 70y_2$$

subject to

$$2y_1 + y_2 \geq 8$$
$$y_1 + 2y_2 \geq 10$$
$$y_1 \geq 0, \quad y_2 \geq 0.$$

Consider the jth constraint of the dual general problem (2):

$$a_{1j}y_1 + a_{2j}y_2 + \cdots + a_{mj}y_m \geq c_j. \tag{7}$$

Since a_{ij} represents amount of resource i per unit of output j, and c_j is the value per unit of output j, we now show from Equation (7) that the dual variable y_i represents "value per unit of resource i" as follows: The dimensions of c_j/a_{ij} are

$$\frac{\text{value/\cancel{unit of output } j}}{\text{amount of resource } i/\cancel{\text{unit of output } j}} = \frac{\text{value}}{\text{amount of resource } i}.$$

Thus the dual variable y_i acts as a price or cost of one unit of resource i. The dual variables are called **shadow prices, fictitious prices,** or **accounting prices.**

First Interpretation. From (7) we see that the left side of this equation is the contribution to the total profit z of the resources that are used in making 1 unit of the jth product. Since the profit of 1 unit of the jth product is c_j, Equation (7) merely says that the contribution of the resources to the total profit has to be at least c_j, for otherwise, we should be using the resources in some better way. In our example, the optimal solution to the dual problem obtained from Tableau 4 in Section 6.2 is

$$y_1 = 2, \qquad y_2 = 4.$$

This means that a shadow price of 2 cents has been attached to each ounce of solution A, and a shadow price of 4 cents has been attached to each unit of solution B. Thus the contribution of the resources used in making 1 quart of Fine to the profit is (in cents)

$$2y_1 + y_2 = 2(2) + 4 = 8,$$

so that the first constraint of the dual has been satisfied. Of course, the constraints

$$y_1 \geqslant 0, \qquad y_2 \geqslant 0$$

in the dual merely state that the contribution of each resource to the total profit is nonnegative; if some y_1 were negative, we would be better off not using this resource. Finally, the objective function

$$z' = b_1 y_1 + b_2 y_2 + \cdots + b_m y_m$$

is to be minimized; this can be interpreted to be minimizing the total shadow value of the resources that are being used to make the products.

Second Interpretation. If $\mathbf{X}_0$ and $\mathbf{Y}_0$ are optimal solutions to the primal and dual problems, respectively, then the maximum profit z_0 satisfies the equation

$$z_0 = \mathbf{B}^T \mathbf{Y}_0 = b_1 y_1^0 + b_2 y_2^0 + \cdots + b_m y_m^0, \tag{8}$$

where

$$\mathbf{Y}_0 = \begin{bmatrix} y_1^0 \\ \vdots \\ y_m^0 \end{bmatrix}.$$

From Equation (8) we see that the manufacturer can increase the profit by increasing the available amount of at least one of the resources. If b_i is increased by 1 unit, the profit will increase by y_i. Thus y_i represents the **marginal value** of the ith resource. Similarly, y_i represents the loss incurred if 1 unit of the ith resource is not used. Thus it can be considered as a replacement value of the ith resource for insurance purposes. Then the objective function of the dual problem

$$z' = b_1 y_1 + b_2 y_2 + \cdots + b_m y_m$$

is the total replacement value. Thus an insurance company would want to minimize this objective function, since it would want to pay out as little as possible to settle a claim.

APPLICATIONS.   In the Theory of Games, Section 7.4, the strategies for two competing players are obtained by solving two dual linear programming problems.

FURTHER READING

Garvin, W. W. *Introduction to Linear Programming*. New York: McGraw-Hill Book Company, 1960.

Gass, Saul I. *Linear Programming*. 4th ed. New York: McGraw-Hill Book Company, 1975.

Kolman, Bernard, and Robert E. Beck. *Elementary Linear Programming with Applications*. New York: Academic Press, Inc., 1980.

Kuester, James L., and Joe H. Mize. *Optimization Techniques with FORTRAN*. New York: McGraw-Hill Book Company, 1974.

Strang, Gilbert, *Introduction to Applied Mathematics*. Wellesley, Mass.: Wellesley-Cambridge Press, 1985.

Wolfe, Carvel S. *Linear Programming with FORTRAN*. Glenview, Ill.: Scott, Foresman and Company, 1970.

6.3 EXERCISES

In Exercises 1 through 4, state the dual of the given linear programming problem.

1. Maximize $z = 3x_1 + 2x_2$
   subject to

$$4x_1 + 3x_2 \leq 7$$
$$5x_1 - 2x_2 \leq 6$$
$$6x_1 + 8x_2 \leq 9$$
$$x_1 \geq 0, \quad x_2 \geq 0.$$

2. Maximize $z = 10x_1 + 12x_2 + 15x_3$
   subject to

$$x_1 + 3x_2 + 4x_3 \leq 5$$
$$2x_1 + 4x_2 - 5x_3 \leq 6$$
$$x_1 \geq 0, \quad x_2 \geq 0, \quad x_3 \geq 0.$$

3. Minimize $z = 3x_1 + 5x_2$
   subject to

$$2x_1 + 3x_2 \geq 7$$
$$8x_1 - 9x_2 \geq 12$$
$$10x_1 + 15x_2 \geq 18$$
$$x_1 \geq 0, \quad x_2 \geq 0.$$

4. Minimize $z' = 14x_1 + 12x_2 + 18x_3$
   subject to

$$3x_1 + 5x_2 - 4x_3 \geq 9$$
$$5x_1 + 2x_2 + 7x_3 \geq 12$$
$$x_1 \geq 0, \quad x_2 \geq 0, \quad x_3 \geq 0.$$

5. Verify that the dual of the dual of the linear programming problem maximize $z = 3x_1 + 6x_2 + 9x_3$ subject to

$$3x_1 + 2x_2 - 3x_3 \leq 12$$
$$5x_1 + 4x_2 + 7x_3 \leq 18$$
$$x_1 \geq 0, \quad x_2 \geq 0, \quad x_3 \geq 0,$$

is the given problem.

In Exercises 6 through 9, solve the dual of the indicated problem by the method of Example 3.

6. Exercise 5 in Section 6.2.

7. Exercise 6 in Section 6.2.

8. Exercise 9 in Section 6.2.

9. Exercise 10 in Section 6.2.

10. A natural cereal consists of dates, nuts, and raisins. Suppose that each ounce of dates contains 24 units of protein, 6 units of iron, and costs 15 cents; each ounce of nuts contains 2 units of protein, 6 units of iron, and costs 18 cents; and each ounce of raisins contains 4 units of protein, 2 units of iron, and costs 12 cents. If each box of cereal is to contain at least 24 units of protein and at least 36 units of iron, how many ounces of each ingredient should be used to minimize the cost of a box of cereal? (*Hint:* Solve the dual problem.)

KEY IDEAS FOR REVIEW

□ **Linear programming problem.** See page 273.
□ **Theorem 6.1.** Let S be the feasible region of a linear programming problem.
  (a) If S is bounded, then the objective function $z = ax + by$ assumes both a maximum and a minimum value on S, these values occur at extreme points of S.
  (b) If S is unbounded, then there may or may not be a maximum or minimum value on S. If a maximum or minimum value does exist on S, it occurs at an extreme point.

□ **Theorem 6.2.** If a linear programming problem has an optimal solution, then it has a basic feasible solution that is optimal.
□ **Simplex method.** See pages 293–299.
□ **Primal and dual problems.** See page 305.
□ **Theorem 6.4 (Duality Theorem).** If either the primal problem or dual problem has an optimal solution with finite objective value, then the other problem also has an optimal solution. Moreover, the objective values of the two problems are equal.

SUPPLEMENTARY EXERCISES

1. Solve the following linear programming problem geometrically.

$$\text{Maximize} \quad z = 2x + 3y$$
subject to
$$3x + y \leq 6$$
$$x + y \leq 4$$
$$x + 2y \leq 6$$
$$x \geq 0, \quad y \geq 0.$$

2. Solve the following linear programming problem geometrically.

A microprocessor manufacturer makes two types of microprocessors, model A and model B. The size of the work force limits the total daily production to at most 600 microprocessors. On the other hand, the suppliers of components limit production to at most 400 model A units and 500 model B units. If the net profit on each model A unit is $80 and it is $100 on each model B unit, how many microprocessors of each type should the manufacturer produce daily to maximize the profit?

3. Solve the following linear programming problem by the simplex method.

$$\text{Maximize} \quad z = 50x + 100y$$
subject to
$$x + 2y \leq 16$$
$$3x + 2y \leq 24$$
$$2x + 2y \leq 18$$
$$x \geq 0, \quad y \geq 0.$$

4. Determine the dual of the following linear programming problem.

$$\text{Minimize} \quad z = 6x_1 + 5x_2$$

subject to

$$2x_1 + 3x_2 \geqslant 6$$
$$5x_1 + 2x_2 \geqslant 10$$
$$x_1 \geqslant 0, \quad x_2 \geqslant 0.$$

5. Solve the problem in Exercise 4 by solving its dual.

CHAPTER TEST

1. Solve the following linear programming problem geometrically.

A farmer who has a 120-acre farm plants corn and wheat. The expenses are $12 for each acre of corn planted and $24 for each acre of wheat planted. Each acre of corn requires 32 bushels of storage and yields a profit of $40; each acre of wheat requires 8 bushels of storage and yields a profit of $50. If the total amount of storage available is 160 bushels and the farmer has $1200 of capital, how many acres of corn and how many acres of wheat should be planted to maximize profit?

2. Solve the following linear programming problem by the simplex method.

$$\text{Maximize} \quad z = 8x_1 + 9x_2 + 5x_3$$

subject to

$$x_1 + x_2 + 2x_3 \leqslant 2$$
$$2x_1 + 3x_2 + 4x_3 \leqslant 3$$
$$3x_1 + 3x_2 + x_3 \leqslant 4$$
$$x_1 \geqslant 0, \quad x_2 \geqslant 0, \quad x_3 \geqslant 0.$$

3. Determine the dual of the following linear programming problem.

$$\text{Minimize} \quad z = 3x_1 + 4x_2$$

subject to

$$x_1 + 4x_2 \geqslant 8$$
$$2x_1 + 3x_2 \geqslant 12$$
$$2x_1 + x_2 \geqslant 6$$
$$x_1 \geqslant 0, \quad x_2 \geqslant 0.$$

7

Applications

7.1

LINES AND PLANES

Prerequisites. Chapter 2, Determinants. Chapter 3, Vectors and Vector Spaces.

Lines in R^2

Any two distinct points $P_1(x_1, y_1)$ and $P_2(x_2, y_2)$ in R^2 (Figure 7.1) determine a straight line whose equation is

$$ax + by + c = 0, \tag{1}$$

where a, b, and c are real numbers, and a and b are not both zero. Since P_1 and P_2 lie on the line, their coordinates satisfy Equation (1):

$$ax_1 + by_1 + c = 0 \tag{2}$$
$$ax_2 + by_2 + c = 0. \tag{3}$$

We now write (1), (2), and (3) as a linear system in the unknowns a, b, and c, obtaining

$$\begin{aligned} xa + yb + c &= 0 \\ x_1a + y_1b + c &= 0 \\ x_2a + y_2b + c &= 0. \end{aligned} \tag{4}$$

314

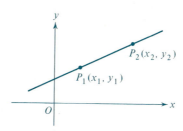

Figure 7.1

We seek a condition on the values x and y that allows (4) to have a nontrivial solution a, b, and c. Since (4) is a homogeneous system, it has a nontrivial solution if and only if the determinant of the coefficient matrix is zero, that is, if and only if

$$\begin{vmatrix} x & y & 1 \\ x_1 & y_1 & 1 \\ x_2 & y_2 & 1 \end{vmatrix} = 0. \tag{5}$$

Thus every point $P(x, y)$ on the line satisfies (5) and, conversely, a point satisfying (5) lies on the line.

Example 1 Find the equation of the line determined by the points $P_1(-1, 3)$ and $P_2(4, 6)$.

Solution
Substituting in (5), we obtain

$$\begin{vmatrix} x & y & 1 \\ -1 & 3 & 1 \\ 4 & 6 & 1 \end{vmatrix} = 0.$$

Expanding this determinant in cofactors about the first row, we have (verify)

$$-3x + 5y - 18 = 0.$$

Lines in R^3

We may recall that in R^2 a line is determined by specifying its slope and one of its points. In R^3 a line is determined by specifying its direction and one of its points. Let $\mathbf{U} = (u, v, w)$ be a nonzero vector in R^3, and let $P_0 = (x_0, y_0, z_0)$ be a point in R^3. Let $\mathbf{X}_0$ be the position vector of P_0. Then the line L through P_0 and parallel to $\mathbf{U}$ consists of the points $P(x, y, z)$ whose position vector $\mathbf{X}$ satisfies (Figure 7.2)

$$\mathbf{X} = \mathbf{X}_0 + t\mathbf{U} \qquad (-\infty < t < \infty), \tag{6}$$

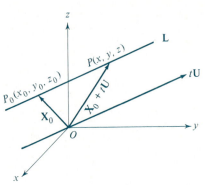

Figure 7.2

Equation (6) is called a **parametric equation** of L, since it contains the parameter t, which can be assigned any real number. Equation (6) can also be written in terms of the components as

$$\begin{aligned} x &= x_0 + tu \\ y &= y_0 + tv \qquad (-\infty < t < \infty) \\ z &= z_0 + tw, \end{aligned} \tag{7}$$

which are called **parametric equations** of L.

Example 2 Parametric equations of the line through the point $P_0(-3, 2, 1)$, which is parallel to the vector $\mathbf{U} = (2, -3, 4)$, are

$$\begin{aligned} x &= -3 + 2t \\ y &= 2 - 3t \qquad (-\infty < t < \infty) \\ z &= 1 + 4t. \end{aligned}$$

Example 3 Find parametric equations of the line L through the points $P_0(2, 3, -4)$ and $P_1(3, -2, 5)$.

Solution
The desired line is parallel to the vector $\mathbf{U} = \overrightarrow{P_0P_1}$. Now

$$\mathbf{U} = (3 - 2, -2 - 3, 5 - (-4)) = (1, -5, 9).$$

Since P_0 lies on the line, we can write the parametric equations of L as

$$\begin{aligned} x &= 2 + t \\ y &= 3 - 5t \qquad (-\infty < t < \infty) \\ z &= -4 + 9t. \end{aligned}$$

In Example 3 we could have used the point P_2 instead of P_1. In fact, we could use any point on the line in the parametric equation of L. Thus a line can be represented in infinitely many ways in parametric form. If u, v, and w are all nonzero in (7), we can solve each equation for t and equate the results to obtain the equations in **symmetric form** of the line through P_0 and parallel to $\mathbf{U}$:

$$\frac{x - x_0}{u} = \frac{y - y_0}{v} = \frac{z - z_0}{w}.$$

The equations in symmetric form of a line are useful in some analytic geometry applications.

Example 4 The equations in symmetric form of the line in Example 3 are

$$\frac{x - 2}{1} = \frac{y - 3}{-5} = \frac{z + 4}{9}.$$

Exercises T.2 and T.3 discuss the intersection of two lines in R^3.

Planes in R^3

A plane in R^3 can be determined by specifying a point in the plane and a vector perpendicular to it. A vector perpendicular to a plane is called a **normal** to the plane.

To obtain an equation of the plane passing through the point $P_0(x_0, y_0, z_0)$ and having the nonzero vector $\mathbf{N} = (a, b, c)$ as a normal, we proceed as follows. A point $P(x, y, z)$ lies in the plane if and only if the vector $\overrightarrow{P_0P}$ is perpendicular to $\mathbf{N}$ (Figure 7.3). Thus $P(x, y, z)$ lies in the plane if and only if

$$\mathbf{N} \cdot \overrightarrow{P_0P} = 0. \tag{8}$$

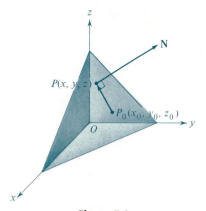

Figure 7.3

Since

$$\overrightarrow{P_0P} = (x - x_0, y - y_0, z - z_0),$$

we can write (8) as

$$a(x - x_0) + b(y - y_0) + c(z - z_0) = 0. \qquad (9)$$

Example 5 Find an equation of the plane passing through the point $(3, 4, -3)$ and perpendicular to the vector $\mathbf{N} = (5, -2, 4)$.

Solution
Substituting in (9), we obtain the equation of the plane as

$$5(x - 3) - 2(y - 4) + 4(z + 3) = 0. \qquad (10)$$

A plane is also determined by three noncollinear points, as we show in the following example.

Example 6 Find an equation of the plane passing through the points $P_1(2, -2, 1)$, $P_2(-1, 0, 3)$, and $P_3(5, -3, 4)$.

Solution
The nonparallel vectors $\overrightarrow{P_1P_2} = (-3, 2, 2)$ and $\overrightarrow{P_1P_3} = (3, -1, 3)$ lie in the plane, since the points P_1, P_2, and P_3 lie in the plane. The vector

$$\mathbf{N} = \overrightarrow{P_1P_2} \times \overrightarrow{P_1P_3} = (8, 15, -3)$$

is then perpendicular to both $\overrightarrow{P_1P_2}$ and $\overrightarrow{P_1P_3}$ and is thus a normal to the plane. Using the vector $\mathbf{N}$ and the point $P_1(2, -2, 1)$ in (9), we obtain an equation of the plane as

$$8(x - 2) + 15(y + 2) - 3(z - 1) = 0.$$

If we multiply out and simplify, (9) can be rewritten as

$$ax + by + cz + d = 0. \qquad (11)$$

Example 7 The equation for the plane in Example 6 can be rewritten in the form given by Equation (11) as

$$8x + 15y - 3z + 17 = 0. \qquad (12)$$

It is not difficult to show (Exercise T.1) that the graph of an equation of the form given in (11), where a, b, c, and d are constants, is a plane with normal $\mathbf{N} = (a, b, c)$ provided that a, b, and c are not all zero.

Example 8 A second solution to Example 6 is as follows. Let the equation of the desired plane be

$$ax + by + cz + d = 0, \qquad (13)$$

where a, b, c, and d are to be determined. Since P_1, P_2, and P_3 lie in the plane, their coordinates satisfy (13). Thus we obtain the linear system

$$2a - 2b + c + d = 0$$
$$-a + 3c + d = 0$$
$$5a - 3b + 4c + d = 0.$$

Solving this system, we have

$$a = \tfrac{8}{17}r, \qquad b = \tfrac{15}{17}r, \qquad c = -\tfrac{3}{17}r, \qquad d = r,$$

where r is any real number. Letting $r = 17$, we obtain

$$a = 8, \qquad b = 15, \qquad c = -3, \qquad d = 17,$$

which yields (12) as in the first solution.

Example 9   A third solution to Example 6 is as follows. Proceeding as in the case of a line in R^2 determined by two distinct points P_1 and P_2, it is not difficult to show (Exercise T.7) that an equation of the plane through the noncollinear points $P_1(x_1, y_1, z_1)$, $P_2(x_2, y_2, z_2)$, and $P_3(x_3, y_3, z_3)$ is

$$\begin{vmatrix} x & y & z & 1 \\ x_1 & y_1 & z_1 & 1 \\ x_2 & y_2 & z_2 & 1 \\ x_3 & y_3 & z_3 & 1 \end{vmatrix} = 0.$$

In our example, the equation of the desired plane is

$$\begin{vmatrix} x & y & z & 1 \\ 2 & -2 & 1 & 1 \\ -1 & 0 & 3 & 1 \\ 5 & -3 & 4 & 1 \end{vmatrix} = 0.$$

Expanding this determinant in cofactors about the first row, we obtain Equation (12).

The equations of a line in symmetric form can be used to determine two planes whose intersection is the given line.

Example 10   Find two planes whose intersection is the line

$$x = -2 + 3t$$
$$y = 3 - 2t \qquad (-\infty < t < \infty)$$
$$z = 5 + 4t.$$

Solution
First, find equations of the line in symmetric form as

$$\frac{x + 2}{3} = \frac{y - 3}{-2} = \frac{z - 5}{4}.$$

The given line is then the intersection of the planes

$$\frac{x + 2}{3} = \frac{y - 3}{-2} \quad \text{and} \quad \frac{x + 2}{3} = \frac{z - 5}{4}.$$

Thus the given line is the intersection of the planes

$$2x + 3y - 5 = 0 \quad \text{and} \quad 4x - 3z + 23 = 0.$$

Two planes are either parallel or they intersect in a straight line. They are parallel if their normals are parallel. In the following example we determine the line of intersection of two planes.

Example 11 Find parametric equations of the line of intersection of the planes

$$\pi_1 : 2x + 3y - 2z + 4 = 0 \quad \text{and} \quad \pi_2 : x - y + 2z + 3 = 0.$$

Solution
Solving the linear system consisting of the equations of π_1 and π_2, we obtain (verify)

$$
\begin{aligned}
x &= -\tfrac{13}{5} - \tfrac{4}{5}t \\
y &= \tfrac{2}{5} + \tfrac{6}{5}t \qquad (-\infty < t < \infty) \\
z &= \quad 0 + t
\end{aligned}
$$

as parametric equations of the line L of intersection of the planes (see Figure 7.4).

Three planes in R^3 may intersect in a plane, in a line, in a unique point, or may not intersect at all. These possibilities can be detected by solving the linear system consisting of their equations.

Consider now the subspaces of R^2. First, we have $\{0\}$ and R^2, the trivial subspaces of dimensions 0 and 2, respectively. Now the subspace V of R^2 spanned by a nonzero vector $\mathbf{X}$ is a one-dimensional subspace of R^2, and V is a line through the origin. We have thus classified all the subspaces of $R^2 : R^2$, the origin, and all the lines through the origin. In Exercise T.5 the reader is asked to carry out a similar discussion for the classification of all the subspaces of R^3.

Figure 7.4

7.1 EXERCISES

1. In each of the following, find an equation of the line in R^2 determined by the given points.
   (a) $P_1(-2, -3)$, $P_2(3, 4)$.
   (b) $P_1(2, -5)$, $P_2(-3, 4)$.
   (c) $P_1(0, 0)$, $P_2(-3, 5)$.
   (d) $P_1(-3, -5)$, $P_2(0, 2)$.

2. In each of the following, find the equation of the line in R^2 determined by the given points.
   (a) $P_1(1, 1)$, $P_2(2, 2)$.
   (b) $P_1(1, 2)$, $P_2(1, 3)$.
   (c) $P_1(2, -4)$, $P_2(-3, -4)$.
   (d) $P_1(2, -3)$, $P_2(3, -2)$.

3. State which of the following points are on the line
   $$\begin{aligned} x &= 3 + 2t \\ y &= -2 + 3t \qquad (-\infty < t < \infty) \\ z &= 4 - 3t. \end{aligned}$$
   (a) $(1, 1, 1)$.   (b) $(1, -1, 0)$.
   (c) $(1, 0, -2)$.   (d) $(4, -\frac{1}{2}, \frac{5}{2})$.

4. State which of the following points are on the line
   $$\frac{x - 4}{-2} = \frac{y + 3}{2} = \frac{z - 4}{-5}.$$
   (a) $(0, 1, -6)$.   (b) $(1, 2, 3)$.
   (c) $(4, -3, 4)$.   (d) $(0, 1, -1)$.

5. In each of the following, find the parametric equations of the line through the point $P_0(x_0, y_0, z_0)$, which is parallel to the vector $\mathbf{U}$.
   (a) $P_0 = (3, 4, -2)$, $\mathbf{U} = (4, -5, 2)$.
   (b) $P_0 = (3, 2, 4)$, $\mathbf{U} = (-2, 5, 1)$.
   (c) $P_0 = (0, 0, 0)$, $\mathbf{U} = (2, 2, 2)$.
   (d) $P_0 = (-2, -3, 1)$, $\mathbf{U} = (2, 3, 4)$.

6. In each of the following, find the parametric equations of the line through the given points.
   (a) $(2, -3, 1)$, $(4, 2, 5)$.
   (b) $(-3, -2, -2)$, $(5, 5, 4)$.
   (c) $(-2, 3, 4)$, $(2, -3, 5)$.
   (d) $(0, 0, 0)$, $(4, 5, 2)$.

7. For each of the lines in Exercise 6, find the equations in symmetric form.

8. State which of the following points are on the plane
   $$3(x - 2) + 2(y + 3) - 4(z - 4) = 0.$$
   (a) $(0, -2, 3)$.   (b) $(1, -2, 3)$.
   (c) $(1, -1, 3)$.   (d) $(0, 0, 4)$.

9. In each of the following, find an equation of the plane passing through the given point and perpendicular to the given vector $\mathbf{N}$.
   (a) $(0, 2, -3)$, $\mathbf{N} = (3, -2, 4)$.
   (b) $(-1, 3, 2)$, $\mathbf{N} = (0, 1, -3)$.
   (c) $(-2, 3, 4)$, $\mathbf{N} = (0, 0, -4)$.
   (d) $(5, 2, 3)$, $\mathbf{N} = (-1, -2, 4)$.

10. In each of the following, find an equation of the plane passing through the given three points.
    (a) $(0, 1, 2)$, $(3, -2, 5)$, $(2, 3, 4)$.
    (b) $(2, 3, 4)$, $(-1, -2, 3)$, $(-5, -4, 2)$.
    (c) $(1, 2, 3)$, $(0, 0, 0)$, $(-2, 3, 4)$.
    (d) $(1, 1, 1)$, $(2, 3, 4)$, $(-5, 3, 2)$.

11. In each of the following, find parametric equations of the line of intersection of the given planes.
    (a) $2x + 3y - 4z + 5 = 0$ and
        $-3x + 2y + 5z + 6 = 0$.
    (b) $3x - 2y - 5z + 4 = 0$ and
        $2x + 3y + 4z + 8 = 0$.
    (c) $-x + 2y + z = 0$ and
        $2x - y + 2z + 8 = 0$.

12. In each of the following, find a pair of planes whose intersection is the given line.
    (a) $\begin{aligned} x &= 2 - 3t \\ y &= 3 + t \\ z &= 2 - 4t. \end{aligned}$
    (b) $\dfrac{x - 2}{-2} = \dfrac{y - 3}{4} = \dfrac{z + 4}{3}.$
    (c) $\begin{aligned} x &= 4t \\ y &= 1 + 5t \\ z &= 2 - t. \end{aligned}$

13. Are the points $(2, 3, -2)$, $(4, -2, -3)$, and $(0, 8, -1)$ on the same line?

14. Are the points $(-2, 4, 2)$, $(3, 5, 1)$, and $(4, 2, -1)$ on the same line?

15. Find the point of intersection of the lines

$$x = 2 - 3s \qquad x = 5 + 2t$$
$$y = 3 + 2s \quad \text{and} \quad y = 1 - 3t$$
$$z = 4 + 2s \qquad z = 2 + \ t.$$

16. Which of the following pairs of lines are perpendicular?

(a)
$$x = \ \ \ 2 + 2t \qquad x = 2 + t$$
$$y = -3 - 3t \quad \text{and} \quad y = 4 - t$$
$$z = \ \ \ 4 + 4t \qquad z = 5 - t.$$

(b)
$$x = 3 - \ \ t \qquad x = \ \ \ 2t$$
$$y = 4 + \ \ t \quad \text{and} \quad y = 3 - 2t$$
$$z = 2 + 2t \qquad z = 4 + 2t.$$

17. Show that the following parametric equations define the same line.

$$x = \ \ \ 2 + 3t \qquad x = -1 - \ \ 9t$$
$$y = \ \ \ 3 - 2t \quad \text{and} \quad y = \ \ \ 5 + \ \ 6t$$
$$z = -1 + 4t \qquad z = -5 - 12t.$$

18. Find the parametric equations of the line passing through the point $(3, -1, -3)$ and perpendicular to the line passing through the points $(3, -2, 4)$ and $(0, 3, 5)$.

19. Find an equation of the plane passing through the point $(-2, 3, 4)$ and perpendicular to the line passing through the points $(4, -2, 5)$ and $(0, 2, 4)$.

20. Find the point of intersection of the line

$$x = 2 - 3t$$
$$y = 4 + 2t$$
$$z = 3 - 5t$$

and the plane $2x + 3y + 4z + 8 = 0$.

21. Find a plane containing the lines

$$x = 3 + 2t \qquad x = 1 - 2t$$
$$y = 4 - 3t \quad \text{and} \quad y = 7 + 4t$$
$$z = 5 + 4t \qquad z = 1 - 3t.$$

22. Find a plane that passes through the point $(2, 4, -3)$ and is parallel to the plane

$$-2x + 4y - 5z + 6 = 0.$$

23. Find a line that passes through the point $(-2, 5, -3)$ and is perpendicular to the plane $2x - 3y + 4z + 7 = 0$.

24. **(Uses optional material of Section 3.8).** Find the distance from the point $(2, 3, -1)$ to the plane $3x - 2y + z = 0$.

(*Hint:* First find an orthonormal basis for the plane.)

THEORETICAL EXERCISES

T.1. Show that the graph of the equation $ax + by + cz + d = 0$, where a, b, c, and d are constants with a, b, and c not all zero, is a plane with normal $\mathbf{N} = (a, b, c)$.

T.2. Let the lines L_1 and L_2 be given parametrically by

$$L_1 : \mathbf{X} = \mathbf{X}_0 + s\mathbf{U} \qquad \text{and} \qquad L_2 : \mathbf{X} = \mathbf{X}_1 + t\mathbf{V}.$$

Prove that

(a) L_1 and L_2 are parallel if and only if $\mathbf{U} = a\mathbf{V}$ for some scalar a.

(b) L_1 and L_2 are identical if and only if $\mathbf{X}_1 - \mathbf{X}_0$ and $\mathbf{U}$ are both parallel to $\mathbf{V}$.

(c) L_1 and L_2 are perpendicular if and only if $\mathbf{U} \cdot \mathbf{V} = 0$.

(d) L_1 and L_2 intersect if and only if $\mathbf{X}_1 - \mathbf{X}_0$ is a linear combination of $\mathbf{U}$ and $\mathbf{V}$.

T.3. The lines L_1 and L_2 are said to be **skew** if they are not parallel and do not intersect. Give an example of skew lines L_1 and L_2.

T.4. Consider the planes $a_1x + b_1y + c_1z + d_1 = 0$ and $a_2x + b_2y + c_2z + d_2 = 0$ with normals $\mathbf{N}_1$ and $\mathbf{N}_2$, respectively. Prove that if the planes are identical, then $\mathbf{N}_2 = a\mathbf{N}_1$ for some scalar a.

T.5. Classify all the subspaces of R^3.

T.6. (a) Show that the set of all points on the plane $ax + by + cz = 0$ is a subspace of R^3.

(b) Find a basis for the plane $2x - 3y + 4z = 0$.

T.7. Show that an equation of the plane through the noncollinear points $P_1(a_1, b_1, c_1)$, $P_2(a_2, b_2, c_2)$, and $P_3(a_3, b_3, c_3)$ is

$$\begin{vmatrix} x & y & z & 1 \\ a_1 & b_1 & c_1 & 1 \\ a_2 & b_2 & c_2 & 1 \\ a_3 & b_3 & c_3 & 1 \end{vmatrix} = 0.$$

7.2

QUADRATIC FORMS

Prerequisite. Section 5.2, Diagonalization of Symmetric Matrices.

Let

$$ax^2 + 2bxy + cy^2 = d, \tag{1}$$

where a, b, c, and d are real numbers which are not all zero, be the equation of a conic centered at the origin of a rectangular Cartesian coordinate system in two-dimensional space. In many applications we are interested in identifying the type of conic represented by the equation; that is, we have to decide whether the equation represents an ellipse, a hyperbola, a parabola, or any degenerate forms of the first two of these, such as straight lines. The approach frequently used is to rotate the x- and y-axes (or change coordinates) so as to obtain a new set of axes, x' and y', in which the $x'y'$ term is no longer present; we can then identify the type of conic rather easily. The analogous problem can be studied for quadric surfaces in three-dimensional space, that is, for a surface whose equation is

$$ax^2 + 2dxy + 2exz + by^2 + 2fyz + cz^2 = g, \tag{2}$$

where a, b, c, d, e, f, and g are real numbers that are not all zero. The solution is to again change the x-, y-, and z-axes to x'-, y'-, and z'-axes so as to eliminate the $x'y'$, $x'z'$, and $y'z'$ terms. The expressions on the left sides of (1) and (2) are quadratic forms. Quadratic forms arise in statistics, mechanics, and in other problems in physics; in quadratic programming; in the study of maxima and minima of functions of several variables; and in other applied problems. In this section we apply our results on eigenvalues and eigenvectors of matrices to give a brief treatment of real quadratic forms in n variables, generalizing the techniques used for quadratic forms in two and three variables.

Definition   If $\mathbf{A}$ is a symmetric matrix ($\mathbf{A} = \mathbf{A}^T$), then the function $\mathbf{Q} \colon R^n \to R^1$ (a real-valued function defined on R^n) defined by

$$\mathbf{Q(X)} = \mathbf{X}^T\mathbf{AX},$$

where

$$\mathbf{X} = \begin{bmatrix} x_1 \\ x_2 \\ \vdots \\ x_n \end{bmatrix}$$

is called a **quadratic form in the n variables** $x_1, x_2, \ldots, x_n$. The matrix $\mathbf{A}$ is called the **matrix of the quadratic** form $\mathbf{Q}$. We shall also denote the quadratic form by $\mathbf{Q(X)}$.

Example 1 The left side of Equation (1) is the quadratic form in the variables x and y:

$$Q(X) = X^T A X,$$

where

$$X = \begin{bmatrix} x \\ y \end{bmatrix} \quad \text{and} \quad A = \begin{bmatrix} a & b \\ b & c \end{bmatrix}.$$

Example 2 The left side of Equation (2) is the quadratic form

$$Q(X) = X^T A X,$$

where

$$X = \begin{bmatrix} x \\ y \\ z \end{bmatrix} \quad \text{and} \quad A = \begin{bmatrix} a & d & e \\ d & b & f \\ e & f & c \end{bmatrix}.$$

Example 3 The following expressions are quadratic forms:

(a) $3x^2 - 5xy - 7y^2 = \begin{bmatrix} x & y \end{bmatrix} \begin{bmatrix} 3 & -\frac{5}{2} \\ -\frac{5}{2} & -7 \end{bmatrix} \begin{bmatrix} x \\ y \end{bmatrix}$.

(b) $3x^2 - 7xy + 5xz + 4y^2 - 4yz - 3z^2$

$$= \begin{bmatrix} x & y & z \end{bmatrix} \begin{bmatrix} 3 & -\frac{7}{2} & \frac{5}{2} \\ -\frac{7}{2} & 4 & -2 \\ \frac{5}{2} & -2 & -3 \end{bmatrix} \begin{bmatrix} x \\ y \\ z \end{bmatrix}.$$

Suppose now that $Q(X) = X^T A X$ is a quadratic form. To simplify the quadratic form, we change from the variables $x_1, x_2, \ldots, x_n$ to the variables $y_1, y_2, \ldots, y_n$, where we assume that the old variables are related to the new variables by $X = PY$ for some orthogonal matrix P. Then

$$Q(X) = X^T A X = (PY)^T A (PY) = Y^T (P^T A P) Y = Y^T B Y,$$

where $B = P^T A P$. We shall let the reader verify the fact that if A is a symmetric matrix, then $P^T A P$ is also symmetric (Exercise T.1). Thus

$$Q'(Y) = Y^T B Y$$

is another quadratic form and $Q(X) = Q'(Y)$.

This situation is important enough to formulate the following definitions.

Definition Two $n \times n$ matrices A and B are said to be **congruent** if $B = P^T A P$ for a nonsingular matrix P.

Definition Two quadratic forms **Q** and **Q**′ with matrices **A** and **B**, respectively, are said to be **equivalent** if **A** and **B** are congruent.

By the definition above, if we wish to show that the $n \times n$ matrices **A** and **B** are congruent, we must show that $\mathbf{B} = \mathbf{P}^T\mathbf{A}\mathbf{P}$, for a nonsingular matrix **P**. However, in this section one of the matrices will be symmetric and the other diagonal and it will be simpler to find an orthogonal matrix **P** (hence nonsingular) such that $\mathbf{B} = \mathbf{P}^T\mathbf{A}\mathbf{P}$.

Example 4 Consider the quadratic form in the variables x and y defined by

$$\mathbf{Q}(\mathbf{X}) = 2x^2 + 2xy + 2y^2 = [x \quad y]\begin{bmatrix} 2 & 1 \\ 1 & 2 \end{bmatrix}\begin{bmatrix} x \\ y \end{bmatrix}. \tag{3}$$

We now change from the variables x and y to the variables x' and y'. Suppose that the old variables are related to the new variables by the equations

$$x = \frac{1}{\sqrt{2}}x' - \frac{1}{\sqrt{2}}y' \quad \text{and} \quad y = \frac{1}{\sqrt{2}}x' + \frac{1}{\sqrt{2}}y', \tag{4}$$

which can be written in matrix form as

$$\mathbf{X} = \begin{bmatrix} x \\ y \end{bmatrix} = \begin{bmatrix} \dfrac{1}{\sqrt{2}} & -\dfrac{1}{\sqrt{2}} \\ \dfrac{1}{\sqrt{2}} & \dfrac{1}{\sqrt{2}} \end{bmatrix}\begin{bmatrix} x' \\ y' \end{bmatrix} = \mathbf{PY},$$

where the orthogonal (hence nonsingular) matrix

$$\mathbf{P} = \begin{bmatrix} \dfrac{1}{\sqrt{2}} & -\dfrac{1}{\sqrt{2}} \\ \dfrac{1}{\sqrt{2}} & \dfrac{1}{\sqrt{2}} \end{bmatrix} \quad \text{and} \quad \mathbf{Y} = \begin{bmatrix} x' \\ y' \end{bmatrix}.$$

We shall soon see why and how this particular matrix **P** was selected. Substituting in (3), we obtain

$$\mathbf{Q}(\mathbf{X}) = \mathbf{X}^T\mathbf{A}\mathbf{X} = (\mathbf{PY})^T\mathbf{A}(\mathbf{PY}) = \mathbf{Y}^T\mathbf{P}^T\mathbf{A}\mathbf{P}\mathbf{Y}$$

$$= [x' \quad y']\begin{bmatrix} \dfrac{1}{\sqrt{2}} & -\dfrac{1}{\sqrt{2}} \\ \dfrac{1}{\sqrt{2}} & \dfrac{1}{\sqrt{2}} \end{bmatrix}^T\begin{bmatrix} 2 & 1 \\ 1 & 2 \end{bmatrix}\begin{bmatrix} \dfrac{1}{\sqrt{2}} & -\dfrac{1}{\sqrt{2}} \\ \dfrac{1}{\sqrt{2}} & \dfrac{1}{\sqrt{2}} \end{bmatrix}\begin{bmatrix} x' \\ y' \end{bmatrix}$$

$$= [x' \quad y']\begin{bmatrix} 3 & 0 \\ 0 & 1 \end{bmatrix}\begin{bmatrix} x' \\ y' \end{bmatrix} = \mathbf{Q}'(\mathbf{Y}).$$

Thus the matrices

$$\begin{bmatrix} 2 & 1 \\ 1 & 2 \end{bmatrix} \quad \text{and} \quad \begin{bmatrix} 3 & 0 \\ 0 & 1 \end{bmatrix}$$

are congruent and the quadratic forms $\mathbf{Q}$ and $\mathbf{Q}'$ are equivalent.

The equation

$$Q(\mathbf{X}) = 2x^2 + 2xy + 2y^2 = 9 \tag{5}$$

represents a conic section. Since $\mathbf{Q}$ is the quadratic form defined in Example 4, it is equivalent to the quadratic form

$$\mathbf{Q}'(\mathbf{Y}) = 3x'^2 + y'^2.$$

Now the equation

$$\mathbf{Q}'(\mathbf{Y}) = 3x'^2 + y'^2 = 9 \tag{6}$$

is the equation of an ellipse.

Since the change of coordinates from the (x, y)-system to the (x', y')-system was by means of an orthogonal transformation, we can then conclude (why?) that the conic section whose equation is (6) is the same ellipse as that represented by (5).

We now turn to the question of how to select the matrix $\mathbf{P}$.

Theorem 7.1 *Any quadratic form in n variables $\mathbf{Q}(\mathbf{X}) = \mathbf{X}^T\mathbf{A}\mathbf{X}$ is equivalent by means of an orthogo-*
(Principal *nal matrix $\mathbf{P}$ to a quadratic form, $\mathbf{Q}'(\mathbf{Y}) = \lambda_1 y_1^2 + \lambda_2 y_2^2 + \cdots + \lambda_n y_n^2$, where*
Axes Theorem)

$$\mathbf{Y} = \begin{bmatrix} y_1 \\ y_2 \\ \vdots \\ y_n \end{bmatrix}$$

and $\lambda_1, \lambda_2, \ldots, \lambda_n$ are the eigenvalues of the matrix $\mathbf{A}$ of $\mathbf{Q}$.

Proof

If $\mathbf{A}$ is the matrix of $\mathbf{Q}$, then, since $\mathbf{A}$ is symmetric, we know by Theorem 5.7 that $\mathbf{A}$ can be diagonalized by an orthogonal matrix. This means that there exists an orthogonal matrix $\mathbf{P}$ such that $\mathbf{B} = \mathbf{P}^{-1}\mathbf{A}\mathbf{P}$ is a diagonal matrix. Since $\mathbf{P}$ is orthogonal, $\mathbf{P}^{-1} = \mathbf{P}^T$, so $\mathbf{B} = \mathbf{P}^T\mathbf{A}\mathbf{P}$. Moreover, the elements on the main diagonal of $\mathbf{B}$ are the eigenvalues λ_1, $\lambda_2, \ldots, \lambda_n$ of $\mathbf{A}$. The quadratic form $\mathbf{Q}'$ with matrix $\mathbf{B}$ is given by

$$\mathbf{Q}'(\mathbf{Y}) = \lambda_1 y_1^2 + \lambda_2 y_2^2 + \cdots + \lambda_n y_n^2;$$

$\mathbf{Q}$ and $\mathbf{Q}'$ are equivalent.

Example 5 Consider the quadratic form $\mathbf{Q}$ in the variables x, y, and z, defined by

$$\mathbf{Q}(\mathbf{X}) = 2x^2 + 4y^2 + 6yz - 4z^2.$$

The matrix of $\mathbf{Q}$ is

$$\mathbf{A} = \begin{bmatrix} 2 & 0 & 0 \\ 0 & 4 & 3 \\ 0 & 3 & -4 \end{bmatrix}$$

and the eigenvalues of $\mathbf{A}$ are (verify)

$$\lambda_1 = 2, \qquad \lambda_2 = 5, \qquad \text{and} \qquad \lambda_3 = -5.$$

Let $\mathbf{Q}'$ be the quadratic form in the variables x', y', and z' defined by

$$\mathbf{Q}'(\mathbf{Y}) = 2x'^2 + 5y'^2 - 5z'^2.$$

Then $\mathbf{Q}$ and $\mathbf{Q}'$ are equivalent by means of some orthogonal matrix. Since

$$\mathbf{D} = \begin{bmatrix} 2 & 0 & 0 \\ 0 & 5 & 0 \\ 0 & 0 & -5 \end{bmatrix}$$

is the matrix of $\mathbf{Q}'$, $\mathbf{A}$ and $\mathbf{D}$ are congruent matrices.

Note that to apply Theorem 7.1 to diagonalize a given quadratic form, as shown in Example 5, we do not need to know the eigenvectors of $\mathbf{A}$ (nor the matrix $\mathbf{P}$); we only require the eigenvalues of $\mathbf{A}$.

We now return to quadratic forms in two and three variables. The equation $\mathbf{X}^T\mathbf{A}\mathbf{X} = 1$, where $\mathbf{X}$ is a vector in R^2 is the locus of a central conic in the plane. From Theorem 7.1 it follows that there is a Cartesian coordinate system in the plane with respect to which the equation of the central conic is $ax^2 + by^2 = 1$, where a and b are real numbers. Similarly, the equation $\mathbf{X}^T\mathbf{A}\mathbf{X} = 1$, where $\mathbf{X}$ is a vector in R^3, is the locus of a central quadric surface in R^3. From Theorem 7.1 it follows that there is a Cartesian coordinate system in 3-space with respect to which the equation of the quadric surface is $ax^2 + by^2 + cz^2 = 1$, where a, b, and c are real numbers. The principal axes of the conic or surface lie along the new coordinate axes and this is the reason for calling Theorem 7.1 the **principal axes theorem**.

Example 6 Consider the conic section whose equation is (5),

$$\mathbf{Q}(\mathbf{X}) = 2x^2 + 2xy + 2y^2 = 9.$$

This conic section can also be described by Equation (6)

$$\mathbf{Q}'(\mathbf{Y}) = 3x'^2 + y'^2 = 9,$$

which can be rewritten as

$$\frac{x'^2}{3} + \frac{y'^2}{9} = 1.$$

This is the equation of an ellipse (Figure 7.5) whose major axis is along the y'-axis. The semimajor axis is of length 6; the semiminor axis of length $2\sqrt{3}$. We now note that there is

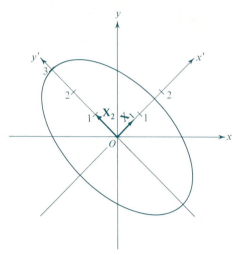

Figure 7.5

a very close connection between the eigenvectors of the matrix $\mathbf{Q}$ in (5) and the location of the x'- and y'-axes.

Since $\mathbf{X} = \mathbf{PY}$, we have $\mathbf{Y} = \mathbf{P}^{-1}\mathbf{X} = \mathbf{P}^T\mathbf{X}$ ($\mathbf{P}$ is orthogonal). Thus

$$x' = \frac{1}{\sqrt{2}}x + \frac{1}{\sqrt{2}}y \quad \text{and} \quad y' = -\frac{1}{\sqrt{2}}x + \frac{1}{\sqrt{2}}y.$$

This means that, in terms of the x- and y-axes, the x'-axis lies along the vector

$$\mathbf{X}_1 = \begin{bmatrix} \dfrac{1}{\sqrt{2}} \\ \dfrac{1}{\sqrt{2}} \end{bmatrix}$$

and the y'-axis lies along the vector

$$\mathbf{X}_2 = \begin{bmatrix} -\dfrac{1}{\sqrt{2}} \\ \dfrac{1}{\sqrt{2}} \end{bmatrix}.$$

Now $\mathbf{X}_1$ and $\mathbf{X}_2$ are the columns of the matrix

$$\mathbf{P} = \begin{bmatrix} \dfrac{1}{\sqrt{2}} & -\dfrac{1}{\sqrt{2}} \\ \dfrac{1}{\sqrt{2}} & \dfrac{1}{\sqrt{2}} \end{bmatrix},$$

which in turn are eigenvectors of the matrix $\mathbf{A}$. Thus the x'- and y'-axes lie along the eigenvectors of the matrix $\mathbf{A}$ (see Figure 7.5).

The situation described in Example 6 is true in general. Thus the principal axes of a conic or surface lie along the eigenvectors of the matrix of the quadratic form.

Let $\mathbf{Q}(\mathbf{X}) = \mathbf{X}^T\mathbf{A}\mathbf{X}$ be a quadratic form in n variables. Then we know that $\mathbf{Q}$ is equivalent to the quadratic form $\mathbf{Q}'(\mathbf{Y}) = \lambda_1 y_1^2 + \lambda_2 y_2^2 + \cdots + \lambda_n y_n^2$, where $\lambda_1, \lambda_2, \ldots, \lambda_n$ are the eigenvalues of the matrix $\mathbf{A}$ of $\mathbf{Q}$. We can label the eigenvalues $\lambda_1, \lambda_2, \ldots, \lambda_n$ so that all the positive eigenvalues of $\mathbf{A}$, if any, are listed first, followed by all the negative eigenvalues of $\mathbf{A}$, if any, followed by the zero eigenvalues, if any. Thus let $\lambda_1, \lambda_2, \ldots, \lambda_p$ be positive, $\lambda_{p+1}, \lambda_{p+2}, \ldots, \lambda_r$ be negative, and $\lambda_{r+1}, \lambda_{r+2}, \ldots, \lambda_n$ be zero. We now define the diagonal matrix $\mathbf{H}$ whose entries on the main diagonal are

$$\frac{1}{\sqrt{\lambda_1}}, \frac{1}{\sqrt{\lambda_2}}, \ldots, \frac{1}{\sqrt{\lambda_p}}, \frac{1}{\sqrt{-\lambda_{p+1}}}, \frac{1}{\sqrt{-\lambda_{p+2}}}, \ldots, \frac{1}{\sqrt{-\lambda_r}}, 1, 1, \ldots, 1,$$

with $n - r$ ones. Let $\mathbf{D}$ be the diagonal matrix whose entries on the main diagonal are $\lambda_1, \lambda_2, \ldots, \lambda_p, \lambda_{p+1}, \ldots, \lambda_r, \lambda_{r+1}, \ldots, \lambda_n$; $\mathbf{A}$ and $\mathbf{D}$ are congruent. Let $\mathbf{D}_1 = \mathbf{H}^T\mathbf{D}\mathbf{H}$ be the matrix whose diagonal elements are $1, 1, \ldots, 1, -1, \ldots, -1, 0, 0, \ldots, 0$ (p ones, $n - r$ zeros); $\mathbf{D}$ and $\mathbf{D}_1$ are then congruent. From Exercise T.2 it follows that $\mathbf{A}$ and $\mathbf{D}_1$ are congruent. In terms of quadratic forms, we have established Theorem 7.2.

Theorem 7.2 *A quadratic form $\mathbf{Q}(\mathbf{X}) = \mathbf{X}^T\mathbf{A}\mathbf{X}$ in n variables is equivalent to a quadratic form $\mathbf{Q}'(\mathbf{Y}) = y_1^2 + y_2^2 + \cdots + y_p^2 - y_{p+1}^2 - y_{p+2}^2 - \cdots - y_r^2$ for some $0 \leq p, r \leq n$.*

It is clear that the rank of the matrix $\mathbf{D}_1$ is r, the number of nonzero entries on its main diagonal. Now it can be shown that congruent matrices have equal ranks. Since the rank of $\mathbf{D}_1$ is r, the rank of $\mathbf{A}$ is also r. We also refer to r as the **rank** of the quadratic form $\mathbf{Q}$ whose matrix is $\mathbf{A}$. It can be shown that the number p of positive terms in the quadratic form $\mathbf{Q}'$ of Theorem 7.2 is unique; that is, no matter how we simplify the given quadratic form $\mathbf{Q}$ to obtain an equivalent quadratic form, the latter will always have p positive terms. Hence the quadratic form $\mathbf{Q}'$ in Theorem 7.2 is unique; it is often called the **canonical form** of a quadratic form in n variables. The difference between the number of positive eigenvalues and the number of negative eigenvalues is $s = p - (r - p) = 2p - r$ and is called the **signature** of the quadratic form. Thus, if $\mathbf{Q}$ and $\mathbf{Q}'$ are equivalent quadratic forms, then they have equal ranks and signatures. Conversely, it can also be shown that if $\mathbf{Q}$ and $\mathbf{Q}'$ have equal ranks and signatures, then they are equivalent.

Example 7 Consider the quadratic form

$$\mathbf{Q}(\mathbf{X}) = 3x_2^2 + 8x_2x_3 - 3x_3^2 = \mathbf{X}^T\mathbf{A}\mathbf{X}$$

$$= [x_1 \quad x_2 \quad x_3] \begin{bmatrix} 0 & 0 & 0 \\ 0 & 3 & 4 \\ 0 & 4 & -3 \end{bmatrix} \begin{bmatrix} x_1 \\ x_2 \\ x_3 \end{bmatrix}.$$

The eigenvalues of **A** are (verify)

$$\lambda_1 = 5, \qquad \lambda_2 = -5, \qquad \text{and} \qquad \lambda_3 = 0.$$

In this case **A** is congruent to

$$\mathbf{D} = \begin{bmatrix} 5 & 0 & 0 \\ 0 & -5 & 0 \\ 0 & 0 & 0 \end{bmatrix}.$$

If we let

$$\mathbf{H} = \begin{bmatrix} \dfrac{1}{\sqrt{5}} & 0 & 0 \\ 0 & \dfrac{1}{\sqrt{5}} & 0 \\ 0 & 0 & 1 \end{bmatrix},$$

then

$$\mathbf{D}_1 = \mathbf{H}^T \mathbf{D} \mathbf{H} = \begin{bmatrix} 1 & 0 & 0 \\ 0 & -1 & 0 \\ 0 & 0 & 0 \end{bmatrix}$$

and **A** are congruent, and the given quadratic form is equivalent to the canonical form

$$\mathbf{Q}'(\mathbf{Y}) = y_1^2 - y_2^2.$$

The rank of **Q** is 2, and since $p = 1$, the signature $s = 2p - r = 0$.

As a final application of quadratic forms we consider positive-definite symmetric matrices.

Definition A symmetric $n \times n$ matrix **A** is called **positive definite** if $\mathbf{X}^T \mathbf{A} \mathbf{X} > 0$ for every nonzero vector **X** in R^n.

If **A** is a symmetric matrix, then $\mathbf{X}^T \mathbf{A} \mathbf{X}$ is a quadratic form $\mathbf{Q}(\mathbf{X}) = \mathbf{X}^T \mathbf{A} \mathbf{X}$ and, by Theorem 7.1, **Q** is equivalent to **Q**', where $\mathbf{Q}'(\mathbf{Y}) = \lambda_1 y_1^2 + \lambda_2 y_2^2 + \cdots + \lambda_p y_p^2 + \lambda_{p+1} y_{p+1}^2 + \lambda_{p+2} y_{p+2}^2 + \cdots + \lambda_r y_r^2$. Now **A** is positive definite if and only if $\mathbf{Q}'(\mathbf{Y}) > 0$ for each $\mathbf{Y} \neq \mathbf{0}$. However, this can happen if and only if all the summands in $\mathbf{Q}'(\mathbf{Y})$ are positive and $r = n$. These remarks have established the following theorem.

Theorem 7.3 *A symmetric matrix **A** is positive definite if and only if all the eigenvalues of **A** are positive.*

A quadratic form is then called **positive definite** if its matrix is positive definite.

7.2 EXERCISES

1. Write each of the following quadratic forms as $\mathbf{X}^T\mathbf{A}\mathbf{X}$.
   (a) $-3x^2 + 5xy - 2y^2$.
   (b) $2x_1^2 + 3x_1x_2 - 5x_1x_3 + 7x_2x_3$.
   (c) $3x_1^2 + x_1x_2 - 2x_1x_3 + x_2^2 - 4x_2x_3 - 2x_3^2$.

2. Write each of the following quadratic forms as $\mathbf{X}^T\mathbf{A}\mathbf{X}$.
   (a) $x^2 - 4xy - 3y^2 + 6yz + 4z^2$.
   (b) $4x^2 - 6xy + 2y^2$.
   (c) $-2x_1x_2 + 4x_1x_3 + 6x_2x_3$.

3. For each of the following symmetric matrices $\mathbf{A}$, find a diagonal matrix $\mathbf{D}$ that is congruent to $\mathbf{A}$.

   (a) $\mathbf{A} = \begin{bmatrix} -1 & 0 & 0 \\ 0 & 1 & 1 \\ 0 & 1 & 1 \end{bmatrix}$.

   (b) $\mathbf{A} = \begin{bmatrix} 1 & 1 & 1 \\ 1 & 1 & 1 \\ 1 & 1 & 1 \end{bmatrix}$.

In Exercises 4 through 8, find a quadratic form of the type in Theorem 7.1 that is equivalent to the given quadratic form.

4. $2x^2 - 4xy - y^2$.

5. $x_1^2 + x_2^2 + 2x_2x_3 + x_3^2$.

6. $2xz$.

7. $2x_2^2 + 4x_2x_3 + 2x_3^2$.

8. $-2x_1^2 - 4x_2^2 - 6x_2x_3 + 4x_3^2$.

In Exercises 9 through 13, find a quadratic form of the type in Theorem 7.2 that is equivalent to the given quadratic form.

9. $2x^2 + 4xy + 2y^2$.

10. $x_1^2 + 2x_1x_2 + x_2^2 + x_3^2$.

11. $2x_1^2 + 4x_2^2 + 10x_2x_3 + 4x_3^2$.

12. $2x^2 + 3y^2 + 4yz + 3z^2$.

13. $-3x_1^2 + 2x_2^2 + 4x_2x_3 + 2x_3^2$.

14. Let $\mathbf{Q}(\mathbf{X}) = 4x^2 - 10xy + 4y^2 = 1$ be the equation of a central conic. Identify the conic by finding a quadratic form of the type in Theorem 7.2 that is equivalent to $\mathbf{Q}$.

15. Let $\mathbf{Q}(\mathbf{X}) = x^2 + 2y^2 + 2yz + 2z^2 = 1$ be the equation of a central quadric surface. Identify the quadric surface by finding a quadratic form of the type in Theorem 7.2 that is equivalent to $\mathbf{Q}$.

16. Let $\mathbf{Q}(\mathbf{X}) = 4x_2^2 - 10x_2x_3 + 4x_3^2$ be a quadratic form in three variables. Find a quadratic form of the type in Theorem 7.2 that is equivalent to $\mathbf{Q}$. What is the rank of $\mathbf{Q}$? What is the signature of $\mathbf{Q}$?

17. Which of the following quadratic forms in three variables are equivalent?

$$\mathbf{Q}_1(\mathbf{X}) = x_1^2 + 2x_1x_2 + x_2^2 + x_3^2$$
$$\mathbf{Q}_2(\mathbf{X}) = 2x_2^2 + 2x_2x_3 + 2x_3^2$$
$$\mathbf{Q}_3(\mathbf{X}) = 3x_2^2 + 8x_2x_3 - 3x_3^2$$
$$\mathbf{Q}_4(\mathbf{X}) = 3x_2^2 - 4x_2x_3 + 3x_3^2.$$

18. Which of the following matrices are positive definite?

   (a) $\begin{bmatrix} 2 & -1 \\ -1 & 2 \end{bmatrix}$.   (b) $\begin{bmatrix} 2 & 1 \\ 1 & 2 \end{bmatrix}$.

   (c) $\begin{bmatrix} 0 & -1 \\ -1 & 0 \end{bmatrix}$.   (d) $\begin{bmatrix} 1 & 1 \\ 1 & 1 \end{bmatrix}$.

   (e) $\begin{bmatrix} 2 & 2 \\ 2 & 2 \end{bmatrix}$.

THEORETICAL EXERCISES

T.1. Prove that if $\mathbf{A}$ is a symmetric matrix, then $\mathbf{P}^T\mathbf{AP}$ is also symmetric.

T.2. If $\mathbf{A}$, $\mathbf{B}$, and $\mathbf{C}$ are $n \times n$ symmetric matrices, prove that

(a) $\mathbf{A}$ and $\mathbf{A}$ are congruent.

(b) If $\mathbf{A}$ and $\mathbf{B}$ are congruent, then $\mathbf{B}$ and $\mathbf{A}$ are congruent.

(c) If $\mathbf{A}$ and $\mathbf{B}$ are congruent and if $\mathbf{B}$ and $\mathbf{C}$ are congruent, then $\mathbf{A}$ and $\mathbf{C}$ are congruent.

T.3. Prove that if $\mathbf{A}$ is symmetric, then $\mathbf{A}$ is congruent to a diagonal matrix $\mathbf{D}$.

T.4. Find all quadratic forms $\mathbf{Q(X)} = \mathbf{X}^T\mathbf{AX}$ in two variables of the type described in Theorem 7.2.

What conics do the equations $\mathbf{X}^T\mathbf{AX} = 1$ represent?

T.5. Find all quadratic forms $\mathbf{Q(X)} = \mathbf{X}^T\mathbf{AX}$ in two variables of rank 1 of the type described in Theorem 7.2. What conics do the equations $\mathbf{X}^T\mathbf{AX} = 1$ represent?

T.6. Let $\mathbf{A} = \begin{bmatrix} a & b \\ b & d \end{bmatrix}$ be a 2×2 symmetric matrix. Prove that $\mathbf{A}$ is positive definite if and only if $|\mathbf{A}| > 0$ and $a > 0$.

T.7. Prove that a symmetric matrix $\mathbf{A}$ is positive definite if and only if $\mathbf{A} = \mathbf{P}^T\mathbf{P}$ for a nonsingular matrix $\mathbf{P}$.

7.3

GRAPH THEORY

Prerequisite. Section 1.3, Properties of Matrix Operations.

Graph theory is a new area of applied mathematics that is being widely used in formulating models in many problems in business, the social sciences, and the physical sciences. These applications include communications problems and the study of organizations and social structures. In this section we present a very brief introduction to the subject as it relates to matrices and show how these elementary and simple notions can be used in formulating models of some important problems.

Graphs

Definition   A **graph** G is a finite set of points, called **vertices** or **nodes**, together with a finite set of **edges**, each of which joins a pair of vertices. An edge joining a vertex to itself is called a **loop**.

The vertices of a graph are represented by dots and the edges by straight or curved line segments.

Example 1   Figure 7.6 shows examples of graphs. The graph in Figure 7.6(a) has four vertices, P_1, P_2, P_3, and P_4, and six edges, P_1P_2, P_1P_3, P_1P_4, P_2P_3, P_2P_4, and P_3P_4. In this representation the edges P_1P_4 and P_2P_3 intersect at a point other than a vertex, so they are drawn as shown. The graph in Figure 7.6(b) has four vertices, P_1, P_2, P_3, and P_4, and two edges,

(a)   (b)   (c)

Figure 7.6

P_1P_2 and P_1P_3. Vertices P_2 and P_3 are not joined by an edge. Moreover, vertex P_4 is not joined to any vertex. The graph in Figure 7.6(c) has three vertices, P_1, P_2, and P_3, and three edges, P_1P_2, P_1P_3, and P_3P_3. The edge P_3P_3 is a **loop**.

A graph is called **complete** if every pair of vertices is joined by an edge. The complete graph with n vertices is denoted by K_n. The graph in Figure 7.6(a) is the complete graph K_4.

Example 2   A bowling league consists of seven teams: T_1, T_2, T_3, T_4, T_5, T_6, and T_7. Suppose that after a number of games have been played we have the following situation:

T_1 has played T_2, T_3, and T_5.
T_2 has played T_1, T_3, and T_5.
T_3 has played T_1, T_2, and T_4.
T_4 has played T_3 and T_7.
T_5 has played T_1 and T_2.
T_6 has not played anyone.
T_7 has played T_4.

This situation can be described by a graph in which the teams are represented by the vertices, and an edge joining two vertices means that the corresponding teams have played each other. We thus obtain the graph in Figure 7.7.

Example 3   A pair of vertices in a graph may be joined by more than one edge, as shown in Figure 7.8. In this case we say that we have a **multiple edge**. Thus, in a communication network there might be two lines between P_1 and P_2 and one line between P_2 and P_3.

Figure 7.7

Figure 7.8

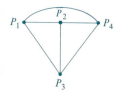

Figure 7.9

A graph is determined by the set of vertices and by the set of edges joining the vertices, not by the particular appearance of the configuration. Thus two graphs G and G' are said to be **equal** or **isomorphic** if they have the same number of vertices, the same number of edges, and if the vertices (respectively, edges) of G may be put into one-to-one correspondences with the vertices (respectively, edges) of G' in such a way that if edge e of G corresponds to edge e' of G', and the end points of e are P_i and P_j, then the end points of e' are the vertices corresponding to P_i and P_j.

Example 4 The graph in Figure 7.9 is equal to the graph in Figure 7.6(a) (verify).

Matrices provide a convenient way of describing graphs, and since matrices lend themselves well to computer use, they make it possible to use the computer for extensive computational work in graph theory.

If we are given a graph G with n vertices, we can associate a symmetric $n \times n$ matrix $\mathbf{M}(G)$ with it by letting the i, jth element be the number of edges from the vertex P_i to vertex P_j. The matrix $\mathbf{M}(G)$ is called the **matrix representing** G.

Example 5 The matrix of the graph in Figure 7.10 is $\mathbf{M}(G)$.

$$
\mathbf{M}(G) = \begin{array}{c} \\ P_1 \\ P_2 \\ P_3 \\ P_4 \\ P_5 \\ P_6 \end{array} \begin{array}{cccccc} P_1 & P_2 & P_3 & P_4 & P_5 & P_6 \\ \left[\begin{array}{cccccc} 1 & 1 & 0 & 0 & 0 & 0 \\ 1 & 0 & 0 & 1 & 0 & 0 \\ 0 & 0 & 0 & 2 & 1 & 0 \\ 0 & 1 & 2 & 0 & 0 & 1 \\ 0 & 0 & 1 & 0 & 0 & 2 \\ 0 & 0 & 0 & 1 & 2 & 0 \end{array}\right] \end{array}.
$$

G :

Figure 7.10

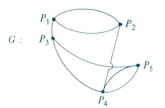

G :

Figure 7.11

Conversely, a given $n \times n$ symmetric matrix with nonnegative integer entries gives rise to a graph G with the given matrix representing G.

Example 6 The matrix

$$\begin{bmatrix} 0 & 2 & 1 & 0 & 0 \\ 2 & 0 & 0 & 1 & 0 \\ 1 & 0 & 0 & 1 & 1 \\ 0 & 1 & 1 & 0 & 2 \\ 0 & 0 & 1 & 2 & 0 \end{bmatrix}$$

determines the graph G, shown in Figure 7.11.

Of course, it is easy to see that graphs G and G' are equal if and only if there is a labeling of the vertices for which the matrices representing them are equal.

There is another matrix associated with a graph, which we now define.

Definition If G is a graph that has n vertices, then the matrix $\mathbf{A}(G)$, whose i, jth element is 1 if there is at least one edge between P_i and P_j and zero otherwise. is called the **adjacency matrix** of G. Note that $\mathbf{A}(G)$ is a symmetric matrix.

Example 7 The adjacency matrix of the graph in Figure 7.10 is

$$\mathbf{A}(G) = \begin{bmatrix} 1 & 1 & 0 & 0 & 0 & 0 \\ 1 & 0 & 0 & 1 & 0 & 0 \\ 0 & 0 & 0 & 1 & 1 & 0 \\ 0 & 1 & 1 & 0 & 0 & 1 \\ 0 & 0 & 1 & 0 & 0 & 1 \\ 0 & 0 & 0 & 1 & 1 & 0 \end{bmatrix}.$$

Of course, from a symmetric matrix whose entries are zeros and ones we can obtain a graph whose adjacency matrix is the given matrix. However, this graph need not be unique.

(a)   (b)

Figure 7.12

Example 8 The matrix

$$\begin{bmatrix} 0 & 1 & 0 & 1 \\ 1 & 0 & 1 & 0 \\ 0 & 1 & 1 & 1 \\ 1 & 0 & 1 & 0 \end{bmatrix}$$

gives rise to the graphs in Figure 7.12.

Sometimes the adjacency matrix of G may equal the matrix representing G. This will occur if G has no multiple edges.

Figure 7.7 indicated the teams that played each other. However, if we wanted to indicate the winner in each competition, we need to develop the notion of a directed graph.

Digraphs

A **directed graph**, or a **digraph**, is a finite set of points, called **vertices** or **nodes**, together with a finite set of **directed edges**, each of which joins an ordered pair of distinct vertices. Thus a digraph contains no loops. Let us also assume that there are no multiple edges. Moreover, the directed edge P_iP_j is now different from the directed edge P_jP_i. The matrix $\mathbf{A}(G)$, whose i, jth element is 1 if there is a directed edge from P_i to P_j and zero otherwise, is called the **adjacency matrix of G**. The adjacency matrix of a digraph need not be symmetric.

Example 9 In Figure 7.13 are shown four examples of digraphs. The digraph in Figure 7.13(a) has vertices P_1, P_2, and P_3 and directed edges P_1P_2 and P_2P_3; the digraph in Figure 7.13(b) has vertices P_1, P_2, P_3, and P_4 and directed edges P_1P_2 and P_1P_3; the digraph in Figure 7.13(c) has vertices P_1, P_2, and P_3 and directed edges P_1P_2, P_1P_3, and P_3P_1; the digraph in Figure 7.13(d) has vertices P_1, P_2, and P_3 and directed edges P_2P_1, P_2P_3, P_1P_3, and P_3P_1. The pair of directed edges P_1P_3 and P_3P_1 are indicated by a curved or straight segment with a double arrow, $P_1 \longleftrightarrow P_3$.

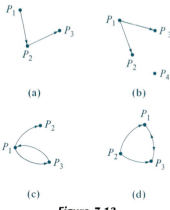

Figure 7.13

Example 10 Consider Example 2 again. Suppose we now know that

T_1 has defeated T_2 and T_5 and lost to T_3;
T_2 has defeated T_5 and lost to T_1 and T_3;
T_3 has defeated T_1 and T_2 and lost to T_4;
T_4 has defeated T_3 and lost to T_7;
T_5 has lost to T_1 and T_2;
T_6 has not played anyone;
T_7 has defeated T_4.

We now obtain the directed graph in Figure 7.14, where $T_i \to T_j$ means that T_i defeated T_j.

Of course, digraphs can be used in a great many situations including communications problems, family relationships, social structures, street maps, flow charts, transportation problems, electrical circuits, and ecological chains. We shall deal with some of these below and in the exercises.

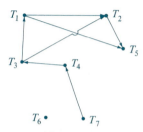

Figure 7.14

Models in Sociology and in Communications

Suppose that we have n individuals $P_1, P_2 \ldots, P_n$, some of whom are associated with each other. We assume that no one is associated with himself. Examples of such relations are

1. P_i has access to P_j. In this case it may or may not be the case that if P_i has access to P_j, then P_j has access to P_i. For example, many emergency telephones on turnpikes allow a distressed traveler to contact a nearby emergency station but make no provision for the station to contact the traveler. This model can thus be represented by a digraph G as follows. Let $P_1, P_2, \ldots, P_n$ be the vertices of G and draw a directed edge from P_i to P_j if P_i has access to P_j. It is important to observe that this relation need not be transitive. That is, P_i may have access to P_j and P_j may have access to P_k, but P_i need not have access to P_k.

2. P_i influences P_j. This situation is identical to that in 1: if P_i influences P_j, then it may or may not happen that P_j influences P_i.

3. For every pair of individuals P_i, P_j, either P_i dominates P_j or P_j dominates P_i, but not both. The graph representing this situation is the complete directed graph with n vertices. Such graphs are often called **dominance digraphs**.

Example 11 Suppose that six individuals have been meeting in group therapy for a long time and their leader, who is not part of the group, has drawn the digraph G in Figure 7.15 to describe the influence relations among the various individuals. The adjacency matrix of G is

$$\mathbf{A}(G) = \begin{array}{c} \\ P_1 \\ P_2 \\ P_3 \\ P_4 \\ P_5 \\ P_6 \end{array} \begin{array}{c} \begin{array}{cccccc} P_1 & P_2 & P_3 & P_4 & P_5 & P_6 \end{array} \\ \left[\begin{array}{cccccc} 0 & 0 & 0 & 0 & 1 & 0 \\ 0 & 0 & 0 & 0 & 1 & 0 \\ 1 & 1 & 0 & 0 & 1 & 0 \\ 0 & 1 & 0 & 0 & 1 & 0 \\ 0 & 0 & 0 & 0 & 0 & 0 \\ 0 & 1 & 0 & 1 & 0 & 0 \end{array} \right] \end{array}.$$

Looking at the rows of $\mathbf{A}(G)$, we see that P_3 has three 1's in its row so that P_3

Figure 7.15

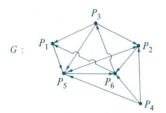

Figure 7.16

influences three people—more than any other individual. Thus P_3 would be declared as the leader of the group. On the other hand, P_5 influences no one.

Example 12   Consider a communication network whose digraph G is shown in Figure 7.16. The adjacency matrix of G is

$$\mathbf{A}(G) = \begin{array}{c} \\ P_1 \\ P_2 \\ P_3 \\ P_4 \\ P_5 \\ P_6 \end{array} \begin{array}{c} \begin{array}{cccccc} P_1 & P_2 & P_3 & P_4 & P_5 & P_6 \end{array} \\ \left[\begin{array}{cccccc} 0 & 0 & 0 & 0 & 1 & 0 \\ 0 & 0 & 0 & 0 & 1 & 1 \\ 1 & 1 & 0 & 0 & 1 & 0 \\ 0 & 1 & 0 & 0 & 1 & 1 \\ 0 & 1 & 0 & 0 & 0 & 1 \\ 1 & 0 & 1 & 0 & 0 & 0 \end{array} \right]. \end{array}$$

Although the relation "P_i has access to P_j" need not be transitive, we can speak of two-stage access. We say that P_i has **two-stage access** to P_k if we can find an individual P_j such that P_i has access to P_j and P_j has access to P_k. Similarly, P_i has r-stage access to P_k if we can find $r - 1$ individuals $P_{j_1}, \ldots, P_{j_{r-1}}$, such that P_i has access to P_{j_1}, P_{j_1} has access to $P_{j_2}, \ldots, P_{j_{r-2}}$ has access to $P_{j_{r-1}}$ and $P_{j_{r-1}}$ has access to P_k. Some of the $r + 1$ individuals $P_i, P_{j_1}, \ldots, P_{j_{r-1}}, P_k$ may be the same.

The following theorem, whose proof we omit, can be established.

Theorem 7.4   *Let* $\mathbf{A}(G)$ *be the adjacency matrix of a digraph* G *and let the* rth *power of* $\mathbf{A}(G)$ *be* $\mathbf{B}_r$:

$$[\mathbf{A}(G)]^r = \mathbf{B}_r = [b_{ij}^{(r)}].$$

Then the i, jth *element in* $\mathbf{B}_r$, $b_{ij}^{(r)}$, *is the number of ways in which* P_i *has access to* P_j *in* r *stages.*

The sum of the elements in the jth column of $[\mathbf{A}(G)]^r$ gives the number of ways in which P_j is reached by all the other individuals in r stages.

If we let

$$\mathbf{A}(G) + [\mathbf{A}(G)]^2 + \cdots + [\mathbf{A}(G)]^r = \mathbf{C} = [c_{ij}], \qquad (1)$$

then c_{ij} is the number of ways in which P_i has access to P_j in one, two, . . . , or r stages.

Similarly, we speak of r-stage dominance, r-stage influence, and so forth. We can also use this model to study the spread of a rumor. Thus c_{ij} in (1) is the number of ways in which P_i has spread the rumor to P_j in one, two, . . . , or r stages. In the influence relation, r-stage influence shows the effect of indirect influence.

Example 13   If G is the digraph in Figure 7.16, then we find that

$$[\mathbf{A}(G)]^2 = \begin{array}{c} \\ P_1 \\ P_2 \\ P_3 \\ P_4 \\ P_5 \\ P_6 \end{array} \begin{array}{cccccc} P_1 & P_2 & P_3 & P_4 & P_5 & P_6 \\ \begin{bmatrix} 0 & 1 & 0 & 0 & 0 & 1 \\ 1 & 1 & 1 & 0 & 0 & 1 \\ 0 & 1 & 0 & 0 & 2 & 2 \\ 1 & 1 & 1 & 0 & 1 & 2 \\ 1 & 0 & 1 & 0 & 1 & 1 \\ 1 & 1 & 0 & 0 & 2 & 0 \end{bmatrix} \end{array}$$

and

$$\mathbf{A}(G) + [\mathbf{A}(G)]^2 = \mathbf{C} = \begin{array}{c} \\ P_1 \\ P_2 \\ P_3 \\ P_4 \\ P_5 \\ P_6 \end{array} \begin{array}{cccccc} P_1 & P_2 & P_3 & P_4 & P_5 & P_6 \\ \begin{bmatrix} 0 & 1 & 0 & 0 & 1 & 1 \\ 1 & 1 & 1 & 0 & 1 & 2 \\ 1 & 2 & 0 & 0 & 3 & 2 \\ 1 & 2 & 1 & 0 & 2 & 3 \\ 1 & 1 & 1 & 0 & 1 & 2 \\ 2 & 1 & 1 & 0 & 2 & 0 \end{bmatrix} \end{array}.$$

Since $c_{35} = 3$, there are three ways in which P_3 has access to P_5 in one or two stages: $P_3 \rightarrow P_5$, $P_3 \rightarrow P_2 \rightarrow P_5$, and $P_3 \rightarrow P_1 \rightarrow P_5$.

In studying organizational structures, we often find subsets of people in which any pair of individuals are related. This is an example of a clique, which we now define.

Definition   A **clique** in a digraph is a subset S of the vertices satisfying the following properties:
(a) S contains three or more vertices.
(b) If P_i and P_j are in S, then there is a directed edge from P_i to P_j and a directed edge from P_j to P_i.
(c) There is no larger subset T of the vertices that satisfies property (b) and contains S [that is, S is a maximal subset satisfying (b)].

Example 14   Consider the digraph in Figure 7.17. The set $\{P_1, P_2, P_3\}$ satisfies conditions (a) and (b) for a clique, but it is not a clique, since it fails to satisfy condition (c). That is, $\{P_1, P_2, P_3\}$ is contained in $\{P_1, P_2, P_3, P_4\}$, which satisfies conditions (a), (b), and (c). Thus the only clique in this digraph is $\{P_1, P_2, P_3, P_4\}$.

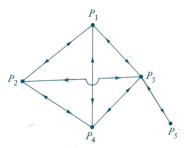

Figure 7.17

Example 15 Consider the digraph in Figure 7.18. In this case we have two cliques: $\{P_1, P_2, P_3, P_4\}$ and $\{P_4, P_5, P_6\}$. Moreover, P_4 belongs to both cliques.

   For large digraphs it is difficult to determine cliques. The following approach provides a useful method for detecting cliques that can easily be implemented on a computer. If $\mathbf{A}(G) = [a_{ij}]$ is the given adjacency matrix of a digraph, form a new matrix $\mathbf{S} = [s_{ij}]$:

$$s_{ij} = s_{ji} = 1 \quad \text{if} \quad a_{ij} = a_{ji} = 1;$$

otherwise, let $s_{ij} = s_{ji} = 0$. Thus $s_{ij} = 1$ if P_i and P_j have access to each other; otherwise, $s_{ij} = 0$. It should be noted that $\mathbf{S}$ is a symmetric matrix ($\mathbf{S} = \mathbf{S}^T$).

Example 16 Consider a digraph with adjacency matrix

$$\mathbf{A}(G) = \begin{array}{c} \\ P_1 \\ P_2 \\ P_3 \\ P_4 \\ P_5 \\ P_6 \end{array} \begin{array}{c} \begin{array}{cccccc} P_1 & P_2 & P_3 & P_4 & P_5 & P_6 \end{array} \\ \left[\begin{array}{cccccc} 0 & 0 & 1 & 1 & 1 & 0 \\ 1 & 0 & 1 & 1 & 1 & 1 \\ 0 & 1 & 0 & 1 & 1 & 1 \\ 1 & 0 & 1 & 0 & 0 & 1 \\ 1 & 1 & 0 & 1 & 0 & 1 \\ 0 & 1 & 1 & 1 & 1 & 0 \end{array} \right]. \end{array}$$

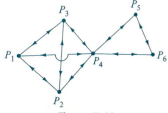

Figure 7.18

Then

$$
\mathbf{S} = \begin{array}{c} \\ P_1 \\ P_2 \\ P_3 \\ P_4 \\ P_5 \\ P_6 \end{array}
\begin{array}{c}
\begin{array}{cccccc} P_1 & P_2 & P_3 & P_4 & P_5 & P_6 \end{array} \\
\left[\begin{array}{cccccc}
0 & 0 & 0 & 1 & 1 & 0 \\
0 & 0 & 1 & 0 & 1 & 1 \\
0 & 1 & 0 & 1 & 0 & 1 \\
1 & 0 & 1 & 0 & 0 & 1 \\
1 & 1 & 0 & 0 & 0 & 1 \\
0 & 1 & 1 & 1 & 1 & 0
\end{array}\right].
\end{array}
$$

The following theorem can be proved.

Theorem 7.5 *Let $\mathbf{A}(G)$ be the adjacency matrix of a digraph and let $\mathbf{S} = [s_{ij}]$ be the symmetric matrix defined above, with $\mathbf{S}^3 = [s_{ij}^{(3)}]$, where $s_{ij}^{(3)}$ is the i, jth element in $\mathbf{S}^3$. Then P_i belongs to a clique if and only if the diagonal entry $s_{ii}^{(3)}$ is positive.*

Let us briefly consider why we examine the diagonal entries of $\mathbf{S}^3$ in Theorem 7.5. First, note that the diagonal entry $s_{ii}^{(3)}$ of $\mathbf{S}^3$ gives the number of ways in which P_i has access to himself in three stages. If $s_{ii}^{(3)} > 0$, then there is at least one way in which P_i has access to himself. Since a digraph has no loops, this access must occur through two individuals: $P_i \rightarrow P_j \rightarrow P_k \rightarrow P_i$. Thus $s_{ij} \neq 0$. But $s_{ij} \neq 0$ implies that $s_{ji} \neq 0$; so $P_j \rightarrow P_i$. Similarly, any two of the individuals in $\{P_i, P_j, P_k\}$ have access to each other. This means that P_i, P_j, and P_k all belong to the same clique. The opposite direction (if P_i is in a clique, then $s_{ii}^{(3)} > 0$) is left as an exercise.

The procedure for determining a clique in a digraph is as follows.

STEP 1. If $\mathbf{A} = [a_{ij}]$ is the adjacency matrix of the given digraph, compute the symmetric matrix $\mathbf{S} = [s_{ij}]$, where

$$s_{ij} = s_{ji} = 1 \qquad \text{if} \qquad a_{ij} = a_{ji} = 1$$

otherwise, $s_{ij} = 0$.

STEP 2. Compute $\mathbf{S}^3 = [s_{ij}^{(3)}]$

STEP 3. P_i belongs to a clique if and only if $s_{ii}^{(3)}$ is positive.

Example 17 Consider the digraph in Figure 7.19, whose adjacency matrix is

$$
\mathbf{A}(G) = \begin{array}{c} \\ P_1 \\ P_2 \\ P_3 \\ P_4 \\ P_5 \end{array}
\begin{array}{c}
\begin{array}{ccccc} P_1 & P_2 & P_3 & P_4 & P_5 \end{array} \\
\left[\begin{array}{ccccc}
0 & 1 & 1 & 0 & 1 \\
1 & 0 & 0 & 0 & 1 \\
1 & 0 & 0 & 1 & 0 \\
0 & 1 & 1 & 0 & 0 \\
1 & 0 & 0 & 0 & 0
\end{array}\right].
\end{array}
$$

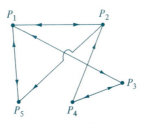

Figure 7.19

Then (verify)

$$
\mathbf{S} = \begin{array}{c}
 \\
P_1 \\
P_2 \\
P_3 \\
P_4 \\
P_5
\end{array}
\begin{array}{c}
\begin{array}{ccccc}
P_1 & P_2 & P_3 & P_4 & P_5
\end{array} \\
\left[\begin{array}{ccccc}
0 & 1 & 1 & 0 & 1 \\
1 & 0 & 0 & 0 & 0 \\
1 & 0 & 0 & 1 & 0 \\
0 & 0 & 1 & 0 & 0 \\
1 & 0 & 0 & 0 & 0
\end{array}\right]
\end{array}
$$

and

$$
\mathbf{S}^3 = \begin{array}{c}
 \\
P_1 \\
P_2 \\
P_3 \\
P_4 \\
P_5
\end{array}
\begin{array}{c}
\begin{array}{ccccc}
P_1 & P_2 & P_3 & P_4 & P_5
\end{array} \\
\left[\begin{array}{ccccc}
0 & 3 & 4 & 0 & 3 \\
3 & 0 & 0 & 1 & 0 \\
4 & 0 & 0 & 2 & 0 \\
0 & 1 & 2 & 0 & 1 \\
3 & 0 & 0 & 1 & 0
\end{array}\right]
\end{array}.
$$

Since every diagonal entry in $\mathbf{S}^3$ is zero, we conclude that there are no cliques.

Example 18 Consider the digraph in Figure 7.20, whose adjacency matrix is

$$
\mathbf{A}(G) = \begin{array}{c}
 \\
P_1 \\
P_2 \\
P_3 \\
P_4 \\
P_5
\end{array}
\begin{array}{c}
\begin{array}{ccccc}
P_1 & P_2 & P_3 & P_4 & P_5
\end{array} \\
\left[\begin{array}{ccccc}
0 & 0 & 1 & 1 & 1 \\
0 & 0 & 1 & 0 & 1 \\
1 & 0 & 0 & 0 & 1 \\
0 & 0 & 1 & 0 & 1 \\
1 & 1 & 1 & 0 & 0
\end{array}\right]
\end{array}.
$$

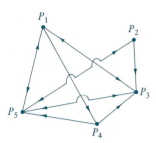

Figure 7.20

Then (verify)

$$\mathbf{S} = \begin{array}{c@{\quad}c} & \begin{array}{ccccc} P_1 & P_2 & P_3 & P_4 & P_5 \end{array} \\ \begin{array}{c} P_1 \\ P_2 \\ P_3 \\ P_4 \\ P_5 \end{array} & \left[\begin{array}{ccccc} 0 & 0 & 1 & 0 & 1 \\ 0 & 0 & 0 & 0 & 1 \\ 1 & 0 & 0 & 0 & 1 \\ 0 & 0 & 0 & 0 & 0 \\ 1 & 1 & 1 & 0 & 0 \end{array}\right] \end{array}$$

and

$$\mathbf{S}^3 = \begin{array}{c@{\quad}c} & \begin{array}{ccccc} P_1 & P_2 & P_3 & P_4 & P_5 \end{array} \\ \begin{array}{c} P_1 \\ P_2 \\ P_3 \\ P_4 \\ P_5 \end{array} & \left[\begin{array}{ccccc} 2 & 1 & 3 & 0 & 4 \\ 1 & 0 & 1 & 0 & 3 \\ 3 & 1 & 2 & 0 & 4 \\ 0 & 0 & 0 & 0 & 0 \\ 4 & 3 & 4 & 0 & 2 \end{array}\right] \end{array}.$$

Since s_{11}, s_{33}, and s_{55} are positive, we conclude that P_1, P_3, and P_5 belong to cliques and in fact they form the only clique in this digraph.

We now consider the notion of a strongly connected digraph.

Definition A **path** joining two individuals P_i and P_k in a digraph is a sequence of distinct vertices P_i, P_a, P_b, P_c, . . . , P_r, P_k and directed edges P_iP_a, P_aP_b, . . . , P_rP_k.

Example 19 Consider the digraph in Figure 7.21. The sequence

$$P_2 \to P_3 \to P_4 \to P_5$$

is a path. The sequence

$$P_2 \to P_3 \to P_4 \to P_2 \to P_5$$

is not a path, since the vertex P_2 is repeated.

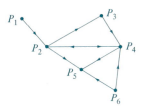

Figure 7.21

Definition The digraph G is said to be **strongly connected** if for every two distinct vertices P_i and P_j there is a path from P_i to P_j and a path from P_j to P_i. Otherwise, G is said to be **not strongly connected**.

An example of a strongly connected digraph is provided by the streets of a city.

For many digraphs it is a tedious task to determine whether or not they are strongly connected. First, observe that if our digraph has n vertices, then the number of edges in a path from P_i to P_j cannot exceed $n - 1$, since all the vertices of a path are distinct and we cannot go through more than the n vertices in the digraph. If $[\mathbf{A}(G)]^r = [b_{ij}^{(r)}]$, then $b_{ij}^{(r)}$ is the number of ways of getting from P_i to P_j in r stages. A way of getting from P_i to P_j in r stages need not be a path, since it may contain repeated vertices. If these repeated vertices and all edges between repeated vertices are deleted, we do obtain a path between P_i and P_j with at most r edges. For example, if we have $P_1 \rightarrow P_2 \rightarrow P_4 \rightarrow P_3 \rightarrow P_2 \rightarrow P_5$, we can eliminate the second P_2 and all edges between the P_2's and obtain the path $P_1 \rightarrow P_2 \rightarrow P_5$. Hence, if the i, jth element in

$$[\mathbf{A}(G)] + [\mathbf{A}(G)]^2 + \cdots + [\mathbf{A}(G)]^{n-1}$$

Theorem 7.6 *A digraph with n vertices is strongly connected if and only if its adjacency matrix $\mathbf{A}(G)$ has the property that*

$$\mathbf{A}(G) + [\mathbf{A}(G)]^2 + \cdots + [\mathbf{A}(G)]^{n-1} = \mathbf{E}$$

has no zero entries.

The procedure for determining if a digraph G with n vertices is strongly connected is as follows.

STEP 1. If $\mathbf{A}(G)$ is the adjacency matrix of the digraph, compute

$$\mathbf{A}(G) + [\mathbf{A}(G)]^2 + \cdots + [\mathbf{A}(G)]^{n-1} = \mathbf{E}$$

STEP 2. G is strongly connected if and only if $\mathbf{E}$ has no zero entries.

Figure 7.22

Example 20 Consider the digraph in Figure 7.22. The adjacency matrix is

$$\mathbf{A}(G) = \begin{array}{c} \\ P_1 \\ P_2 \\ P_3 \\ P_4 \\ P_5 \end{array} \begin{array}{ccccc} P_1 & P_2 & P_3 & P_4 & P_5 \\ \begin{bmatrix} 0 & 1 & 0 & 1 & 0 \\ 0 & 0 & 1 & 1 & 0 \\ 0 & 0 & 0 & 1 & 1 \\ 1 & 0 & 1 & 0 & 0 \\ 0 & 1 & 0 & 1 & 0 \end{bmatrix} \end{array}.$$

Then (verify)

$$\mathbf{A}(G) + [\mathbf{A}(G)]^2 + [\mathbf{A}(G)]^3 + [\mathbf{A}(G)]^4 = \mathbf{E}$$

$$= \begin{array}{c} \\ P_1 \\ P_2 \\ P_3 \\ P_4 \\ P_5 \end{array} \begin{array}{ccccc} P_1 & P_2 & P_3 & P_4 & P_5 \\ \begin{bmatrix} 5 & 5 & 7 & 10 & 3 \\ 5 & 4 & 8 & 10 & 3 \\ 5 & 4 & 7 & 10 & 4 \\ 5 & 4 & 7 & 10 & 4 \\ 5 & 5 & 7 & 10 & 3 \end{bmatrix} \end{array}.$$

Since all the entries in **E** are positive, the given digraph is strongly connected.

This approach can be used to trace the spread of a contaminant in a group of individuals; if there is a path from P_i to P_j, then the contaminant can spread from P_i to P_j.

FURTHER READING

Berge, C. *The Theory of Graphs and Its Applications.* London: Methuen & Company Ltd., 1962.
Busacker, R. G., and T. L. Saaty. *Finite Graphs and Networks: An Introduction with Applications.* New York: McGraw-Hill Book Company, 1965.
Even, S. *Algorithmic Combinatorics.* New York: Macmillan Publishing Co., Inc., 1973.
Johnston, J. B., G. Price, and F. S. van Vleck. *Linear Equations and Matrices.* Reading, Mass.: Addison-Wesley Publishing Company, Inc., 1966.
Kolman, B. and R. C. Busby, *Discrete Mathematical Structures for Computer Science,* Englewood Cliffs, Prentice-Hall, Inc., 1984.
Liu, C. L. *Introduction to Combinatorial Mathematics.* New York: McGraw-Hill Book Company, 1968.
Ore, O. *Graphs and Their Uses.* New York: Random House, Inc., 1963.

7.3 EXERCISES

1. Draw the complete graph on five vertices.

2. Write the matrices $\mathbf{M}(g)$ representing digraphs (2a) and (2b), as in Example 5.

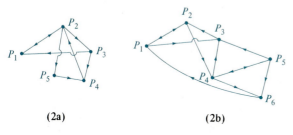

(2a)   (2b)

3. Which of the digraphs (3a), (3b), and (3c) are equal?

(*Hint:* If two digraphs are equal, the number of directed edges emanating from corresponding vertices must be the same.)

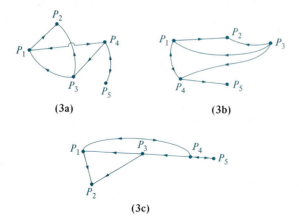

(3a)   (3b)

(3c)

4. Draw a digraph determined by the given matrix.

(a) $\begin{bmatrix} 0 & 1 & 0 & 0 & 0 \\ 0 & 0 & 1 & 1 & 0 \\ 1 & 0 & 0 & 1 & 1 \\ 1 & 1 & 0 & 0 & 1 \\ 1 & 0 & 1 & 1 & 0 \end{bmatrix}$.

(b) $\begin{bmatrix} 0 & 1 & 1 & 1 \\ 1 & 0 & 0 & 1 \\ 1 & 1 & 0 & 0 \\ 0 & 1 & 0 & 0 \end{bmatrix}$.

5. Write the adjacency matrices of the digraphs in Exercise 2.

6. Consider a group of five people, P_1, P_2, P_3, P_4, and P_5, who have been stationed on a remote island to operate a weather station. The following social interactions have been observed:

P_1 gets along with P_2, P_3, and P_4.
P_2 gets along with P_1 and P_5.
P_3 gets along with P_1, P_2, and P_4.
P_4 gets along with P_2, P_3, and P_5.
P_5 gets along with P_4.

Draw a digraph G describing the situation. Write the adjacency matrix representing G.

7. Which of the following matrices can be adjacency matrices for a dominance digraph?

(a) $\begin{bmatrix} 0 & 1 & 1 & 0 \\ 0 & 0 & 1 & 1 \\ 0 & 0 & 0 & 0 \\ 0 & 0 & 1 & 0 \end{bmatrix}$.

(b) $\begin{bmatrix} 0 & 0 & 1 & 0 & 0 \\ 1 & 0 & 1 & 0 & 1 \\ 0 & 0 & 0 & 0 & 1 \\ 1 & 1 & 1 & 0 & 1 \\ 1 & 0 & 0 & 0 & 0 \end{bmatrix}$.

8. The following data have been obtained by studying a group of six individuals in the course of a sociological study:

Carter influences Smith and Gordon.
Gordon influences Jones.
Smith is influenced by Peters.
Russell is influenced by Carter, Smith, and Gordon.
Peters is influenced by Russell.
Jones influences Carter and Russell.
Smith influences Jones.
Carter is influenced by Peters.
Peters influences Jones and Gordon.

(a) Who influences the most people?
(b) Who is influenced by the most people?

9. Consider a communication network among five individuals with adjacency matrix

$$\begin{array}{c c c c c c} & P_1 & P_2 & P_3 & P_4 & P_5 \\ \begin{matrix} P_1 \\ P_2 \\ P_3 \\ P_4 \\ P_5 \end{matrix} & \begin{bmatrix} 0 & 1 & 1 & 1 & 1 \\ 0 & 0 & 0 & 1 & 1 \\ 0 & 1 & 0 & 1 & 0 \\ 0 & 0 & 0 & 0 & 1 \\ 1 & 0 & 1 & 0 & 0 \end{bmatrix} \end{array}.$$

(a) In how many ways does P_2 have access to P_1 through one individual?

(b) What is the smallest number of individuals through which P_2 has access to himself?

10. Consider the following influence relation among five individuals?

P_1 influences P_2, P_4, and P_5.
P_2 influences P_3 and P_4.
P_3 influences P_1 and P_4.
P_4 influences P_5.
P_5 influences P_2 and P_3.

(a) Can P_4 influence P_1 in at most two stages?

(b) In how many ways does P_1 influence P_4 in exactly three stages?

(c) In how many ways does P_1 influence P_4 in one, two, or three stages?

In Exercises 11 through 13, determine a clique, if there is one, for the digraph with given adjacency matrix.

11. $\begin{bmatrix} 0 & 0 & 0 & 0 & 0 \\ 1 & 0 & 1 & 1 & 1 \\ 0 & 1 & 0 & 1 & 0 \\ 1 & 1 & 1 & 0 & 0 \\ 0 & 0 & 1 & 1 & 0 \end{bmatrix}.$

12. $\begin{bmatrix} 0 & 1 & 1 & 0 & 1 & 1 \\ 1 & 0 & 0 & 0 & 0 & 0 \\ 0 & 0 & 0 & 1 & 0 & 0 \\ 1 & 0 & 0 & 0 & 1 & 1 \\ 1 & 0 & 1 & 1 & 0 & 1 \\ 1 & 0 & 1 & 1 & 1 & 0 \end{bmatrix}.$

13. $\begin{bmatrix} 0 & 1 & 1 & 0 & 1 \\ 1 & 0 & 1 & 1 & 1 \\ 0 & 0 & 0 & 1 & 0 \\ 1 & 0 & 0 & 0 & 1 \\ 0 & 1 & 0 & 1 & 0 \end{bmatrix}.$

14. Consider a communication network among five individuals with adjacency matrix

$$\begin{bmatrix} 0 & 1 & 0 & 0 & 0 \\ 0 & 0 & 1 & 0 & 1 \\ 0 & 0 & 0 & 1 & 0 \\ 1 & 1 & 0 & 0 & 0 \\ 0 & 0 & 1 & 1 & 0 \end{bmatrix}.$$

(a) Can P_3 get a message to P_5 in at most two stages?

(b) What is the minimum number of stages that will guarantee that every person can get a message to any other (different) person?

(c) What is the minimum number of stages that will guarantee that every person can get a message to any person (including himself)?

15. Determine whether the digraph with given adjacency matrix is strongly connected.

(a) $\begin{bmatrix} 0 & 1 & 1 & 1 \\ 0 & 0 & 1 & 1 \\ 1 & 0 & 0 & 1 \\ 0 & 0 & 1 & 0 \end{bmatrix}.$

(b) $\begin{bmatrix} 0 & 1 & 0 & 0 & 0 \\ 0 & 0 & 1 & 0 & 0 \\ 1 & 0 & 0 & 0 & 0 \\ 0 & 0 & 0 & 0 & 1 \\ 0 & 0 & 1 & 1 & 0 \end{bmatrix}.$

16. Determine whether the digraph with given adjacency matrix is strongly connected.

(a) $\begin{bmatrix} 0 & 0 & 1 & 1 & 1 \\ 1 & 0 & 1 & 1 & 0 \\ 0 & 1 & 0 & 0 & 0 \\ 0 & 1 & 0 & 0 & 1 \\ 1 & 1 & 0 & 0 & 0 \end{bmatrix}.$

$$\text{(b)} \begin{bmatrix} 0 & 0 & 0 & 0 & 1 \\ 0 & 0 & 1 & 1 & 0 \\ 0 & 1 & 0 & 0 & 1 \\ 1 & 0 & 0 & 0 & 0 \\ 0 & 0 & 0 & 1 & 0 \end{bmatrix}.$$

17. A group of five acrobats performs a pyramiding act in which there must be a supporting path between any two different persons or the pyramid collapses.

In the following adjacency matrix, $a_{ij} = 1$ means that P_i supports P_j.

$$\begin{bmatrix} 0 & 0 & 1 & 1 & 0 \\ 1 & 0 & 0 & 0 & 1 \\ 1 & 0 & 0 & 1 & 0 \\ 1 & 0 & 0 & 0 & 1 \\ 0 & 1 & 1 & 0 & 0 \end{bmatrix}.$$

Who can be left out of the pyramid without having a collapse occur?

THEORETICAL EXERCISES

T.1. Show that a digraph which has the directed edges P_iP_j and P_jP_i cannot be a dominance graph.

T.2. Prove Theorem 7.4.

T.3. Prove Theorem 7.5.

7.4

THE THEORY OF GAMES

Prerequisite. Chapter 6, Linear Programming.

There are many problems in economics, politics, warfare, business, and so on, that require decisions to be made in conflicting or competitive situations. The theory of games is a rather new area of applied mathematics that attempts to analyze conflict situations and provides a basis for rational decision making.

The theory of games was developed in the 1920s by John von Neumann[†] and E. Borel,[‡] but the subject did not come to fruition until the publication, in 1944, of the

---

  † John von Neumann (1903–1957) was born in Hungary and came to the United States in 1930. His talents in mathematics were recognized early, and he is considered one of the greatest mathematicians of the twentieth century. He made fundamental contributions to the foundations of mathematics, quantum mechanics, operator calculus, and the theory of computing machines and automata. He also participated in many defense projects during World War II and helped to develop the atomic bomb. Beginning in 1926 he developed the theory of games, culminating in his joint work with Oskar Morgenstern in 1944, *Theory of Games and Economic Behavior*.

  ‡ Emile Borel (1871–1956) was born in Saint Affrique, France. A child prodigy and prolific author of mathematical papers, he married an author, was part of French literary circles, and was active in politics, serving in the National Assembly for a time and joining in the Resistance during World War II. In mathematics he contributed to the theory of functions, measure theory, and probability theory. In the 1920s he wrote the first papers in game theory, defining the basic terms, stating the minimax theorem, and discussing applications to war and economics.

decisive book *Theory of Games and Economic Behavior* by John von Neumann and Oskar Morgenstern.†

A **game** is a competitive situation in which each of a number of players is pursuing his objective in direct conflict with the other players. Each player is doing everything he can to gain as much as possible for himself. Essentially, games are of two types. First, there are **games of chance**, such as roulette, which require no skill on the part of the players; the outcomes and winnings are determined solely by the laws of probability and can in no way be affected by any actions of the players. Second, there are **games of strategy**, such as chess, checkers, bridge, and poker, which require skill on the part of the players; the outcomes and winnings are determined by the skills of the players. By **game**, we mean a game of strategy. In addition to such parlor games there are many games in economic competition, warfare, geological exploration, farming, administration of justice, and so on, in which each player (competitor) can choose one of a set of possible moves and the outcomes depend upon the player's (competitor's) skill. The theory of games attempts to determine the best course of action for each player. The subject is still being developed and many new theoretical and applied results are needed for handling the complex games that occur in everyday problems.

We shall limit our treatment to games played by two players, usually denoted by R and C. We shall also assume that R has m possible moves (or courses of action) and that C has n moves. We now form an $m \times n$ matrix by labeling its rows, from top to bottom, with the moves of R, and labeling its columns, from left to right, with the moves of C. Entry a_{ij}, in row i and column j, indicates the amount (money or some other valuable item) received by R if R makes his ith move and C makes his jth move. The entry a_{ij} is called a **payoff** and the matrix $\mathbf{A} = [a_{ij}]$ is called the **payoff matrix** (for R). Such games are called **two-person games**. We also refer to them as **matrix games**.

We could also, if we wished, construct a second matrix that represents payoffs for C. However, in a class of games called **constant-sum games**, the sum of the payoff to R and the payoff to C is constant for all mn pairs of RC moves. The more R gains, the less C gains (or the more C loses), and vice versa. A special kind of constant-sum game is the **zero-sum game** in which the amount won by one player is exactly the amount lost by the other player. Because of this strict interrelation between the payoff matrix for C and that for R, in the case of a constant-sum game it is sufficient to study only the payoff matrix for R. In the following discussion we study only constant-sum games, and the matrix involved is the payoff matrix for R.

In the study of matrix games, it is always assumed that both players are equally capable, that each is playing as well as he can possibly play, and that each player makes his move without knowing what his opponent's move will be.

Example 1   Consider the game of matching pennies, consisting of two players, R and C, each of whom has a penny in his hand. Each player shows one side of the coin without knowing

---

† Oskar Morgenstern (1902–1977) was born in Germany and came to the United States in 1938. He was a professor of economics at Princeton University until 1970 and at New York University until his death. In 1944 he collaborated with von Neumann on the influential book *Theory of Games and Economic Behavior*.

his opponent's choice. If both players are showing the same side of the coin, then R wins $1 from C; otherwise, C wins $1 from R.

In this two-person, zero-sum game each player has two possible moves: he can show a tail or he can show a head. The payoff matrix is thus

$$
\begin{array}{c}
 \\
R
\end{array}
\begin{array}{c}
 \\
\begin{array}{cc} H & T \end{array} \\
\begin{array}{c} H \\ T \end{array}
\left[
\begin{array}{cc}
1 & -1 \\
-1 & 1
\end{array}
\right].
\end{array}
$$

over C

Example 2 There are two suppliers, firms R and C, of a new specialized type of tire that has 100,000 customers. Each company can advertise its product on TV or in the newspapers. A marketing firm determines that if both firms advertise on TV, then firm R gets 40,000 customers (and firm C gets 60,000 customers). If they both use newspapers, then each gets 50,000 customers. If R uses newspapers and C uses TV, then R gets 60,000 customers. If R uses TV and C uses newspapers, they each get 50,000 customers.

We can consider this situation as a game between firms R and C, with the payoff matrix as shown in Figure 7.23. The entries in the matrix indicate the number of customers secured by firm R. This is a constant-sum game, for the sum of the R customers and the C customers is always the total population of 100,000 tire customers.

Consider now a two-person, constant-sum game with the $m \times n$ payoff matrix $\mathbf{A} = [a_{ij}]$, so that player R has m moves and player C has n moves. If player R plays his ith move, he is assured of winning at least the smallest entry in the ith row of $\mathbf{A}$, no matter what C does. Thus R's best course of action is to choose that move which will maximize his assured winnings in spite of C's best countermove. Player R will get his largest payoff by maximizing his smallest gain. Player C's goals are in direct conflict with those of player R: he is trying to keep R's winnings to a minimum. If C plays his jth move, he is assured of losing no more than the largest entry in the jth column of $\mathbf{A}$, no matter what R does. Thus C's best course of action is to choose that move which will minimize his assured losses in spite of R's best countermove. Player C will do his best by minimizing his largest loss.

Definition If the payoff matrix of a matrix game contains an entry a_{rs}, which is at the same time the minimum of row r and the maximum of column s, then a_{rs} is called a **saddle point**. Also, a_{rs} is called the **value** of the game. If the value of a zero-sum game is zero, the game is said to be **fair**.

		Firm C	
		T.V.	Newspapers
Firm R	T.V.	40,000	50,000
	Newspapers	60,000	50,000

Figure 7.23

Definition   A matrix game is said to be **strictly determined** if its payoff matrix has a saddle point.

If a_{rs} is a saddle point for a matrix game, then player R will be assured of winning at least a_{rs} by playing his rth move and player C will be guaranteed that he will lose no more than a_{rs} by playing his sth move. This is the best that each player can do.

Example 3   Consider a game with payoff matrix

$$R \begin{bmatrix} 0 & -3 & -1 & 3 \\ 3 & 2 & 2 & 4 \\ 1 & 4 & 0 & 6 \end{bmatrix}.$$

$$\overset{C}{}$$

To determine whether this game has a saddle point, we write the minimum of each row to the right of the row and the maximum of each column at the bottom of each column. Thus we have

	C				Row minima
R	0	−3	−1	3	−3
	3	2	2	4	2
	1	4	0	6	0
Column maxima	3	4	2	6	

Entry $a_{23} = 2$ is both the least entry in the second row and the largest entry in the third column. Hence it is a saddle point for the game, which is then a strictly determined game. The value of the game is 2 and player R has an advantage. The best course of action for R is to play his second move; he will win at least 2 units from C, no matter what C does. The best course of action for C is to play his third move; he will limit his loss to not more than 2 units, no matter what R does.

Example 4   Consider the advertising game of Example 2. The payoff matrix is shown in Figure 7.24. Thus entry $a_{22} = 50,000$ is a saddle point. The best course of action for both firms is to advertise in newspapers. The game is strictly determined with value 50,000.

There are many games that are not strictly determined.

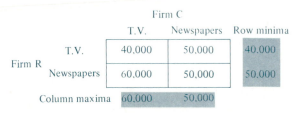

		Firm C		Row minima
		T.V.	Newspapers	
Firm R	T.V.	40,000	50,000	40,000
	Newspapers	60,000	50,000	50,000
	Column maxima	60,000	50,000	

Figure 7.24

Example 5 Consider the game with payoff matrix

$$\begin{bmatrix} 1 & 6 & -1 \\ 3 & -2 & 4 \\ 4 & 5 & -3 \end{bmatrix} \quad \begin{matrix} -1 \\ -2 \\ -3 \end{matrix}.$$

Column maxima $\quad 4 \quad 6 \quad 4$

It is clear that there is no saddle point.

On the other hand, a game may have more than one saddle point. However, it can be proved that all saddle points must have the same value.

Example 6 Consider the game with payoff matrix

$$\begin{bmatrix} 5 & 4 & 5 & 4 \\ 6 & -1 & 3 & 2 \\ 6 & 4 & 6 & 4 \end{bmatrix} \quad \begin{matrix} 4 \\ -1 \\ 4 \end{matrix}.$$

Column maxima $\quad 6 \quad 4 \quad 6 \quad 4$

Entries a_{12}, a_{14}, a_{32}, and a_{34} are all saddle points and have the same value, 4. They appear shaded in the payoff matrix. The value of the game is also 4.

Consider now the penny-matching game of Example 1 with payoff matrix

$$\begin{array}{cc} & C \\ & \begin{array}{cc} H & T \end{array} \quad \text{Row minima} \\ R \begin{array}{c} H \\ T \end{array} \begin{bmatrix} 1 & -1 \\ -1 & 1 \end{bmatrix} & \begin{matrix} -1 \\ -1 \end{matrix}. \end{array}$$

Column maxima $\quad 1 \quad 1$

It is clear that this game is not strictly determined; that is, it has no saddle point.

To analyze this type of situation, we assume that a game is played repeatedly and that each player is trying to determine his best course of action. Thus player R tries to maximize his winnings while player C tries to minimize his losses. A **strategy** for a player is a decision for choosing his moves.

Consider now the above penny-matching game. Suppose that in the repeated play of the game, player R always chooses the first row (he chooses to show heads), in the hope that player C will always choose the first column (play heads), thereby ensuring a win of \$1 for himself. However, as player C begins to notice that player R always chooses his first row, then player C will choose his second column, resulting in a loss of \$1 for R. Similarly, if R always chooses the second row, then C will choose the first column, resulting in a loss of \$1 for R. We can thus conclude that each player must somehow keep the other player from anticipating his choice of moves. This situation is in marked contrast with the case in strictly determined games. In a strictly determined game each player will make the same move whether or not he has advanced knowledge of his opponent's move. Thus, in a nonstrictly determined game, each player will make each move with a certain relative frequency.

Definition   Suppose that we have a matrix game with an $m \times n$ payoff matrix $\mathbf{A}$. Let p_i, $1 \le i \le m$, be the probability that R chooses the ith row of $\mathbf{A}$ (that is, chooses his ith move). Let q_j, $1 \le j \le n$, be the probability that C chooses the jth column of $\mathbf{A}$. The vector $\mathbf{P} = [p_1 \, p_2 \, \cdots \, p_m]$ is called a **strategy** for player R; the vector

$$\mathbf{Q} = \begin{bmatrix} q_1 \\ q_2 \\ \vdots \\ q_n \end{bmatrix}$$

is called a **strategy** for player C.

Of course, the probabilities p_i and q_j in the definition satisfy

$$p_1 + p_2 + \cdots + p_m = 1$$
$$q_1 + q_2 + \cdots + q_n = 1.$$

If a matrix game is strictly determined, then optimal strategies for R and C are strategies having 1 as one component and zero for all other components. Such strategies are called **pure strategies**. A strategy that is not pure is called a **mixed strategy**. Thus, in Example 3, the pure strategy for R is

$$\mathbf{p} = [0 \quad 1 \quad 0],$$

and the pure strategy for C is

$$\mathbf{Q} = \begin{bmatrix} 0 \\ 0 \\ 1 \\ 0 \end{bmatrix}.$$

Consider now a matrix game with payoff matrix

$$\mathbf{A} = \begin{bmatrix} a_{11} & a_{12} \\ a_{21} & a_{22} \end{bmatrix}. \tag{1}$$

Suppose that $\mathbf{P} = [p_1 \quad p_2]$ and $\mathbf{Q} = \begin{bmatrix} q_1 \\ q_2 \end{bmatrix}$ are strategies for R and C, respectively. Then if R plays his first row with probability p_1 and if C plays his first column with probability q_1, then R's expected payoff is $p_1 q_1 a_{11}$. Similarly, we can examine the remaining three possibilities, obtaining Table 7.1. The expected payoff $E(\mathbf{P}, \mathbf{Q})$ of the game to R is then the sum of the four quantities in the rightmost column. We obtain

$$E(\mathbf{P}, \mathbf{Q}) = p_1 q_1 a_{11} + p_1 q_2 a_{12} + p_2 q_1 a_{21} + p_2 q_2 a_{22},$$

which can be written in matrix form (verify) as

$$E(\mathbf{P}, \mathbf{Q}) = \mathbf{PAQ}. \tag{2}$$

Table 7.1

Moves				
Player R	Player C	Probability	Payoff to Player R	Expected Payoff to Player R
Row 1	Column 1	$p_1 q_1$	a_{11}	$p_1 q_1 a_{11}$
Row 1	Column 2	$p_1 q_2$	a_{12}	$p_1 q_2 a_{12}$
Row 2	Column 1	$p_2 q_1$	a_{21}	$p_2 q_1 a_{21}$
Row 2	Column 2	$p_2 q_2$	a_{22}	$p_2 q_2 a_{22}$

The same analysis applies to a matrix game with an $m \times n$ payoff matrix $\mathbf{A}$. Thus, if

$$\mathbf{P} = [p_1 \quad p_2 \quad \cdots \quad p_m] \quad \text{and} \quad \mathbf{Q} = \begin{bmatrix} q_1 \\ q_2 \\ \vdots \\ q_n \end{bmatrix}$$

are strategies for R and C, respectively, then the expected payoff to player R is given by (2).

Example 7 Consider a matrix game with payoff matrix

$$\mathbf{A} = \begin{bmatrix} 2 & -2 & 3 \\ 4 & 0 & -3 \end{bmatrix}.$$

If

$$\mathbf{P} = [\tfrac{1}{4} \quad \tfrac{3}{4}] \quad \text{and} \quad \mathbf{Q} = \begin{bmatrix} \tfrac{1}{3} \\ \tfrac{1}{3} \\ \tfrac{1}{3} \end{bmatrix}$$

are strategies for R and C, respectively, then the expected payoff to R is

$$E(\mathbf{P}, \mathbf{Q}) = \mathbf{PAQ}$$

$$= [\tfrac{1}{4} \quad \tfrac{3}{4}] \begin{bmatrix} 2 & -2 & 3 \\ 4 & 0 & -3 \end{bmatrix} \begin{bmatrix} \tfrac{1}{3} \\ \tfrac{1}{3} \\ \tfrac{1}{3} \end{bmatrix} = \tfrac{1}{2}.$$

If

$$\mathbf{P} = [\tfrac{3}{4} \quad \tfrac{1}{4}] \quad \text{and} \quad \mathbf{Q} = \begin{bmatrix} \tfrac{1}{3} \\ \tfrac{2}{3} \\ 0 \end{bmatrix}$$

are strategies for R and C, respectively, then the expected payoff to R is $-\tfrac{1}{6}$. Thus, in the first case, R gains $\tfrac{1}{2}$ from C, whereas in the second case R loses $\tfrac{1}{6}$ to C.

A strategy for player R is said to be **optimal** if it guarantees R the largest possible payoff no matter what his opponent may do. Similarly, a strategy for player C is said to be **optimal** if it guarantees the smallest possible payoff to R no matter what R may do.

If $\mathbf{P}$ and $\mathbf{Q}$ are optimal strategies for R and C, respectively, then the expected payoff to R, $v = E(\mathbf{P}, \mathbf{Q})$, is called the **value** of the game. Although $E(\mathbf{P}, \mathbf{Q})$ is a 1×1 matrix, we think of it merely as a number v. If the value of a zero-sum game is zero, the game is said to be **fair**. The principal task of the theory of games is the determination of optimal strategies for each player.

Consider again a matrix game with the 2×2 payoff matrix (1) and suppose that the game is not strictly determined. It can then be shown that

$$a_{11} + a_{22} - a_{12} - a_{21} \neq 0.$$

To determine an optimal strategy for R, we proceed as follows. Suppose that R's strategy is $[p_1 \quad p_2]$. Then if C plays his first column, the expected payoff to R is

$$a_{11}p_1 + a_{21}p_2. \tag{3}$$

If C plays his second column, the expected payoff to R is

$$a_{12}p_1 + a_{22}p_2. \tag{4}$$

If v is the minimum of the expected payoffs (3) and (4), then R expects to gain at least v units from C no matter what C does. Thus we have

$$a_{11}p_1 + a_{21}p_2 \geqslant v \tag{5}$$
$$a_{12}p_1 + a_{22}p_2 \geqslant v. \tag{6}$$

Moreover, player R seeks to make v as large as possible. Thus player R seeks to find p_1, p_2, and v such that

$$v \text{ is a maximum}$$

and

$$\begin{aligned} a_{11}p_1 + a_{21}p_2 - v &\geqslant 0 \\ a_{12}p_1 + a_{22}p_2 - v &\geqslant 0 \\ p_1 + p_2 &= 1 \\ p_1 \geqslant 0, \quad p_2 \geqslant 0, \quad v &\geqslant 0. \end{aligned} \tag{7}$$

We shall see below (in a more general situation) that problem (7) is a linear programming problem. It can be shown that a solution to (7), giving an optimal strategy for R, is

$$p_1 = \frac{a_{22} - a_{21}}{a_{11} + a_{22} - a_{12} - a_{21}}, \qquad p_2 = \frac{a_{11} - a_{12}}{a_{11} + a_{22} - a_{12} - a_{21}} \qquad (8)$$

and

$$v = \frac{a_{11}a_{22} - a_{12}a_{21}}{a_{11} + a_{22} - a_{12} - a_{21}}. \qquad (9)$$

We now find an optimal strategy for C. Suppose that C's strategy is $\begin{bmatrix} q_1 \\ q_2 \end{bmatrix}$. If R plays the first row, then the expected payoff to R is

$$a_{11}q_1 + a_{12}q_2, \qquad (10)$$

while if R plays the second row, the expected payoff to R is

$$a_{21}q_1 + a_{22}q_2. \qquad (11)$$

If v' is the maximum of the expected payoffs (10) and (11), then

$$a_{11}q_1 + a_{12}q_2 \leq v'$$
$$a_{21}q_1 + a_{22}q_2 \leq v'.$$

Since player C wishes to lose as little as possible, he seeks to make v' as small as possible. Thus C wants to find q_1, q_2, and v such that

$$v' \text{ is a minimum}$$

and

$$
\begin{aligned}
a_{11}q_1 + a_{12}q_2 - v' &\leq 0 \\
a_{21}q_1 + a_{22}q_2 - v' &\leq 0 \\
q_1 + q_2 &= 1 \\
q_1 \geq 0, \quad q_2 \geq 0, \quad v' &\geq 0.
\end{aligned}
\qquad (12)
$$

Problem (12) is also a linear programming problem. It can be shown that a solution to (12), giving an optimal strategy for C, is

$$q_1 = \frac{a_{22} - a_{12}}{a_{11} + a_{22} - a_{12} - a_{21}}, \qquad q_2 = \frac{a_{11} - a_{21}}{a_{11} + a_{22} - a_{12} - a_{21}} \qquad (13)$$

and

$$v' = \frac{a_{11}a_{22} - a_{12}a_{21}}{a_{11} + a_{22} - a_{12} - a_{21}}. \qquad (14)$$

Thus $v = v'$ when both players use their optimal strategies.

Example 8 For the zero-sum penny-matching game of Example 1, we have upon substituting in (8), (9), and (13),

$$p_1 = p_2 = \tfrac{1}{2} \quad \text{and} \quad q_1 = q_2 = \tfrac{1}{2}, \quad v = 0,$$

so that optimal strategies for R and C are

$$[\tfrac{1}{2} \quad \tfrac{1}{2}] \quad \text{and} \quad \begin{bmatrix} \tfrac{1}{2} \\ \tfrac{1}{2} \end{bmatrix},$$

respectively. This means that half the time R should show heads and half the time he should show tails; likewise for player C. The value of the game is zero, so the game is fair.

Example 9 Consider a matrix game with payoff matrix

$$\begin{bmatrix} 2 & -5 \\ 1 & 3 \end{bmatrix}.$$

Again substituting in (8), (9), and (13), we obtain

$$p_1 = \frac{3 - 1}{2 + 3 - 1 + 5} = \frac{2}{9}, \qquad p_2 = \frac{2 + 5}{2 + 3 - 1 + 5} = \frac{7}{9},$$

$$q_1 = \frac{3 + 5}{2 + 3 - 1 + 5} = \frac{8}{9}, \qquad q_2 = \frac{2 - 1}{2 + 3 - 1 + 5} = \frac{1}{9},$$

$$v = \frac{6 + 5}{2 + 3 - 1 + 5} = \frac{11}{9}.$$

Thus optimal strategies for R and C are

$$[\tfrac{2}{9} \quad \tfrac{7}{9}] \quad \text{and} \quad \begin{bmatrix} \tfrac{8}{9} \\ \tfrac{1}{9} \end{bmatrix},$$

respectively; when both players use their optimal strategies, the value of the game (the expected payoff to R) is $\tfrac{11}{9}$. If this matrix represents a zero-sum game, the game is not fair and in the long run favors player R.

We can now generalize our discussion to a game with an $m \times n$ payoff matrix $\mathbf{A} = [a_{ij}]$. First, let us observe that if we add a constant r to every entry of $\mathbf{A}$, then the optimal strategies for R and C do not change, and the value of the new game is r plus the value of the old game (Exercise T.2). Thus we can assume that, by adding a suitable constant to every entry of the payoff matrix, every entry of $\mathbf{A}$ is positive.

Player R seeks to find $p_1, p_2, \ldots, p_m$, and v such that

$$v \text{ is a maximum}$$

subject to

$$
\begin{aligned}
a_{11}p_1 + a_{21}p_2 + \cdots + a_{m1}p_m - v &\geq 0 \\
a_{12}p_1 + a_{22}p_2 + \cdots + a_{m2}p_m - v &\geq 0 \\
\vdots \qquad \vdots \qquad\qquad \vdots \qquad \vdots \\
a_{1n}p_1 + a_{2n}p_2 + \cdots + a_{mn}p_m - v &\geq 0 \\
p_1 + p_2 + \cdots + p_m &= 1 \\
p_1 \geq 0,\ p_2 \geq 0,\ \ldots,\ p_m \geq 0,\ v &\geq 0.
\end{aligned}
\tag{15}
$$

Since every entry of $\mathbf{A}$ is positive, we may assume that $v > 0$. Now divide each of the constraints in (15) by v, and let

$$
y_i = \frac{p_i}{v}.
$$

Observe that

$$
\begin{aligned}
y_1 + y_2 + \cdots + y_m &= \frac{p_1}{v} + \frac{p_2}{v} + \cdots + \frac{p_m}{v} \\
&= \frac{1}{v}(p_1 + p_2 + \cdots + p_m) = \frac{1}{v}.
\end{aligned}
$$

Thus v is a maximum if and only if $y_1 + y_2 + \cdots + y_m$ is a minimum. We can now restate problem (15), R's problem, as follows:

$$\text{Minimize} \qquad y_1 + y_2 + \cdots + y_m$$

subject to

$$
\begin{aligned}
a_{11}y_1 + a_{21}y_1 + \cdots + a_{m1}y_m &\geq 1 \\
a_{12}y_1 + a_{22}y_2 + \cdots + a_{m2}y_m &\geq 1 \\
\vdots \qquad \vdots \qquad\qquad \vdots \qquad \vdots \\
a_{1n}y_1 + a_{2n}y_2 + \cdots + a_{mn}y_m &\geq 1 \\
y_1 \geq 0,\ y_2 \geq 0,\ \ldots,\ y_m &\geq 0.
\end{aligned}
\tag{16}
$$

Observe that (16) is a linear programming problem and that it has one fewer constraint and one fewer variable than (15).

Turning next to C's problem, we note that he seeks to find $q_1, q_2, \ldots, q_n$ and v' such that

$$v' \text{ is a minimum}$$

subject to

$$
\begin{aligned}
a_{11}q_1 + a_{12}q_2 + \cdots + a_{1n}q_n - v' &\leq 0 \\
a_{21}q_1 + a_{22}q_2 + \cdots + a_{2n}q_n - v' &\leq 0 \\
\vdots \qquad \vdots \qquad\qquad \vdots \qquad \vdots \\
a_{m1}q_1 + a_{m2}q_2 + \cdots + a_{mn}q_n - v' &\leq 0 \\
q_1 + q_2 + \cdots + q_n &= 1 \\
q_1 \geq 0,\quad q_2 \geq 0,\ \ldots,\ q_n \geq 0,\ v' &\geq 0.
\end{aligned}
\tag{17}
$$

The fundamental theorem of matrix games, which we now state, says that every matrix game has a solution.

Theorem 7.7 *Every matrix game has a solution. That is, there are optimal strategies for R and C.*
(Fundamental *Moreover, $v = v'$.*
Theorem of
Matrix Games)

Since $v = v'$, we can divide each of the constraints in (17) by $v' = v$ and let

$$x_i = \frac{q_i}{v}.$$

Now

$$x_1 + x_2 + \cdots + x_n = \frac{1}{v},$$

so v is a minimum if and only if $x_1 + x_2 + \cdots + x_n$ is a maximum. We can now restate problem (17), C's problem, as follows:

$$\text{Maximize} \quad x_1 + x_2 + \cdots + x_n$$

subject to

$$
\begin{aligned}
a_{11}x_1 + a_{12}x_2 + \cdots + a_{1n}x_n &\leq 1 \\
a_{21}x_1 + a_{22}x_2 + \cdots + a_{2n}x_n &\leq 1 \\
&\;\;\vdots \\
a_{m1}x_1 + a_{m2}x_2 + \cdots + a_{mn}x_n &\leq 1 \\
x_1 \geq 0,\; x_2 \geq 0,\; \ldots,\; x_n &\geq 0.
\end{aligned}
\tag{18}
$$

Observe that (18) is a linear programming problem in standard form, which is the dual of (16). From the results of Section 6.3, it follows that when (18) is solved by the simplex method, the final tableau will contain the optimal strategies for R in the objective row under the columns of the slack variables. That is, y_1 is found in the objective row under the first slack variable, y_2 is found in the objective row under the second slack variable, and so on.

Example 10 Consider a game with payoff matrix

$$
\begin{bmatrix}
2 & -3 & 0 \\
3 & 1 & -2
\end{bmatrix}.
$$

Adding 4 to each element of the matrix, we obtain a matrix **A** with positive entries.

$$
\mathbf{A} =
\begin{bmatrix}
6 & 1 & 4 \\
7 & 5 & 2
\end{bmatrix}.
$$

We now find optimal strategies for the game with payoff matrix **A**. Problem (18), C's problem, becomes:

$$\text{Maximize} \quad x_1 + x_2 + x_3$$

subject to

$$6x_1 + x_2 + 4x_3 \leq 1$$
$$7x_1 + 5x_2 + 2x_3 \leq 1$$
$$x_1 \geq 0, \quad x_2 \geq 0, \quad x_3 \geq 0.$$

If we introduce the slack variables x_4 and x_5, our problem becomes:

$$\text{Maximize} \quad x_1 + x_2 + x_3$$

subject to

$$6x_1 + x_2 + 4x_3 + x_4 \qquad = 1$$
$$7x_1 + 5x_2 + 2x_3 \qquad + x_5 = 1$$
$$x_1 \geq 0, \quad x_2 \geq 0, \quad x_3 \geq 0, \quad x_4 \geq 0, \quad x_5 \geq 0.$$

Using the simplex method, we obtain

	x_1	x_2	x_3	x_4	x_5	z	
← x_4	6	1	④	1	0	0	1
x_5	7	5	2	0	1	0	1
	−1	−1	−1	0	0	1	0

	x_1	x_2	x_3	x_4	x_5	z	
x_3	$\frac{3}{2}$	$\frac{1}{4}$	1	$\frac{1}{4}$	0	0	$\frac{1}{4}$
← x_5	4	$\frac{9}{2}$	0	$-\frac{1}{2}$	1	0	$\frac{1}{2}$
	$\frac{1}{2}$	$-\frac{3}{4}$	0	$\frac{1}{4}$	0	1	$\frac{1}{4}$

	x_1	x_2	x_3	x_4	x_5	z	
x_3	$\frac{23}{18}$	0	1	$\frac{5}{18}$	$-\frac{1}{18}$	0	$\frac{2}{9}$
x_2	$\frac{8}{9}$	1	0	$-\frac{1}{9}$	$\frac{2}{9}$	0	$\frac{1}{9}$
	$\frac{7}{6}$	0	0	$\frac{1}{6}$	$\frac{1}{6}$	1	$\frac{1}{3}$

Thus we have

$$x_1 = 0, \qquad x_2 = \tfrac{1}{9}, \qquad \text{and} \qquad x_3 = \tfrac{2}{9}.$$

The maximum value of $x_1 + x_2 + x_3$ is $\tfrac{1}{3}$, so the minimum value of v is 3. Hence

$$q_1 = x_1 v = 0, \qquad q_2 = x_2 v = (\tfrac{1}{9})(3) = \tfrac{1}{3},$$

and

$$q_3 = x_3 v = (\tfrac{2}{9})(3) = \tfrac{2}{3}.$$

Thus an optimal strategy for C is

$$\mathbf{Q} = \begin{bmatrix} 0 \\ \tfrac{1}{3} \\ \tfrac{2}{3} \end{bmatrix}.$$

An optimal solution to (16), R's problem, is found in the objective row under the columns of the slack variables. Thus, under the slack variables x_4 and x_5, we find

$$y_1 = \tfrac{1}{6} \qquad \text{and} \qquad y_2 = \tfrac{1}{6}.$$

Since $v = 3$,

$$p_1 = y_1 v = (\tfrac{1}{6})(3) = \tfrac{1}{2} \qquad \text{and} \qquad p_2 = y_2 v = (\tfrac{1}{6})(3) = \tfrac{1}{2}.$$

Thus an optimal strategy for R is

$$\mathbf{P} = [\tfrac{1}{2} \quad \tfrac{1}{2}].$$

The zero-sum game with payoff matrix $\mathbf{A}$ is not fair, since the value is 3, with player R having the advantage. Since matrix $\mathbf{A}$ was obtained from the game as initially proposed in Example 10 by adding $r = 4$ to all matrix entries, the value of the *initial* game is $3 - 4 = -1$. That initial game is to C's advantage.

Sometimes, it is possible to solve a matrix game by reducing the size of the payoff matrix $\mathbf{A}$. If each element of the rth *row* of $\mathbf{A}$ is *less than or equal* to the corresponding element of the sth row of $\mathbf{A}$, then the rth row is called **recessive** and the sth row is said to **dominate** the rth row. If each element of the rth *column* of $\mathbf{A}$ is *greater than or equal* to the corresponding element in the sth column of $\mathbf{A}$, then the rth column is called **recessive** and the sth column is said to **dominate** the rth column.

Example 11 In the payoff matrix

$$\begin{bmatrix} 2 & -1 & 3 \\ 0 & 3 & 4 \\ 3 & 2 & 4 \end{bmatrix},$$

the first row is recessive; the third row dominates the first row.

In the payoff matrix

$$\begin{bmatrix} -2 & 4 & 3 \\ 3 & -3 & -3 \\ 5 & 2 & 1 \end{bmatrix},$$

the second column is recessive; the third column dominates the second column.

Consider a matrix game in which the rth row is recessive and the sth row dominates the rth row. It is then obvious that player R will always tend to choose the sth row rather than the rth row, since he will be guaranteed a gain equal to or greater than the gain realized by choosing the rth row. Thus, since the rth row will never be chosen, it can be dropped from further consideration. Suppose now that the rth column is recessive and that the sth column dominates the rth column. Since player C wishes to keep his losses to a minimum, by choosing the sth column he will be guaranteed a loss equal to or smaller than the loss incurred by choosing the rth column. Since the rth column will never be chosen, it can be dropped from further consideration. These techniques, when applicable, result in a smaller payoff matrix.

Example 12 Consider the matrix game with payoff matrix

$$\mathbf{A} = \begin{bmatrix} 2 & -1 & 3 \\ -2 & 2 & 4 \\ 3 & 0 & 4 \end{bmatrix}.$$

Since the third row of $\mathbf{A}$ dominates its first row, the latter can be dropped, obtaining

$$\mathbf{A}_1 = \begin{bmatrix} -2 & 2 & 4 \\ 3 & 0 & 4 \end{bmatrix}.$$

Since the second column of $\mathbf{A}_1$ dominates its third column, the latter can be dropped, obtaining

$$\mathbf{A}_2 = \begin{bmatrix} -2 & 2 \\ 3 & 0 \end{bmatrix},$$

which has no saddle point. The solution to the matrix game with payoff matrix $\mathbf{A}_2$ can be obtained from Equations (8), (9), and (13). We have

$$p_1 = \frac{0 - 3}{-2 + 0 - 2 - 3} = \frac{-3}{-7} = \frac{3}{7},$$

$$p_2 = \frac{-2 - 2}{-2 + 0 - 2 - 3} = \frac{-4}{-7} = \frac{4}{7},$$

$$q_1 = \frac{0 - 2}{-2 + 0 - 2 - 3} = \frac{-2}{-7} = \frac{2}{7},$$

$$q_2 = \frac{-2 - 3}{-2 + 0 - 2 - 3} = \frac{-5}{-7} = \frac{5}{7},$$

and

$$v = \frac{0 - 6}{-2 + 0 - 2 - 3} = \frac{-6}{-7} = \frac{6}{7}.$$

Since in $\mathbf{A}$, the original payoff matrix, the first row and third column were dropped, we obtain

$$\mathbf{P} = [0 \quad \tfrac{3}{7} \quad \tfrac{4}{7}]$$

as an optimal strategy for player R. Similarly,

$$\mathbf{Q} = \begin{bmatrix} \tfrac{2}{7} \\ \tfrac{5}{7} \\ 0 \end{bmatrix}$$

is an optimal strategy for player C.

FURTHER READING

Luce, R. D., and **Howard Raiffa.** *Games and Decisions.* New York: John Wiley & Sons, Inc., 1957.

McKinsey, J. C. C. *Introduction to the Theory of Games.* New York: McGraw-Hill Book Company, 1952.

Owen, G. *Game Theory.* Philadelphia: W. B. Saunders Company, 1968.

7.4 EXERCISES

In Exercises 1 through 4, write the payoff matrix for the given game.

1. Each of two players shows two or three fingers. If the sum of the fingers shown is even, then R pays C an amount equal to the sum of the numbers shown; if the sum is odd, then C pays R an amount equal to the sum of the numbers shown.

2. **(Stone, Scissors, Paper).** Each of two players selects one of the words *stone, scissors, paper*. Stone beats scissors, scissors beats paper, and paper beats stone. In case of a tie, there is no payoff. In case of a win, the winner collects $1.

3. Firms A and B, both handling specialized sporting equipment, are planning to locate in either Abington or Wyncote. If they both locate in the same town, each will capture 50 percent of the trade. If A locates in Abington and B locates in Wyncote, then A will capture 60 percent of the business (and B will keep 40 percent); if A locates in Wyncote and B locates in Abington, then A will hold on to 25 percent of the business (and B to 75 percent).

4. Player R has a nickel and a dime with him. He chooses one of the coins and player C must guess R's choice. If C guesses correctly, he keeps the coin; if he guesses incorrectly, he must give R an amount equal to the coin shown.

5. Find all saddle points for the following matrix games.

   (a) $\begin{bmatrix} 5 & 4 \\ 3 & -2 \end{bmatrix}$.

(b) $\begin{bmatrix} 2 & 1 & 0 \\ 3 & 1 & -2 \\ 4 & 2 & -4 \end{bmatrix}$.

(c) $\begin{bmatrix} 3 & 4 & 5 \\ -2 & 5 & 1 \\ -1 & 0 & 1 \end{bmatrix}$.

(d) $\begin{bmatrix} 5 & 2 & 4 & 2 \\ 0 & -1 & 2 & 0 \\ 3 & 2 & 3 & 2 \\ 1 & 0 & -1 & -1 \end{bmatrix}$.

6. Find optimal strategies for the following strictly determined games. Give the payoff for R.

(a) $\begin{bmatrix} -3 & 4 \\ 3 & 5 \end{bmatrix}$.

(b) $\begin{bmatrix} -1 & -3 & -2 \\ 3 & -1 & 4 \\ -1 & -2 & 5 \end{bmatrix}$.

(c) $\begin{bmatrix} -2 & 3 & -2 & 4 \\ -1 & 2 & -2 & 4 \\ -2 & 3 & -3 & 5 \\ -1 & 2 & -3 & 1 \end{bmatrix}$.

7. Find optimal strategies for the following strictly determined games. Give the payoff for R.

(a) $\begin{bmatrix} 2 & 1 & 3 \\ -2 & 0 & 2 \end{bmatrix}$.

(b) $\begin{bmatrix} -2 & -2 & 4 & 5 \\ -2 & -2 & 1 & 0 \\ 0 & 1 & 1 & 2 \end{bmatrix}$.

(c) $\begin{bmatrix} 6 & 4 \\ 7 & 4 \end{bmatrix}$.

8. Consider a matrix game with payoff matrix

$$\begin{bmatrix} 2 & -3 & -2 \\ -4 & 5 & 6 \end{bmatrix}.$$

Find $E(\mathbf{P}, \mathbf{Q})$, the expected payoff to R, if

(a) $\mathbf{P} = [\frac{1}{4} \ \frac{3}{4}]$ and $\mathbf{Q} = \begin{bmatrix} \frac{1}{3} \\ \frac{1}{6} \\ \frac{1}{2} \end{bmatrix}$.

(b) $p_1 = \frac{2}{3}$, $p_2 = \frac{1}{3}$; $q_1 = \frac{1}{2}$, $q_2 = \frac{1}{4}$, and $q_3 = \frac{1}{4}$.

9. Consider a matrix game with payoff matrix

$$\begin{bmatrix} 3 & -3 \\ 2 & 5 \\ 1 & 0 \end{bmatrix}.$$

Find $E(\mathbf{P}, \mathbf{Q})$, the expected payoff to R, if

(a) $\mathbf{P} = [\frac{1}{2} \ \frac{1}{3} \ \frac{1}{6}]$ and $\mathbf{Q} = \begin{bmatrix} \frac{1}{6} \\ \frac{5}{6} \end{bmatrix}$.

(b) $p_1 = 0$, $p_2 = 0$, $p_3 = 1$; $q_1 = \frac{1}{7}$ and $q_2 = \frac{6}{7}$.

In Exercises 10 and 11, solve the given matrix game using (8) and (13). Find the value of the game using (9).

10. $\begin{bmatrix} 4 & 8 \\ 6 & -2 \end{bmatrix}$.   11. $\begin{bmatrix} -3 & 2 \\ 4 & -5 \end{bmatrix}$.

In Exercises 12 and 13, solve the given matrix game by linear programming.

12. $\begin{bmatrix} -2 & 3 \\ 4 & 5 \\ 5 & 2 \end{bmatrix}$.   13. $\begin{bmatrix} 2 & -3 & 4 \\ 4 & 0 & 1 \\ 3 & 2 & -2 \end{bmatrix}$.

In Exercises 14 and 15, solve the given matrix game by using the method of Example 11.

14. $\begin{bmatrix} -3 & 1 & 3 \\ 1 & -2 & 2 \\ 2 & -1 & 3 \end{bmatrix}$.

15. $\begin{bmatrix} 0 & -4 & 3 & 0 \\ 2 & -3 & 4 & 1 \\ -1 & 2 & 2 & 2 \\ 1 & -4 & 3 & 0 \end{bmatrix}$.

16. Solve Exercise 1.

17. Solve Exercise 2.

18. Solve Exercise 3.

19. Solve Exercise 4.

20. In a labor–management dispute, labor can make one of three different moves L_1, L_2, and L_3 while management can make one of two moves, M_1 and M_2. Suppose that the following payoff matrix is obtained (the entries represent millions of dollars).

Determine the best courses of action for both labor and management.

$$
L \begin{array}{c} \\ L_1 \\ L_2 \\ L_3 \end{array}
\begin{array}{cc}
M_1 & M_2 \\
\end{array}
\left[\begin{array}{cc}
2 & 4 \\
3 & 2 \\
2 & 5
\end{array} \right].
$$

THEORETICAL EXERCISES

T.1. Consider a matrix game with $m \times n$ payoff matrix **A**. Verify that if player R uses strategy **P** and player C uses strategy **Q**, then the expected payoff to R is **PAQ**.

T.2. Consider a matrix game with payoff matrix **A**.

Show that if a constant r is added to each entry of **A**, then we get a new game whose optimal strategies are the same as those for the original game, and the value of the new game is r plus the value of the old game.

7.5

LEAST SQUARES

Prerequisites. Section 1.4, Solutions of Equations. Section 1.5, The Inverse of a Matrix. Calculus (needed for the derivation).

The problem of gathering and analyzing data is one that is present in many facets of human activity. Frequently, we measure a value of y for a given value of x and then plot the points (x, y) on graph paper. From the resulting graph we try to develop a relationship between the variables x and y that can then be used to predict new values of y for given values of x.

Example 1 In the manufacture of product XXX, the amount of the compound beta present in the product is controlled by the amount of the ingredient alpha used in the process. In manufacturing a gallon of XXX, the amount of alpha used and the amount of beta present are recorded. The following data were obtained:

Alpha Used (ounces/gallon)	3	4	5	6	7	8	9	10	11	12
Beta Present (ounces/gallon)	4.5	5.5	5.7	6.6	7.0	7.7	8.5	8.7	9.5	9.7

The points in this table are plotted in Figure 7.25.

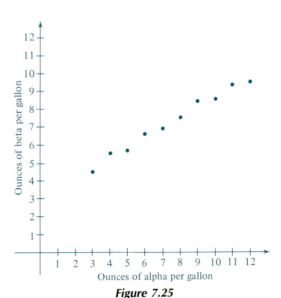

Figure 7.25

Suppose it is known that the relationship between the amount of alpha used and the amount of beta present is given by a linear equation so that the graph is a straight line. It would thus not be reasonable to connect the plotted points by drawing a curve that passes through every point. Moreover, the data are of a *probabilistic* nature. That is, they are not *deterministic* in the sense that if we repeated the experiment, we would expect slightly different values of beta for the same values of alpha, since all measurements are subject to experimental error. The technique for drawing the straight line that "best fits" the given data is called the method of **least squares**, which we now consider.

Suppose that we are given n points (x_1, y_1), (x_2, y_2), . . . , (x_n, y_n) with $x_1 < x_2 < \cdots < x_n$. We are interested in finding the line

$$y = b_0 + b_1 x, \tag{1}$$

which "best fits the data." This is the line which has the property that the sum of the squares of the vertical distances from the points (x_i, y_i) to the line will be as small as possible. This line is called the **line of best fit**.

In Figure 7.26 we show five data points (x_1, y_1), (x_2, y_2), (x_3, y_3), (x_4, y_4), and (x_5, y_5). Let the signed vertical deviation from the point (x_i, y_i) to the (desired) line of best fit be d_i. Then the line of best fit is chosen so that

$$d_1^2 + d_2^2 + d_3^2 + d_4^2 + d_5^2 \tag{2}$$

is a minimum.

It is natural to wonder why we minimize the sum of the squares of the d_i and not some other expression. For example, if we try to minimize the expression

$$d_1 + d_2 + d_3 + d_4 + d_5,$$

Figure 7.26

then we may find several candidates for the line of best fit, some of which will be better than others. If we were to minimize the expression

$$|d_1| + |d_2| + |d_3| + |d_4| + |d_5|,$$

then the analysis becomes very complicated. It turns out that the most successful approach involves minimizing the sum of the squares.

To determine the line of best fit, (1), we must find its y-intercept b_0 and its slope b_1. The y-coordinate of the point on the line of best fit with x-coordinate x_i is

$$b_0 + b_1 x_i,$$

by Equation (1). Hence the signed vertical deviation d_i from the data point (x_i, y_i) to the point on the line of best fit with x-coordinate x_i is

$$d_i = y_i - b_0 - b_1 x_i, \tag{3}$$

We are trying to determine b_0 and b_1 so that the sum of the squares of these vertical deviations will be minimized, that is, so that

$$E = \sum_{i=1}^{n} (y_i - b_0 - b_1 x_i)^2 \tag{4}$$

is minimized.

Since we are to minimize (4), we compute (using calculus) and equate to zero the partial derivatives

$$\frac{\partial E}{\partial b_0} = 0 \qquad \text{and} \qquad \frac{\partial E}{\partial b_1} = 0,$$

yielding

$$-2 \sum_{i=1}^{n} (y_i - b_0 - b_1 x_i) = 0$$

and

$$-2 \sum_{i=1}^{n} x_i(y_i - b_0 - b_1 x_i) = 0.$$

Simplifying and expanding these equations, we are led to having to solve the linear system

$$n b_0 + \left(\sum_{i=1}^{n} x_i \right) b_1 = \sum_{i=1}^{n} y_i,$$

(5)

$$\left(\sum_{i=1}^{n} x_i \right) b_0 + \left(\sum_{i=1}^{n} x_i^2 \right) b_1 = \sum_{i=1}^{n} x_i y_i.$$

The equations in linear system (5) are called the **normal equations** for the least-squares problem. The desired coefficients b_0 and b_1 are the solution to the normal equations.

The procedure for computing the line of best fit $y = b_0 + b_1 x$ to the data $(x_1, y_1), (x_2, y_2), \ldots, (x_n, y_n)$ is as follows.

STEP 1. Compute $\displaystyle\sum_{i=1}^{n} x_i$.

STEP 2. Compute $\displaystyle\sum_{i=1}^{n} x_i^2$.

STEP 3. Compute $\displaystyle\sum_{i=1}^{n} y_i$.

STEP 4. Compute $\displaystyle\sum_{i=1}^{n} x_i y_i$.

STEP 5. Solve the normal equations (5) for b_0 and b_1.

Example 2 Find the line of best fit for the data (1, 2), (2, 5), (3, 3), and (4, 6).

Solution
The given data points are $(x_1, y_1) = (1, 2)$; $(x_2, y_2) = (2, 5)$; $(x_3, y_3) = (3, 3)$; and $(x_4, y_4) = (4, 6)$. Since $n = 4$, we have

$$\sum_{i=1}^{4} x_i = x_1 + x_2 + x_3 + x_4 = 1 + 2 + 3 + 4 = 10$$

$$\sum_{i=1}^{4} x_i^2 = x_1^2 + x_2^2 + x_3^2 + x_4^2 = 1^2 + 2^2 + 3^2 + 4^2 = 30$$

$$\sum_{i=1}^{4} y_i = y_1 + y_2 + y_3 + y_4 = 2 + 5 + 3 + 6 = 16$$

$$\sum_{i=1}^{4} x_i y_i = x_1 y_1 + x_2 y_2 + x_3 y_3 + x_4 y_4 = 2 + 10 + 9 + 24 = 45.$$

Substituting these values into the system (5), we obtain

$$b_0 = \tfrac{3}{2} \quad \text{and} \quad b_1 = 1.$$

Thus the line of best fit is

$$y = x + \tfrac{3}{2},$$

which is shown in Figure 7.27 along with the given four data points.

We now rewrite the foregoing results in matrix notation so that we can even further simplify our approach. From Equation (3) we may write

$$y_i = b_0 + b_1 x_i + d_i, \qquad i = 1, 2, \ldots, n. \tag{6}$$

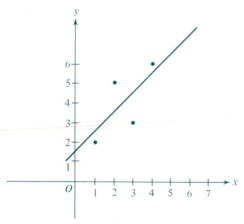

Figure 7.27

If we let

$$\mathbf{Y} = \begin{bmatrix} y_1 \\ y_2 \\ \vdots \\ y_n \end{bmatrix}, \qquad \mathbf{A} = \begin{bmatrix} 1 & x_1 \\ 1 & x_2 \\ \vdots & \vdots \\ 1 & x_n \end{bmatrix}, \qquad \mathbf{X} = \begin{bmatrix} b_0 \\ b_1 \end{bmatrix}, \qquad \text{and} \qquad \mathbf{D} = \begin{bmatrix} d_1 \\ d_2 \\ \vdots \\ d_n \end{bmatrix},$$

then we may write the n equations in (6) as a single matrix equation

$$\mathbf{Y} = \mathbf{AX} + \mathbf{D}.$$

The linear system (5) can be written in matrix form as

$$\mathbf{A}^T\mathbf{AX} = \mathbf{A}^T\mathbf{Y}. \tag{7}$$

It can be shown that if the given n data points (x_1, y_1), (x_2, y_2), . . . , (x_n, y_n) do not all lie on a vertical line, then $\mathbf{A}^T\mathbf{A}$ is a nonsingular matrix. In this case we can solve (7) for $\mathbf{X}$, obtaining

$$\mathbf{X} = (\mathbf{A}^T\mathbf{A})^{-1}\mathbf{A}^T\mathbf{Y}. \tag{8}$$

REMARK If $n > 2$, then $\mathbf{A}$ is not a square matrix and we cannot simplify $(\mathbf{A}^T\mathbf{A})^{-1}$ in Equation (8).

Example 3 Consider the data in Example 2. We have

$$\mathbf{Y} = \begin{bmatrix} 2 \\ 5 \\ 3 \\ 6 \end{bmatrix}, \qquad \mathbf{A} = \begin{bmatrix} 1 & 1 \\ 1 & 2 \\ 1 & 3 \\ 1 & 4 \end{bmatrix}, \qquad \mathbf{X} = \begin{bmatrix} b_0 \\ b_1 \end{bmatrix}.$$

Then

$$\mathbf{A}^T\mathbf{A} = \begin{bmatrix} 4 & 10 \\ 10 & 30 \end{bmatrix}, \qquad \mathbf{A}^T\mathbf{Y} = \begin{bmatrix} 16 \\ 45 \end{bmatrix}.$$

The system (7) in this case is

$$\begin{bmatrix} 4 & 10 \\ 10 & 30 \end{bmatrix} \begin{bmatrix} b_0 \\ b_1 \end{bmatrix} = \begin{bmatrix} 16 \\ 45 \end{bmatrix},$$

or

$$4b_0 + 10b_1 = 16$$
$$10b_0 + 30b_1 = 45,$$

which represents system (5) for Example 2. Substituting in Equation (8), we have

$$\mathbf{X} = \begin{bmatrix} 4 & 10 \\ 10 & 30 \end{bmatrix}^{-1} \begin{bmatrix} 16 \\ 45 \end{bmatrix} = \begin{bmatrix} \frac{3}{2} & -\frac{1}{2} \\ -\frac{1}{2} & \frac{1}{5} \end{bmatrix} \begin{bmatrix} 16 \\ 45 \end{bmatrix} = \begin{bmatrix} \frac{3}{2} \\ 1 \end{bmatrix}.$$

Thus

$$b_0 = \tfrac{3}{2}, \qquad b_1 = 1,$$

agreeing with the values obtained in Example 2.

Example 4   (a) Find the equation of the line of best fit for the data in Example 1.
(b) Use the equation obtained in part (a) to predict the number of ounces of beta present in a gallon if 30 ounces of alpha are used per gallon.

Solution
(a) We have

$$\mathbf{Y} = \begin{bmatrix} 4.5 \\ 5.5 \\ 5.7 \\ 6.6 \\ 7.0 \\ 7.7 \\ 8.5 \\ 8.7 \\ 9.5 \\ 9.7 \end{bmatrix}, \qquad \mathbf{A} = \begin{bmatrix} 1 & 3 \\ 1 & 4 \\ 1 & 5 \\ 1 & 6 \\ 1 & 7 \\ 1 & 8 \\ 1 & 9 \\ 1 & 10 \\ 1 & 11 \\ 1 & 12 \end{bmatrix}, \qquad \mathbf{X} = \begin{bmatrix} b_0 \\ b_1 \end{bmatrix}.$$

Then

$$\mathbf{A}^T\mathbf{A} = \begin{bmatrix} 10 & 75 \\ 75 & 645 \end{bmatrix} \quad \text{and} \quad \mathbf{A}^T\mathbf{Y} = \begin{bmatrix} 73.4 \\ 598.6 \end{bmatrix}.$$

Substituting in Equation (8), we obtain

$$\mathbf{X} = \begin{bmatrix} 10 & 75 \\ 75 & 645 \end{bmatrix}^{-1} \begin{bmatrix} 73.4 \\ 598.6 \end{bmatrix}$$

$$= \begin{bmatrix} \frac{645}{825} & -\frac{75}{825} \\ -\frac{75}{825} & \frac{10}{825} \end{bmatrix} \begin{bmatrix} 73.4 \\ 598.6 \end{bmatrix} = \begin{bmatrix} 2.967 \\ 0.583 \end{bmatrix}.$$

Then

$$b_0 = 2.967 \qquad \text{and} \qquad b_1 = 0.583.$$

Hence the line of best fit, shown in Figure 7.28, is

$$y = 0.583x + 2.967, \tag{9}$$

where y is the amount of beta present and x is the amount of alpha used.

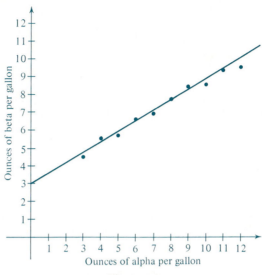

Figure 7.28

(b) If $x = 30$, then substituting in (9), we obtain

$$y = 20.457.$$

Thus there would be 20.457 ounces of beta present in a gallon of the compound.

7.5 EXERCISES

In Exercises 1 through 4, find the line of best fit for the given data points.

1. (2, 1), (3, 2), (4, 3), (5, 2).

2. (3, 2), (4, 3), (5, 2), (6, 4), (7, 3).

3. (2, 3), (3, 4), (4, 3), (5, 4), (6, 3), (7, 4).

4. (3, 3), (4, 5), (5, 4), (6, 5), (7, 5), (8, 6), (9, 5), (10, 6).

5. In an experiment designed to determine the extent of a person's natural orientation, a subject is put in a special room and kept there for a certain length of time. He is then asked to find his way out of a maze and a record is made of the time it takes him to accomplish this task. The following data are obtained:

Time in Room (hours)	1	2	3	4	5	6
Time to Find Way out of Maze (minutes)	0.8	2.1	2.6	2.0	3.1	3.3

Let x denote the number of hours in the room and let y denote the number of minutes that it takes the subject to find his way out.

(a) Find the line of best fit relating x and y.

(b) Use the equation obtained in (a) to estimate the time it will take the subject to find his way out of the maze after 10 hours in the room.

6. A steel producer gathers the following data:

Year	1965	1966	1967	1968	1969	1970
Annual Sales (millions of dollars)	1.2	2.3	3.2	3.6	3.8	5.1

Represent the years 1965, . . . , 1970 as 0, 1, 2, 3, 4, 5, respectively, and let x denote the year. Let y denote the annual sales (in millions of dollars).
(a) Find the line of best fit relating x and y.
(b) Use the equation obtained in (a) to estimate the annual sales for the year 1974.

7. A sales organization obtains the following data relating the number of salespersons to annual sales.

Number of Salespersons	5	6	7	8	9	10
Annual Sales (millions of dollars)	2.3	3.2	4.1	5.0	6.1	7.2

Let x denote the number of salespersons and let y denote the annual sales (in millions of dollars).
(a) Find the line of best fit relating x and y.
(b) Use the equation obtained in (a) to estimate annual sales when there are 14 salespersons.

THEORETICAL EXERCISES

T.1. For the matrices $\mathbf{A}$, $\mathbf{X}$, and $\mathbf{Y}$ as defined in the text prior to Equation (7), verify that Equation (7) is indeed a matrix formulation of the system of normal equations (5).

T.2. Show that if all n x-coordinates x_i are equal, then the matrix $\mathbf{A}^T\mathbf{A}$ is singular [and thus $(\mathbf{A}^T\mathbf{A})^{-1}$ does not exist]. See Equation (8).

7.6

LINEAR ECONOMIC MODELS

Prerequisite. Chapter 1, Linear Equations and Matrices.

As society has grown more complex, increasing attention has been paid to the analysis of economic behavior. Problems of economic behavior are more difficult to handle than problems in the physical sciences for many reasons. For example, we may not know all the factors or variables that must be considered, what data need to be gathered, when we have enough data, or the resulting mathematical problem may be too difficult to solve.

In the 1930s Wassily Leontief, a Harvard University economics professor, developed a pioneering approach to the mathematical analysis of economic behavior. He was awarded the 1973 Nobel Prize in Economics for his work. In this section we give a very brief introduction to the application of linear algebra to economics.

Our approach leans heavily on the presentations in the books by Gale and Johnston, Price and van Vleck given in Further Reading, which may be consulted by the reader for a more extensive treatment of this material.

The Leontief Closed Model

Example 1† Consider a simple society consisting of a farmer who produces all the food, and makes nothing else, a carpenter who builds all the homes, and produces nothing else, and a tailor who makes all the clothes, and makes nothing else. For convenience, we may select our units so that each individual produces one unit of each commodity during the year. Suppose that during the year the portion of each commodity that is consumed by each individual is given in Table 7.2.

Table 7.2

Goods Consumed by:	Goods Produced by:		
	Farmer	Carpenter	Tailor
Farmer	$\frac{7}{16}$	$\frac{1}{2}$	$\frac{3}{16}$
Carpenter	$\frac{5}{16}$	$\frac{1}{6}$	$\frac{5}{16}$
Tailor	$\frac{1}{4}$	$\frac{1}{3}$	$\frac{1}{2}$

Thus the farmer consumes $\frac{7}{16}$ of his own produce while the carpenter consumes $\frac{5}{16}$ of the farmer's produce, the carpenter consumes $\frac{5}{16}$ of the clothes made by the tailor, and so on. Let p_1 be the price per unit of food, p_2 the price per unit of housing, and p_3 the price per unit of clothes. We assume that every one pays the same price for a commodity. Thus the farmer pays the same price for his food as the tailor and carpenter, although he grew the food himself. We are interested in determining the prices p_1, p_2, and p_3 so that we have a state of equilibrium, which is defined as: *No one makes money or loses money.*

The farmer's expenditures are

$$\tfrac{7}{16}p_1 + \tfrac{1}{2}p_2 + \tfrac{3}{16}p_3,$$

while his income is p_1 because he produces one unit of food. Since expenditures must equal income, we have

$$\tfrac{7}{16}p_1 + \tfrac{1}{2}p_2 + \tfrac{3}{16}p_3 = p_1. \tag{1}$$

Similarly, for the carpenter we obtain

$$\tfrac{5}{16}p_1 + \tfrac{1}{6}p_2 + \tfrac{5}{16}p_3 = p_2, \tag{2}$$

and for the tailor we have

$$\tfrac{1}{4}p_1 + \tfrac{1}{3}p_2 + \tfrac{1}{2}p_3 = p_3. \tag{3}$$

Equations (1), (2), and (3) can be rewritten as

---

† This example is due to Johnston, Price, and van Vleck and is described in their book given in Further Reading. The general model for this example is also presented by Gale.

$$AP = P, \tag{4}$$

where

$$A = \begin{bmatrix} \frac{7}{16} & \frac{1}{2} & \frac{3}{16} \\ \frac{5}{16} & \frac{1}{6} & \frac{5}{16} \\ \frac{1}{4} & \frac{1}{3} & \frac{1}{2} \end{bmatrix}, \qquad P = \begin{bmatrix} p_1 \\ p_2 \\ p_3 \end{bmatrix}.$$

Equation (4) can be rewritten as

$$(I_n - A)P = 0, \tag{5}$$

which is a homogeneous system.

Our problem is that of finding a solution P to (5) whose components p_i will be nonnegative with at least one positive p_i, since $P = 0$ means that all the prices are zero, which makes no sense.

Solving (5), we obtain

$$P = r \begin{bmatrix} 4 \\ 3 \\ 4 \end{bmatrix},$$

where r is any real number. Letting r be a positive number, we determine the *relative* prices of the commodities. For example, letting $r = 1000$, we find that food costs \$4000 per unit, housing costs \$3000 per unit, and clothes cost \$4000 per unit.

**Example 2
(The Exchange
Model)** Consider now the general problem where we have n manufacturers $M_1, M_2, \ldots, M_n$, and n goods $G_1, G_2, \ldots, G_n$, with M_i making only G_i. Consider a fixed interval of time, say one year, and suppose that M_i only makes one unit of G_i during the year.

In producing the goods G_i, manufacturer M_i may consume amounts of goods G_1, $G_2, \ldots, G_i, \ldots, G_n$. For example, iron, along with many other ingredients, is used in the manufacture of iron. Let a_{ij} be the amount of good G_j consumed by manufacturer M_i. Then

$$0 \le a_{ij} \le 1.$$

Suppose that the model is **closed**; that is, no goods leave or enter the system. This means that the total consumption of each good must equal its total production. Since the total production of G_j is 1, we obtain

$$a_{1j} + a_{2j} + \cdots + a_{nj} = 1 \qquad (1 \le j \le n).$$

If the price per unit of G_k is p_k, then manufacturer M_i pays

$$a_{i1}p_1 + a_{i2}p_2 + \cdots + a_{in}p_n \tag{6}$$

for the goods he uses.

Our problem is that of determining the prices $p_1, p_2, \ldots, p_n$ so that no manufacturer makes money or loses money. That is, so that each manufacturer's income will equal his expenses. Since M_i only manufactures one unit, his income is p_i. Thus from (6) we have

$$
\begin{aligned}
a_{11}p_1 + a_{12}p_2 + \cdots + a_{1n}p_n &= p_1 \\
a_{21}p_1 + a_{22}p_2 + \cdots + a_{2n}p_n &= p_2 \\
&\ \ \vdots \\
a_{n1}p_1 + a_{n2}p_2 + \cdots + a_{nn}p_n &= p_n,
\end{aligned}
$$

which can be written in matrix form as

$$\mathbf{AP} = \mathbf{P}, \tag{7}$$

where

$$\mathbf{A} = [a_{ij}] \qquad \text{and} \qquad \mathbf{P} = \begin{bmatrix} p_1 \\ p_2 \\ \vdots \\ p_n \end{bmatrix}.$$

Equation (7) can be rewritten as

$$(\mathbf{I}_n - \mathbf{A})\mathbf{P} = \mathbf{0}. \tag{8}$$

Thus our problem is that of finding a vector

$$\mathbf{P} \geq \mathbf{0},$$

with at least one component that is positive and satisfies Equation (8).

Definition An $n \times n$ matrix $\mathbf{A} = [a_{ij}]$ is called an **exchange matrix** if it satisfies the following two properties:
  (a) $a_{ij} \geq 0$ (each entry is nonnegative).
  (b) $a_{1j} + a_{2j} + \cdots + a_{nj} = 1$, for $j = 1, 2, \ldots, n$ (the entries in each column add up to 1).

Example 3 Matrix $\mathbf{A}$ of Example 1 is an exchange matrix, as is the matrix $\mathbf{A}$ of Example 2.

Our general problem can now be stated as: Given as exchange matrix $\mathbf{A}$, find a vector $\mathbf{P} \geq \mathbf{0}$ with at least one positive component, satisfying Equation (8). It can be shown that this problem always has a solution (see the book by Gale, p. 264, given in Further Reading).

In our general problem we required that each manufacturer's income equal his expenses. Instead, we could have required that each manufacturer's expenses not exceed his income. This would have led to

$$\mathbf{AP} \leq \mathbf{P} \tag{9}$$

instead of $\mathbf{AP} = \mathbf{P}$. However, it can be shown (Exercise T.1) that if Equation (9) holds, then Equation (7) will hold. Thus, if no manufacturer spends more than he earns, then everyone's income equals his expenses. An economic interpretation of this statement is that in the closed Leontief model if some manufacturer is making a profit, then at least one manufacturer is taking a loss.

REMARK   (For those who have read Section 5.1). We also observe that a price vector **P** for a Leontief closed model with exchange matrix **A** is an eigenvector for $\lambda = 1$ that has nonnegative components.

An International Trade Model

Example 4   Suppose that n countries $C_1, C_2, \ldots, C_n$ are engaged in trading with each other and that a common currency is in use. We assume that prices are fixed throughout the discussion and that C_j's income y_j comes entirely from selling its goods either internally or to other countries. We also assume that the fraction of C_j's income that is spent on imports from C_i is a fixed number a_{ij} which does not depend upon C_j's income y_j. Since the a_{ij}'s are fractions of y_j, we have

$$a_{ij} \geq 0$$
$$a_{1j} + a_{2j} + \cdots + a_{nj} = 1,$$

so that $\mathbf{A} = [a_{ij}]$ is an exchange matrix. We now wish to determine the total income y_i for each country C_i. Since the value of C_i's exports to C_j is $a_{ij}y_j$, the total income of C_i is

$$a_{i1}y_1 + a_{i2}y_2 + \cdots + a_{in}y_n.$$

Hence, we must have

$$a_{i1}y_1 + a_{i2}y_2 + \cdots + a_{in}y_n = y_i.$$

In matrix notation we must find $\mathbf{P} = \begin{bmatrix} y_1 \\ y_2 \\ \vdots \\ y_n \end{bmatrix} \geq \mathbf{0}$, with at least one $y_i > 0$, so that

$$\mathbf{AP} = \mathbf{P},$$

which is our earlier problem.

The Leontief Open Model

Suppose that we have n goods $G_1, G_2, \ldots, G_n$ and n activities $M_1, M_2, \ldots, M_n$. Assume that each activity M_i produces only one good G_i and that G_i is only produced by M_i. Let $c_{ij} \geq 0$ be the dollar value of G_i that has to be consumed to produce one dollar's worth of G_j. The matrix $\mathbf{C} = [c_{ij}]$ is called the **consumption matrix**. Observe that c_{ii} may be positive, which means that we may require some amount of G_i to make one dollar's worth of G_i.

Let x_i be the dollar value of G_i produced in a fixed period of time, say, one year. The vector

$$\mathbf{X} = \begin{bmatrix} x_1 \\ x_2 \\ \vdots \\ x_n \end{bmatrix} \qquad (x_i \geq 0) \tag{10}$$

is called the **production vector**. The expression

$$c_{i1}x_1 + c_{i2}x_2 + \cdots + c_{in}x_n$$

is the total value of the product G_i that is consumed, as determined by the production vector to make x_1 dollar's worth of G_1, x_2 dollar's worth of G_2, and so on. Observe that the expression given by Equation (10) is the ith entry of the matrix product $\mathbf{CX}$. The difference between the dollar value of G_i that is produced and the total dollar value of G_i that is consumed

$$x_i - (c_{i1}x_1 + c_{i2}x_2 + \cdots + c_{in}x_n) \tag{11}$$

is called the **net production**.

Observe that the expression in Equation (11) is the ith entry in

$$\mathbf{X} - \mathbf{CX} = (\mathbf{I}_n - \mathbf{C})\mathbf{X}$$

Now let d_i represent the dollar value of the outside demand for G_i and let

$$\mathbf{D} = \begin{bmatrix} d_1 \\ d_2 \\ \vdots \\ d_n \end{bmatrix} \qquad (d_i \geq 0)$$

be the **demand vector**.

Our problem can now be stated: Given a demand vector $\mathbf{D} \geq \mathbf{0}$, can we find a production vector $\mathbf{X}$ such that the outside demand $\mathbf{D}$ is met without any surplus? That is, can we find a vector $\mathbf{X} \geq \mathbf{0}$, so that the following equation is satisfied?

$$(\mathbf{I}_n - \mathbf{C})\mathbf{X} = \mathbf{D}. \tag{12}$$

Example 5 Let

$$\mathbf{C} = \begin{bmatrix} \frac{1}{4} & \frac{1}{2} \\ \frac{2}{3} & \frac{1}{3} \end{bmatrix}$$

be a consumption matrix. Then

$$\mathbf{I}_2 - \mathbf{C} = \begin{bmatrix} 1 & 0 \\ 0 & 1 \end{bmatrix} - \begin{bmatrix} \frac{1}{4} & \frac{1}{2} \\ \frac{2}{3} & \frac{1}{3} \end{bmatrix} = \begin{bmatrix} \frac{3}{4} & -\frac{1}{2} \\ -\frac{2}{3} & \frac{2}{3} \end{bmatrix}.$$

Equation (12) becomes

$$\begin{bmatrix} \frac{3}{4} & -\frac{1}{2} \\ -\frac{2}{3} & \frac{2}{3} \end{bmatrix} \begin{bmatrix} x_1 \\ x_2 \end{bmatrix} = \begin{bmatrix} d_1 \\ d_2 \end{bmatrix}$$

so

$$\begin{bmatrix} x_1 \\ x_2 \end{bmatrix} = \begin{bmatrix} \frac{3}{4} & -\frac{1}{2} \\ -\frac{2}{3} & \frac{2}{3} \end{bmatrix}^{-1} \begin{bmatrix} d_1 \\ d_2 \end{bmatrix}$$

$$= \begin{bmatrix} 4 & 3 \\ 4 & \frac{9}{2} \end{bmatrix} \begin{bmatrix} d_1 \\ d_2 \end{bmatrix} \geqslant \mathbf{0},$$

since $d_1 \geqslant 0$ and $d_2 \geqslant 0$. Thus we can obtain a production vector for any given demand vector.

In general, if $(\mathbf{I}_n - \mathbf{C})^{-1}$ exists and is $\geqslant \mathbf{0}$ then $\mathbf{X} = (\mathbf{I}_n - \mathbf{C})^{-1}\mathbf{D} \geqslant \mathbf{0}$ is a production vector for any given demand vector. However, for a given consumption matrix, there may be no solution to Equation (12).

Example 6 Consider the consumption matrix

$$\mathbf{C} = \begin{bmatrix} \frac{1}{2} & \frac{1}{2} \\ \frac{1}{2} & \frac{3}{4} \end{bmatrix}.$$

Then

$$\mathbf{I}_2 - \mathbf{C} = \begin{bmatrix} \frac{1}{2} & -\frac{1}{2} \\ -\frac{1}{2} & \frac{1}{4} \end{bmatrix}$$

and

$$(\mathbf{I}_2 - \mathbf{C})^{-1} = \begin{bmatrix} -2 & -4 \\ -4 & -4 \end{bmatrix},$$

so that

$$\mathbf{X} = (\mathbf{I}_2 - \mathbf{C})^{-1}\mathbf{D}$$

is not a production vector for $\mathbf{D} \neq \mathbf{0}$, since all of its components are negative. Thus the problem has no solution. If $\mathbf{D} = \mathbf{0}$, we do have a solution, namely, $\mathbf{X} = \mathbf{0}$, which means that if there is no outside demand, then nothing is produced.

Definition An $n \times n$ consumption matrix $\mathbf{C}$ is called **productive** if $(\mathbf{I}_n - \mathbf{C})^{-1}$ exists and $(\mathbf{I}_n - \mathbf{C})^{-1} \geqslant \mathbf{0}$. That is, $\mathbf{C}$ is productive if $(\mathbf{I}_n - \mathbf{C})$ is nonsingular and every entry of $(\mathbf{I}_n - \mathbf{C})^{-1}$ is nonnegative. In this case, the model is also called **productive**.

It then follows that if $\mathbf{C}$ is productive, then for any demand vector $\mathbf{D} \geqslant \mathbf{0}$, the equation

$$(\mathbf{I}_n - \mathbf{C})\mathbf{X} = \mathbf{D}$$

has a unique solution $\mathbf{X} \geqslant \mathbf{0}$.

Example 7 Consider the consumption matrix

$$C = \begin{bmatrix} \frac{1}{2} & \frac{1}{3} \\ \frac{1}{4} & \frac{1}{3} \end{bmatrix}.$$

Then

$$(I_2 - C) = \begin{bmatrix} \frac{1}{2} & -\frac{1}{3} \\ -\frac{1}{4} & \frac{2}{3} \end{bmatrix},$$

and

$$(I_2 - C)^{-1} = 4\begin{bmatrix} \frac{2}{3} & \frac{1}{3} \\ \frac{1}{4} & \frac{1}{2} \end{bmatrix}.$$

Thus C is productive. If $D \geqslant 0$ is a demand vector, then the equation $(I_n - C)X = D$ has the unique solution $X = (I_n - C)^{-1}D \geqslant 0$.

Results are given in more advanced books (see the book by Johnson, p. 251, given in Further Reading) for telling when a given consumption matrix is productive.

FURTHER READING

Gale, David. *The Theory of Linear Economic Models.* New York: McGraw-Hill Book Company, 1960.

Johnston, B., **G. Price,** and **F. S. van Vleck.** *Linear Equations and Matrices.* Reading, Mass.: Addison-Wesley Publishing Co., Inc.

7.6 EXERCISES

1. Which of the following matrices are exchange matrices?

(a) $\begin{bmatrix} \frac{1}{3} & 0 & 1 \\ \frac{2}{3} & 1 & 0 \\ \frac{1}{2} & -\frac{1}{2} & 0 \end{bmatrix}.$ (b) $\begin{bmatrix} \frac{1}{2} & \frac{1}{3} & \frac{3}{4} \\ \frac{1}{2} & \frac{1}{3} & \frac{1}{4} \\ 0 & \frac{1}{3} & 0 \end{bmatrix}.$

(c) $\begin{bmatrix} \frac{1}{3} & -\frac{2}{3} & \frac{1}{2} \\ \frac{2}{3} & \frac{2}{3} & \frac{1}{2} \\ 0 & 1 & 0 \end{bmatrix}.$ (d) $\begin{bmatrix} 1 & \frac{1}{4} & \frac{5}{6} \\ 0 & \frac{1}{4} & \frac{1}{6} \\ 0 & \frac{1}{2} & 0 \end{bmatrix}.$

In Exercises 2 through 4, find a vector $P \geqslant 0$, with at least one positive component, satisfying Equation (8) for the given exchange matrix.

2. $\begin{bmatrix} \frac{1}{3} & \frac{2}{3} & 0 \\ \frac{1}{3} & 0 & \frac{1}{4} \\ \frac{1}{3} & \frac{1}{3} & \frac{3}{4} \end{bmatrix},$ **3.** $\begin{bmatrix} \frac{1}{2} & 1 & \frac{2}{3} \\ 0 & 0 & 0 \\ \frac{1}{2} & 0 & \frac{1}{3} \end{bmatrix}.$

4. $\begin{bmatrix} 0 & \frac{1}{3} & 1 \\ \frac{1}{6} & \frac{1}{6} & 0 \\ \frac{5}{6} & \frac{1}{2} & 0 \end{bmatrix}.$

5. Consider the simple economy of Example 1. Suppose that the farmer consumes $\frac{2}{5}$ of the food, $\frac{1}{3}$ of the shelter, and $\frac{1}{2}$ of the clothes; that the carpenter consumes $\frac{2}{5}$ of the food, $\frac{1}{3}$ of the shelter, and $\frac{1}{2}$ of the clothes; that the tailor consumes $\frac{1}{5}$ of the food, $\frac{1}{3}$ of the shelter, and none of the clothes. Find the exchange matrix A for this problem and a vector $P \geqslant 0$, with at least one positive component satisfying Equation (8).

6. Consider the international trade model consisting of three countries, C_1, C_2, and C_3. Suppose that the fraction of C_1's income spent on imports from C_1 is

$\frac{1}{4}$, from C_2 is $\frac{1}{2}$, and from C_3 is $\frac{1}{4}$; that the fraction of C_2's income spent on imports from C_1 is $\frac{2}{5}$, and C_2 is $\frac{1}{5}$, and from C_3 is $\frac{2}{5}$; that the fraction of C_3's income spent on imports from C_1 is $\frac{1}{2}$, from C_2 is $\frac{1}{2}$, and from C_3 is 0. Find the income of each country.

In Exercises 7 through 10, determine which matrices are productive.

7. $\begin{bmatrix} \frac{1}{2} & \frac{1}{3} & 0 \\ 0 & \frac{2}{3} & 0 \\ 1 & 0 & 2 \end{bmatrix}$.
8. $\begin{bmatrix} 0 & \frac{2}{3} & 0 \\ \frac{1}{2} & 0 & 0 \\ 0 & \frac{1}{4} & 0 \end{bmatrix}$.

9. $\begin{bmatrix} 0 & \frac{1}{3} & \frac{1}{2} \\ \frac{1}{2} & 0 & \frac{1}{4} \\ \frac{1}{4} & \frac{1}{3} & 0 \end{bmatrix}$.
10. $\begin{bmatrix} 0 & \frac{1}{3} & \frac{1}{3} \\ \frac{1}{4} & 0 & \frac{1}{6} \\ \frac{1}{3} & \frac{2}{3} & 0 \end{bmatrix}$.

11. Suppose that the consumption matrix for the linear production model is

$$\begin{bmatrix} \frac{1}{2} & \frac{1}{2} \\ \frac{1}{2} & \frac{1}{4} \end{bmatrix}.$$

(a) Find the production vector for the demand vector $\begin{bmatrix} 1 \\ 3 \end{bmatrix}$.

(b) Find the production vector for the demand vector $\begin{bmatrix} 2 \\ 0 \end{bmatrix}$.

12. A small town has three primary industries: a copper mine, a railroad, and an electric utility. To produce $1 of copper, the copper mine uses $0.20 of copper, $0.10 of transportation, and $0.20 of electric power. To provide $1 of transportation, the railroad uses $0.10 of copper, $0.10 of transportation, and $0.40 of electric power. To provide $1 of electric power, the electric utility uses $0.20 of copper, $0.20 of transportation, and $0.30 of electric power. Suppose that during the year there is an outside demand of 1.2 million dollars for copper, 0.8 million dollars for transportation, and 1.5 million dollars for electric power. How much should each industry produce to satisfy the demands?

THEORETICAL EXERCISE

T.1. In the exchange model (Example 3), show that $AP \le P$ implies that $AP = P$.

7.7

MARKOV† CHAINS

Prerequisites. Section 1.4, Solutions of Equations. Basic Ideas of Probability. The Notion of a Limit.

Consider a system that is, at any one time, in one and only one of a finite number of states. For example, the weather in a certain area is either rainy or dry; a person is either a smoker or a nonsmoker; we either go or do not go to college; we live in an urban or suburban area; we are in the lower-, middle-, or upper-income bracket; we buy a Chevrolet, Ford, or other make of car. As times goes on, the system may move from one state to another, and we assume that the state of the system is observed at fixed time periods (for example, each day, each hour, and so on). In many applications we know the present state of the system and wish to predict the state at the next observation period, or at some future observation period. We often can predict the probability of the system's being in a particular state at a future observation period from its past history. The applications mentioned above are of this type.

Definition A **Markov chain** or **Markov process** is a process in which the probability of the system's being in a particular state at a given observation period depends only on its state at the immediately preceding observation period.

Suppose that the system has n possible states. For each $i = 1, 2, \ldots, n$, $j = 1, 2, \ldots, n$ let t_{ij} be the probability that if the system is in state j at a certain observation period, it will be in state i at the next observation period; t_{ij} is called a **transition probability**. Moreover, t_{ij} applies to every time period; that is, it does not change with time.

Since t_{ij} is a probability, we must have

$$0 \le t_{ij} \le 1 \qquad (1 \le i, j \le n).$$

Also, if the system is in state j at a certain observation period, then it must be in one of the n states (it may remain in state j) at the next observation period. Thus we have

$$t_{1j} + t_{2j} + \cdots + t_{nj} = 1. \tag{1}$$

It is convenient to arrange the transition probabilities as the $n \times n$ matrix $\mathbf{T} = [t_{ij}]$, which is called the **transition matrix** of the Markov chain. Other names for a transition

---

† Andrei Andreevitch Markov (1856–1922) lived most of his life in St. Petersburg (now Leningrad), where his father was an official of the Russian forestry department. He was first a student and then a professor at St. Petersburg University. A political liberal, he participated in the protests against the Czarist rule in the first decade of the twentieth century. Although he was interested in many aspects of mathematical analysis, his most important work was in helping to lay the foundations of modern probability theory. His ideas on Markov processes were motivated by a desire to give a strictly rigorous proof of the law of large numbers and to extend the applicability of this law; these ideas appeared in a series of papers between 1906 and 1912.

matrix are **Markov matrix, stochastic matrix**, and **probability matrix**. We see that the entries in each column of $\mathbf{T}$ are nonnegative and, from Equation (1), add up to 1.

Example 1   Suppose that the weather in a certain city is either rainy or dry. As a result of extensive record keeping it has been determined that the probability of a rainy day following a dry day is $\frac{1}{3}$, and the probability of a rainy day following a rainy day is $\frac{1}{2}$. Let state 1 be a dry day and state 2 be a rainy day. Then the transition matrix of this Markov chain is

$$\mathbf{T} = \begin{matrix} & 1 & 2 & \\ \begin{bmatrix} \frac{2}{3} & \frac{1}{2} \\ \frac{1}{3} & \frac{1}{2} \end{bmatrix} & & & \begin{matrix} 1 \\ 2 \end{matrix} \end{matrix}$$

Example 2   A market research organization is studying a large group of coffee buyers who buy a can of coffee each week. It is found that 50 percent of those presently using brand A will again buy brand A next week, 25 percent will switch to brand B, and 25 percent will switch to another brand. Of those using brand B now, 30 percent will again buy brand B next week, 60 percent will switch to brand A and 10 percent will switch to another brand. Of those using another brand now, 30 percent will again buy another brand next week, 40 percent will switch to brand A, and 30 percent will switch to brand B. Let states 1, 2, and 3 denote brand A, brand B, and another brand, respectively. The probability that a person presently using brand A will switch to brand B is 0.25, the probability that a person presently using brand B will again buy brand B is 0.3, and so on. Thus the transition matrix of this Markov chain is

$$\mathbf{T} = \begin{matrix} & 1 & 2 & 3 & \\ \begin{bmatrix} 0.50 & 0.60 & 0.40 \\ 0.25 & 0.30 & 0.30 \\ 0.25 & 0.10 & 0.30 \end{bmatrix} & & & & \begin{matrix} 1 \\ 2 \\ 3. \end{matrix} \end{matrix}$$

We shall now use the transition matrix of the Markov process to determine the probability of the system being in any of the n states at future times.

Let

$$\mathbf{P}^{(k)} = \begin{bmatrix} p_1^{(k)} \\ p_2^{(k)} \\ \vdots \\ p_n^{(k)} \end{bmatrix} \qquad (k \geq 0)$$

denote the **state vector** of the Markov process at the observation period k, where $p_j^{(k)}$ is the probability that the system is in state j at the observation period k. The state vector $\mathbf{P}^{(0)}$, at the observation period 0, is called the **initial state vector**.

The following theorem, whose proof we omit, is proved by using the basic ideas of probability theory.

Theorem 7.8 *If* **T** *is the transition matrix of a Markov process, then the state vector* $\mathbf{P}^{(k+1)}$, *at the* $(k+1)$th *observation period, can be determined by from the state vector* $\mathbf{P}^{(k)}$, *at the* kth *observation period, as*

$$\mathbf{P}^{(k+1)} = \mathbf{T}\mathbf{P}^{(k)}. \tag{2}$$

From (2) we have

$$\mathbf{P}^{(1)} = \mathbf{T}\mathbf{P}^{(0)}$$
$$\mathbf{P}^{(2)} = \mathbf{T}\mathbf{P}^{(1)} = \mathbf{T}(\mathbf{T}\mathbf{P}^{(0)}) = \mathbf{T}^2\mathbf{P}^{(0)}$$
$$\mathbf{P}^{(3)} = \mathbf{T}\mathbf{P}^{(2)} = \mathbf{T}(\mathbf{T}^2\mathbf{P}^0) = \mathbf{T}^3\mathbf{P}^{(0)},$$

and in general

$$\mathbf{P}^{(n)} = \mathbf{T}^n\mathbf{P}^{(0)}.$$

Thus the transition matrix and the initial state vector completely determine every other state vector.

Example 3 Consider Example 1 again. Suppose that when we begin our observations (day 0), it is dry, so the initial state vector is

$$\mathbf{P}^{(0)} = \begin{bmatrix} 1 \\ 0 \end{bmatrix}.$$

Then the state vector on day 1 (the day after we begin our observation) is

$$\mathbf{P}^{(1)} = \mathbf{T}\mathbf{P}^{(0)} = \begin{bmatrix} 0.67 & 0.5 \\ 0.33 & 0.5 \end{bmatrix}\begin{bmatrix} 1 \\ 0 \end{bmatrix} = \begin{bmatrix} 0.67 \\ 0.33 \end{bmatrix}.$$

Thus the probability of no rain on day 1 is 0.67 and the probability of rain on that day is 0.33. Similarly,

$$\mathbf{P}^{(2)} = \mathbf{T}\mathbf{P}^{(1)} = \begin{bmatrix} 0.67 & 0.5 \\ 0.33 & 0.5 \end{bmatrix}\begin{bmatrix} 0.67 \\ 0.33 \end{bmatrix} = \begin{bmatrix} 0.614 \\ 0.386 \end{bmatrix}$$

$$\mathbf{P}^{(3)} = \mathbf{T}\mathbf{P}^{(2)} = \begin{bmatrix} 0.67 & 0.5 \\ 0.33 & 0.5 \end{bmatrix}\begin{bmatrix} 0.614 \\ 0.386 \end{bmatrix} = \begin{bmatrix} 0.604 \\ 0.396 \end{bmatrix}$$

$$\mathbf{P}^{(4)} = \mathbf{T}\mathbf{P}^{(3)} = \begin{bmatrix} 0.67 & 0.5 \\ 0.33 & 0.5 \end{bmatrix}\begin{bmatrix} 0.604 \\ 0.396 \end{bmatrix} = \begin{bmatrix} 0.603 \\ 0.397 \end{bmatrix}$$

$$\mathbf{P}^{(5)} = \mathbf{T}\mathbf{P}^{(4)} = \begin{bmatrix} 0.67 & 0.5 \\ 0.33 & 0.5 \end{bmatrix}\begin{bmatrix} 0.603 \\ 0.397 \end{bmatrix} = \begin{bmatrix} 0.603 \\ 0.397 \end{bmatrix}.$$

From the fourth day on, the state vector of the system is always the same, $\begin{bmatrix} 0.603 \\ 0.397 \end{bmatrix}$. This means that from the fourth day on, it is dry about 60 percent of the time, and it rains about 40 percent of the time.

Example 4 Consider Example 2 again. Suppose that when our survey begins, we find that brand A has 20 percent of the market, brand B has 20 percent of the market, and the other brands have 60 percent of the market. Thus

$$\mathbf{P}^{(0)} = \begin{bmatrix} 0.2 \\ 0.2 \\ 0.6 \end{bmatrix}.$$

The state vector after the first week is

$$\mathbf{P}^{(1)} = \mathbf{TP}^{(0)} = \begin{bmatrix} 0.50 & 0.60 & 0.40 \\ 0.25 & 0.30 & 0.30 \\ 0.25 & 0.10 & 0.30 \end{bmatrix} \begin{bmatrix} 0.2 \\ 0.2 \\ 0.6 \end{bmatrix} = \begin{bmatrix} 0.4600 \\ 0.2900 \\ 0.2500 \end{bmatrix}.$$

Similarly,

$$\mathbf{P}^{(2)} = \mathbf{TP}^{(1)} = \begin{bmatrix} 0.50 & 0.60 & 0.40 \\ 0.25 & 0.30 & 0.30 \\ 0.25 & 0.10 & 0.30 \end{bmatrix} \begin{bmatrix} 0.4600 \\ 0.2900 \\ 0.2500 \end{bmatrix} = \begin{bmatrix} 0.5040 \\ 0.2770 \\ 0.2190 \end{bmatrix}$$

$$\mathbf{P}^{(3)} = \mathbf{TP}^{(2)} = \begin{bmatrix} 0.50 & 0.60 & 0.40 \\ 0.25 & 0.30 & 0.30 \\ 0.25 & 0.10 & 0.30 \end{bmatrix} \begin{bmatrix} 0.5040 \\ 0.2770 \\ 0.2190 \end{bmatrix} = \begin{bmatrix} 0.5058 \\ 0.2748 \\ 0.2194 \end{bmatrix}.$$

$$\mathbf{P}^{(4)} = \mathbf{TP}^{(3)} = \begin{bmatrix} 0.50 & 0.60 & 0.40 \\ 0.25 & 0.30 & 0.30 \\ 0.25 & 0.10 & 0.30 \end{bmatrix} \begin{bmatrix} 0.5058 \\ 0.2748 \\ 0.2194 \end{bmatrix} = \begin{bmatrix} 0.5055 \\ 0.2747 \\ 0.2198 \end{bmatrix}.$$

$$\mathbf{P}^{(5)} = \mathbf{TP}^{(4)} = \begin{bmatrix} 0.50 & 0.60 & 0.40 \\ 0.25 & 0.30 & 0.30 \\ 0.25 & 0.10 & 0.30 \end{bmatrix} \begin{bmatrix} 0.5055 \\ 0.2747 \\ 0.2198 \end{bmatrix} = \begin{bmatrix} 0.5055 \\ 0.2747 \\ 0.2198 \end{bmatrix}.$$

Thus as n increases, the state vectors approach the fixed vector

$$\begin{bmatrix} 0.5055 \\ 0.2747 \\ 0.2198 \end{bmatrix}.$$

This means that in the long run, brand A will command about 51 percent of the market, brand B will retain about 27 percent, and the other brands will command about 22 percent.

In the last two examples we have seen that as the number of observation periods increases, the state vectors converge to a fixed vector. In this case we say that the Markov process has reached equilibrium. The fixed vector is called the **steady-state vector**. Markov processes are generally used to determine the behavior of a system in the long run; for example, the share of the market that a certain manufacturer can expect to retain on a somewhat permanent basis. Thus the question of whether or not a Markov process reaches equilibrium is of paramount importance. The following example shows that not every Markov process reaches equilibrium.

Example 5 Let

$$\mathbf{T} = \begin{bmatrix} 0 & 1 \\ 1 & 0 \end{bmatrix} \qquad \text{and} \qquad \mathbf{P}^{(0)} = \begin{bmatrix} \frac{1}{3} \\ \frac{2}{3} \end{bmatrix}.$$

Then

$$\mathbf{P}^{(1)} = \begin{bmatrix} \frac{2}{3} \\ \frac{1}{3} \end{bmatrix}, \qquad \mathbf{P}^{(2)} = \begin{bmatrix} \frac{1}{3} \\ \frac{2}{3} \end{bmatrix}, \qquad \mathbf{P}^{(3)} = \begin{bmatrix} \frac{2}{3} \\ \frac{1}{3} \end{bmatrix}, \qquad \mathbf{P}^{(4)} = \begin{bmatrix} \frac{1}{3} \\ \frac{2}{3} \end{bmatrix}, \ \ldots.$$

Thus the state vectors oscillate between the vectors

$$\begin{bmatrix} \frac{2}{3} \\ \frac{1}{3} \end{bmatrix} \qquad \text{and} \qquad \begin{bmatrix} \frac{1}{3} \\ \frac{2}{3} \end{bmatrix}$$

and do not converge to a fixed vector.

However, if we demand that the transition matrix of a Markov process satisfy a rather reasonable property, we obtain a large class of Markov processes, many of which arise in practical applications, that *do* reach equilibrium. We now proceed to formulate these ideas.

Definition The vector

$$\mathbf{U} = \begin{bmatrix} u_1 \\ u_2 \\ \vdots \\ u_n \end{bmatrix}$$

is called a **probability vector** if $u_i \geq 0$ $(1 \leq i \leq n)$ and

$$u_1 + u_2 + \cdots + u_n = 1.$$

Example 6 The vectors

$$\begin{bmatrix} \frac{1}{2} \\ \frac{1}{4} \\ \frac{1}{4} \end{bmatrix} \qquad \text{and} \qquad \begin{bmatrix} \frac{1}{3} \\ \frac{2}{3} \\ 0 \end{bmatrix}$$

are probability vectors; the vectors

$$\begin{bmatrix} \frac{1}{5} \\ \frac{1}{5} \\ \frac{2}{5} \end{bmatrix} \qquad \text{and} \qquad \begin{bmatrix} \frac{1}{3} \\ \frac{1}{2} \\ \frac{1}{2} \end{bmatrix}$$

are not probability vectors. (Why not?)

Definition A transition matrix $\mathbf{T}$ of a Markov process is called **regular** if all the entries in some power of $\mathbf{T}$ are positive. A Markov process is called **regular** if its transition matrix is regular.

The Markov processes in Examples 1 and 2 are regular, since all the entries in the transition matrices themselves are positive.

Example 7 The transition matrix

$$\mathbf{T} = \begin{bmatrix} 0.2 & 1 \\ 0.8 & 0 \end{bmatrix}$$

is regular, since

$$\mathbf{T}^2 = \begin{bmatrix} 0.84 & 0.2 \\ 0.16 & 0.8 \end{bmatrix}.$$

We now state the following fundamental theorem of regular Markov processes; the proof, which we omit, can be found in the book by Kemeny and Snell given in Further Reading at the end of the section.

Theorem 7.9 *If $\mathbf{T}$ is the transition matrix of a regular Markov process, then:*
 (a) *As $n \to \infty$, $\mathbf{T}^n$ approaches a matrix*

$$\mathbf{A} = \begin{bmatrix} u_1 & u_1 & \cdots & u_1 \\ u_2 & u_2 & \cdots & u_2 \\ \vdots & \vdots & & \vdots \\ u_n & u_n & \cdots & u_n \end{bmatrix},$$

 all of whose columns are identical.
 (b) *Every column*

$$\mathbf{U} = \begin{bmatrix} u_1 \\ u_2 \\ \vdots \\ u_n \end{bmatrix}$$

 of $\mathbf{A}$ is a probability vector all of whose components are positive. That is, $u_i > 0$ $(1 \leq i \leq n)$ and

$$u_1 + u_2 + \cdots + u_n = 1.$$

We now establish the following result.

Theorem 7.10 *If* **T** *is a regular transition matrix and* **A** *and* **U** *are as in Theorem 7.9, then:*

(a) *For any probability vector* **X**, $\mathbf{T}^n\mathbf{X} \to \mathbf{U}$ *as* $n \to \infty$, *so that* **U** *is a steady-state vector.*

(b) *The steady-state vector* **U** *is the unique probability vector satisfying the matrix equation* $\mathbf{TU} = \mathbf{U}$.

Proof

(a) Let

$$\mathbf{X} = \begin{bmatrix} x_1 \\ x_2 \\ \vdots \\ x_n \end{bmatrix}$$

be a probability vector. Since $\mathbf{T}^n \to \mathbf{A}$ as $n \to \infty$, we have

$$\mathbf{T}^n\mathbf{X} \to \mathbf{AX}.$$

Now

$$\mathbf{AX} = \begin{bmatrix} u_1 & u_1 & \cdots & u_1 \\ u_2 & u_2 & \cdots & u_2 \\ \vdots & \vdots & & \vdots \\ u_n & u_n & \cdots & u_n \end{bmatrix}\begin{bmatrix} x_1 \\ x_2 \\ \vdots \\ x_n \end{bmatrix}$$

$$= \begin{bmatrix} u_1x_1 + u_1x_2 + \cdots + u_1x_n \\ u_2x_1 + u_2x_2 + \cdots + u_2x_n \\ \vdots \\ u_nx_1 + u_nx_2 + \cdots + u_nx_n \end{bmatrix}$$

$$= \begin{bmatrix} u_1(x_1 + x_2 + \cdots + x_n) \\ u_2(x_1 + x_2 + \cdots + x_n) \\ \vdots \\ u_n(x_1 + x_2 + \cdots + x_n) \end{bmatrix} = \begin{bmatrix} u_1 \\ u_2 \\ \vdots \\ u_n \end{bmatrix},$$

since $x_1 + x_2 + \cdots + x_n = 1$. Hence $\mathbf{T}^n\mathbf{X} \to \mathbf{U}$.

(b) Since $\mathbf{T}^n \to \mathbf{A}$, we also have $\mathbf{T}^{n+1} \to \mathbf{A}$. However,

$$\mathbf{T}^{n+1} = \mathbf{TT}^n,$$

so $\mathbf{T}^{n+1} \to \mathbf{TA}$. Hence $\mathbf{TA} = \mathbf{A}$. Equating corresponding columns of this equation (using Exercise T.5 of Section 1.2), we have $\mathbf{TU} = \mathbf{U}$. To show that **U** is unique, we let **V** be another probability vector such that $\mathbf{TV} = \mathbf{V}$. From (a), $\mathbf{T}^n\mathbf{V} \to \mathbf{U}$, and since $\mathbf{TV} = \mathbf{V}$, we have $\mathbf{T}^n\mathbf{V} = \mathbf{V}$ for all n. Hence $\mathbf{V} = \mathbf{U}$.

REMARK   (For those who have read Section 5.1). Thus a steady-state vector **U** of a regular matrix **T** is that eigenvector for $\lambda = 1$ which is also a probability vector.

In Examples 3 and 4 we calculated steady-state vectors by computing the powers T^nX. An alternative way of computing the steady-state vector of a regular transition matrix is as follows. From (b) of Theorem 7.10, we write

$$TU = U$$

as

$$TU = I_nU$$

or

$$(I_n - T)U = 0. \tag{3}$$

We have shown that the homogeneous system (3) has a unique solution U which is a probability vector, so that

$$u_1 + u_2 + \cdots + u_n = 1. \tag{4}$$

The first procedure for computing the steady-state vector U of a regular transition matrix T is as follows.

STEP 1. Compute the powers T^nX, where X is any probability vector.

STEP 2. U is the limit of the powers T^nX.

The second procedure for computing the steady-state vector U of a regular transition matrix T is as follows.

STEP 1. Solve the homogeneous system

$$(I_n - T)U = 0$$

STEP 2. From the infinitely many solutions obtained in Step 1, determine a unique solution U by requiring that its components satisfy Equation (4).

Example 8 Consider the matrix of Example 2. The homogeneous system (3) is (verify)

$$\begin{bmatrix} 0.50 & -0.60 & -0.40 \\ -0.25 & 0.70 & -0.30 \\ -0.25 & -0.10 & 0.70 \end{bmatrix} \begin{bmatrix} u_1 \\ u_2 \\ u_3 \end{bmatrix} = \begin{bmatrix} 0 \\ 0 \\ 0 \end{bmatrix}.$$

The reduced row echelon form of the augmented matrix is (verify)

$$\begin{bmatrix} 1 & 0 & -2.30 & \vdots & 0 \\ 0 & 1 & -1.25 & \vdots & 0 \\ 0 & 0 & 0.00 & \vdots & 0 \end{bmatrix}.$$

Hence a solution is

$$u_1 = 2.3r$$
$$u_2 = 1.25r$$
$$u_3 = r,$$

where r is any real number. From (4), we have

$$2.3r + 1.25r + r = 1,$$

or

$$r = \frac{1}{4.55} = 0.2198.$$

Hence

$$u_1 = 0.5055$$
$$u_2 = 0.2747$$
$$u_3 = 0.2198.$$

These results agree with those obtained in Example 4.

FURTHER READING

Kemeny, John G., and **J. Laurie Snell.** *Finite Markov Chains.* New York: Springer-Verlag, 1976.

Maki, D. P., and **M. Thompson.** *Mathematical Models and Applications: With Emphasis on the Social, Life, and Management Sciences.* Englewood Cliffs, N.J.: Prentice-Hall, Inc., 1973.

Roberts, Fred S. *Discrete Mathematical Models with Applications to Social, Biological, and Environmental Problems.* Englewood Cliffs, N.J.: Prentice-Hall, Inc., 1976.

7.7 EXERCISES

1. Which of the following can be transition matrices of a Markov process?

(a) $\begin{bmatrix} 0.3 & 0.7 \\ 0.4 & 0.6 \end{bmatrix}.$

(b) $\begin{bmatrix} 0.2 & 0.3 & 0.1 \\ 0.8 & 0.5 & 0.7 \\ 0.0 & 0.2 & 0.2 \end{bmatrix}.$

(c) $\begin{bmatrix} 0.55 & 0.33 \\ 0.45 & 0.67 \end{bmatrix}.$

(d) $\begin{bmatrix} 0.3 & 0.4 & 0.2 \\ 0.2 & 0.0 & 0.8 \\ 0.1 & 0.3 & 0.6 \end{bmatrix}.$

2. Which of the following are probability vectors?

(a) $\begin{bmatrix} \frac{1}{2} \\ \frac{1}{3} \\ \frac{2}{3} \end{bmatrix}.$ (b) $\begin{bmatrix} 0 \\ 1 \\ 0 \end{bmatrix}.$

(c) $\begin{bmatrix} \frac{1}{4} \\ \frac{1}{6} \\ \frac{1}{3} \\ \frac{1}{4} \end{bmatrix}.$ (d) $\begin{bmatrix} \frac{1}{5} \\ \frac{2}{5} \\ \frac{1}{10} \\ \frac{2}{10} \end{bmatrix}.$

3. Consider the transition matrix $\mathbf{T} = \begin{bmatrix} 0.7 & 0.4 \\ 0.3 & 0.6 \end{bmatrix}.$

(a) If $\mathbf{P} = \begin{bmatrix} 1 \\ 0 \end{bmatrix}$, compute $\mathbf{P}^{(1)}$, $\mathbf{P}^{(2)}$, and $\mathbf{P}^{(3)}$ to three decimal places.

(b) Show that **T** is regular and find its steady-state vector.

4. Consider the transition matrix

$$\mathbf{T} = \begin{bmatrix} 0 & 0.2 & 0.0 \\ 0 & 0.3 & 0.3 \\ 1 & 0.5 & 0.7 \end{bmatrix}.$$

(a) If $\mathbf{P} = \begin{bmatrix} 0 \\ 1 \\ 0 \end{bmatrix}$, compute $\mathbf{P}^{(1)}$, $\mathbf{P}^{(2)}$, $\mathbf{P}^{(3)}$, and $\mathbf{P}^{(4)}$

to three decimal places.

(b) Show that **T** is regular and find its steady-state vector.

5. Which of the following transition matrices are regular?

(a) $\begin{bmatrix} 0 & \frac{1}{2} \\ 1 & \frac{1}{2} \end{bmatrix}.$ (b) $\begin{bmatrix} \frac{1}{2} & 0 & 0 \\ 0 & 1 & \frac{1}{2} \\ \frac{1}{2} & 0 & \frac{1}{2} \end{bmatrix}.$

(c) $\begin{bmatrix} 1 & \frac{1}{3} & 0 \\ 0 & \frac{1}{3} & 1 \\ 0 & \frac{1}{3} & 0 \end{bmatrix}.$ (d) $\begin{bmatrix} \frac{1}{4} & \frac{3}{5} & \frac{1}{2} \\ \frac{1}{2} & 0 & 0 \\ \frac{1}{4} & \frac{2}{5} & \frac{1}{2} \end{bmatrix}.$

6. Show that each of the following transition matrices reaches a state of equilibrium.

(a) $\begin{bmatrix} \frac{1}{2} & 1 \\ \frac{1}{2} & 0 \end{bmatrix}.$ (b) $\begin{bmatrix} 0.4 & 0.2 \\ 0.6 & 0.8 \end{bmatrix}.$

(c) $\begin{bmatrix} \frac{1}{3} & 1 & \frac{1}{2} \\ \frac{1}{3} & 0 & \frac{1}{4} \\ \frac{1}{3} & 0 & \frac{1}{4} \end{bmatrix}.$

(d) $\begin{bmatrix} 0.3 & 0.1 & 0.4 \\ 0.2 & 0.4 & 0.0 \\ 0.5 & 0.5 & 0.6 \end{bmatrix}.$

7. Let

$$\mathbf{T} = \begin{bmatrix} \frac{1}{2} & 0 \\ \frac{1}{2} & 1 \end{bmatrix}.$$

(a) Show that **T** is not regular.

(b) Show that $\mathbf{T}^n\mathbf{X} \to \begin{bmatrix} 0 \\ 1 \end{bmatrix}$ for any probability vector **X**. Thus a Markov chain may have a unique steady-state vector even though its transition matrix is not regular.

8. Find the steady-state vector of each of the following regular matrices.

(a) $\begin{bmatrix} \frac{1}{3} & \frac{1}{2} \\ \frac{2}{3} & \frac{1}{2} \end{bmatrix}.$ (b) $\begin{bmatrix} 0.3 & 0.1 \\ 0.7 & 0.9 \end{bmatrix}.$

(c) $\begin{bmatrix} \frac{1}{4} & \frac{1}{2} & \frac{1}{3} \\ 0 & \frac{1}{2} & \frac{2}{3} \\ \frac{3}{4} & 0 & 0 \end{bmatrix}.$

(d) $\begin{bmatrix} 0.4 & 0.0 & 0.1 \\ 0.2 & 0.5 & 0.3 \\ 0.4 & 0.5 & 0.6 \end{bmatrix}.$

9. **(Psychology).** A behavioral psychologist places a rat each day in a cage with two doors, A and B. The rat can go through door A, where it receives an electric shock, or through door B, where it receives some food. A record is made of the door through which the rat passes. At the start of the experiment, on a Monday, the rat is equally likely to go through door A as through door B. After going through door A, and receiving a shock, the probability of going through the same door on the next day is 0.3. After going through door B, and receiving food, the probability of going through the same door on the next day is 0.6.

(a) Write the transition matrix for the Markov process.

(b) What is the probability of the rat going through door A on Thursday (the third day after starting the experiment)?

(c) What is the steady-state vector?

10. **(Business).** The subscription department of a magazine sends out a letter to a large mailing list inviting subscriptions for the magazine. Some of the people receiving this letter already subscribe to the magazine, while others do not. From this mailing list, 60 percent of those who already subscribe will subscribe again, while 25 percent of those who do not now subscribe will subscribe.

(a) Write the transition matrix for this Markov process.

(b) On the last letter it was found that 40 percent of those receiving it, ordered a subscription. What percentage of those receiving the current letter can be expected to order a subscription?

11. **(Sociology).** A study has determined that the occupation of a boy, as an adult, depends upon the occupation of his father and is given by the following transition matrix

Father's occupation

		Profes-sional	Farmer	Laborer
Son's	Profes-sional	0.8	0.3	0.2
occupa-	Farmer	0.1	0.5	0.2
tion	Laborer	0.1	0.2	0.6

Thus the probability that the son of a professional will also be a professional is 0.8, and so on.
   (a) What is the probability that the grandchild of a professional will also be a professional?
   (b) In the long run, what proportion of the population will be farmers?

12. **(Genetics).** Consider a plant that can have red flowers (R), pink flowers (P), or white flowers (W), depending upon the genotypes RR, RW, and WW. When we cross each of these genotypes with a genotype RW, we obtain the transition matrix

Flowers of parent plant

		Red	Pink	White
Flowers of	Red	0.5	0.25	0.0
offspring	Pink	0.5	0.50	0.5
plant	White	0.0	0.25	0.5

Suppose that each successive generation is produced by crossing only with plants of RW genotype. When the process reaches equilibrium, what percentage of the plants will have red, pink, or white flowers?

13. **(Mass transit).** A new mass transit system has just gone into operation. The transit authority has made studies that predict the percentage of commuters who will change to mass transit or continue driving their automobile. The following transition matrix has been obtained:

This year

		Mass transit	Automobile
Next	Mass transit	0.7	0.2
year	Automobile	0.3	0.8

Suppose that the population of the area remains constant, and that initially 30 percent of the commuters use mass transit and 70 percent use their automobiles.
   (a) What percentage of the commuters will be using the mass transit system after 1 year? After 2 years?
   (b) What percentage of the commuters will be using the mass transit system in the long run?

7.8

THE FIBONACCI SEQUENCE

Prerequisite. Section 5.1, Diagonalization.

In 1202, Leonardo of Pisa, also called Fibonacci,† wrote a book on mathematics in which he posed the following problem. A pair of newborn rabbits begins to breed at the age of one month, and thereafter produces one pair of offspring per month. Suppose that we start

---

† Leonardo Fibonacci of Pisa (about 1170–1250) was born and lived most of his life in Pisa, Italy. When he was about 20, his father was appointed director of Pisan commercial interests in northern Africa, now a part of Algeria. Leonardo accompanied his father to Africa and for several years traveled extensively throughout the Mediterranean area on behalf of his father. In these travels he learned the Hindu-Arabic method of numeration and calculation and decided to promote its use in Italy. This was one purpose of his most famous book, *Liber Abaci,* which appeared in 1202 and contained the rabbit problem stated here.

with a pair of newly born rabbits and that none of the rabbits produced from this pair die. How many pairs of rabbits will there be at the beginning of each month?

The breeding pattern of the rabbits is shown in Figure 7.29, where an arrow indicates an offspring pair.

At the beginning of month zero, we have the newly born pair of rabbits P_1. At the beginning of month 1 we still have only the original pair of rabbits P_1, which have not yet produced any offspring. At the beginning of month 2 we have the original pair P_1 and its first pair of offspring, P_2. At the beginning of month 3 we have the original pair P_1, its first pair of offspring P_2 born at the beginning of month 2, and its second pair of offspring, P_3. At the beginning of month 4 we have P_1, P_2, and P_3; P_4, the offspring of P_1; and P_5, the offspring of P_2. Let u_n denote the number of pairs of rabbits at the beginning of month n. We see that

$$u_0 = 1, \quad u_1 = 1, \quad u_2 = 2, \quad u_3 = 3, \quad u_4 = 5, \quad u_5 = 8.$$

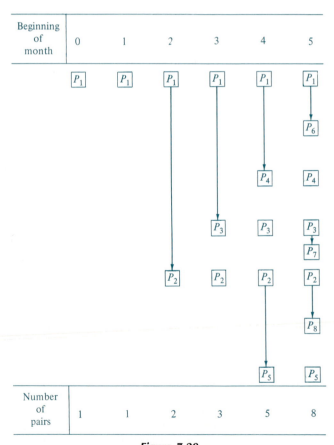

Figure 7.29

To obtain a formula for u_n, we proceed as follows. The number of pairs of rabbits that are alive at the beginning of month n is u_{n-1}, the number of pairs who were alive the previous month, plus the number of pairs newly born at the beginning of month n. The latter number is u_{n-2}, since a pair of rabbits produces a pair of offspring starting with its second month of life. Thus

$$u_n = u_{n-1} + u_{n-2}. \tag{1}$$

That is, each number is the sum of its two predecessors. The resulting sequence of numbers, called a **Fibonacci sequence**, occurs in a remarkably varied number of applications, such as the distribution of leaves on certain trees, the arrangements of seeds on sunflowers, search techniques in numerical analysis, generating random numbers in statistics, and others.

To compute u_n by using the **recursion relation** (or difference equation) (1), we have to compute $u_0, u_1, \ldots, u_{n-2}, u_{n-1}$. This can be rather tedious for large n. We now develop a formula that will enable us to calculate u_n directly.

In addition to Equation (1) we write

$$u_{n-1} = u_{n-1}. \tag{2}$$

We now define

$$\mathbf{U}_k = \begin{bmatrix} u_{k+1} \\ u_k \end{bmatrix} \quad \text{and} \quad \mathbf{A} = \begin{bmatrix} 1 & 1 \\ 1 & 0 \end{bmatrix}, \quad (0 \leq k \leq n-1)$$

so that

$$\mathbf{U}_0 = \begin{bmatrix} u_1 \\ u_0 \end{bmatrix} = \begin{bmatrix} 1 \\ 1 \end{bmatrix}, \quad \mathbf{U}_1 = \begin{bmatrix} u_2 \\ u_1 \end{bmatrix} = \begin{bmatrix} 2 \\ 1 \end{bmatrix}, \ldots,$$

$$\mathbf{U}_{n-2} = \begin{bmatrix} u_{n-1} \\ u_{n-2} \end{bmatrix}, \quad \text{and} \quad \mathbf{U}_{n-1} = \begin{bmatrix} u_n \\ u_{n-1} \end{bmatrix}.$$

Then (1) and (2) together can be written as

$$\mathbf{U}_{n-1} = \mathbf{A}\mathbf{U}_{n-2}.$$

Thus

$$\begin{aligned} \mathbf{U}_1 &= \mathbf{A}\mathbf{U}_0 \\ \mathbf{U}_2 &= \mathbf{A}\mathbf{U}_1 = \mathbf{A}(\mathbf{A}\mathbf{U}_0) = \mathbf{A}^2\mathbf{U}_0 \\ \mathbf{U}_3 &= \mathbf{A}\mathbf{U}_2 = \mathbf{A}(\mathbf{A}^2\mathbf{U}_0) = \mathbf{A}^3\mathbf{U}_0 \\ &\vdots \\ \mathbf{U}_{n-1} &= \mathbf{A}^{n-1}\mathbf{U}_0. \end{aligned} \tag{3}$$

Hence, to find u_n, we merely have to calculate $\mathbf{A}^{n-1}$, which is still rather tedious if n is large. To avoid this difficulty, we find a diagonal matrix $\mathbf{B}$ that is similar to $\mathbf{A}$. The characteristic equation of $\mathbf{A}$ is

$$\begin{vmatrix} \lambda - 1 & -1 \\ -1 & \lambda \end{vmatrix} = \lambda^2 - \lambda - 1 = 0.$$

The eigenvalues of **A** are (verify)

$$\lambda_1 = \frac{1 + \sqrt{5}}{2} \quad \text{and} \quad \lambda_2 = \frac{1 - \sqrt{5}}{2}$$

so that

$$D = \begin{bmatrix} \dfrac{1 + \sqrt{5}}{2} & 0 \\ 0 & \dfrac{1 - \sqrt{5}}{2} \end{bmatrix}.$$

Corresponding eigenvectors are (verify)

$$X_1 = \begin{bmatrix} \dfrac{1 + \sqrt{5}}{2} \\ 1 \end{bmatrix} \quad \text{and} \quad X_2 = \begin{bmatrix} \dfrac{1 - \sqrt{5}}{2} \\ 1 \end{bmatrix}.$$

Thus

$$P = \begin{bmatrix} \dfrac{1 + \sqrt{5}}{2} & \dfrac{1 - \sqrt{5}}{2} \\ 1 & 1 \end{bmatrix}, \quad P^{-1} = \begin{bmatrix} \dfrac{1}{\sqrt{5}} & -\dfrac{1 - \sqrt{5}}{2\sqrt{5}} \\ -\dfrac{1}{\sqrt{5}} & \dfrac{1 + \sqrt{5}}{2\sqrt{5}} \end{bmatrix},$$

and

$$A = PDP^{-1}.$$

Hence (verify) for any nonnegative integer k,

$$A^k = PD^kP^{-1}.$$

Since **D** is diagonal, D^k is easy to calculate; its entries are the diagonal entries of **D** raised to the kth power. From (3) we have

$$U_{n-1} = A^{n-1}U_0 = PD^{n-1}P^{-1}U_0$$

$$= \begin{bmatrix} \dfrac{1 + \sqrt{5}}{2} & \dfrac{1 - \sqrt{5}}{2} \\ 1 & 1 \end{bmatrix} \begin{bmatrix} \left(\dfrac{1 + \sqrt{5}}{2}\right)^{n-1} & 0 \\ 0 & \left(\dfrac{1 - \sqrt{5}}{2}\right)^{n-1} \end{bmatrix} \begin{bmatrix} \dfrac{1}{\sqrt{5}} & -\dfrac{1 - \sqrt{5}}{2\sqrt{5}} \\ -\dfrac{1}{\sqrt{5}} & \dfrac{1 + \sqrt{5}}{2\sqrt{5}} \end{bmatrix} \begin{bmatrix} 1 \\ 1 \end{bmatrix}.$$

This equation gives the formula (verify)

$$u_n = \frac{1}{\sqrt{5}}\left[\left(\frac{1 + \sqrt{5}}{2}\right)^{n+1} - \left(\frac{1 - \sqrt{5}}{2}\right)^{n+1}\right] \tag{4}$$

for calculating u_n directly.

Using a hand calculator, we find that for $n = 50$, u_{50} is approximately 20.365 billion.

FURTHER READING

A Primer for the Fibonacci Numbers. San Jose State University, San Jose, Calif., 1972.

Vorobyov, N. N., *The Fibonacci Numbers*. Boston: D. C. Heath and Company, 1963.

7.8 EXERCISES

1. Compute the eigenvalues and eigenvectors of
$$A = \begin{bmatrix} 1 & 1 \\ 1 & 0 \end{bmatrix}$$
and verify that they are as given in the text.

2. Verify that if $A = PBP^{-1}$ and k is a positive integer, then $A^k = PB^kP^{-1}$.

3. Using a hand calculator, compute
   (a) u_8.   (b) u_{12}.   (c) u_{20}.

4. Consider the rabbit breeding problem but suppose that each pair of rabbits now produces two pairs of rabbits starting with its second month of life and continuing every month thereafter.
   (a) Formulate a recursion relation for the number u_n of rabbits at the beginning of month n.
   (b) Develop a formula for calculating u_n directly.

THEORETICAL EXERCISE

T.1. Let $A = \begin{bmatrix} 1 & 1 \\ 1 & 0 \end{bmatrix}$. It can then be shown that

$$A^{n+1} = \begin{bmatrix} u_{n+1} & u_n \\ u_n & u_{n-1} \end{bmatrix}.$$

Use this result to obtain the formula

$$u_{n+1}u_{n-1} - u_n^2 = (-1)^{n+1}.$$

7.9

DIFFERENTIAL EQUATIONS

Prerequisites. Section 5.1, Diagonalization. Calculus required.

A **differential equation** is an equation that involves an unknown function and its derivatives. An important, simple example of a differential equation is

$$\frac{dy}{dx} = ry,$$

where r is a constant. The idea here is to find a function f that will satisfy the given differential equation; that is, $f' = rf$. This differential equation is further discussed below.

Differential equations occur often in all branches of science and engineering; linear algebra is helpful in the formulation and solution of differential equations. In this section we provide only a brief survey of the approach; books on differential equations deal with the subject in much greater detail and several suggestions for further reading are given at the end of this section.

We consider the **homogeneous linear system** of differential equations

$$
\begin{aligned}
x_1'(t) &= a_{11}x_1(t) + a_{12}x_2(t) + \cdots + a_{1n}x_n(t) \\
x_2'(t) &= a_{21}x_1(t) + a_{22}x_2(t) + \cdots + a_{2n}x_n(t) \\
&\ \ \vdots \qquad\quad \vdots \qquad\qquad\qquad \vdots \\
x_n'(t) &= a_{n1}x_1(t) + a_{n2}x_2(t) + \cdots + a_{nn}x_n(t),
\end{aligned}
\tag{1}
$$

where the a_{ij} are known constants. We seek functions $x_1(t), x_2(t), \ldots, x_n(t)$ defined and differentiable on the real line satisfying (1).

We can write (1) in matrix form by letting

$$
\mathbf{X}(t) = \begin{bmatrix} x_1(t) \\ x_2(t) \\ \vdots \\ x_n(t) \end{bmatrix}, \qquad
\mathbf{A} = \begin{bmatrix}
a_{11} & a_{12} & \cdots & a_{1n} \\
a_{21} & a_{22} & \cdots & a_{2n} \\
\vdots & \vdots & & \vdots \\
a_{n1} & a_{n2} & \cdots & a_{nn}
\end{bmatrix},
$$

and defining

$$
\mathbf{X}'(t) = \begin{bmatrix} x_1'(t) \\ x_2'(t) \\ \vdots \\ x_n'(t) \end{bmatrix}
$$

Then (1) can be written as

$$
\mathbf{X}'(t) = \mathbf{A}\mathbf{X}(t).
\tag{2}
$$

We shall often write (2) more briefly as

$$
\mathbf{X}' = \mathbf{A}\mathbf{X}.
$$

With this notation, a vector function

$$
\mathbf{X}(t) = \begin{bmatrix} x_1(t) \\ x_2(t) \\ \vdots \\ x_n(t) \end{bmatrix}
$$

satisfying (2) is called a **solution** to the given system.

We leave it to the reader to verify that if $\mathbf{X}^{(1)}(t), \mathbf{X}^{(2)}(t), \ldots, \mathbf{X}^{(n)}(t)$ are all solutions to (2), then any linear combination

$$
\mathbf{X}(t) = b_1\mathbf{X}^{(1)}(t) + b_2\mathbf{X}^{(2)}(t) + \cdots + b_n\mathbf{X}^{(n)}(t)
\tag{3}
$$

is also a solution to (2).

A set of vector functions $\{X^{(1)}(t), X^{(2)}(t), \ldots, X^{(n)}(t)\}$ is said to be a **fundamental system** for (1) if every solution to (1) can be written in the form (3). In this case, the right side of (3), where $b_1, b_2, \ldots, b_n$ are arbitrary constants, is said to be the **general solution** to (2).

It can be shown that any system of the form (2) has a fundamental system (in fact, infinitely many).

The problem of finding a solution $X(t)$ to (2) such that $X(0) = X_0$, an **initial condition**, where X_0 is a given vector, is called an **initial value problem**. If the general solution (3) is known, then the initial value problem can be solved by setting $t = 0$ in (3) and determining the constants $b_1, b_2, \ldots, b_n$ so that

$$X_0 = b_1 X^{(1)}(0) + b_2 X^{(2)}(0) + \cdots + b_n X^{(n)}(0).$$

It is easily seen that this is actually an $n \times n$ linear system with unknowns $b_1, b_2, \ldots, b_n$. This linear system can also be written as

$$CB = X_0, \tag{4}$$

where

$$B = \begin{bmatrix} b_1 \\ b_2 \\ \vdots \\ b_n \end{bmatrix}$$

and C is the $n \times n$ matrix whose columns are $X^{(1)}(0), X^{(2)}(0), \ldots, X^{(n)}(0)$, respectively. It can be shown that if $X^{(1)}(t), X^{(2)}(t), \ldots, X^{(n)}(t)$ form a fundamental system for (1), then C is nonsingular, so (4) always has a solution.

Example 1   The simplest system of the form (1) is the single equation

$$\frac{dx}{dt} = ax, \tag{5}$$

where a is a constant. From calculus, the solutions to this equation are of the form

$$x = be^{at}; \tag{6}$$

that is, this is the general solution to (5). To solve the initial value problem

$$\frac{dx}{dt} = ax, \qquad x(0) = x_0,$$

we set $t = 0$ in (6) and obtain $b = x_0$. Thus the solution to the initial value problem is

$$x = x_0 e^{at}.$$

The system (2) is said to be **diagonal** if the matrix A is diagonal. Then (1) can be rewritten as

$$
\begin{aligned}
x_1'(t) &= a_{11}x_1(t) \\
x_2'(t) &= \qquad\quad a_{22}x_2(t) \\
&\vdots \\
x_n'(t) &= \qquad\qquad\qquad a_{nn}x_n(t).
\end{aligned}
\tag{7}
$$

This system is easy to solve, since the equations can be solved separately. Applying the results of Example 1 to each equation in (7) yields

$$
\begin{aligned}
x_1(t) &= b_1 e^{a_{11}t} \\
x_2(t) &= b_2 e^{a_{22}t} \\
&\vdots \qquad \vdots \\
x_n(t) &= b_n e^{a_{nn}t},
\end{aligned}
\tag{8}
$$

where $b_1, b_2, \ldots, b_n$ are arbitrary constants. Writing (8) in vector form yields

$$
\mathbf{X}(t) = \begin{bmatrix} b_1 e^{a_{11}t} \\ b_2 e^{a_{22}t} \\ \vdots \\ b_n e^{a_{nn}t} \end{bmatrix} = b_1 \begin{bmatrix} 1 \\ 0 \\ 0 \\ \vdots \\ 0 \end{bmatrix} e^{a_{11}t} + b_2 \begin{bmatrix} 0 \\ 1 \\ 0 \\ \vdots \\ 0 \end{bmatrix} e^{a_{22}t}
$$

$$
+ \cdots + b_n \begin{bmatrix} 0 \\ 0 \\ \vdots \\ 0 \\ 1 \end{bmatrix} e^{a_{nn}t}.
$$

This implies that the vector functions

$$
\mathbf{X}^{(1)}(t) = \begin{bmatrix} 1 \\ 0 \\ 0 \\ \vdots \\ 0 \end{bmatrix} e^{a_{11}t}, \quad \mathbf{X}^{(2)}(t) = \begin{bmatrix} 0 \\ 1 \\ 0 \\ \vdots \\ 0 \end{bmatrix} e^{a_{22}t}, \quad \ldots, \quad \mathbf{X}^{(n)}(t) = \begin{bmatrix} 0 \\ 0 \\ \vdots \\ 0 \\ 1 \end{bmatrix} e^{a_{nn}t}
$$

form a fundamental system for the diagonal system (7).

Example 2 The diagonal system

$$
\begin{bmatrix} x_1' \\ x_2' \\ x_3' \end{bmatrix} = \begin{bmatrix} 3 & 0 & 0 \\ 0 & -2 & 0 \\ 0 & 0 & 4 \end{bmatrix} \begin{bmatrix} x_1 \\ x_2 \\ x_3 \end{bmatrix}
\tag{9}
$$

can be written as three equations:

$$x_1' = 3x_1$$
$$x_2' = -2x_2$$
$$x_3' = 4x_3.$$

Integrating these, we obtain

$$x_1 = b_1 e^{3t}, \qquad x_2 = b_2 e^{-2t}, \qquad x_3 = b_3 e^{4t},$$

where b_1, b_2, and b_3 are arbitrary constants. Thus

$$\mathbf{X}(t) = \begin{bmatrix} b_1 e^{3t} \\ b_2 e^{-2t} \\ b_3 e^{4t} \end{bmatrix} = b_1 \begin{bmatrix} 1 \\ 0 \\ 0 \end{bmatrix} e^{3t} + b_2 \begin{bmatrix} 0 \\ 1 \\ 0 \end{bmatrix} e^{-2t} + b_3 \begin{bmatrix} 0 \\ 0 \\ 1 \end{bmatrix} e^{4t}$$

is the general solution to (9) and the functions

$$\mathbf{X}^{(1)}(t) = \begin{bmatrix} 1 \\ 0 \\ 0 \end{bmatrix} e^{3t}, \qquad \mathbf{X}^{(2)}(t) = \begin{bmatrix} 0 \\ 1 \\ 0 \end{bmatrix} e^{-2t}, \qquad \mathbf{X}^{(3)}(t) = \begin{bmatrix} 0 \\ 0 \\ 1 \end{bmatrix} e^{4t}$$

form a fundamental system for (9).

If the system (2) is not diagonal, then it cannot be solved as simply as the system in the preceding example. However, there is an extension of this method that yields the general solution in the case where $\mathbf{A}$ is diagonalizable. Suppose that $\mathbf{A}$ is diagonalizable and $\mathbf{P}$ is a nonsingular matrix such that

$$\mathbf{P}^{-1}\mathbf{AP} = \mathbf{D}, \tag{10}$$

where $\mathbf{D}$ is diagonal. Then multiplying the given system

$$\mathbf{X}' = \mathbf{AX}$$

on the left by $\mathbf{P}^{-1}$, we obtain

$$\mathbf{P}^{-1}\mathbf{X}' = \mathbf{P}^{-1}\mathbf{AX}.$$

Since $\mathbf{P}^{-1}\mathbf{P} = \mathbf{I}_n$, we can rewrite the last equation as

$$\mathbf{P}^{-1}\mathbf{X}' = (\mathbf{P}^{-1}\mathbf{AP})(\mathbf{P}^{-1}\mathbf{X}). \tag{11}$$

Temporarily, let

$$\mathbf{U} = \mathbf{P}^{-1}\mathbf{X}. \tag{12}$$

Since $\mathbf{P}^{-1}$ is a constant matrix,

$$\mathbf{U}' = \mathbf{P}^{-1}\mathbf{X}'. \tag{13}$$

Therefore, substituting (10), (12), and (13) into (11), we obtain

$$\mathbf{U}' = \mathbf{DU}. \tag{14}$$

Equation (14) is a diagonal system and can be solved by the methods just discussed.

Before proceeding, however, let us recall from Theorem 5.2 that

$$\mathbf{D} = \begin{bmatrix} \lambda_1 & 0 & \cdots & 0 \\ 0 & \lambda_2 & \cdots & 0 \\ \vdots & \vdots & & \vdots \\ 0 & 0 & \cdots & \lambda_n \end{bmatrix},$$

where $\lambda_1, \lambda_2, \ldots, \lambda_n$ are the eigenvalues of $\mathbf{A}$, and that the columns of $\mathbf{P}$ are linearly independent eigenvectors of $\mathbf{A}$ associated, respectively, with $\lambda_1, \lambda_2, \ldots, \lambda_n$. From the discussion just given for diagonal systems, the general solution to (14) is

$$\mathbf{U}(t) = b_1\mathbf{U}^{(1)}(t) + b_2\mathbf{U}^{(2)}(t) + \cdots + b_n\mathbf{U}^{(n)}(t)$$

$$= \begin{bmatrix} b_1 e^{\lambda_1 t} \\ b_2 e^{\lambda_2 t} \\ \vdots \\ b_n e^{\lambda_n t} \end{bmatrix},$$

where

$$\mathbf{U}^{(1)}(t) = \begin{bmatrix} 1 \\ 0 \\ 0 \\ \vdots \\ 0 \end{bmatrix} e^{\lambda_1 t}, \quad \mathbf{U}^{(2)}(t) = \begin{bmatrix} 0 \\ 1 \\ 0 \\ \vdots \\ 0 \end{bmatrix} e^{\lambda_2 t}, \quad \ldots, \quad \mathbf{U}^{(n)}(t) = \begin{bmatrix} 0 \\ 0 \\ \vdots \\ \vdots \\ 0 \\ 1 \end{bmatrix} e^{\lambda_n t} \quad (15)$$

and $b_1, b_2, \ldots, b_n$ are arbitrary constants. From Equation (12), $\mathbf{X} = \mathbf{PU}$, so the general solution to the given system $\mathbf{X}' = \mathbf{AX}$ is

$$\mathbf{X}(t) = \mathbf{PU}(t) = b_1\mathbf{PU}^{(1)}(t) + b_2\mathbf{PU}^{(2)}(t) + \cdots + b_n\mathbf{PU}^{(n)}(t). \quad (16)$$

However, since the constant vectors in (15) are the columns of the identity matrix and $\mathbf{PI}_n = \mathbf{P}$, (16) can be rewritten as

$$\mathbf{X}(t) = b_1\mathbf{P}_1 e^{\lambda_1 t} + b_2\mathbf{P}_2 e^{\lambda_2 t} + \cdots + b_n\mathbf{P}_n e^{\lambda_n t}, \quad (17)$$

where $\mathbf{P}_1, \mathbf{P}_2, \ldots, \mathbf{P}_n$ are the columns of $\mathbf{P}$, and therefore eigenvectors of $\mathbf{A}$ associated with $\lambda_1, \lambda_2, \ldots, \lambda_n$, respectively.

We summarize the discussion above in the following theorem.

Theorem 7.11 *If the $n \times n$ matrix $\mathbf{A}$ has n linearly independent eigenvectors $\mathbf{P}_1, \mathbf{P}_2, \ldots, \mathbf{P}_n$ associated with the eigenvalues $\lambda_1, \lambda_2, \ldots, \lambda_n$, respectively, then the general solution to the system of differential equations*

$$\mathbf{X}' = \mathbf{AX}$$

is given by (17).

The procedure for obtaining the general solution to the homogeneous linear system $\mathbf{X}'(t) = \mathbf{AX}(t)$, where $\mathbf{A}$ is diagonalizable, is as follows.

STEP 1. Compute the eigenvalues $\lambda_1, \lambda_2, \ldots, \lambda_n$ of $\mathbf{A}$.

STEP 2. Compute eigenvectors $\mathbf{P}_1, \mathbf{P}_2, \ldots, \mathbf{P}_n$ of $\mathbf{A}$ associated, respectively, with $\lambda_1, \lambda_2, \ldots, \lambda_n$.

STEP 3. The general solution is given by Equation (17).

Example 3 For the system

$$\mathbf{X}' = \begin{bmatrix} 1 & -1 \\ 2 & 4 \end{bmatrix} \mathbf{X},$$

the matrix

$$\mathbf{A} = \begin{bmatrix} 1 & -1 \\ 2 & 4 \end{bmatrix}$$

has eigenvalues $\lambda_1 = 2$ and $\lambda_2 = 3$ with associated eigenvectors

$$\mathbf{P}_1 = \begin{bmatrix} 1 \\ -1 \end{bmatrix} \quad \text{and} \quad \mathbf{P}_2 = \begin{bmatrix} 1 \\ -2 \end{bmatrix}.$$

These eigenvectors are automatically linearly independent, since they are associated with distinct eigenvalues (proof of Theorem 5.3). Hence the general solution to the given system is

$$\mathbf{X}(t) = b_1 \begin{bmatrix} 1 \\ -1 \end{bmatrix} e^{2t} + b_2 \begin{bmatrix} 1 \\ -2 \end{bmatrix} e^{3t}.$$

In terms of components, this can be written as

$$x_1(t) = b_1 e^{2t} + b_2 e^{3t}$$
$$x_2(t) = -b_1 e^{2t} - 2b_2 e^{3t}.$$

Example 4 Consider the following linear system of differential equations:

$$\mathbf{X}' = \begin{bmatrix} x_1' \\ x_2' \\ x_3' \end{bmatrix} = \begin{bmatrix} 0 & 1 & 0 \\ 0 & 0 & 1 \\ 8 & -14 & 7 \end{bmatrix} \begin{bmatrix} x_1 \\ x_2 \\ x_3 \end{bmatrix}.$$

The characteristic polynomial of $\mathbf{A}$ is (verify)

$$f(\lambda) = \lambda^3 - 7\lambda^2 + 14\lambda - 8$$

or

$$f(\lambda) = (\lambda - 1)(\lambda - 2)(\lambda - 4),$$

so the eigenvalues of $\mathbf{A}$ are $\lambda_1 = 1$, $\lambda_2 = 2$, and $\lambda_3 = 4$. Associated eigenvectors are (verify)

$$\begin{bmatrix} 1 \\ 1 \\ 1 \end{bmatrix}, \begin{bmatrix} 1 \\ 2 \\ 4 \end{bmatrix}, \begin{bmatrix} 1 \\ 4 \\ 16 \end{bmatrix},$$

respectively. The general solution is then given by

$$\mathbf{X}(t) = b_1 \begin{bmatrix} 1 \\ 1 \\ 1 \end{bmatrix} e^t + b_2 \begin{bmatrix} 1 \\ 2 \\ 4 \end{bmatrix} e^{2t} + b_3 \begin{bmatrix} 1 \\ 4 \\ 16 \end{bmatrix} e^{4t},$$

where b_1, b_2, and b_3 are arbitrary constants.

Example 5 We consider the linear system of Example 4 and now wish to solve the initial value problem determined by the **initial conditions** $x_1(0) = 4$, $x_2(0) = 6$, and $x_3(0) = 8$.

We write our general solution in the form $\mathbf{X} = \mathbf{PU}$ as

$$\mathbf{X}(t) = \begin{bmatrix} 1 & 1 & 1 \\ 1 & 2 & 4 \\ 1 & 4 & 16 \end{bmatrix} \begin{bmatrix} b_1 e^t \\ b_2 e^{2t} \\ b_3 e^{4t} \end{bmatrix}.$$

Now

$$\mathbf{X}(0) = \begin{bmatrix} 4 \\ 6 \\ 8 \end{bmatrix} = \begin{bmatrix} 1 & 1 & 1 \\ 1 & 2 & 4 \\ 1 & 4 & 16 \end{bmatrix} \begin{bmatrix} b_1 e^0 \\ b_2 e^0 \\ b_3 e^0 \end{bmatrix}$$

or

$$\begin{bmatrix} 1 & 1 & 1 \\ 1 & 2 & 4 \\ 1 & 4 & 16 \end{bmatrix} \begin{bmatrix} b_1 \\ b_2 \\ b_3 \end{bmatrix} = \begin{bmatrix} 4 \\ 6 \\ 8 \end{bmatrix}. \tag{18}$$

Solving (18) by Gauss–Jordan reduction, we obtain (verify)

$$b_1 = \tfrac{4}{3}, \qquad b_2 = 3, \qquad b_3 = -\tfrac{1}{3}.$$

Therefore, the solution to the initial value problem is

$$\mathbf{X}(t) = \tfrac{4}{3} \begin{bmatrix} 1 \\ 1 \\ 1 \end{bmatrix} e^t + 3 \begin{bmatrix} 1 \\ 2 \\ 4 \end{bmatrix} e^{2t} - \tfrac{1}{3} \begin{bmatrix} 1 \\ 4 \\ 16 \end{bmatrix} e^{4t}.$$

We now recall several facts from Chapter 5. If $\mathbf{A}$ does not have distinct eigenvalues, then we may or may not be able to diagonalize $\mathbf{A}$. Let λ be an eigenvalue of $\mathbf{A}$ of multiplicity k. Then $\mathbf{A}$ can be diagonalized if and only if the dimension of the eigenspace associated with λ is k, that is, if and only if the rank of the matrix $(\lambda \mathbf{I}_n - \mathbf{A})$ is $n - k$ (verify). If the rank of $(\lambda \mathbf{I}_n - \mathbf{A})$ is $n - k$, then we can find k linearly independent eigenvectors of $\mathbf{A}$ associated with λ.

Example 6 Consider the linear system

$$\mathbf{X}' = \mathbf{AX} = \begin{bmatrix} 1 & 0 & 0 \\ 0 & 3 & -2 \\ 0 & -2 & 3 \end{bmatrix} \mathbf{X}.$$

The eigenvalues of $\mathbf{A}$ are $\lambda_1 = \lambda_2 = 1$ and $\lambda_3 = 5$. The rank of the matrix

$$1\mathbf{I}_3 - \mathbf{A} = \begin{bmatrix} 0 & 0 & 0 \\ 0 & -2 & 2 \\ 0 & 2 & -2 \end{bmatrix}$$

is 1, so the linearly independent eigenvectors

$$\begin{bmatrix} 1 \\ 0 \\ 0 \end{bmatrix} \quad \text{and} \quad \begin{bmatrix} 0 \\ 1 \\ 1 \end{bmatrix}$$

are associated with the eigenvalue 1. The eigenvector

$$\begin{bmatrix} 0 \\ 1 \\ -1 \end{bmatrix}$$

is associated with the eigenvalue 5. The general solution to the given system is then

$$\mathbf{X}(t) = b_1 \begin{bmatrix} 1 \\ 0 \\ 0 \end{bmatrix} e^t + b_2 \begin{bmatrix} 0 \\ 1 \\ 1 \end{bmatrix} e^t + b_3 \begin{bmatrix} 0 \\ 1 \\ -1 \end{bmatrix} e^{5t},$$

where b_1, b_2, and b_3 are arbitrary constants.

If we cannot diagonalize $\mathbf{A}$ as in the examples, we are in a considerably more difficult situation. Methods for dealing with such problems are discussed in more advanced books (see the book by Brauer and Nohel given in Further Reading).

APPLICATION (A DIFFUSION PROCESS).   The following example is a modification of an example presented by Derrick and Grossman in *Elementary Differential Equations with Applications* (see Further Reading).

Example 7 Consider two adjoining cells separated by a permeable membrane and suppose that a fluid flows from the first cell to the second one at a rate (in milliliters per minute) that is numerically equal to three times the volume (in milliliters) of the fluid in the first cell. It then flows out of the second cell at a rate (in milliliters per minute) that is numerically equal to twice the volume in the second cell. Let $x_1(t)$ and $x_2(t)$ denote the volumes of the fluid in the first and second cells at time t, respectively. Assume that initially the first cell has 40 milliliters of fluid, while the second one has 5 milliliters of fluid. Find the volume of fluid in each cell at time t.

Solution

The change in volume of the fluid in each cell is the difference between the amount flowing in and the amount flowing out. Since no fluid flows into the first cell, we have

$$\frac{dx_1(t)}{dt} = -3x_1(t),$$

where the minus sign indicates that the fluid is flowing out of the cell. The flow $3x_1(t)$ from the first cell flows into the second cell. The flow out of the second cell is $2x_2(t)$. Thus the change in volume of the fluid in the second cell is given by

$$\frac{dx_2(t)}{dt} = 3x_1(t) - 2x_2(t).$$

We have thus obtained the linear system

$$\frac{dx_1(t)}{dt} = -3x_1(t)$$

$$\frac{dx_2(t)}{dt} = 3x_1(t) - 2x_2(t),$$

which can be written in matrix form as

$$\begin{bmatrix} x_1'(t) \\ x_2'(t) \end{bmatrix} = \begin{bmatrix} -3 & 0 \\ 3 & -2 \end{bmatrix} \begin{bmatrix} x_1(t) \\ x_2(t) \end{bmatrix}.$$

The eigenvalues of the matrix

$$\mathbf{A} = \begin{bmatrix} -3 & 0 \\ 3 & -2 \end{bmatrix}$$

are (verify)

$$\lambda_1 = -3, \qquad \lambda_2 = -2$$

and corresponding associated eigenvectors are (verify)

$$\begin{bmatrix} 1 \\ -3 \end{bmatrix}, \qquad \begin{bmatrix} 0 \\ 1 \end{bmatrix}.$$

Hence a solution is given by

$$\mathbf{X}(t) = \begin{bmatrix} x_1(t) \\ x_2(t) \end{bmatrix} = b_1 \begin{bmatrix} 1 \\ -3 \end{bmatrix} e^{-3t} + b_2 \begin{bmatrix} 0 \\ 1 \end{bmatrix} e^{-2t}.$$

Using the initial conditions, we find that (verify)

$$b_1 = 40, \qquad b_2 = 125.$$

Thus the volume of fluid in each cell at time t is given by

$$x_1(t) = 40e^{-3t}$$
$$x_2(t) = -120e^{-3t} + 125e^{-2t}.$$

It should also be pointed out that many differential equations cannot be solved in the sense that we can write a formula for the solution. Numerical methods, some of which are studied in numerical analysis, exist for obtaining numerical solutions to differential equations; computer codes for some of these methods are widely available.

FURTHER READING

Bentley, Donald L., and **Kenneth L. Cooke.** *Linear Algebra with Differential Equations.* New York: Holt, Rinehart and Winston, 1973.

Brauer, Fred, and **John A. Nohel.** *Ordinary Differential Equations: A First Course.* Reading, Mass.: Cummings Publishing Company, Inc., 1973.

Derrick, William R., and **Stanley I. Grossman.** *Elementary Differential Equations with Applications.* Reading, Mass.: Addison-Wesley Publishing Company, Inc., 1976.

Dettman, John H. *Introduction to Linear Algebra and Differential Equations.* New York: McGraw-Hill Book Company, 1974.

Rabenstein, Albert L. *Elementary Differential Equations with Linear Algebra,* 2nd ed. New York: Academic Press, Inc., 1975.

Wolfenstein, Samuel. *Introduction to Linear Algebra and Differential Equations.* San Francisco: Holden-Day, Inc., 1969.

7.9 EXERCISES

1. Consider the linear system of differential equations

$$\begin{bmatrix} x_1' \\ x_2' \\ x_3' \end{bmatrix} = \begin{bmatrix} -3 & 0 & 0 \\ 0 & 4 & 0 \\ 0 & 0 & 2 \end{bmatrix} \begin{bmatrix} x_1 \\ x_2 \\ x_3 \end{bmatrix}.$$

(a) Find the general solution.
(b) Find the solution to the initial value problem determined by the initial conditions $x_1(0) = 3$, $x_2(0) = 4$, $x_3(0) = 5$.

2. Consider the linear system of differential equations

$$\begin{bmatrix} x_1' \\ x_2' \\ x_3' \end{bmatrix} = \begin{bmatrix} 1 & 0 & 0 \\ 0 & -2 & 1 \\ 0 & 0 & 3 \end{bmatrix} \begin{bmatrix} x_1 \\ x_2 \\ x_3 \end{bmatrix}.$$

(a) Find the general solution.

(b) Find the solution to the initial value problem determined by the initial conditions $x_1(0) = 2$, $x_2(0) = 7$, $x_3(0) = 20$.

3. Find the general solution to the linear system of differential equations

$$\begin{bmatrix} x_1' \\ x_2' \\ x_3' \end{bmatrix} = \begin{bmatrix} 4 & 0 & 0 \\ 3 & -5 & 0 \\ 2 & 1 & 2 \end{bmatrix} \begin{bmatrix} x_1 \\ x_2 \\ x_3 \end{bmatrix}.$$

4. Find the general solution to the linear system of differential equations

$$\begin{bmatrix} x_1' \\ x_2' \\ x_3' \end{bmatrix} = \begin{bmatrix} 2 & 3 & 0 \\ 0 & 1 & 0 \\ 0 & 0 & 2 \end{bmatrix} \begin{bmatrix} x_1 \\ x_2 \\ x_3 \end{bmatrix}.$$

5. Find the general solution to the linear system of differential equations

$$\begin{bmatrix} x_1' \\ x_2' \\ x_3' \end{bmatrix} = \begin{bmatrix} 5 & 0 & 0 \\ 0 & -4 & 3 \\ 0 & 3 & 4 \end{bmatrix} \begin{bmatrix} x_1 \\ x_2 \\ x_3 \end{bmatrix}.$$

6. Find the general solution to the linear system of differential equations

$$\begin{bmatrix} x_1' \\ x_2' \end{bmatrix} = \begin{bmatrix} 3 & -2 \\ -2 & 3 \end{bmatrix} \begin{bmatrix} x_1 \\ x_2 \end{bmatrix}.$$

7. Find the general solution to the linear system of differential equations

$$\begin{bmatrix} x_1' \\ x_2' \\ x_3' \end{bmatrix} = \begin{bmatrix} 1 & 2 & 3 \\ 0 & 1 & 0 \\ 2 & 1 & 2 \end{bmatrix} \begin{bmatrix} x_1 \\ x_2 \\ x_3 \end{bmatrix}.$$

8. Find the general solution to the linear system of differential equations

$$\begin{bmatrix} x_1' \\ x_2' \\ x_3' \end{bmatrix} = \begin{bmatrix} 1 & 1 & 2 \\ 0 & 1 & 0 \\ 0 & 1 & 3 \end{bmatrix} \begin{bmatrix} x_1 \\ x_2 \\ x_3 \end{bmatrix}.$$

9. Consider two competing species that live in the same forest and let $x_1(t)$ and $x_2(t)$ denote the respective populations of the species at time t. Suppose that the initial populations are $x_1(0) = 500$ and $x_2(0) = 200$. If the growth rates of the species are given by

$$x_1'(t) = -3x_1(t) + 6x_2(t)$$
$$x_2'(t) = x_1(t) - 2x_2(t),$$

what is the population of each species at time t?

THEORETICAL EXERCISE

T.1. Prove that the set of all solutions to a linear system of differential equations $\mathbf{X}' = \mathbf{AX}$ is a vector space.

KEY IDEAS FOR REVIEW

☐ The **parametric equation** of the line through P_0 and parallel to $\mathbf{U}$ is

$$\mathbf{X} = \mathbf{X}_0 + t\mathbf{U} \quad (-\infty < t < \infty),$$

where $\mathbf{X}_0$ is the position vector of P_0.

☐ The equation of the plane with normal $\mathbf{N} = (a, b, c)$ and passing through the point $P_0 = (x_0, y_0, z_0)$ is

$$a(x - x_0) + b(y - y_0) + c(z - z_0) = 0.$$

☐ **Theorem 7.1 (Principal Axes Theorem).** Any quadratic form in n variables $Q(\mathbf{X}) = \mathbf{X}^T\mathbf{AX}$ is equivalent to a quadratic form,

$$Q'(\mathbf{Y}) = \lambda_1 y_1^2 + \lambda_2 y_2^2 + \cdots + \lambda_n y_n^2,$$

where

$$\mathbf{Y} = \begin{bmatrix} y_1 \\ y_2 \\ \vdots \\ y_n \end{bmatrix}$$

and $\lambda_1, \lambda_2, \ldots, \lambda_n$ are the eigenvalues of the matrix $\mathbf{A}$ of Q.

☐ **Theorem 7.2.** A quadratic form $Q(\mathbf{X}) = \mathbf{X}^T\mathbf{AX}$ in n variables is equivalent to a quadratic form $Q'(\mathbf{Y}) = y_1^2 + y_2^2 + \cdots + y_p^2 - y_{p+1}^2 - y_{p+2}^2 - \cdots - y_r^2$ for some $0 \leqslant p, \ r \leqslant n$.

☐ **Theorem 7.4.** Let $\mathbf{A}(G)$ be the adjacency matrix of a digraph G and let the rth power of $\mathbf{A}(G)$ be $\mathbf{B}_r$:

$$[\mathbf{A}(G)]^r = \mathbf{B}_r = [b_{ij}^{(r)}].$$

Then the i, jth element in $\mathbf{B}_r$, $b_{ij}^{(r)}$, is the number of ways in which P_i has access to P_j in r stages.

☐ **Theorem 7.5.** Let $\mathbf{A}(G)$ be the adjacency matrix of a digraph and let $\mathbf{S} = [s_{ij}]$ be the symmetric matrix defined above, with $\mathbf{S}^3 = [s_{ij}^{(3)}]$, where $s_{ij}^{(3)}$ is the i, jth element in $\mathbf{S}^3$. Then P_i belongs to a clique if and only if the diagonal entry $s_{ii}^{(3)}$ is positive.

☐ **Theorem 7.6.** A digraph with n vertices is strongly connected if and only if its adjacency matrix $\mathbf{A}(G)$

has the property that

$$A(G) + [A(G)]^2 + \cdots + [A(G)]^{n-1} = E$$

has no zero entries.

☐ If a_{rs} is a saddle point for a matrix game, then the optimal strategy for player R is his rth move, the optimal strategy for player C is his sth move. The value of the game is a_{rs}.

☐ The optimal strategies $P = [p_1 \quad p_2]$ and $Q = \begin{bmatrix} q_1 \\ q_2 \end{bmatrix}$ for players R and C, respectively, in a 2×2 matrix game are given by

$$p_1 = \frac{a_{22} - a_{21}}{a_{11} + a_{22} - a_{12} - a_{21}},$$

$$p_2 = \frac{a_{11} - a_{12}}{a_{11} + a_{22} - a_{12} - a_{21}}$$

$$q_1 = \frac{a_{22} - a_{12}}{a_{11} + a_{22} - a_{12} - a_{21}},$$

$$q_2 = \frac{a_{11} - a_{21}}{a_{11} + a_{22} - a_{12} - a_{21}}$$

The value of the game is

$$v = \frac{a_{11}a_{22} - a_{12}a_{21}}{a_{11} + a_{22} - a_{12} - a_{21}} = v'.$$

☐ **Theorem 7.7 (Fundamental Theorem of Matrix Games).** Every matrix game has a solution. That is, there are optimal strategies for R and C. Moreover, $v = v'$.

☐ **Method of least squares** [for determining the straight line $y = b_0 + b_1 x$ that best fits the data $(x_1, y_1), (x_2, y_2), \ldots, (x_n, y_n)$]. Let

$$Y = \begin{bmatrix} y_1 \\ y_2 \\ \vdots \\ y_n \end{bmatrix}, \quad A = \begin{bmatrix} 1 & x_1 \\ 1 & x_2 \\ \vdots & \vdots \\ 1 & x_n \end{bmatrix}, \quad \text{and} \quad X = \begin{bmatrix} b_0 \\ b_1 \end{bmatrix}.$$

Then b_0 and b_1 are found as follows:

$$X = (A^T A)^{-1} A^T Y,$$

☐ **Leontief closed model:** Given an exchange matrix A, find a vector $P \geq 0$ with at least one positive component satisfying $(I_n - A)P = 0$.

☐ **Leontief open model:** Given a consumption matrix C, and a demand vector D, find a production vector $X \geq 0$, satisfying the equation $(I_n - C)X = D$.

☐ **Theorem 7.8.** If T is the transition matrix of a Markov process, then the state vector $P^{(k+1)}$, at the $(k + 1)$th observation period, can be determined by the state vector $P^{(k)}$, at the kth observation period, as

$$P^{(k+1)} = TP^{(k)}.$$

☐ **Theorem 7.9.** If T is the transition matrix of a regular Markov process, then:

(a) As $n \to \infty$, T^n approaches a matrix

$$A = \begin{bmatrix} u_1 & u_1 & \cdots & u_1 \\ u_2 & u_2 & \cdots & u_2 \\ \vdots & \vdots & & \vdots \\ u_n & u_n & \cdots & u_n \end{bmatrix},$$

all of whose columns are identical.

(b) Every column

$$U = \begin{bmatrix} u_1 \\ u_2 \\ \vdots \\ u_n \end{bmatrix}$$

of A is a probability vector all of whose components are positive. That is, $u_i > 0$ $(1 \leq i \leq n)$ and

$$u_1 + u_2 + \cdots + u_n = 1.$$

☐ **Theorem 7.10.** If T is a regular transition matrix and A and U are as in Theorem 7.9, Then:

(a) For any probability vector X, $T^n X \to U$ as $n \to \infty$, so that U is a steady-state vector.

(b) The steady-state vector U is the unique probability vector satisfying the matrix equation $TU = U$.

☐ **Fibonacci sequence.**

$$u_n = \frac{1}{\sqrt{5}} \left[\left(\frac{1 + \sqrt{5}}{2} \right)^{n+1} - \left(\frac{1 - \sqrt{5}}{2} \right)^{n+1} \right]$$

☐ **Theorem 7.11.** If the $n \times n$ matrix A has n linearly independent eigenvectors $P_1, P_2, \ldots, P_n$ associated with the eigenvalues $\lambda_1, \lambda_2, \ldots, \lambda_n$, respectively, then the general solution to the system of differential equations

$$X' = AX$$

is given by

$$X(t) = b_1 P_1 e^{\lambda_1 t} + b_2 P_2 e^{\lambda_2 t} + \cdots + b_n P_n e^{\lambda_n t}.$$

SUPPLEMENTARY EXERCISES

1. Which of the following points are on the line

$$\frac{x - 3}{2} = \frac{y + 3}{4} = \frac{z + 5}{-4}.$$

(a) $(1, 2, 3)$.   (b) $(5, 1, -9)$.   (c) $(1, -7, -1)$.

2. Find the parametric equations of the line of intersection of the planes $x - 2y + z + 3 = 0$ and $2x - y + 3z + 4 = 0$.

3. Find a quadratic form of the type in Theorem 7.2 that is equivalent to the quadratic form $x^2 + 2y^2 + z^2 - 2xy - 2yz$.

4. Consider a communication network among six individuals with adjacency matrix

$$\begin{array}{c} \\ P_1 \\ P_2 \\ P_3 \\ P_4 \\ P_5 \\ P_6 \end{array} \begin{array}{cccccc} P_1 & P_2 & P_3 & P_4 & P_5 & P_6 \\ \begin{bmatrix} 0 & 1 & 0 & 0 & 0 & 1 \\ 0 & 0 & 1 & 1 & 0 & 0 \\ 0 & 0 & 0 & 1 & 0 & 0 \\ 1 & 0 & 1 & 0 & 1 & 0 \\ 0 & 0 & 0 & 0 & 0 & 1 \\ 0 & 0 & 0 & 1 & 0 & 0 \end{bmatrix} \end{array}.$$

In how many ways does P_1 have access to P_3 through two individuals?

5. Solve the following matrix game.

$$\begin{bmatrix} 6 & 2 & 3 \\ 3 & 4 & 2 \\ 4 & 1 & 2 \end{bmatrix}.$$

6. Find the line of best fit for the following data points.

$$(0, 1), \quad (3, 2), \quad (5, 4), \quad (8, 10).$$

7. Consider an isolated village in a remote part of Australia consisting of three households: a rancher who raises all the cattle needed and nothing else, a dairy farmer who produces all the dairy products needed and nothing else, and a vegetable farmer who produces all the necessary vegetables and nothing else. Suppose that the units are selected so that each person produces one unit of each commodity. Suppose that during the year the portion of each commodity that is consumed by each individual is given in Table 7.3.

Table 7.3

Goods Consumed by:	Goods Produced by:		
	Rancher	Dairy Farmer	Vegetable Farmer
Rancher	$\frac{1}{2}$	$\frac{3}{8}$	$\frac{1}{3}$
Dairy farmer	$\frac{1}{4}$	$\frac{1}{4}$	$\frac{1}{3}$
Vegetable farmer	$\frac{1}{4}$	$\frac{3}{8}$	$\frac{1}{3}$

Let p_1, p_2, and p_3 be the prices per unit of cattle, dairy produce, and vegetable produce, respectively. Suppose that every one pays the same price for a commodity. What prices p_1, p_2, and p_3 should be assigned to the commodities so that we have a state of equilibrium?

8. A marketing research organization has determined the following behavior of the average student at a certain college. If the student plays a video game on a given day, there is a probability of 0.2 that he or she will play a game on the following day, whereas if the student has not played on a given day, there is a probability of 0.6 that he or she will play on the following day.
(a) Write the transition matrix for the Markov process.
(b) If the average student plays a video game on Monday, what is the probability that he or she will play on Friday of the same week?
(c) In the long run, what is the probability that the average student will play a video game?

9. Using a hand calculator, compute the Fibonacci number u_{25}.

10. Consider the linear system of differential equations

$$\begin{bmatrix} x_1' \\ x_2' \end{bmatrix} = \begin{bmatrix} 1 & 1 \\ 3 & -1 \end{bmatrix} \begin{bmatrix} x_1 \\ x_2 \end{bmatrix}.$$

(a) Find the general solution.
(b) Find the solution to the initial value problem determined by the initial value conditions $x_1(0) = 4$, $x_2(0) = 6$.

CHAPTER TEST

1. Find parametric equations of the line through the point $(5, -2, 1)$, which is parallel to the vector $\mathbf{U} = (3, -2, 5)$.

2. Find an equation of the plane passing through the points $(1, 2, -1)$, $(3, 4, 5)$, $(0, 1, 1)$.

3. Let $\mathbf{Q(X)} = 2x^2 + 6xy + 2y^2 = 1$ be the equation of a central conic. Identify the conic by finding a quadratic form of the type in Theorem 7.2 that is equivalent to $\mathbf{Q}$.

4. Determine a clique, if there is one, for the digraph with the following adjacency matrix.

$$\begin{bmatrix} 0 & 1 & 1 & 0 & 1 & 1 \\ 0 & 0 & 0 & 1 & 0 & 1 \\ 1 & 0 & 0 & 0 & 1 & 1 \\ 0 & 1 & 0 & 0 & 1 & 0 \\ 1 & 0 & 1 & 1 & 0 & 0 \\ 1 & 1 & 1 & 0 & 0 & 0 \end{bmatrix}.$$

5. Solve the following matrix game by linear programming.

$$\begin{bmatrix} -3 & 2 & 4 \\ 4 & 1 & 5 \end{bmatrix}.$$

6. Show that the following matrix game is strictly determined regardless of the value of a.

$$\begin{bmatrix} 2 & 3 \\ 1 & a \end{bmatrix}.$$

7. A physical equipment center determines the relationship between the length of time spent on a certain piece of equipment and the number of calories lost. The following data are obtained:

Time on Equipment (minutes)	5	8	10	12	15
Lost Calories (hundreds)	3.2	5.5	6.8	7.8	9.2

Let x denote the number of minutes on the equipment, and let y denote the number of calories lost (in hundreds).
(a) Find the line of best fit relating x and y.
(b) Use the equation obtained in part (a) to estimate the number of calories lost after 20 minutes on the equipment.

8. Consider a city that has three basic industries: a steel plant, a coal mine, and a railroad. To produce $1 of steel, the steel plant uses $0.50 of steel, $0.30 of coal, and $0.10 of transportation. To mine $1 of coal, the coal mine uses $0.10 of steel, $0.20 of coal, and $0.30 of transportation. To provide $1 of transportation, the railroad uses $0.10 of steel, $0.40 of coal, and $0.05 of transportation. Suppose that during the month of December there is an outside demand of 2 million dollars for steel, 1.5 million dollars for coal, and 0.5 million dollars for transportation. How much should each industry produce to satisfy the demands?

9. A political analyst who has been studying the voting patterns of nonregistered voters in a certain city has come to the following conclusions. If on a given year, a voter votes Republican, then the probability that on the following year he or she will again vote Republican is 0.4. If on a given year, a voter votes Democratic, then the probability that on the following year he or she will again vote Democratic is 0.5.
(a) Write the transition matrix for the Markov process.
(b) If a voter votes Republican in 1984, what is the probability that he or she will vote Republican in 1988?
(c) In the long run, what is the probability of a voter voting Democratic?

10. Using a hand calculator, compute the Fibonacci number u_{30}.

11. Find the general solution to the linear system of differential equations.

$$\begin{bmatrix} x_1' \\ x_2' \end{bmatrix} = \begin{bmatrix} 3 & 2 \\ 6 & -1 \end{bmatrix} \begin{bmatrix} x_1 \\ x_2 \end{bmatrix}.$$

NUMERICAL

LINEAR

ALGEBRA

8

Numerical Linear Algebra

Almost all computational work in applied linear algebra using large matrices is done on a computer. This is necessary because the matrices encountered in practice are seldom 2×2 or 3×3. These efforts have given added impetus to the relatively new area of **numerical linear algebra**, which seeks to evaluate various methods, improve existing methods, and find new methods. In this chapter we give a very brief sketch of some widely used numerical methods for solving linear systems and for finding eigenvalues and eigenvectors. Computer programs that implement these methods are widely available.

8.1

ERROR ANALYSIS

Before discussing the evaluation of different numerical methods, we look at the way in which numbers are handled in a computer. Most of the computational work done by the computer is carried out in **floating-point arithmetic**. In this system each number is written as

$$\pm .d_1 d_2 \cdots d_n \times 10^e, \tag{1}$$

415

where e is an integer that satisfies

$$-M \leq e \leq M,$$

and the digits d_i satisfy

$$1 \leq d_1 \leq 9 \quad \text{and} \quad 0 \leq d_i \leq 9 \qquad (2 \leq i \leq n).$$

The number M depends upon the computer being used. For the IBM 4300 and 3030 series of computers, $M = 75$, while for the Univac 1100 series of computers, $M = 38$. If $M = 75$, the smallest positive number that can be stored in the computer is 10^{-75}, and the largest is 10^{75}. The number n of digits in (1) is called the **number of significant digits**. In the IBM 4300 and 3030 series of computers, seven significant digits are used; in the UNIVAC 1100 series of computers, eight digits are used. The fraction $.d_1 d_2 \cdots d_n$ in (1) is called the **mantissa** and e is called the **exponent**. A number of the form in (1) is said to be a **floating-point number**.

Examples of floating-point numbers for $n = 4$ are

$$+0.2310 \times 10^2, \quad -0.6844 \times 10^4, \quad -0.7510 \times 10^{-2},$$
$$+0.4000 \times 10^2, \quad +0.6300 \times 10^0, \quad +0.3142 \times 10^1.$$

Suppose that we now want to represent a number N in the computer. If N has a nonterminating decimal representation (for example, $N = \frac{7}{3} = 2.33 \ldots$), then we must approximate the number as it is entered into the computer, for we only have n significant digits at our disposal. First, we write the number as

$$\pm .d_1 d_2 \cdots d_n d_{n+1} \cdots \times 10^e, \tag{2}$$

$1 \leq d_1 \leq 9$, and $0 \leq d_i \leq 9$ $(i \geq 2)$. Now we proceed in one of two ways.

Truncation or Chopping. Keep the first n digits in (2). Thus $\frac{7}{3} = 2.333 \ldots =$ $0.233 \ldots \times 10^1$, which when truncated to four significant figures becomes 0.2333×10^1. Also, $\frac{8}{3} = 2.666 \ldots = 0.2666 \ldots \times 10^1$, which when truncated to four significant figures becomes 0.2666×10^1.

Rounding. If $d_{n+1} \geq 5$, add 1 to d_n and truncate. Otherwise, merely truncate. Another way of stating this procedure is: Add 5 to d_{n+1} and then truncate. Thus $\frac{7}{3}$ rounded to four significant figures is 0.2333×10^1 and $\frac{8}{3}$ rounded to four significant figures is 0.2667×10^1. Some computers deal with the problem of representing numbers by rounding, others by truncating. Also, many computers make available **double-precision arithmetic** (through either hardware or software), thereby providing twice as many digits as in single-precision floating-point arithmetic.

Numerical errors that occur due to either rounding or truncating are called **roundoff errors**. There are two types of errors that can be studied. Let N be a real number and let $\hat{N}$ be an approximation to it. The **absolute error** ε_a is defined as

$$\varepsilon_a = \hat{N} - N,$$

and the **relative error** ε_r is

$$\varepsilon_r = \frac{\hat{N} - N}{N} \qquad \text{if } N \neq 0.$$

The relative error is not defined if $N = 0$. The relative error is the more important quantity, since a small error in a large number may not be too harmful, while the same error in a small number can be disastrous. Thus an error of 0.1234 in the number 20,642.3217 is rather trivial, but the same error in 0.8266 poses some serious problems. The three major causes of roundoff error are (1) adding a large number to a relatively small number, (2) subtracting two numbers that are almost equal, and (3) dividing a number by a very small number. We might also note that today's microcomputers and hand-held calculators have, except for storage and speed, many of the capabilities of a large computer. They also have the same problems with roundoff.

Much effort in numerical linear algebra goes into the development of methods that reduce roundoff error and methods that estimate the roundoff error incurred in a numerical procedure.

8.1 EXERCISES

In Exercises 1 through 4, write the given number as a floating-point number with four significant digits.

1. 34.7213   **2.** −0.0002135

3. −284   **4.** 24.00

In Exercises 5 through 8, write the given number as a floating-point number with four significant digits, first by rounding and then by truncating.

5. 1.230   **6.** −4.25678   **7.** $\frac{17}{3}$ **8.** $\frac{13}{6}$

In Exercises 9 through 12, find the absolute error and the relative error.

9. $N = 12.341$, $\hat{N} = 12.362$.

10. $N = -0.4821$, $\hat{N} = -0.4215$.

11. $N = 6482.0$, $\hat{N} = 6483.1$.

12. $N = 0.00724$, $\hat{N} = 0.00742$.

8.2

LINEAR SYSTEMS

Prerequisite. Section 1.4, Solutions of Equations.

In Section 1.4 we discussed the Gauss–Jordan reduction method for solving a linear system of m equations in n unknowns by transforming the augmented matrix to reduced row echelon form. We first present here a similar method, which is more efficient from a computational point of view.

Definition An $m \times n$ matrix **A** is said to be in **row echelon form** when it satisfies the following
properties:

    (a) All rows consisting entirely of zeros, if any, are at the bottom of the matrix.

    (b) The first nonzero entry in each row that does not consist entirely of zeros is a 1,
         called the **leading entry** of its row.

    (c) If rows i and $i + 1$ are two successive rows that do not consist entirely of zeros,
         then the leading entry of row $i + 1$ must be to the right of the leading entry of
         row i.

Row echelon form differs from the earlier concept of reduced row echelon form in that
in a column that contains the leading entry of some row, the entries *above* the leading
entry need not be zero.

Example 1 The matrices

$$
\begin{bmatrix} 1 & -3 & 2 & 4 \\ 0 & 1 & 4 & 5 \\ 0 & 0 & 1 & 2 \end{bmatrix},
\qquad
\begin{bmatrix} 1 & 2 & 3 & 1 & 2 \\ 0 & 1 & -2 & 4 & 3 \\ 0 & 0 & 1 & -2 & 6 \\ 0 & 0 & 0 & 0 & 0 \\ 0 & 0 & 0 & 0 & 0 \end{bmatrix}
$$

are in row echelon form.

It is not difficult to see that using the first six steps of Theorem 1.5 of Section 1.4, we
can show that every $m \times n$ matrix is row equivalent to a matrix in row echelon form,
which need not be unique. We transform a matrix to row echelon form by using the first
six steps of Theorem 1.5 in the following example.

Example 2 Let **A** be the matrix of Example 5 in Section 1.4.

$$
\mathbf{A} = \begin{bmatrix} 0 & 2 & 3 & -4 & 1 \\ 0 & 0 & 2 & 3 & 4 \\ 2 & 2 & -5 & 2 & 4 \\ 2 & 0 & -6 & 9 & 7 \end{bmatrix}.
$$

Proceeding as far as Step 6, we have

$$
\mathbf{B}_3 = \begin{bmatrix} 1 & 1 & -\tfrac{5}{2} & 1 & 2 \\ 0 & 1 & \tfrac{3}{2} & -2 & \tfrac{1}{2} \\ 0 & 0 & 2 & 3 & 4 \\ 0 & 0 & 2 & 3 & 4 \end{bmatrix}.
$$

We now identify **C** by deleting but not erasing the first row of $\mathbf{B}_3$.

$$\mathbf{C} = \begin{bmatrix} 1 & 1 & -\frac{5}{2} & 1 & 2 \\ 0 & 1 & \frac{3}{2} & -1 & \frac{1}{2} \\ 0 & 0 & ② & 3 & 4 \\ 0 & 0 & 2 & 3 & 4 \end{bmatrix}$$

$$\mathbf{C}_1 = \mathbf{C}_2 = \begin{bmatrix} 1 & 1 & -\frac{5}{2} & 1 & 2 \\ 0 & 1 & \frac{3}{2} & -2 & \frac{1}{2} \\ 0 & 0 & 1 & \frac{3}{2} & 2 \\ 0 & 0 & 2 & 3 & 4 \end{bmatrix}$$

No row of **C** had to be interchanged. The first row of **C** was divided by 2.

$$\mathbf{C}_3 = \begin{bmatrix} 1 & 1 & -\frac{5}{2} & 1 & 2 \\ 0 & 1 & \frac{3}{2} & -2 & \frac{1}{2} \\ 0 & 0 & 1 & \frac{3}{2} & 2 \\ 0 & 0 & 0 & 0 & 0 \end{bmatrix}$$

-2 times the first row of $\mathbf{C}_2$ was added to its second row.

The final matrix

$$\begin{bmatrix} 1 & 1 & -\frac{5}{2} & 1 & 2 \\ 0 & 1 & \frac{3}{2} & -2 & \frac{1}{2} \\ 0 & 0 & 1 & \frac{3}{2} & 2 \\ 0 & 0 & 0 & 0 & 0 \end{bmatrix}$$

is in row echelon form.

If we transform the augmented matrix $[\mathbf{A} \mid \mathbf{B}]$ of the linear system $\mathbf{AX} = \mathbf{B}$ to row echelon form, we get another method for solving linear systems, which is called **Gaussian elimination**.

The Gaussian elimination method for solving the linear system $\mathbf{AX} = \mathbf{B}$ is as follows.

STEP 1. Form the augmented matrix $[\mathbf{A} \mid \mathbf{B}]$.

STEP 2. Transform the augmented matrix to row echelon form by using elementary row operations.

STEP 3. Solve the linear system that corresponds to the matrix in row echelon form that has been obtained in Step 2 by using back substitution.

Example 3 Consider Example 6 of Section 1.4:

$$\begin{aligned} x + 2y + 3z &= 9 \\ 2x - y + z &= 8 \\ 3x \quad - z &= 3. \end{aligned}$$

The augmented matrix is row equivalent to the matrix (verify),

$$\left[\begin{array}{ccc|c} 1 & 2 & 3 & 9 \\ 0 & 1 & 1 & 2 \\ 0 & 0 & 1 & 3 \end{array}\right], \tag{1}$$

which is in row echelon form. The solution can be obtained by using **back substitution** as follows.

From the last row of (1) we have

$$0x + 0y + 1z = 3$$

so that $z = 3$. From the second row of (1),

$$0x + 1y + 1z = 2,$$

so

$$y = 2 - z = 2 - 3 = -1.$$

Now from the first row of (1),

$$x + 2y + 3z = 9,$$

so

$$x = 9 - 2y - 3z = 9 - 2(-1) - 3(3) = 2.$$

Thus the solution is $x = 2$, $y = -1$, $z = 3$.

Throughout the rest of this section we limit our discussion to linear systems of n equations with n unknowns.

One of the troubles with Gaussian elimination is that we have to divide by the pivot; if the latter is a very small number, then the roundoff error can cast considerable doubt upon the final answer. A partial remedy to this problem consists in using the technique called **partial pivoting**. This method calls for choosing the largest (in absolute value) nonzero entry in the pivotal column as the pivot element.

Example 4 To solve the linear system

$$\begin{array}{rcl} 14x_1 + 2x_2 + 4x_3 & = & -10 \\ 16x_1 + 40x_2 - 4x_3 & = & 55 \\ -2x_1 + 4x_2 - 16x_3 & = & -38 \end{array} \tag{2}$$

by Gaussian elimination with partial pivoting, we proceed as follows. Form the augmented matrix

$$\left[\begin{array}{ccc|c} 14 & 2 & 4 & -10 \\ 16 & 40 & -4 & 55 \\ -2 & 4 & -16 & -38 \end{array}\right].$$

Our computations will be carried out to three decimal places and we round after each multiplication and division.

$$\begin{bmatrix} 16 & 40 & -4 & \vdots & 55 \\ 14 & 2 & 4 & \vdots & -10 \\ -2 & 4 & -16 & \vdots & -38 \end{bmatrix}$$

The first and second rows were interchanged.

$$\begin{bmatrix} 1.000 & 2.500 & -0.250 & \vdots & 3.438 \\ 14.000 & 2.000 & 4.000 & \vdots & -10.000 \\ -2.000 & 4.000 & -16.000 & \vdots & -38.000 \end{bmatrix}$$

The first row was divided by the pivot 16.

$$\begin{bmatrix} 1.000 & 2.500 & -0.250 & \vdots & 3.438 \\ 0.000 & -33.000 & 7.500 & \vdots & -58.132 \\ 0.000 & 9.000 & -16.500 & \vdots & -31.124 \end{bmatrix}$$

-14 times the first row was added to the second row; 2 times the first row was added to its third row.

$$\begin{bmatrix} 1.000 & 2.500 & -0.250 & \vdots & 3.438 \\ 0.000 & 1.000 & -0.227 & \vdots & 1.762 \\ 0.000 & 9.000 & -16.500 & \vdots & -31.124 \end{bmatrix}$$

The second row was divided by -33; that is, -33 is the pivot for the 2×4 submatrix obtained by removing the first row.

$$\begin{bmatrix} 1.000 & 2.500 & -0.250 & \vdots & 3.438 \\ 0.000 & 1.000 & -0.227 & \vdots & 1.762 \\ 0.000 & 0.000 & -14.457 & \vdots & -46.982 \end{bmatrix}$$

-9 times the second row was added to the third row.

$$\begin{bmatrix} 1.000 & 2.500 & -0.250 & \vdots & 3.438 \\ 0.000 & 1.000 & -0.227 & \vdots & 1.762 \\ 0.000 & 0.000 & 1.000 & \vdots & 3.250 \end{bmatrix}$$

The third row was divided by -14.457; that is, -14.457 is the pivot for the 1×4 submatrix obtained by removing the first two rows.

Then

$$x_3 = 3.250$$
$$x_2 = 1.762 + (0.227)(3.250) = 2.500$$
$$x_1 = 3.438 + (0.250)(3.250) - (2.500)(2.500) = -1.999.$$

Thus the solution obtained is

$$x_1 = -1.999, \qquad x_2 = 2.500, \qquad x_3 = 3.250,$$

which agrees rather well with the exact solution:

$$x_1 = -2, \qquad x_2 = 2.5, \qquad x_3 = 3.25.$$

Full pivoting consists in choosing the largest nonzero entry in the entire augmented matrix as the pivot. Although this variant sometimes gives better results than partial pivoting, it requires a relabeling of the variables. The additional programming and book-keeping required make Gaussian elimination with partial pivoting a more popular method than Gaussian elimination with full pivoting.

Ill-Conditioned Systems

There are linear systems, $\mathbf{AX} = \mathbf{B}$, where $\mathbf{A}$ is $n \times n$, in which relatively small changes in some of the elements of the augmented matrix $[\mathbf{A} \mid \mathbf{B}]$ may lead to relatively large changes in the solution. Such systems are called **ill-conditioned systems**. Gaussian elimination with partial pivoting or full pivoting may reduce the roundoff error for some of these systems. Sometimes **double-precision arithmetic** (requiring twice as much storage as single-precision and longer running times) may provide some help. However, for some ill-conditioned systems there may be no possible remedy.

Example 5 Consider the linear system

$$\begin{aligned} x - y &= 1 \\ x - 1.01y &= 0 \end{aligned} \qquad (3)$$

whose exact solution is

$$x = 101, \qquad y = 100.$$

Suppose that the coefficient of y in the second equation of (3) now changes from 1.01 to 0.99. The resulting system

$$\begin{aligned} x - y &= 1 \\ x - 0.99y &= 0 \end{aligned} \qquad (4)$$

has the exact solution

$$x = -99, \qquad y = -100.$$

Thus the systems (3) and (4), which are nearly identical, have vastly different solutions. System (3) is an ill-conditioned system. If we sketch the lines for (3) (Figure 8.1), we see that they are nearly parallel. Thus a minor shift of one of the lines results in a major shift of the point of intersection.

Iterative Methods

Gaussian elimination with partial pivoting is a **direct** method for solving $\mathbf{AX} = \mathbf{B}$, where $\mathbf{A}$ is $n \times n$. That is, the solution will be obtained after a finite number of steps. Moreover, the number of steps required to obtain the solution can be estimated in advance.

We shall now outline two additional methods, one due to Jacobi, the other to Gauss and

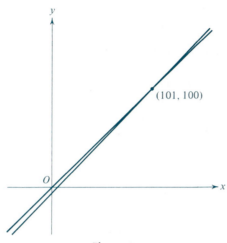

Figure 8.1

Seidel, for solving such linear systems. These methods are **iterative** in nature: that is, we start out with an initial approximation to the solution, which we successively try to improve. If the successive approximations tend to approach the solution, we say that the method **converges**. Otherwise, we say that the method **diverges**.

The Jacobi and Gauss–Seidel methods do not always converge, and we give below a sufficient condition for their convergence. These methods are more efficient, when applicable, than Gaussian elimination with partial pivoting especially for very large matrices that are **sparse** matrices (very large matrices having a large number of zero elements).

To discuss these methods, we start with the linear system $AX = B$, where A is $n \times n$. Assume that the determinant of A is nonzero and that the diagonal entries of A are all nonzero.

We shall illustrate the two methods (see pages 406 and 407) with the linear system (2). Throughout the rest of this section we work with three decimal places and *round* after each multiplication and division.

In Step 3 of the Jacobi iteration method, we substitute the values of the variables calculated in the last iteration into the right side of (5) to obtain a new approximation. However, in the next iteration, once we have a new value for x_1, we can use it to calculate x_2 in Equation (5); once we have new values for x_1 and x_2, we can use them to calculate x_3; and so on. Thus we can replace Step 3 of the Jacobi iteration method by the following procedure.

STEP 3. The kth iteration consists of the following. For the ith equation of (5), substitute the most recently calculated values of the variables into the right side to obtain the next approximation $x_i^{(k)}$. Do this for $i = 1, 2, \ldots, n$. The resulting method is called the **Gauss–Seidel iteration method**.

Jacobi[†] Iteration Method

The procedure for carrying out the Jacobi iteration method for solving n equations in n unknowns is as follows.

STEP 1. Rewrite (2) by expressing x_i in the ith equation in terms of the remaining variables. Thus

$$x_1 = -\tfrac{10}{14} - \tfrac{2}{14}x_2 - \tfrac{4}{14}x_3$$
$$x_2 = \tfrac{55}{40} - \tfrac{16}{40}x_1 + \tfrac{4}{40}x_3$$
$$x_3 = \tfrac{38}{16} - \tfrac{2}{16}x_1 + \tfrac{4}{16}x_2$$

or

$$x_1 = -0.714 - 0.143x_2 - 0.286x_3$$
$$x_2 = 1.375 - 0.400x_1 + 0.100x_3 \tag{5}$$
$$x_3 = 2.375 - 0.125x_1 + 0.250x_2.$$

STEP 2. Choose an initial approximation $x_1^{(0)}, x_2^{(0)}, \ldots, x_n^{(0)}$ to the solution. In the absence of other information, let $x_1^{(0)} = x_2^{(0)} = \cdots = x_n^{(0)} = 0$. In our example, $x_1^{(0)} = x_2^{(0)} = x_3^{(0)} = 0$.

STEP 3. Substitute the values of the variables calculated in the previous iteration $[(k-1)$st iteration] into the right side of (5) to obtain a new approximation, $x_1^{(k)}, x_2^{(k)}, \ldots, x_n^{(k)}$. For our example,

$$x_1^{(1)} = -0.714, \qquad x_2^{(1)} = 1.375, \qquad x_3^{(1)} = 2.375.$$

Using Step 3 again, we now obtain our second approximation,

$$x_1^{(2)} = -1.590, \qquad x_2^{(2)} = 1.898, \qquad x_3^{(2)} = 2.808.$$

The first nine approximations ($k = 8$) to the solution are shown in Table 8.1 on page 408. Thus our approximation to the solution is

$$x_1 \cong -1.998, \qquad x_2 \cong 2.497, \qquad x_3 \cong 3.247$$

and it was obtained in nine iterations. The exact solution is $x_1 = -2$, $x_2 = 2.5$, $x_3 = 3.25$. The symbol $\cong$ means "approximately."

[†] Karl Gustav Jacobi (1804–1851) was born in Potsdam, Germany, and was educated at the University of Berlin. He was a professor at the University of Königsberg and made many important contributions to the theory of elliptic functions, determinants, and differential equations.

[‡] Philipp Ludwig von Seidel (1821–1896) was born in Bavaria, Germany. He studied at Berlin, Königsberg, and Munich, and taught at Munich before he was forced to retire early by failing eyesight. He participated in the general development of analysis during the nineteenth century and also made contributions to astronomy.

Gauss–Seidel‡ Iteration Method

The procedure for carrying out the Gauss–Seidel iteration method for solving n equations in n unknowns is as follows.

STEP 1. Rewrite the given system by expressing x_i in the ith equation in terms of the remaining variables. Thus

$$x_1 = -\tfrac{10}{14} - \tfrac{2}{14}x_2 - \tfrac{4}{14}x_3$$
$$x_2 = \tfrac{55}{40} - \tfrac{16}{40}x_1 + \tfrac{4}{40}x_3$$
$$x_3 = \tfrac{38}{16} - \tfrac{2}{16}x_1 + \tfrac{4}{16}x_2$$

STEP 2. Choose an initial approximation $x_1^{(0)} = x_2^{(0)} = \cdots = x_n^{(0)}$ to the solution. In the absence of other information, let $x_1^{(0)} = x_2^{(0)} = \cdots = x_n^{(0)} = 0$. In our example, $x_1^{(0)} = x_2^{(0)} = x_3^{(0)} = 0$.

STEP 3. Substitute the most recently calculated values of the variables into the right side of the ith equation in the equations in Step 1 to obtain the new approximation $x_i^{(k)}$. For our example, we substitute these values into the right side of the first equation of (5), and obtain

$$x_1^{(1)} = -0.714.$$

We next substitute

$$x_1^{(1)} = -0.714 \qquad \text{and} \qquad x_3^{(0)} = 0$$

into the right side of the second equation of (5) to obtain

$$x_2^{(1)} = 1.661.$$

Now substitute

$$x_1^{(1)} = -0.714 \qquad \text{and} \qquad x_2^{(1)} = 1.661$$

into the right side of the third equation of (5) to obtain

$$x_3^{(1)} = 2.880.$$

We next use the values of $x_2^{(1)} = 1.661$ and $x_3^{(1)} = 2.880$ in the first equation of (5) to obtain $x_1^{(2)} = -1.775$. Now use $x_1^{(2)} = -1.775$ and $x_3^{(1)} = 2.880$ in the second equation of (5) to obtain $x_2^{(2)} = 2.373$. The first six approximations ($k = 5$) to the solution are shown in Table 8.2 on page 426. Thus our approximation to the solution is

$$x_1 \cong -2.000, \qquad x_2 \cong 2.500, \qquad x_3 \cong 3.250,$$

and it was obtained in six iterations.

Table 8.1

Iteration	$x_1^{(k)}$	$x_2^{(k)}$	$x_3^{(k)}$
0	0	0	0
1	−0.714	1.375	2.375
2	−1.590	1.898	2.808
3	−1.789	2.292	3.048
4	−1.913	2.395	3.172
5	−1.964	2.457	3.213
6	−1.984	2.482	3.235
7	−1.994	2.492	3.244
8	−1.998	2.497	3.247

Table 8.2

Iteration	$x_1^{(k)}$	$x_2^{(k)}$	$x_3^{(k)}$
0	0	0	0
1	−0.714	1.661	2.880
2	−1.775	2.373	3.190
3	−1.966	2.480	3.241
4	−1.996	2.498	3.249
5	−2.000	2.500	3.250

Although Gauss–Seidel iteration took fewer iterations than the Jacobi iteration method for our illustrative problem, there are problems where the opposite is true. Moreover, one cannot predict, in advance, which method is better for a given problem.

Convergence

Consider the linear system $\mathbf{AX} = \mathbf{B}$, where $\mathbf{A}$ is $n \times n$. A sufficient condition for the convergence of the Jacobi and Gauss–Seidel iteration methods is

$$|a_{ii}| > \sum_{\substack{j=1 \\ j \neq i}}^{n} |a_{ij}| \qquad (1 \leq i \leq n). \tag{6}$$

Equation (6) means that every diagonal element is in absolute value larger than the sum of the absolute values of the other entries in its row. A matrix $\mathbf{A}$ satisfying (6) is called **diagonally dominant**.

If $\mathbf{A}$ is diagonally dominant, then the Jacobi and Gauss–Seidel methods converge. Of course, the methods may converge even if this condition fails to hold (see Exercise 15).

Example 6 The matrix

$$\begin{bmatrix} 3 & -2 & 0 \\ 2 & -6 & 3 \\ 5 & 2 & 8 \end{bmatrix}$$

is diagonally dominant, whereas

$$\begin{bmatrix} 4 & 2 & -2 \\ 2 & 4 & 1 \\ 5 & 2 & 8 \end{bmatrix}$$

is not, since $|4|$ is not greater than $|2| + |-2|$.

Sometimes it is necessary to interchange the equations so that the coefficient matrix will be diagonally dominant.

Example 7 Consider the linear system

$$\begin{align*} 12x_1 - 14x_2 + x_3 &= 4 \\ -6x_1 + 4x_2 + 11x_3 &= -6 \\ 6x_1 + 3x_2 + 2x_3 &= 8. \end{align*}$$

The coefficient matrix is obviously not diagonally dominant. However, if we interchange the first and second rows and then the first and third rows of the resulting system (verify), we obtain the coefficient matrix

$$\begin{bmatrix} 6 & 3 & 2 \\ 12 & -14 & 1 \\ -6 & 4 & 11 \end{bmatrix},$$

which is diagonally dominant: $|6| > |3| + |2|$; $|-14| > |12| + |1|$; $|11| > |-6| + |4|$.

8.2 EXERCISES

In Exercises 1 and 2, transform the given matrix to row echelon form.

1. $\begin{bmatrix} 1 & -2 & 0 \\ 2 & -3 & -1 \\ 1 & 3 & 2 \end{bmatrix}$.

2. $\begin{bmatrix} 0 & -1 & 2 & 3 \\ 2 & 3 & 4 & 5 \\ 1 & 3 & -1 & 2 \\ 3 & 2 & 4 & 1 \end{bmatrix}$.

In Exercises 3 and 4, solve the given linear system by Gaussian elimination.

3. $\begin{align*} x_1 + 2x_2 + x_3 &= 0 \\ -3x_1 + 3x_2 + 2x_3 &= -7 \\ 4x_1 - 2x_2 - 3x_3 &= 2. \end{align*}$

4. $\begin{align*} 3x_1 - 3x_2 + 2x_3 &= 12 \\ 4x_1 + 6x_2 - 4x_3 &= -2 \\ 5x_1 - 9x_2 - 6x_3 &= 6. \end{align*}$

In Exercises 5 through 8, solve the given linear system by Gaussian elimination with partial pivoting. Carry out your computations to three decimal places and round after each multiplication and division.

5. $\begin{align*} 3x_1 - 2x_2 + 3x_3 &= -8 \\ 6x_1 - 4x_2 + 5x_3 &= -14 \\ -12x_1 + 6x_2 + 7x_3 &= -8. \end{align*}$

6. $\begin{align*} -2x_1 + 3x_2 + 4x_3 &= 8 \\ 3x_1 + 6x_2 - x_3 &= -8 \\ x_1 + 2x_2 + x_3 &= 4. \end{align*}$

7. $\begin{align*} 2.5x_1 + 3.5x_2 - 4.25x_3 &= 37.3 \\ 3.4x_1 + 2.5x_2 - 2.01x_3 &= 26.8 \\ 5.3x_1 - 2.4x_2 + 6.21x_3 &= -20.68. \end{align*}$

8. $\begin{align*} 2.2x_1 - 3.5x_2 + 1.2x_3 &= -18.3 \\ 4.4x_1 + 4.6x_2 - 3.8x_3 &= -25.6 \\ -6.3x_1 + 3.8x_2 + 2.4x_3 &= 57.1. \end{align*}$

9. Verify that the linear system

$$2.121x + 3.421y = 13.205$$
$$2.12x + 3.42y = 13.200$$

is ill-conditioned by solving the linear system

$$2.121x + 3.421y = 13.205$$
$$2.12x + 3.42y = 13.203.$$

Carry out your computations to *four* decimal places and round after each multiplication and division.

10. Which of the following matrices are diagonally dominant?

(a) $\begin{bmatrix} 2 & 1 \\ 4 & -5 \end{bmatrix}$.   (b) $\begin{bmatrix} 3 & 2 \\ -4 & 3 \end{bmatrix}$.

(c) $\begin{bmatrix} 3 & 1 & 2 \\ 0 & 4 & -2 \\ 1 & -2 & 4 \end{bmatrix}$.   (d) $\begin{bmatrix} -4 & 3 & 0 \\ 4 & 6 & 1 \\ 2 & -2 & 8 \end{bmatrix}$.

(e) $\begin{bmatrix} 3 & -2 & -1 \\ 4 & 5 & 1 \\ -2 & -4 & 6 \end{bmatrix}$.

In Exercises 11 through 14, solve the given linear system by (1) Jacobi's iteration method and (2) the Gauss–Seidel iteration method. Carry out the computations to three decimal places for six iterations ($k = 5$) and round after each multiplication and division.

11. $16x + 5y = -7$
$4x + 15y = 67.$

12. $6x_1 - x_2 - 2x_3 = 12$
$3x_1 + 5x_2 + x_3 = 3$
$2x_1 - x_2 + 7x_3 = 36.$

13. $9x_1 - 2x_2 + 6x_3 = 9$
$3x_1 + 6x_2 - 2x_3 = 15$
$6x_1 - x_2 + 8x_3 = 4.5.$

14. $15x_1 - 4x_2 - 2x_3 = -2$
$5x_1 - 8x_2 + x_3 = -9$
$10x_1 - 2x_2 + 15x_3 = -50.$

15. (a) Consider the linear system

$$2x + 4y = 8$$
$$x - y = -5.$$

The coefficient matrix is not diagonally dominant. Show that the Gauss–Seidel iteration method diverges. Use seven iterations ($k = 6$).

(b) Consider the linear system

$$2x - y = -8$$
$$x + y = -1.$$

The coefficient matrix is not diagonally dominant. Show that the Gauss–Seidel iteration method converges. Use seven iterations ($k = 6$).

8.3

LU-FACTORIZATION (OPTIONAL)

Prerequisites. Sections 1.4 and 8.2.

Gaussian elimination with partial pivoting, as discussed in Section 8.2, is more efficient than Gauss–Jordan elimination. In the special case that **A** is *square, upper triangular,* and has *all its diagonal entries different from zero,* the linear system $\mathbf{AX} = \mathbf{B}$ can be solved directly. There is no need to obtain the reduced row echelon form or even the row echelon form of the augmented matrix $[\mathbf{A} \vdots \mathbf{B}]$. The augmented matrix of such a system is given by

$$\begin{bmatrix} a_{11} & a_{12} & a_{13} & \cdots & a_{1n} & \vdots & b_1 \\ 0 & a_{22} & a_{23} & \cdots & a_{2n} & \vdots & b_2 \\ 0 & 0 & a_{33} & \cdots & a_{3n} & \vdots & b_3 \\ \vdots & \vdots & \vdots & \cdots & \vdots & \vdots & \vdots \\ 0 & 0 & 0 & \cdots & a_{nn} & \vdots & b_n \end{bmatrix}.$$

The solution is obtained by the following algorithm:

$$x_n = \frac{b_n}{a_{nn}}$$

$$x_{n-1} = \frac{b_{n-1} - a_{n-1\,n}x_n}{a_{n-1\,n-1}}$$

$$\vdots$$

$$x_j = \frac{b_j - \sum_{k=n}^{j-1} a_{jk}x_k}{a_{jj}} \qquad j = n,\, n-1,\, \ldots,\, 1.$$

This procedure is **back substitution,** which we used in conjunction with Gaussian elimination in Section 8.2.

In a similar manner, if $\mathbf{A}$ is *square, lower triangular,* and has *all its diagonal entries different from zero,* the system $\mathbf{AX} = \mathbf{B}$ can be solved by **forward substitution,** which is the algorithm we present next. The augmented matrix $[\mathbf{A} \mid \mathbf{B}]$ has the form

$$\begin{bmatrix} a_{11} & 0 & 0 & \cdots & 0 & \vdots & b_1 \\ a_{21} & a_{22} & 0 & \cdots & 0 & \vdots & b_2 \\ a_{31} & a_{32} & a_{33} & \cdots & 0 & \vdots & b_3 \\ \vdots & \vdots & \vdots & \cdots & \vdots & \vdots & \vdots \\ a_{n1} & a_{n2} & a_{n3} & \cdots & a_{nn} & \vdots & b_n \end{bmatrix}$$

and the solution is given by

$$x_1 = \frac{b_1}{a_{11}}$$

$$x_2 = \frac{b_2 - a_{21}x_1}{a_{22}}$$

$$\vdots$$

$$x_j = \frac{b_j - \sum_{k=1}^{j-1} a_{jk}x_k}{a_{jj}} \qquad j = 1,\, 2,\, \ldots,\, n.$$

We illustrate forward substitution in the following example.

Example 1 To solve the linear system

$$\begin{aligned} 5x_1 &= 10 \\ 4x_1 - 2x_2 &= 28 \\ 2x_1 + 3x_2 + 4x_3 &= 26 \end{aligned}$$

we use forward substitution. Hence we obtain from the previous algorithm

$$x_1 = \frac{10}{5} = 2$$

$$x_2 = \frac{28 - 4x_1}{-2} = -10$$

$$x_3 = \frac{26 - 2x_1 - 3x_2}{4} = 13,$$

which implies that the solution to the given lower triangular system of equations is

$$\mathbf{X} = \begin{bmatrix} 2 \\ -10 \\ 13 \end{bmatrix}.$$

The ease with which systems of equations with upper or lower triangular coefficient matrices can be solved is quite attractive. The forward substitution and back substitution algorithms are fast and simple to use. In order to introduce another important numerical procedure for solving linear systems of equations, we begin with the following definition.

Definition *An $n \times n$ matrix $\mathbf{A}$ is said to be in* **triangular form** *when it satisfies the following properties:*
   (a) $\mathbf{A}$ *is either an upper or a lower triangular matrix.*
   (b) *All the diagonal entries of $\mathbf{A}$ are different from zero.*

We use the terms *upper triangular form* and *lower triangular form* to describe particular matrices in triangular form. We also say that a linear system $\mathbf{AX} = \mathbf{B}$ is in **triangular form** if its coefficient matrix $\mathbf{A}$ is in triangular form. Any linear system of equations that is in triangular form can be solved directly by either forward substitution or back substitution.

Suppose that a nonsingular matrix $\mathbf{A}$ can be written as a product of a matrix $\mathbf{L}$ in lower triangular form and a matrix $\mathbf{U}$ in upper triangular form; that is,

$$\mathbf{A} = \mathbf{LU}.$$

In this case we say that $\mathbf{A}$ has an **LU-factorization** or an **LU-decomposition.** The LU-factorization of a matrix $\mathbf{A}$ can be used to efficiently solve a linear system $\mathbf{AX} = \mathbf{B}$. It follows that $\mathbf{AX} = \mathbf{B}$ is equivalent to $(\mathbf{LU})\mathbf{X} = \mathbf{B}$. Let $\mathbf{UX} = \mathbf{Z}$. Then we have

$$\mathbf{LZ} = \mathbf{B}.$$

Since $\mathbf{L}$ is in lower triangular form, we solve directly for $\mathbf{Z}$ by forward substitution. Next, since $\mathbf{U}$ is in upper triangular form we solve $\mathbf{UX} = \mathbf{Z}$ for $\mathbf{X}$ by back substitution. In summary, *if a nonsingular matrix* **A** *has an* **LU**-*factorization, then the solution of* **AX** = **B** *can be determined by a forward substitution followed by a back substitution.* We illustrate this procedure in the next example.

Example 2   Solve the following linear system using the **LU**-factorization, given below, of its coefficient matrix.

$$\begin{aligned} 6x_1 - 2x_2 - 4x_3 + 4x_4 &= 2 \\ 3x_1 - 3x_2 - 6x_3 + x_4 &= -4 \\ -12x_1 + 8x_2 + 21x_3 - 8x_4 &= 8 \\ -6x_1 \qquad\quad - 10x_3 + 7x_4 &= -43 \end{aligned}$$

The coefficient matrix

$$\mathbf{A} = \begin{bmatrix} 6 & -2 & -4 & 4 \\ 3 & -3 & -6 & 1 \\ -12 & 8 & 21 & -8 \\ -6 & 0 & -10 & 7 \end{bmatrix}$$

has an **LU**-factorization where

$$\mathbf{L} = \begin{bmatrix} 2 & 0 & 0 & 0 \\ 1 & -1 & 0 & 0 \\ -4 & 2 & 1 & 0 \\ -2 & -1 & -2 & 2 \end{bmatrix} \quad \text{and} \quad \mathbf{U} = \begin{bmatrix} 3 & -1 & -2 & 2 \\ 0 & 2 & 4 & 1 \\ 0 & 0 & 5 & -2 \\ 0 & 0 & 0 & 4 \end{bmatrix}$$

(verify). Let

$$\mathbf{B} = \begin{bmatrix} 2 \\ -4 \\ 8 \\ -43 \end{bmatrix}.$$

Then we solve **AX** = **B** by writing **LUX** = **B**. First, let **UX** = **Z** and now solve **LZ** = **B**:

$$\begin{bmatrix} 2 & 0 & 0 & 0 \\ 1 & -1 & 0 & 0 \\ -4 & 2 & 1 & 0 \\ -2 & -1 & -2 & 2 \end{bmatrix} \begin{bmatrix} z_1 \\ z_2 \\ z_3 \\ z_4 \end{bmatrix} = \begin{bmatrix} 2 \\ -4 \\ 8 \\ -43 \end{bmatrix}$$

by forward substitution. We obtain

$$z_1 = \frac{2}{2} = 1$$

$$z_2 = \frac{-4 - z_1}{-1} = 5$$

$$z_3 = \frac{8 + 4z_1 - 2z_2}{1} = 2$$

$$z_4 = \frac{-43 + 2z_1 + z_2 + 2z_3}{2} = -16.$$

Next we solve $\mathbf{UX} = \mathbf{Z}$,

$$\begin{bmatrix} 3 & -1 & -2 & 2 \\ 0 & 2 & 4 & 1 \\ 0 & 0 & 5 & -2 \\ 0 & 0 & 0 & 4 \end{bmatrix} \begin{bmatrix} x_1 \\ x_2 \\ x_3 \\ x_4 \end{bmatrix} = \begin{bmatrix} 1 \\ 5 \\ 2 \\ -16 \end{bmatrix},$$

by back substitution. We obtain

$$x_4 = \frac{-16}{4} = -4$$

$$x_3 = \frac{2 + 2x_4}{5} = -1.2$$

$$x_2 = \frac{5 - 4x_3 - x_4}{2} = 6.9$$

$$x_1 = \frac{1 + x_2 + 2x_3 - 2x_4}{3} = 4.5.$$

Thus the solution to the linear system is

$$\mathbf{X} = \begin{bmatrix} 4.5 \\ 6.9 \\ -1.2 \\ -4 \end{bmatrix}.$$

Next we show how to obtain an **LU**-factorization of a matrix by modifying the Gaussian elimination procedure from Section 8.2. No row interchanges will be permitted and we do not require that the pivots have value 1. The following **LU**-factorization procedure may fail in the same way that Gaussian elimination may fail to find nonzero pivots when row interchanges are not permitted. At the end of this section we discuss how to enhance the **LU**-factorization scheme presented to prevent such failures. We observe that the only elementary row operation permitted is the one that adds a multiple of one row to a different row.

To describe the **LU**-factorization we present a step-by-step procedure in the next example.

Example 3 Let **A** be the coefficient matrix of the linear system of Example 2.

$$\mathbf{A} = \begin{bmatrix} 6 & -2 & -4 & 4 \\ 3 & -3 & -6 & 1 \\ -12 & 8 & 21 & -8 \\ -6 & 0 & -10 & 7 \end{bmatrix}.$$

We proceed to "zero out" entries below the diagonal entries using only the row operation that adds a multiple of one row to a different row.

<table>
<tr><td align="center">**PROCEDURE**</td><td align="center">**MATRICES USED**</td></tr>
</table>

STEP 1. "Zero out" below the first diagonal entry of $\mathbf{A}$. Add $-\frac{1}{2}$ times the first row of $\mathbf{A}$ to the second row of $\mathbf{A}$. Add 2 times the first row of $\mathbf{A}$ to the third row of $\mathbf{A}$. Add 1 times the first row of $\mathbf{A}$ to the fourth row of $\mathbf{A}$. Call the new resulting matrix $\mathbf{U}_1$.

$$\mathbf{U}_1 = \begin{bmatrix} 6 & -2 & -4 & 4 \\ 0 & -2 & -4 & -1 \\ 0 & 4 & 13 & 0 \\ 0 & -2 & -14 & 11 \end{bmatrix}$$

We begin building a lower triangular matrix, $\mathbf{L}_1$, with 1's on the main diagonal, to record the row operations. Enter the *negatives of the multipliers* used in the row operations in the first column of $\mathbf{L}_1$, below the first diagonal entry of $\mathbf{L}_1$.

$$\mathbf{L}_1 = \begin{bmatrix} 1 & 0 & 0 & 0 \\ \frac{1}{2} & 1 & 0 & 0 \\ -2 & * & 1 & 0 \\ -1 & * & * & 1 \end{bmatrix}$$

STEP 2. "Zero out" below the second diagonal entry of $\mathbf{U}_1$. Add 2 times the second row of $\mathbf{U}_1$ to the third row of $\mathbf{U}_1$. Add -1 times the second row of $\mathbf{U}_1$ to the fourth row of $\mathbf{U}_1$. Call the new resulting matrix $\mathbf{U}_2$.

$$\mathbf{U}_2 = \begin{bmatrix} 6 & -2 & -4 & 4 \\ 0 & -2 & -4 & -1 \\ 0 & 0 & 5 & -2 \\ 0 & 0 & -10 & 12 \end{bmatrix}$$

Enter the negatives of the multipliers from the row operations below the second diagonal entry of $\mathbf{L}_1$. Call the new matrix $\mathbf{L}_2$.

$$\mathbf{L}_2 = \begin{bmatrix} 1 & 0 & 0 & 0 \\ \frac{1}{2} & 1 & 0 & 0 \\ -2 & -2 & 1 & 0 \\ -1 & 1 & * & 1 \end{bmatrix}$$

STEP 3. "Zero out" below the third diagonal entry of $\mathbf{U}_2$. Add 2 times the third row of $\mathbf{U}_2$ to the fourth row. Call the new resulting matrix $\mathbf{U}_3$.

$$\mathbf{U}_3 = \begin{bmatrix} 6 & -2 & -4 & 4 \\ 0 & -2 & -4 & -1 \\ 0 & 0 & 5 & -2 \\ 0 & 0 & 0 & 8 \end{bmatrix}$$

Enter the negative of the multiplier below the third diagonal entry of $\mathbf{L}_2$. Call the new matrix $\mathbf{L}_3$.

$$\mathbf{L}_3 = \begin{bmatrix} 1 & 0 & 0 & 0 \\ \frac{1}{2} & 1 & 0 & 0 \\ -2 & -2 & 1 & 0 \\ -1 & 1 & -2 & 1 \end{bmatrix}$$

Let $\mathbf{L} = \mathbf{L}_3$ and $\mathbf{U} = \mathbf{U}_3$. Then the product $\mathbf{LU}$ gives the original matrix $\mathbf{A}$ (verify). The linear system of equations in Example 2 can be solved using the $\mathbf{LU}$-factorization just constructed. Use a forward substitution followed by a back substitution.

In the preceding discussions we have asserted that Gaussian elimination and **LU**-factor-ization are more efficient than Gauss–Jordan reduction. A way to measure efficiency of techniques for solving systems of equations is to count the number of arithmetic opera-tions (additions, subtractions, multiplications, and divisions) required to compute the solution. The usual procedure is to count only the multiplications and divisions involved. This custom arose because the first digital computers required significantly more time to perform multiplications and divisions than additions and subtractions. Also, the number of additions and subtractions is approximately the same as the number of multiplications for methods based on row operations. We call the number of multiplications and divisions the **operation count.**

We consider the operation count for back substitution. To compute x_n, one division is required. To compute x_{n-1}, one division and one multiplication are required. In general, to compute x_j, one division and $n - j$ multiplications are required. Thus the operation count for back substitution is

$$1 + 2 + \cdots + (n - j) + 1 + \cdots + (n - 1) + 1.$$

Using summation notation, we have that the operation count for back substitution is

$$\sum_{k=1}^{n} k$$

which simplifies to $n(n + 1)/2$. Forward substitution has the same operation count.

Next consider the operation count for a row operation that adds a multiple of one row to a different row. In the elimination-type methods used to solve linear systems, this type of row operation is specifically used to "zero out" entries. Consider a pair of rows from an augmented matrix that has undergone a number of elimination steps with $q_1 \neq 0$:

$$
\begin{array}{ccccccc|c}
0 & \cdots & 0 & q_1 & q_2 & \cdots & q_k & q_{k+1} \\
0 & \cdots & 0 & s_1 & s_2 & \cdots & s_k & s_{k+1}
\end{array}.
$$

Gaussian elimination as described in Section 8.2 used $k + 1$ divisions to get to the form

$$
\begin{array}{ccccccc|c}
0 & \cdots & 0 & 1 & Q_2 & \cdots & Q_k & Q_{k+1} \\
0 & \cdots & 0 & s_1 & s_2 & \cdots & s_k & s_{k+1}
\end{array},
$$

where $Q_j = q_j/q_1$ for $j = 2, \cdots, k + 1$. Then k multiplications are required to obtain the form

$$
\begin{array}{ccccccc|c}
0 & \cdots & 0 & 1 & Q_2 & \cdots & Q_k & Q_{k+1} \\
0 & \cdots & 0 & 0 & S_2 & \cdots & S_k & S_{k+1}
\end{array},
$$

where $S_j = s_j - s_1 Q_j$ for $j = 2, \cdots, k + 1$. The entry that is "zeroed out" is set to zero, so no multiplication is required. Thus one such step of Gaussian elimination has an operation count of $2k + 1$, except the last step, which requires only two divisions. Care-fully totaling the operations involved, we find that the operation count to obtain upper triangular form is

$$\frac{2n^3 + 3n^2 + 7n}{6}.$$

Since the diagonal extries are 1, back substitution for an upper triangular matrix requires no divisions; hence the operation count is $(n(n + 1)/2) - n$. Thus Gaussian elimination with back substitution has an operation count of

$$\frac{2n^3 + 3n^2 + 7n}{6} + \frac{n(n + 1)}{2} - n = \frac{2n^3 + 6n^2 + 4n}{6}.$$

A similar development shows that Gauss–Jordan has an operation count of

$$\frac{n^3 + 2n^2 - n}{2}.$$

It can be shown that the method of solving a linear system by **LU**-factorization using storage of multipliers followed by a forward and back substitution has an operation count of

$$\frac{n^3 + 3n^2 - n}{3}.$$

Table 8.3 gives a sample of comparisons for the number of multiplications and divisions required for the three solution methods of Gaussian elimination, **LU**-factorization, and Gauss–Jordan reduction. These results clearly imply that Gauss–Jordan requires more operations. Hence Gauss–Jordan is seldom used in computer algorithms.

Table 8.3 Operation Count Sample

n	Gaussian Elimination	Gauss–Jordan Reduction	LU-Factorization
3	20	21	17
4	40	46	36
5	70	85	65
10	440	595	430
15	1,360	1,905	1,345
20	3,080	4,390	3,060
50	44,200	64,975	44,150

Note that a matrix **A** can have more than one **LU**-factorization. This is illustrated by Examples 2 and 3. There are many methods for obtaining an **LU**-factorization besides the scheme for **storage of multipliers** described in Example 3. It is important to note that if $a_{11} = 0$, then the procedure in Example 3 fails. Also, if the second diagonal entry of U_1 is zero or if the third diagonal entry of U_2 is zero, the procedure fails. In such cases we can try rearranging the equations of the system and beginning again, or using one of the other methods for **LU**-factorization. Most computer programs for **LU**-factorization incorporate row interchanges into the storage of multipliers scheme and use a partial pivoting strategy to help control round-off error. In this case the product of **L** and **U** is not necessarily **A**. The resulting matrix consists of a permutation of the rows of **A**. The book by Hill (see Further Reading, below) explores such a modification of the procedure for **LU**-factorization which has been presented here.

FURTHER READING

Hill, **David R.** *Experiments in Computational Matrix Algebra*. New York: Random House/ Birkhäuser, 1987.

8.3 EXERCISES

In Exercises 1 through 4, solve the linear system **AX** = **B** with the given **LU**-factorization of the coefficient matrix **A**. Solve the linear system using a forward substitution followed by a back substitution.

1. A $= \begin{bmatrix} 2 & 8 & 0 \\ 2 & 2 & -3 \\ 1 & 2 & 7 \end{bmatrix}$, **B** $= \begin{bmatrix} 18 \\ 3 \\ 12 \end{bmatrix}$,

L $= \begin{bmatrix} 2 & 0 & 0 \\ 2 & -3 & 0 \\ 1 & -1 & 4 \end{bmatrix}$, **U** $= \begin{bmatrix} 1 & 4 & 0 \\ 0 & 2 & 1 \\ 0 & 0 & 2 \end{bmatrix}$.

2. A $= \begin{bmatrix} 8 & 12 & -4 \\ 6 & 5 & 7 \\ 2 & 1 & 6 \end{bmatrix}$, **B** $= \begin{bmatrix} -36 \\ 11 \\ 16 \end{bmatrix}$,

L $= \begin{bmatrix} 4 & 0 & 0 \\ 3 & 2 & 0 \\ 1 & 1 & 1 \end{bmatrix}$, **U** $= \begin{bmatrix} 2 & 3 & -1 \\ 0 & -2 & 5 \\ 0 & 0 & 2 \end{bmatrix}$.

3. A $= \begin{bmatrix} 2 & 3 & 0 & 1 \\ 4 & 5 & 3 & 3 \\ -2 & -6 & 7 & 7 \\ 8 & 9 & 5 & 21 \end{bmatrix}$, **B** $= \begin{bmatrix} -2 \\ -2 \\ -16 \\ -66 \end{bmatrix}$,

L $= \begin{bmatrix} 1 & 0 & 0 & 0 \\ 2 & 1 & 0 & 0 \\ -1 & 3 & 1 & 0 \\ 4 & 3 & 2 & 1 \end{bmatrix}$,

U $= \begin{bmatrix} 2 & 3 & 0 & 1 \\ 0 & -1 & 3 & 1 \\ 0 & 0 & -2 & 5 \\ 0 & 0 & 0 & 4 \end{bmatrix}$.

4. A $= \begin{bmatrix} 4 & 2 & 1 & 0 \\ -4 & -6 & 1 & 3 \\ 8 & 16 & -3 & -4 \\ 20 & 10 & 4 & -3 \end{bmatrix}$, **B** $= \begin{bmatrix} 6 \\ 13 \\ -20 \\ 15 \end{bmatrix}$,

L $= \begin{bmatrix} 1 & 0 & 0 & 0 \\ -1 & 1 & 0 & 0 \\ 2 & -3 & 1 & 0 \\ 5 & 0 & -1 & 1 \end{bmatrix}$,

U $= \begin{bmatrix} 4 & 2 & 1 & 0 \\ 0 & -4 & 2 & 3 \\ 0 & 0 & 1 & 5 \\ 0 & 0 & 0 & 2 \end{bmatrix}$.

In Exercises 5 through 10, find an **LU**-factorization of the coefficient matrix of the given linear system **AX** = **B**. Solve the linear system using a forward substitution followed by a back substitution.

5. A $= \begin{bmatrix} 2 & 3 & 4 \\ 4 & 5 & 10 \\ 4 & 8 & 2 \end{bmatrix}$, **B** $= \begin{bmatrix} 6 \\ 16 \\ 2 \end{bmatrix}$.

6. A $= \begin{bmatrix} -3 & 1 & -2 \\ -12 & 10 & -6 \\ 15 & 13 & 12 \end{bmatrix}$, **B** $= \begin{bmatrix} 15 \\ 82 \\ -5 \end{bmatrix}$.

7. A $= \begin{bmatrix} 4 & 2 & 3 \\ 2 & 0 & 5 \\ 1 & 2 & 1 \end{bmatrix}$, **B** $= \begin{bmatrix} 1 \\ -1 \\ -3 \end{bmatrix}$.

8. A $= \begin{bmatrix} -5 & 4 & 0 & 1 \\ -30 & 27 & 2 & 7 \\ 5 & 2 & 0 & 2 \\ 10 & 1 & -2 & 1 \end{bmatrix}$, **B** $= \begin{bmatrix} -17 \\ -102 \\ -7 \\ -6 \end{bmatrix}$.

9. $\mathbf{A} = \begin{bmatrix} 2 & 1 & 0 & -4 \\ 1 & 0 & 0.25 & -1 \\ -2 & -1.1 & 0.25 & 6.2 \\ 4 & 2.2 & 0.3 & -2.4 \end{bmatrix}$,

10. $\mathbf{A} = \begin{bmatrix} 4 & 1 & 0.25 & -0.5 \\ 0.8 & 0.6 & 1.25 & -2.6 \\ -1.6 & -0.08 & 0.01 & 0.2 \\ 8 & 1.52 & -0.6 & -1.3 \end{bmatrix}$,

$\mathbf{B} = \begin{bmatrix} -3 \\ -1.5 \\ 5.6 \\ 2.2 \end{bmatrix}$.

$\mathbf{B} = \begin{bmatrix} -0.15 \\ 9.77 \\ 1.69 \\ -4.576 \end{bmatrix}$.

8.4

EIGENVALUES AND EIGENVECTORS

Prerequisite. Chapter 5, Eigenvalues and Eigenvectors.

In this section we present two numerical methods for finding eigenvalues and eigenvectors of matrices. The first method finds the eigenvalue of largest absolute value and an associated eigenvector. The second method finds all the eigenvalues and associated eigenvectors of a symmetric matrix.

The Power Method

Suppose that $\mathbf{A}$ is an $n \times n$ matrix which is diagonalizable and that the eigenvalues λ_i have been indexed in such a way that

$$|\lambda_1| \geq |\lambda_2| \geq |\lambda_3| \geq \cdots \geq |\lambda_n|.$$

Further, assume that $|\lambda_1| > |\lambda_2|$: the magnitude of λ_1 is strictly greater than that of λ_2. The power method, an easily implemented technique, computes λ_1 and an associated eigenvector.

The key idea in the power method is the following result, whose proof we omit.

If $\mathbf{U}_0$ is an arbitrary nonzero vector, then the vectors $\mathbf{A}^k\mathbf{U}_0$ (k a nonnegative integer) will usually form a sequence of vectors whose directions converge to the direction of $\mathbf{X}_1$, an eigenvector of $\mathbf{A}$ associated with the dominant eigenvalue λ_1.

If $\mathbf{X}_1$ were exactly known, then using the inner product in R^n, we form the ratio

$$\frac{(\mathbf{A}\mathbf{X}_1, \mathbf{X}_1)}{(\mathbf{X}_1, \mathbf{X}_1)} = \frac{(\lambda_1\mathbf{X}_1, \mathbf{X}_1)}{(\mathbf{X}_1, \mathbf{X}_1)} = \frac{\lambda_1(\mathbf{X}_1, \mathbf{X}_1)}{(\mathbf{X}_1, \mathbf{X}_1)} = \lambda_1,$$

and thus obtain λ_1.

Suppose now that for k sufficiently large, we have a good approximation $\hat{\mathbf{X}}_1$ to $\mathbf{X}_1$. Then

$$\frac{(\mathbf{A}\hat{\mathbf{X}}_1, \hat{\mathbf{X}}_1)}{(\hat{\mathbf{X}}_1, \hat{\mathbf{X}}_1)}$$

is an approximation to λ_1.

The components of $\mathbf{A}^k\mathbf{U}_0$ may become increasingly large numbers as k increases which leads to a large roundoff error. This problem can be avoided by multiplying $\mathbf{A}^k\mathbf{U}_0$ by a suitable scalar for each k.

The procedure for carrying out the power method is as follows.

STEP 1. Choose an arbitrary nonzero vector $\mathbf{U}_0$ as an initial approximation to an eigenvector of $\mathbf{A}$ associated with λ_1. For example, $\mathbf{U}_0$ can be taken as the vector all of whose components are 1.

STEP 2. Compute $\mathbf{A}\mathbf{U}_0 = \mathbf{U}_1$. Let $k = 1$.

STEP 3. Let u_{kr} be the component of $\mathbf{U}_k$ with largest absolute value. Define

$$\mathbf{V}_k = \frac{1}{|u_{kr}|}\mathbf{U}_k \qquad (k \geq 1).$$

Thus $\mathbf{V}_k$ is parallel to and in the same direction as $\mathbf{U}_k$, and is an approximation to an eigenvector of $\mathbf{A}$ associated with λ_1.

STEP 4. Form an approximation to λ_1:

$$\frac{(\mathbf{A}\mathbf{V}_k, \mathbf{V}_k)}{(\mathbf{V}_k, \mathbf{V}_k)} \cong \lambda_1 \qquad (k \geq 1).$$

STEP 5. Increase k by 1, let $\mathbf{U}_{k+1} = \mathbf{A}\mathbf{V}_k$, and return to Step 3.

Example 1 Let

$$\mathbf{A} = \begin{bmatrix} 4 & 1 \\ 2 & 5 \end{bmatrix}.$$

If we carry three decimal places and round after each multiplication and division, our results for nine iterations ($k = 8$) are shown in Table 8.4.

We conclude that the dominant eigenvalue is approximately 6.001 and that an associated eigenvector is approximately

$$\hat{\mathbf{X}}_1 = \begin{bmatrix} 0.501 \\ 1 \end{bmatrix}.$$

Table 8.4

Iteration	V_k	Approximate Value of λ_1
0	$[1 \quad 1]^T$	—
1	$[0.714 \quad 1]^T$	6.080
2	$[0.600 \quad 1]^T$	6.059
3	$[0.548 \quad 1]^T$	6.035
4	$[0.524 \quad 1]^T$	6.016
5	$[0.512 \quad 1]^T$	6.010
6	$[0.506 \quad 1]^T$	6.005
7	$[0.503 \quad 1]^T$	6.002
8	$[0.501 \quad 1]^T$	6.001

The exact answers are

$$\lambda_1 = 6 \quad \text{and} \quad \mathbf{X}_1 = \begin{bmatrix} \frac{1}{2} \\ 1 \end{bmatrix}.$$

The rate of convergence of the power method depends upon the ratio $|\lambda_2/\lambda_1|$ and upon the choice of the vector $\mathbf{U}_0$. If the ratio $|\lambda_2/\lambda_1|$ is close to 1, then convergence is very slow. There is no way of knowing how to choose a good $\mathbf{U}_0$. There are many ways of extending the power method to handle cases excluded in our discussion. These and methods for improving the rate of convergence are discussed in the book by Fadeev and Fadeeva listed in Further Reading at the end of the chapter.

If $\mathbf{A}$ is an $n \times n$ symmetric matrix with eigenvalues $\lambda_1, \lambda_2, \ldots, \lambda_n$, where

$$|\lambda_1| > |\lambda_2| > |\lambda_3| \geq \cdots \geq |\lambda_n|,$$

it is possible to find λ_2 and an associated eigenvector $\mathbf{X}_2$ by a method called **deflation**, which applies the power method to the matrix $\mathbf{A}_1 = \mathbf{A} - \lambda_1 \mathbf{X}_1 \mathbf{X}_1^T$. This procedure works very well if we have a *very good* approximation to λ_1 and $\mathbf{X}_1$. Otherwise, the results can be rather poor. A much better procedure for finding all the eigenvalues and associated eigenvectors of a *symmetric* matrix is Jacobi's method, which is widely used.

Jacobi's Method

Let $\mathbf{A}$ be a symmetric $n \times n$ matrix and let $\mathbf{E}_{pq}$ $(p \neq q)$ be the $n \times n$ matrix obtained from the identity matrix $\mathbf{I}_n$ by replacing the following elements of $\mathbf{I}_n$ as indicated: p, pth element replaced by $\cos \theta$; p, qth element replaced by $\sin \theta$; q, pth element replaced by $-\sin \theta$; q, qth element replaced by $\cos \theta$. Thus, if we start with $\mathbf{I}_4$, then

$$\mathbf{E}_{24} = \begin{bmatrix} 1 & 0 & 0 & 0 \\ 0 & \cos \theta & 0 & \sin \theta \\ 0 & 0 & 1 & 0 \\ 0 & -\sin \theta & 0 & \cos \theta \end{bmatrix}.$$

The $\mathbf{E}_{pq}$ matrices are called **rotation matrices** because, for $n = 2$, they represent a rotation of axes through the angle θ.

The matrices $\mathbf{E}_{pq}$ are orthogonal; that is, $\mathbf{E}_{pq}^T \mathbf{E}_{pq} = \mathbf{I}_n$ (Exercise T.1). If we now form the matrix

$$\mathbf{B}_1 = \mathbf{E}_{pq}^T \mathbf{A} \mathbf{E}_{pq}, \tag{1}$$

then the entries of $\mathbf{B}_1$ differ from those of $\mathbf{A}$ only in the pth and qth rows and in the pth and qth columns. More specifically, the p, qth element of $\mathbf{B}_1$ is

$$(a_{pp} - a_{qq}) \cos \theta \sin \theta + a_{pq}(\cos^2 \theta - \sin^2 \theta). \tag{2}$$

Since

$$\cos \theta \sin \theta = \tfrac{1}{2} \sin 2\theta \quad \text{and} \quad \cos^2 \theta - \sin^2 \theta = \cos 2\theta,$$

we can write (2) as

$$\tfrac{1}{2}(a_{pp} - a_{qq}) \sin 2\theta + a_{pq} \cos 2\theta. \tag{3}$$

Let a_{pq} be in absolute value the largest nonzero element of $\mathbf{A}$ that is not on the main diagonal, and choose θ so that the p, qth element of $\mathbf{B}_1$ is zero. If there are several candidates for a_{pq}, choose any one. From (3) we see that θ is chosen so that

$$\tan 2\theta = -\frac{2a_{pq}}{a_{pp} - a_{qq}} \quad \text{for } a_{pp} - a_{qq} \neq 0. \tag{4}$$

If $a_{pp} - a_{qq} = 0$, then from (3) the p, qth element of $\mathbf{B}_1$ is

$$a_{pq}(\cos^2 \theta - \sin^2 \theta) = a_{pq} \cos 2\theta,$$

which can be made zero by taking $\theta = \pi/4$. We now repeat the same procedure with $\mathbf{B}_1$ taking the place of $\mathbf{A}$. That is, if b_{rs} is in absolute value the largest off-diagonal element of $\mathbf{B}_1$, then we form the matrix

$$\mathbf{B}_2 = \mathbf{E}_{rs}^T \mathbf{B}_1 \mathbf{E}_{rs},$$

where $\mathbf{E}_{rs}$ is the rotation matrix chosen so that the r, sth element of $\mathbf{B}_2$ will be zero; that is, θ is chosen so that

$$\tan 2\theta = -\frac{2a_{rs}}{a_{rr} - a_{ss}} \quad \text{for } a_{rr} - a_{ss} \neq 0.$$

If $a_{rr} - a_{ss} = 0$, let $\theta = \pi/4$.

It can then be proved that the sequence of matrices $\mathbf{B}_1, \mathbf{B}_2, \ldots, \mathbf{B}_t, \ldots$, converges to a diagonal matrix $\mathbf{D}$, whose diagonal elements are the eigenvalues of $\mathbf{A}$.

Thus, when we stop our computations, we have

$$\mathbf{E}_{k+1}^T \cdots \mathbf{E}_2^T \mathbf{E}_1^T \mathbf{A} \mathbf{E}_1 \mathbf{E}_2 \cdots \mathbf{E}_{k+1} = \mathbf{B}_{k+1}, \tag{5}$$

where $\mathbf{E}_1, \mathbf{E}_2, \ldots, \mathbf{E}_{k+1}$ are rotation matrices. If we let

$$\mathbf{E}_1 \mathbf{E}_2 \cdots \mathbf{E}_{k+1} = \mathbf{P}, \tag{6}$$

then we can write (5) as

$$\mathbf{P}^T\mathbf{A}\mathbf{P} = \mathbf{B}_{k+1}. \tag{7}$$

Now $\mathbf{P}$ is an orthogonal matrix and we may recall from Section 5.2 that the columns of $\mathbf{P}$ are eigenvectors of $\mathbf{A}$. Thus, to find approximations to the eigenvectors of $\mathbf{A}$ associated with $\lambda_1, \lambda_2, \ldots, \lambda_n$, we have to form the products of the rotation matrices as in (6).

From (7) it follows that

$$\mathbf{A} = \mathbf{P}\mathbf{B}_{k+1}\mathbf{P}^T.$$

As an additional check on the accuracy of our results, we may compute $\mathbf{P}\mathbf{B}_{k+1}\mathbf{P}^T$ and see how close it is to the given matrix $\mathbf{A}$.

In using the Jacobi method, it is not necessary to evaluate trigonometric functions to compute $\sin\theta$ and $\cos\theta$ from (4); instead we proceed as follows. Let

$$f = -a_{pq} \quad \text{and} \quad g = \tfrac{1}{2}(a_{pp} - a_{qq}).$$

Compute

$$h = \text{sign}\,(g)\,\frac{f}{\sqrt{f^2 + g^2}}. \tag{8}$$

Then

$$\sin\theta = \frac{h}{\sqrt{2(1 + \sqrt{1 - h^2})}}, \qquad \cos\theta = \sqrt{1 - \sin^2\theta}. \tag{9}$$

Observe that if $a_{pp} - a_{qq} = 0$, then taking sign (g) as $+$, we obtain $h = 1$ in (8), and from (9) $\sin\theta = 1/\sqrt{2}$, so that $\theta = \pi/4$, as has already been mentioned earlier. Thus we can use (8) and (9) in all cases, without having to check whether $a_{pp} - a_{qq} \neq 0$.

For a detailed discussion of a computer implementation of Jacobi's method, the reader is referred to the article by Greenstadt listed in Further Reading at the end of the chapter.

The procedure for carrying out Jacobi's method for diagonalizing a symmetric matrix $\mathbf{A}$ is as follows.

STEP 1. Identify a_{pq} as the largest (in absolute value) nonzero element of $\mathbf{A}$ that is not on the main diagonal.

STEP 2. Compute

$$f = -a_{pq} \qquad g = \frac{1}{2}(a_{pp} - a_{qq}) \qquad h = \text{sign}\,(g)\,\frac{f}{\sqrt{f^2 + g^2}}$$

STEP 3. Compute

$$\sin\theta = \frac{h}{\sqrt{2(1 + \sqrt{1 - h^2})}} \qquad \cos\theta = \sqrt{1 - \sin^2\theta}$$

STEP 4. Compute the $n \times n$ matrix $\mathbf{E}_{pq}$ from $\mathbf{I}_n$ by replacing the following elements of $\mathbf{I}_n$ as indicated:

p, pth element by $\cos \theta$ p, qth element by $\sin \theta$
q, pth element by $-\sin \theta$ q, qth element by $\cos \theta$

STEP 5. Compute $\mathbf{B}_1 = \mathbf{E}_{pq}^T \mathbf{A} \mathbf{E}_{pq}$.

STEP 6. Repeat Steps 1–5 with $\mathbf{B}_1$ taking the place of $\mathbf{A}$, obtaining matrices $\mathbf{B}_1, \mathbf{B}_2, \ldots, \mathbf{B}_{k+1}$. Then

$$\mathbf{P} = \mathbf{E}_1 \mathbf{E}_2 \cdots \mathbf{E}_{k+1}$$

is orthogonal and

$$\mathbf{B} = \mathbf{P}^T \mathbf{A} \mathbf{P}$$

is approximately diagonal.

Example 2 Consider the symmetric matrix

$$\mathbf{A} = \begin{bmatrix} -1 & 1 & -2 \\ 1 & -1 & -2 \\ -2 & -2 & -2 \end{bmatrix}.$$

We now find the eigenvalues and associated eigenvectors of $\mathbf{A}$ by Jacobi's method. We carry three decimal places and round after each multiplication and division. Moreover, the matrix product $\mathbf{ABC}$ will be calculated by first computing $\mathbf{BC}$ and then $\mathbf{A}(\mathbf{BC})$. Proceed as follows.

We can reduce the 1, 3; 2, 3; 3, 1; or 3, 2 element of $\mathbf{A}$ to zero. We choose to reduce the 1, 3 element to zero, obtaining

$$\mathbf{E}_{13} = \begin{bmatrix} 0.789 & 0.000 & 0.615 \\ 0.000 & 1.000 & 0.000 \\ -0.615 & 0.000 & 0.789 \end{bmatrix} = \mathbf{E}_1 \qquad \begin{matrix} f = 2.000 \\ g = 0.500 \\ h = 0.970; \end{matrix}$$

$$\mathbf{B}_1 = \begin{bmatrix} 0.562 & 2.019 & -0.003 \\ 2.019 & -1.000 & -0.963 \\ -0.004 & -0.963 & -3.565 \end{bmatrix}.$$

We now reduce the 2, 1 element of $\mathbf{B}_1$ to zero, obtaining

$$\mathbf{E}_{21} = \begin{bmatrix} 0.825 & -0.565 & 0.000 \\ 0.565 & 0.825 & 0.000 \\ 0.000 & 0.000 & 1.000 \end{bmatrix} = \mathbf{E}_2 \qquad \begin{matrix} f = -2.019 \\ g = -0.781 \\ h = 0.933; \end{matrix}$$

$$\mathbf{B}_2 = \begin{bmatrix} 1.946 & 0.001 & -0.546 \\ 0.001 & -2.384 & -0.792 \\ -0.547 & -0.792 & -3.565 \end{bmatrix}.$$

We next reduce the 2, 3 element of $\mathbf{B}_2$ to zero, obtaining

$$\mathbf{E}_{23} = \begin{bmatrix} 1.000 & 0.000 & 0.000 \\ 0.000 & 0.893 & 0.449 \\ 0.000 & -0.449 & 0.893 \end{bmatrix} = \mathbf{E}_3 \qquad \begin{matrix} f = 0.792 \\ g = 0.591 \\ h = 0.802; \end{matrix}$$

$$\mathbf{B}_3 = \begin{bmatrix} 1.946 & 0.246 & -0.488 \\ 0.247 & -1.984 & 0.002 \\ -0.488 & 0.002 & -3.959 \end{bmatrix}.$$

We now reduce the 1, 3 element of $\mathbf{B}_3$ to zero, obtaining

$$\mathbf{E}_{13} = \begin{bmatrix} 0.996 & 0.000 & 0.082 \\ 0.000 & 1.000 & 0.000 \\ -0.082 & 0.000 & 0.996 \end{bmatrix} = \mathbf{E}_4 \qquad \begin{matrix} f = 0.488 \\ g = 2.953 \\ h = 0.163; \end{matrix}$$

$$\mathbf{B}_4 = \begin{bmatrix} 1.983 & 0.245 & 0.002 \\ 0.246 & -1.984 & 0.022 \\ 0.002 & 0.022 & -3.994 \end{bmatrix}.$$

We next reduce the 2, 1 element of $\mathbf{B}_4$ to zero, obtaining

$$\mathbf{E}_{21} = \begin{bmatrix} 0.998 & -0.062 & 0.000 \\ 0.062 & 0.998 & 0.000 \\ 0.000 & 0.000 & 1.000 \end{bmatrix} = \mathbf{E}_5 \qquad \begin{matrix} f = -0.246 \\ g = -1.984 \\ h = 0.123; \end{matrix}$$

$$\mathbf{B}_5 = \begin{bmatrix} 1.998 & -0.002 & 0.003 \\ -0.001 & -1.999 & 0.022 \\ 0.003 & 0.022 & -3.994 \end{bmatrix}.$$

We now stop; the diagonal elements of $\mathbf{B}_5$ are approximations to the eigenvalues of $\mathbf{A}$. Thus we have $\lambda_1 \cong 1.998$, $\lambda_2 \cong -1.999$, and $\lambda_3 \cong -3.994$, which agree rather well with the exact values;

$$\lambda_1 = 2, \quad \lambda_2 = -2, \quad \lambda_3 = -4.$$

To find associated eigenvectors, we form $\mathbf{E}_1\mathbf{E}_2\mathbf{E}_3\mathbf{E}_4\mathbf{E}_5 = \mathbf{P}$. We have

$$\mathbf{E}_1\mathbf{E}_2\mathbf{E}_3\mathbf{E}_4\mathbf{E}_5 = \mathbf{P} = \begin{bmatrix} 0.576 & -0.707 & 0.409 \\ 0.578 & 0.708 & 0.407 \\ -0.578 & -0.001 & 0.816 \end{bmatrix}.$$

The columns of $\mathbf{P}$ give the approximate eigenvectors

$$\mathbf{X}_1 \cong \begin{bmatrix} 0.576 \\ 0.578 \\ -0.578 \end{bmatrix}, \qquad \mathbf{X}_2 \cong \begin{bmatrix} -0.707 \\ 0.708 \\ -0.001 \end{bmatrix}, \qquad \mathbf{X}_3 \cong \begin{bmatrix} 0.409 \\ 0.407 \\ 0.816 \end{bmatrix},$$

which are associated with λ_1, λ_2, and λ_3, respectively.

As a final check of our results, we compute $\mathbf{PB}_5\mathbf{P}^T$, obtaining

$$\mathbf{PB}_5\mathbf{P}^T = \begin{bmatrix} -1.015 & 1.002 & -2.010 \\ 1.001 & -0.982 & -1.981 \\ -2.010 & -1.981 & -1.995 \end{bmatrix},$$

which is in very close agreement with the given matrix $\mathbf{A}$.

FURTHER READING

Fadeev, D. K., and **V. N. Fadeeva.** *Computational Methods of Linear Algebra.* San Francisco: W. H. Freeman and Company, Publishers, 1963.

Forsythe, G., and **C. B. Moler.** *Computer Solution of Linear Algebraic Systems.* Englewood Cliffs, N.J.: Prentice-Hall, Inc., 1967.

Fox, L. *An Introduction to Numerical Linear Algebra.* New York: Oxford University Press, Inc., 1965.

Franklin, J. N. *Matrix Theory.* Englewood Cliffs, N.J.: Prentice-Hall, Inc., 1968.

Greenstadt, John. "The Determination of the Characteristic Roots of a Matrix by the Jacobi Method," in *Mathematical Methods for Digital Computers,* ed. by A. Ralston and H. S. Wilf. New York: John Wiley & Sons, Inc., 1960.

Noble, Ben. *Applied Linear Algebra,* 2nd ed. Englewood Cliffs, N.J.: Prentice-Hall, Inc., 1977.

Stewart, G. W. *Introduction to Matrix Computations.* New York: Academic Press, Inc., 1973.

Wilkinson, J. H. *The Algebraic Eigenvalue Problem.* New York: Oxford University Press, Inc., 1965.

8.4 EXERCISES

In Exercises 1 through 5, compute an approximation to the eigenvalue of largest magnitude and an associated eigenvector by the power method. Carry out the computations to three decimal places and round after each multiplication and division for six iterations ($k = 5$).

1. $\begin{bmatrix} -4 & 2 \\ 3 & 1 \end{bmatrix}$.   2. $\begin{bmatrix} 0 & -1 \\ 2 & 3 \end{bmatrix}$.

3. $\begin{bmatrix} 4 & 3 \\ 2 & 3 \end{bmatrix}$.   4. $\begin{bmatrix} 6 & 2 \\ -4 & -3 \end{bmatrix}$.

5. $\begin{bmatrix} 8 & 8 \\ 3 & -2 \end{bmatrix}$.

In Exercises 6 through 9, find the eigenvalues and associated eigenvectors by Jacobi's method. Carry out the computations to three decimal places and round after each multiplication and division. In each case calculate $\mathbf{ABC}$ by first calculating $\mathbf{BC}$.

6. $\begin{bmatrix} 2 & 4 \\ 4 & 2 \end{bmatrix}$. Find $\mathbf{B}_3$.

7. $\begin{bmatrix} 2 & 2 \\ 2 & 2 \end{bmatrix}$. Find $\mathbf{B}_3$.

8. $\begin{bmatrix} -1 & -4 & -8 \\ -4 & -7 & 4 \\ -8 & 4 & -1 \end{bmatrix}$. Find $\mathbf{B}_5$.

9. $\begin{bmatrix} 8 & -2 & 0 \\ -2 & 9 & -2 \\ 0 & -2 & 10 \end{bmatrix}$. Find $\mathbf{B}_5$.

THEORETICAL EXERCISE

T.1. Prove that the matrices $\mathbf{E}_{pq}$ are orthogonal.

KEY IDEAS FOR REVIEW

☐ **Gaussian elimination method** (for solving the linear system $\mathbf{AX} = \mathbf{B}$). Transform the augmented matrix $[\mathbf{A} \mid \mathbf{B}]$ to row echelon form and use back substitution.

☐ **Iterative methods** (for solving $\mathbf{AX} = \mathbf{B}$): **Jacobi's iteration method.** See page 424. **Gauss–Seidel iteration method.** See page 425.

☐ **LU factorization** (for writing an $n \times n$ matrix $\mathbf{A}$ as

LU, where $\mathbf{L}$ is in lower triangular form and $\mathbf{U}$ is in upper triangular form). See page 433.

☐ **Power method** (for finding the eigenvalue of largest absolute value and an associated eigenvector). See page 438.

☐ **Jacobi's method** (for finding the eigenvalues and eigenvectors of a symmetric matrix). See page 441.

SUPPLEMENTARY EXERCISES

1. Solve the following linear system by (a) Jacobi's iteration method and (b) the Gauss–Seidel iteration method. Carry out the computations to three decimal places for six iterations ($k = 5$) and round after each multiplication and division.

$$\begin{aligned} 10x + 3y &= -24 \\ 2x + 11y &= 16. \end{aligned}$$

In Exercises 2 and 3, find an **LU**-factorization of the coefficient matrix of the given linear system $\mathbf{AX} = \mathbf{B}$. Solve the linear system using a forward substitution followed by a back substitution.

2. $\mathbf{A} = \begin{bmatrix} 2 & 2 & 3 \\ 6 & 5 & 7 \\ -6 & -8 & -10 \end{bmatrix}$, $\mathbf{B} = \begin{bmatrix} -6 \\ -13 \\ 22 \end{bmatrix}$.

3. $\mathbf{A} = \begin{bmatrix} -2 & 1 & -2 \\ 6 & 1 & 9 \\ -4 & 18 & 5 \end{bmatrix}$, $\mathbf{B} = \begin{bmatrix} -6 \\ 19 \\ -17 \end{bmatrix}$.

4. For the following matrix, compute an approximation to the eigenvalue of largest magnitude and an associated eigenvector by the power method. Carry out the computations to three decimal places and round after each multiplication and division for six iterations ($k = 5$).

$$\begin{bmatrix} 7 & 6 \\ -5 & -4 \end{bmatrix}.$$

5. For the following matrix, find the eigenvalues and associated eigenvectors by Jacobi's method. Carry out the computations to three decimal places and round after each multiplication and division. In each case calculate **ABC** by first calculating **BC**.

$$\begin{bmatrix} 1 & -1 & -1 \\ -1 & 1 & -1 \\ -1 & -1 & 1 \end{bmatrix}.$$

Find $\mathbf{B}_2$.

CHAPTER TEST

1. Solve the following linear system by (a) Jacobi's iteration method and (b) the Gauss–Seidel iteration method. Carry out the computations to three decimal places for six iterations ($k = 5$) and round after each multiplication and division.

$$16x + 3y = 55$$
$$x + 10y = -26.$$

2. Find an **LU** factorization of the coefficient matrix of the linear system $\mathbf{AX} = \mathbf{B}$. Solve the linear system by using a forward substitution followed by a back substitution.

$$\mathbf{A} = \begin{bmatrix} 2 & 2 & -1 \\ -8 & -11 & 5 \\ 4 & 13 & -7 \end{bmatrix}, \quad \mathbf{B} = \begin{bmatrix} 3 \\ -14 \\ -5 \end{bmatrix}.$$

3. For the following matrix, compute an approximation to the eigenvalue of largest magnitude and an associated eigenvector by the power method. Carry out the computations to three decimal places and round after each multiplication and division for six iterations ($k = 5$).

$$\begin{bmatrix} 4 & 3 \\ 5 & 6 \end{bmatrix}.$$

4. For the following matrix, find the eigenvalues and associated eigenvectors by Jacobi's method. Carry out the computations to three decimal places and round after each multiplication and division. In each case calculate **ABC** by first calculating **BC**.

$$\begin{bmatrix} -1 & 2 & 2 \\ 2 & -1 & 2 \\ 2 & 2 & -1 \end{bmatrix}.$$

Find $\mathbf{B}_2$.

Appendix A: Complex Numbers

A.1

COMPLEX NUMBERS

Complex numbers are usually introduced in an algebra course to "complete" the solution of the quadratic equation

$$ax^2 + bx + c = 0.$$

In using the quadratic formula

$$x = \frac{-b \pm \sqrt{b^2 - 4ac}}{2a},$$

the case in which $b^2 - 4ac < 0$ is not resolved unless we can cope with the square roots of negative numbers. In the sixteenth century mathematicians and scientists justified this "completion" of the solution of quadratic equations by intuition. Naturally, a controversy arose, with some mathematicians denying their existence and others using them along with real numbers. The use of complex numbers did not lead to any contradictions, and the idea proved to be an important milestone in the development of mathematics.

A **complex number** c is of the form $c = a + bi$, where a and b are real numbers and where $i = \sqrt{-1}$; a is called the **real part** of c and b is called the **imaginary part** of c. The term "imaginary part" arose from the mysticism surrounding the beginnings of complex numbers; however, these numbers are as "real" as the real numbers.

Example 1 (a) $5 - 3i$ has real part 5 and imaginary part -3; (b) $-6 + \sqrt{2}i$ has real part -6 and imaginary part $\sqrt{2}$.

The symbol $i = \sqrt{-1}$ has the property that $i^2 = -1$ and we can deduce the following relationships:

$$i^3 = -i, \ i^4 = 1, \ i^5 = i, \ i^6 = -1, \ i^7 = -i, \ \ldots.$$

These results will be handy for simplifying operations involving complex numbers.

We say that two complex numbers $c_1 = a_1 + b_1 i$ and $c_2 = a_2 + b_2 i$ are **equal** if their real and imaginary parts are equal, that is, if $a_1 = a_2$ and $b_1 = b_2$. Of course, every real number a is a complex number with its imaginary part zero: $c = a + 0i$.

Operations on Complex Numbers

If $c_1 = a_1 + b_1 i$ and $c_2 = a_2 + b_2 i$ are complex numbers, then their **sum** is

$$c_1 + c_2 = (a_1 + a_2) + (b_1 + b_2)i,$$

and their **difference** is

$$c_1 - c_2 = (a_1 - a_2) + (b_1 - b_2)i.$$

In words, to form the sum of two complex numbers, add the real parts and add the imaginary parts. The **product** of c_1 and c_2 is

$$c_1 c_2 = (a_1 + b_1 i) \cdot (a_2 + b_2 i) = a_1 a_2 + (a_1 b_2 + b_1 a_2)i + b_1 b_2 i^2$$
$$= (a_1 a_2 - b_1 b_2) + (a_1 b_2 + b_1 a_2)i.$$

A special case of multiplication of complex numbers occurs when c_1 is real. In this case we obtain the simple result

$$c_1 c_2 = c_1 \cdot (a_2 + b_2 i) = c_1 a_2 + c_1 b_2 i.$$

If $c = a + bi$ is a complex number, then the **conjugate** of c is the complex number $\bar{c} = a - bi$. It is easy to show that if c and d are complex numbers, then the following basic properties of complex arithmetic hold:

1. $\bar{\bar{c}} = c$.
2. $\overline{c + d} = \bar{c} + \bar{d}$.
3. $\overline{cd} = \bar{c}\bar{d}$.
4. c is a real number if and only if $c = \bar{c}$.
5. $c\bar{c}$ is a nonnegative real number and $c\bar{c} = 0$ if and only if $c = 0$.

We prove property 4 here and leave the others as exercises. Let $c = a + bi$ so that $\bar{c} = a - bi$. If $c = \bar{c}$, then $a + bi = a - bi$, so $b = 0$ and c is real. On the other hand, if c is real, then $c = a$ and $\bar{c} = a$, so $c = \bar{c}$.

Example 2 Let $c_1 = 5 - 3i$, $c_2 = 4 + 2i$, and $c_3 = -3 + i$.
      (a) $c_1 + c_2 = (5 - 3i) + (4 + 2i) = 9 - i$.
      (b) $c_2 - c_3 = (4 + 2i) - (-3 + i) = (4 - (-3)) + (2 - 1)i = 7 + i$.

(c) $c_1c_2 = (5 - 3i) \cdot (4 + 2i) = 20 + 10i - 12i - 6i^2 = 26 - 2i$.

(d) $c_1\bar{c}_3 = (5 - 3i) \cdot \overline{(-3 + i)} = (5 - 3i) \cdot (-3 - i) = -15 - 5i + 9i + 3i^2$
$$= -18 + 4i.$$

(e) $3c_1 + 2\bar{c}_2 = 3(5 - 3i) + 2\overline{(4 + 2i)} = (15 - 9i) + 2(4 - 2i)$
$$= (15 - 9i) + (8 - 4i) = 23 - 13i.$$

(f) $c_1\bar{c}_1 = (5 - 3i)\overline{(5 - 3i)} = (5 - 3i)(5 + 3i) = 34$.

When we consider systems of linear equations with complex coefficients, we will need to divide complex numbers to complete the solution process and obtain a reasonable form for the solution. Let $c_1 = a_1 + b_1i$ and $c_2 = a_2 + b_2i$. If $c_2 \neq 0$, that is, if $a_2 \neq 0$ or $b_2 \neq 0$, then we can **divide** c_1 by c_2:

$$\frac{c_1}{c_2} = \frac{a_1 + b_1i}{a_2 + b_2i}.$$

To conform to our practice of expressing a complex number in the form real part + imaginary part $\cdot\, i$, we must simplify the foregoing expression for c_1/c_2. To simplify this complex fraction, we multiply the numerator and the denominator by the conjugate of the denominator. Thus, dividing c_1 by c_2 gives the complex number

$$\frac{c_1}{c_2} = \frac{a_1 + b_1i}{a_2 + b_2i} = \frac{(a_1 + b_1i)(a_2 - b_2i)}{(a_2 + b_2i)(a_2 - b_2i)} = \frac{a_1a_2 + b_1b_2}{a_2^2 + b_2^2} + \frac{-a_1b_2 + a_2b_1}{a_2^2 + b_2^2}i.$$

Example 3   Let $c_1 = 2 - 5i$ and $c_2 = -3 + 4i$. Then

$$\frac{c_1}{c_2} = \frac{2 - 5i}{-3 + 4i} = \frac{(2 - 5i)(-3 - 4i)}{(-3 + 4i)(-3 - 4i)} = \frac{-26 + 7i}{(-3)^2 + (4)^2} = -\frac{26}{25} + \frac{7}{25}i.$$

Finding the reciprocal of a complex number is a special case of division of complex numbers. If $c = a + bi$, $c \neq 0$, then

$$\frac{1}{c} = \frac{1}{a + bi} = \frac{a - bi}{(a + bi)(a - bi)} = \frac{a - bi}{a^2 + b^2}$$

$$= \frac{a}{a^2 + b^2} - \frac{b}{a^2 + b^2}i.$$

Example 4   (a) $\dfrac{1}{2 + 3i} = \dfrac{2 - 3i}{(2 + 3i)(2 - 3i)} = \dfrac{2 - 3i}{2^2 + 3^2} = \dfrac{2}{13} - \dfrac{3}{13}i$.

(b) $\dfrac{1}{i} = \dfrac{-i}{i(-i)} = \dfrac{-i}{-i^2} = \dfrac{-i}{-(-1)} = -i$.

Summarizing, we can say that complex numbers are mathematical objects for which addition, subtraction, multiplication, and division are defined in such a way that these operations on real numbers can be derived as special cases.

Geometric Representation of Complex Numbers

A complex number $c = a + bi$ may be regarded as an ordered pair (a, b) of real numbers. This ordered pair of real numbers corresponds to a point in the plane. Such a correspondence naturally suggests that we represent $a + bi$ as a point in the **complex plane,** where the horizontal axis is used to represent the real part of c and the vertical axis is used to represent the imaginary part of c. To simplify matters, we call these the **real axis** and **imaginary axis,** respectively (see Figure A.1).

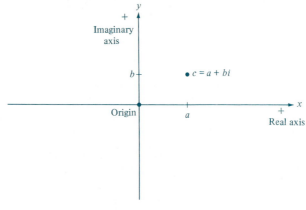

Figure A.1

Example 5 Plot the complex numbers $c = 2 - 3i$, $d = 1 + 4i$, $e = -3$, and $f = 2i$ in the complex plane.

Solution
See Figure A.2.

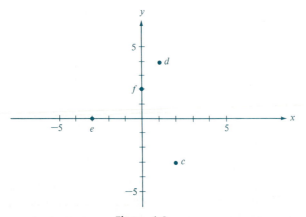

Figure A.2

The rules concerning inequality of real numbers, such as less than and greater than, *do not apply to complex numbers*. There is no way to arrange the complex numbers according to size. However, using the geometric representation from the complex plane, we can attach a notion of size to a complex number by measuring its distance from the origin. The distance from the origin to $c = a + bi$ is called the **absolute value** or **modulus** of the complex number, and is denoted by $|c| = |a + bi|$. Using the formula for the distance between ordered pairs of real numbers, we obtain

$$|c| = |a + bi| = \sqrt{a^2 + b^2}.$$

It follows that $c\bar{c} = |c|^2$ (verify).

Example 6 Referring to Example 5: $|c| = \sqrt{13}$; $|d| = \sqrt{17}$; $|e| = 3$; $|f| = 2$.

A different, but related interpretation of a complex number is obtained if we associate with $c = a + bi$ the vector OP, where O is the origin $(0, 0)$ and P is the point (a, b). There is an obvious correspondence between this representation and vectors in the plane discussed in calculus, which are reviewed at the beginning of Chapter 3. Using a vector representation, addition and subtraction of complex numbers can be viewed as the corresponding vector operations. These are represented in Figures 3.11, 3.12, and 3.16. We will not pursue the manipulation of complex numbers by vector operations here, but such a point of view is important for the development and study of complex variables.

Matrices with Complex Entries

If the entries of a matrix are complex numbers, we can perform the matrix operations of addition, subtraction, multiplication, and scalar multiplication in a manner completely analogous to that for real matrices. The validity of these operations can be verified using properties of complex arithmetic, just imitating the proofs for real matrices presented in the text. We illustrate these concepts in the following example.

Example 7 Let

$$\mathbf{A} = \begin{bmatrix} 4 + i & -2 + 3i \\ 6 + 4i & -3i \end{bmatrix}, \mathbf{B} = \begin{bmatrix} 2 - i & 3 - 4i \\ 5 + 2i & -7 + 5i \end{bmatrix}, \mathbf{C} = \begin{bmatrix} 1 + 2i & i \\ 3 - i & 8 \\ 4 + 2i & 1 - i \end{bmatrix}.$$

(a) $\mathbf{A} + \mathbf{B} = \begin{bmatrix} (4 + i) + (2 - i) & (-2 + 3i) + (3 - 4i) \\ (6 + 4i) + (5 + 2i) & (-3i) + (-7 + 5i) \end{bmatrix} = \begin{bmatrix} 6 & 1 - i \\ 11 + 6i & -7 + 2i \end{bmatrix}.$

(b) $\mathbf{B} - \mathbf{A} = \begin{bmatrix} (2 - i) - (4 + i) & (3 - 4i) - (-2 + 3i) \\ (5 + 2i) - (6 + 4i) & (-7 + 5i) - (-3i) \end{bmatrix} = \begin{bmatrix} -2 - 2i & 5 - 7i \\ -1 - 2i & -7 + 8i \end{bmatrix}.$

(c) $\mathbf{CA} = \begin{bmatrix} 1 + 2i & i \\ 3 - i & 8 \\ 4 + 2i & 1 - i \end{bmatrix} \begin{bmatrix} 4 + i & -2 + 3i \\ 6 + 4i & -3i \end{bmatrix}$

$= \begin{bmatrix} (1 + 2i)(4 + i) + (i)(6 + 4i) & (1 + 2i)(-2 + 3i) + (i)(-3i) \\ (3 - i)(4 + i) + (8)(6 + 4i) & (3 - i)(-2 + 3i) + (8)(-3i) \\ (4 + 2i)(4 + i) + (1 - i)(6 + 4i) & (4 + 2i)(-2 + 3i) + (1 - i)(-3i) \end{bmatrix}$

$= \begin{bmatrix} -2 + 15i & -5 - i \\ 61 + 31i & -3 - 13i \\ 24 + 10i & -17 + 5i \end{bmatrix}$.

(d) $(2 + i)\mathbf{B} = \begin{bmatrix} (2 + i)(2 - i) & (2 + i)(3 - 4i) \\ (2 + i)(5 + 2i) & (2 + i)(-7 + 5i) \end{bmatrix} = \begin{bmatrix} 5 & 10 - 5i \\ 8 + 9i & -19 + 3i \end{bmatrix}$.

Just as we can compute the conjugate of a complex number, we can compute the **conjugate of a matrix** by computing the conjugate of each entry of the matrix. We denote the conjugate of matrix $\mathbf{A}$ by $\overline{\mathbf{A}}$, and write

$$\overline{\mathbf{A}} = [\overline{a_{ij}}].$$

Example 8 Referring to Example 7, we find that

$$\overline{\mathbf{A}} = \begin{bmatrix} 4 - i & -2 - 3i \\ 6 - 4i & 3i \end{bmatrix} \quad \text{and} \quad \overline{\mathbf{B}} = \begin{bmatrix} 2 + i & 3 + 4i \\ 5 - 2i & -7 - 5i \end{bmatrix}.$$

The following properties of the conjugate of a matrix hold:

1. $\overline{\overline{\mathbf{A}}} = \mathbf{A}$.
2. $\overline{\mathbf{A} + \mathbf{B}} = \overline{\mathbf{A}} + \overline{\mathbf{B}}$.
3. $\overline{\mathbf{AB}} = \overline{\mathbf{A}}\,\overline{\mathbf{B}}$.
4. For any real number k, $\overline{k\mathbf{A}} = k\overline{\mathbf{A}}$.
5. For any complex number c, $\overline{c\mathbf{A}} = \overline{c}\overline{\mathbf{A}}$.
6. $(\overline{\mathbf{A}})^T = \overline{\mathbf{A}^T}$.
7. If $\mathbf{A}$ is nonsingular, then $(\overline{\mathbf{A}})^{-1} = \overline{\mathbf{A}^{-1}}$.

We prove properties 5 and 6 here and leave the others as exercises. First property 5: If c is complex, the (i, j) entry of $\overline{c\mathbf{A}}$ is

$$\overline{ca_{ij}} = \overline{c}\,\overline{a}_{ij},$$

which is the (i, j) entry of $\overline{c}\overline{\mathbf{A}}$. Next property 6: The (i, j) entry of $(\overline{\mathbf{A}})^T$ is $\overline{a}_{ji}$, which is the (i, j) entry of $\overline{\mathbf{A}^T}$.

Complex Numbers and Roots of Polynomials

A polynomial of degree n with real coefficients has n complex roots, some, all, or none of which may be real numbers. Thus the polynomial $f_1(x) = x^4 - 1$ has the roots i, $-i$, 1, and -1; the polynomial $f_2(x) = x^2 - 1$ has the roots 1 and -1; and the polynomial $f_3(x) = x^2 + 1$ has the roots i and $-i$. We shall make use of this result later, when we extend the definition of an eigenvalue (Section 5.1) to allow it to be a complex number.

A.1 EXERCISES

1. Let $c_1 = 3 + 4i$, $c_2 = 1 - 2i$, and $c_3 = -1 + i$. Compute each of the following and simplify as much as possible.
   (a) $c_1 + c_2$.　　(b) $c_3 - c_1$.
   (c) $c_1 c_2$.　　(d) $c_2 \bar{c}_3$.
   (e) $4c_3 + \bar{c}_2$.　　(f) $(-i) \cdot c_2$.
   (g) $\overline{3c_1 - ic_2}$.　　(h) $c_1 c_2 c_3$.

2. Write in the form $a + bi$.
   (a) $\dfrac{1 + 2i}{3 - 4i}$.　　(b) $\dfrac{2 - 3i}{3 - i}$.
   (c) $\dfrac{(2 + i)^2}{i}$.　　(d) $\dfrac{1}{(3 + 2i)(1 + i)}$.

3. Represent each complex number as a point and as a vector in the complex plane.
   (a) $4 + 2i$.　　(b) $-3 + i$.　　(c) $3 - 2i$.
   (d) $i(4 + i)$.

4. Find the modulus of each complex number in Exercise 3.

5. In the complex plane sketch the vectors corresponding to c and $\bar{c}$ for $c = 2 + 3i$ and $c = -1 + 4i$. Geometrically, we can say that $\bar{c}$ is the reflection of c with respect to the real axis. (See also Example 9 in Section 4.1.)

6. Let
$$\mathbf{A} = \begin{bmatrix} 2 + 2i & -1 + 3i \\ -2 & 1 - i \end{bmatrix},$$
$$\mathbf{B} = \begin{bmatrix} 2i & 1 + 2i \\ 0 & 3 - i \end{bmatrix}, \quad \mathbf{C} = \begin{bmatrix} 2 + i \\ -i \end{bmatrix}.$$

Compute each of the following and simplify each entry as $a + bi$.
   (a) $\mathbf{A} + \mathbf{B}$.　　(b) $(1 - 2i)\mathbf{C}$.　　(c) $\mathbf{AB}$.
   (d) $\mathbf{BC}$.　　(e) $\mathbf{A} - 2\mathbf{I}_2$.　　(f) $\bar{\mathbf{B}}$.
   (g) $\mathbf{A}\bar{\mathbf{C}}$.　　(h) $(\mathbf{A} + \mathbf{B})\mathbf{C}$.

7. If $\mathbf{A} = \begin{bmatrix} 0 & i \\ i & 0 \end{bmatrix}$, compute $\mathbf{A}^2$, $\mathbf{A}^3$, and $\mathbf{A}^4$. Give a general rule for $\mathbf{A}^n$, n a positive integer.

8. Find all the roots:
   (a) $x^2 + x + 1 = 0$.
   (b) $x^3 + 2x^2 + x + 2 = 0$.
   (c) $x^5 + x^4 - x - 1 = 0$.

9. Let $p(x)$ denote a polynomial and let $\mathbf{A}$ be a square matrix. Then $p(\mathbf{A})$ is called a **matrix polynomial** or a **polynomial in the matrix** A. For $p(x) = 2x^2 + 5x - 3$, compute $p(\mathbf{A}) = 2\mathbf{A}^2 + 5\mathbf{A} - 3\mathbf{I}_n$ for each of the following.
   (a) $\mathbf{A} = \begin{bmatrix} -3 & 0 \\ 0 & -3 \end{bmatrix}$.　　(b) $\mathbf{A} = \begin{bmatrix} 1 & 2 \\ 0 & 1 \end{bmatrix}$.
   (c) $\mathbf{A} = \begin{bmatrix} 0 & i \\ i & 0 \end{bmatrix}$.　　(d) $\mathbf{A} = \begin{bmatrix} 1 & i \\ 0 & 0 \end{bmatrix}$.

10. Let $p(x) = x^2 + 1$.
   (a) Determine two different 2×2 matrices $\mathbf{A}$ of the form $k\mathbf{I}_2$ that satisfy $p(\mathbf{A}) = \mathbf{0}$.
   (b) Verify that $p(\mathbf{A}) = \mathbf{0}$, for $\mathbf{A} = \begin{bmatrix} 1 & 2 \\ -1 & -1 \end{bmatrix}$.

11. Find all the 2×2 matrices $\mathbf{A}$ of the form $k\mathbf{I}_2$ that satisfy $p(\mathbf{A}) = \mathbf{0}$ for $p(x) = x^2 - x - 2$.

12. In Supplementary Exercise 27 in Chapter 1, we introduced the concept of a square root of a matrix with real entries. We can generalize the notion of a square root of a matrix if we permit complex entries.

(a) Compute a complex square root of

$$\mathbf{A} = \begin{bmatrix} -1 & 0 \\ 0 & 0 \end{bmatrix}.$$

(b) Compute a complex square root of

$$\mathbf{A} = \begin{bmatrix} -2 & 2 \\ 2 & -2 \end{bmatrix}.$$

THEORETICAL EXERCISES

T.1. If $c = a + bi$, then we can denote the real part of c by Re (c) and the imaginary part of c by Im (c).

(a) For any complex numbers $c_1 = a_1 + b_1 i$, $c_2 = a_2 + b_2 i$, prove that Re $(c_1 + c_2) =$ Re $(c_1) +$ Re (c_2) and Im $(c_1 + c_2) =$ Im $(c_1) +$ Im (c_2).

(b) For any real number k, prove that Re $(kc) = k$ Re (c) and Im $(kc) = k$ Im (c).

(c) Is part (b) true if k is a complex number?

(d) Prove or disprove:
Re $(c_1 c_2) =$ Re $(c_1) \cdot$ Re (c_2).

T.2. Let $\mathbf{A}$ and $\mathbf{B}$ be $m \times n$ complex matrices, and let $\mathbf{C}$ be an $n \times m$ nonsingular matrix.

(a) Prove that $\overline{\mathbf{A} + \mathbf{B}} = \overline{\mathbf{A}} + \overline{\mathbf{B}}$.

(b) Prove that for any real number k, $\overline{k\mathbf{A}} = k\overline{\mathbf{A}}$.

(c) $(\overline{\mathbf{C}})^{-1} = \overline{\mathbf{C}^{-1}}$.

A.2

COMPLEX NUMBERS IN LINEAR ALGEBRA

Almost everything in the first five chapters of this book remains true if we cross out the word *real* and replace it by the word *complex*. In Section A.1 we introduced matrices with complex entries. Here we will consider linear systems with complex entries, determinants of complex matrices, complex vector spaces, and eigenvalues and eigenvectors of complex matrices. The theory we have developed for these topics using real numbers is immediately applicable to the case with complex numbers. We could have developed Chapters 1 through 5 with complex numbers, but we limited ourselves to the real numbers for the following reasons:

1. The results would not look markedly different.
2. For a good many applications the real number situation is adequate; the transition to complex numbers is, as we have just remarked, not too difficult.
3. The computational effort of complex arithmetic is, in most cases, more than double that resulting with only real numbers.

The primary goal of this appendix is to provide an easy transition to complex numbers in linear algebra. This is of particular importance in Chapter 5, where complex eigenvalues and eigenvectors arise naturally for matrices with real entries. Hence we will not restate our theorems in the complex case, but will provide a discussion and examples of the major ideas needed to accomplish this transition. It will soon be evident that the increased computational effort of complex arithmetic becomes quite tedious for hand computation. However, such calculations can be easily done by computers.

Solving Linear Systems with Complex Entries

The results and techniques dealing with the solution of linear systems that we developed in Chapter 1 carry over directly to linear systems with complex coefficients. We will illustrate row operations and echelon forms for such systems with Gauss–Jordan reduction using complex arithmetic.

Example 1 Solve the linear system

$$(1 + i)x_1 + (2 + i)x_2 = 5$$
$$(2 - 2i)x_1 + \quad (i)x_2 = 1 + 2i.$$

Solution
We form the augmented matrix and use elementary row operations to transform it to reduced row echelon form. For the augmented matrix $[\mathbf{A} \mid \mathbf{B}]$,

$$\begin{bmatrix} 1 + i & 2 + i & \vdots & 5 \\ 2 - 2i & i & \vdots & 1 + 2i \end{bmatrix},$$

multiply the first row by $1/(1 + i)$ to obtain

$$\begin{bmatrix} 1 & \frac{3}{2} - \frac{1}{2}i & \vdots & \frac{5}{2} - \frac{5}{2}i \\ 2 - 2i & i & \vdots & 1 + 2i \end{bmatrix}.$$

We now add $-(2 - 2i)$ times the first row to the second row to get

$$\begin{bmatrix} 1 & \frac{3}{2} - \frac{1}{2}i & \vdots & \frac{5}{2} - \frac{5}{2}i \\ 0 & -2 + 5i & \vdots & 1 + 12i \end{bmatrix}.$$

Multiply the second row by $1/(-2 + 5i)$ to obtain

$$\begin{bmatrix} 1 & \frac{3}{2} - \frac{1}{2}i & \vdots & \frac{5}{2} - \frac{5}{2}i \\ 0 & 1 & \vdots & 2 - i \end{bmatrix},$$

which is in row echelon form. To get to reduced row echelon form, we add $-(\frac{3}{2} - \frac{1}{2}i)$ times the second row to the first row to get

$$\begin{bmatrix} 1 & 0 & \vdots & 0 \\ 0 & 1 & \vdots & 2 - i \end{bmatrix}.$$

Hence the solution is $x_1 = 0$ and $x_2 = 2 - i$.

If you carry out the arithmetic for the row operations in the preceding example, you will feel the burden of the complex arithmetic even though there were just two equations in two unknowns. Gaussian elimination with back substitution can also be used on linear systems with complex coefficients.

Example 2 Suppose that the augmented matrix of a linear system has been transformed to the following matrix in row echelon form:

$$\left[\begin{array}{ccc|c} 1 & 0 & 1+i & -1 \\ 0 & 1 & 3i & 2+i \\ 0 & 0 & 1 & 2i \end{array}\right].$$

The back-substitution procedure gives us

$$x_3 = 2i$$
$$x_2 = 2 + i - 3i(2i) = 2 + i + 6 = 8 + i$$
$$x_1 = -1 - (1+i)(2i) = -1 - 2i + 2 = 3 - 2i.$$

We can alleviate the tediousness of complex arithmetic by using a computer code for linear systems with complex entries. However, we must still pay a high price because the execution time will be approximately twice as long as that for the same size linear system with all real entries. We can illustrate this by showing how to transform an $n \times n$ linear system with complex coefficients to a $2n \times 2n$ linear system with only real coefficients.

Example 3 Consider the linear system

$$(2+i)x_1 + (1+i)x_2 = 3 + 6i$$
$$(3-i)x_1 + (2-2i)x_2 = 7 - i.$$

If we let $x_1 = a_1 + b_1 i$ and $x_2 = a_2 + b_2 i$, with a_1, b_1, a_2, and b_2 real numbers, then we can write this system in matrix form as

$$\left[\begin{array}{cc} 2+i & 1+i \\ 3-i & 2-2i \end{array}\right]\left[\begin{array}{c} a_1 + b_1 i \\ a_2 + b_2 i \end{array}\right] = \left[\begin{array}{c} 3 + 6i \\ 7 - i \end{array}\right].$$

We first rewrite the given linear system as

$$\left(\left[\begin{array}{cc} 2 & 1 \\ 3 & 2 \end{array}\right] + i\left[\begin{array}{cc} 1 & 1 \\ -1 & -2 \end{array}\right]\right)\left(\left[\begin{array}{c} a_1 \\ a_1 \end{array}\right] + i\left[\begin{array}{c} b_1 \\ b_2 \end{array}\right]\right) = \left[\begin{array}{c} 3 \\ 7 \end{array}\right] + i\left[\begin{array}{c} 6 \\ -1 \end{array}\right].$$

Multiplying, we have

$$\left(\left[\begin{array}{cc} 2 & 1 \\ 3 & 2 \end{array}\right]\left[\begin{array}{c} a_1 \\ a_2 \end{array}\right] - \left[\begin{array}{cc} 1 & 1 \\ -1 & -2 \end{array}\right]\left[\begin{array}{c} b_1 \\ b_2 \end{array}\right]\right)$$

$$+ i\left(\left[\begin{array}{cc} 2 & 1 \\ 3 & 2 \end{array}\right]\left[\begin{array}{c} b_1 \\ b_2 \end{array}\right] + \left[\begin{array}{cc} 1 & 1 \\ -1 & -2 \end{array}\right]\left[\begin{array}{c} a_1 \\ a_2 \end{array}\right]\right) = \left[\begin{array}{c} 3 \\ 7 \end{array}\right] + i\left[\begin{array}{c} 6 \\ -1 \end{array}\right].$$

The real and imaginary parts on both sides of the equation must agree, respectively, and so we have

$$\begin{bmatrix} 2 & 1 \\ 3 & 2 \end{bmatrix}\begin{bmatrix} a_1 \\ a_2 \end{bmatrix} - \begin{bmatrix} 1 & 1 \\ -1 & -2 \end{bmatrix}\begin{bmatrix} b_1 \\ b_2 \end{bmatrix} = \begin{bmatrix} 3 \\ 7 \end{bmatrix}$$

and

$$\begin{bmatrix} 2 & 1 \\ 3 & 2 \end{bmatrix}\begin{bmatrix} b_1 \\ b_2 \end{bmatrix} + \begin{bmatrix} 1 & 1 \\ -1 & -2 \end{bmatrix}\begin{bmatrix} a_1 \\ a_2 \end{bmatrix} = \begin{bmatrix} 6 \\ -1 \end{bmatrix}$$

This leads to the linear system

$$\begin{array}{rcrcrcrcr} 2a_1 & + & a_2 & - & b_1 & - & b_2 & = & 3 \\ 3a_1 & + & 2a_2 & + & b_1 & + & 2b_2 & = & 7 \\ a_1 & + & a_2 & + & 2b_1 & + & b_2 & = & 6 \\ -a_1 & - & 2a_2 & + & 3b_1 & + & 2b_2 & = & -1, \end{array}$$

which can be written as

$$\begin{bmatrix} 2 & 1 & -1 & -1 \\ 3 & 2 & 1 & 2 \\ 1 & 1 & 2 & 1 \\ -1 & -2 & 3 & 2 \end{bmatrix}\begin{bmatrix} a_1 \\ a_2 \\ b_1 \\ b_2 \end{bmatrix} = \begin{bmatrix} 3 \\ 7 \\ 6 \\ -1 \end{bmatrix}.$$

This linear system of four equations in four unknowns is now solved as in Chapter 1. The solution is (verify) $a_1 = 1$, $a_2 = 2$, $b_1 = 2$, and $b_2 = -1$. Thus $x_1 = 1 + 2i$ and $x_2 = 2 - i$ is the solution to the given linear system.

Determinants of Complex Matrices

The definition of a determinant and all the properties derived in Chapter 2 apply to matrices with complex entries. The following example is an illustration.

Example 4 Compute $|A|$ for the coefficient matrix A in Example 3.

Solution

$$\begin{vmatrix} 2 + i & 1 + i \\ 3 - i & 2 - 2i \end{vmatrix} = (2 + i)(2 - 2i) - (3 - i)(1 + i)$$
$$= (6 - 2i) - (4 + 2i)$$
$$= 2 - 4i.$$

Complex Vector Spaces

A **complex vector space** is defined exactly as was a real vector space in Definition 1 in Section 3.4, except that the scalars in properties (e) through (h) are permitted to be complex numbers. The terms *complex* vector space and *real* vector space emphasize the set from which the scalars are chosen. It happens that in order to satisfy the closure property of scalar multiplication [Definition 1(β) in Section 3.4] in a complex vector space, we must, in most examples, consider vectors that involve complex numbers.

Most of the real vector spaces of Chapter 3 have complex vector space analogues.

Example 5 (a) Consider C^n, the set of all $n \times 1$ matrices $\begin{bmatrix} a_1 \\ a_2 \\ \vdots \\ a_n \end{bmatrix}$ with complex entries. Let the opera-

tion $\oplus$ be matrix addition and let the operation $\odot$ be multiplication of a matrix by a complex number. We can verify that C^n is a complex vector space by using the properties of matrices established in Section 1.3 and the properties of complex arithmetic established in Section A.1. (Note that if the operation $\odot$ is taken as multiplication of a matrix by a real number, then C^n is a real vector space whose vectors have complex components.)

(b) The set of all $m \times n$ matrices with complex entries with matrix addition as $\oplus$ and multiplication of a matrix by a complex number as $\odot$ is a complex vector space (verify).

(c) The set of polynomials with complex coefficients with polynomial addition as $\oplus$ and multiplication of a polynomial by a complex constant as $\odot$ forms a complex vector space. Verification follows the pattern of Example 6 in Section 3.4.

(d) The set of complex-valued continuous functions defined on an interval $[a, b]$ (i.e., all functions of the form $f(t) = f_1(t) + if_2(t)$, where f_1 and f_2 are real-valued continuous functions on $[a, b]$), with $\oplus$ defined by $(f \oplus g)(t) = f(t) + g(t)$ and $\odot$ defined by $(c \odot f)(t) = c \cdot f(t)$ for a complex scalar c, forms a complex vector space. The corresponding real vector space is given in Example 14 in Section 3.4 for the interval $(-\infty, \infty)$.

A **complex vector subspace** W of a complex vector space V is defined as in Section 3.4, but with real scalars replaced by complex ones. The analogue of Theorem 3.8 can be proved to show that a nonempty subset W of a complex vector space V is a complex vector subspace if and only if the following conditions hold:

(a) If $\mathbf{X}$ and $\mathbf{Y}$ are any vectors in W, then $\mathbf{X} \oplus \mathbf{Y}$ is in W.

(b) If c is any complex number and $\mathbf{X}$ is any vector in W, then $c \odot \mathbf{X}$ is in W.

Example 6 (a) Let W be the set of all vectors in C^3 of the form $(a, 0, b)$ where a and b are complex numbers. It easily follows that

$$(a, 0, b) \oplus (d, 0, e) = (a + d, 0, b + e)$$

belongs to W and for any complex scalar c,

$$c \odot (a, 0, b) = (c \cdot a, 0, c \cdot b)$$

belongs to W. Hence W is a complex vector subspace of C^3.

(b) Let V be the vector space of all $m \times n$ matrices with complex entries, and let W be the set of all vectors in V having only real entries. If $\mathbf{A} = [a_{ij}]$ and $\mathbf{B} = [b_{ij}]$ belong to W, then so will $\mathbf{A} \oplus \mathbf{B}$, because if a_{ij} and b_{ij} are real, then so is their sum. However, if c is any complex scalar and $\mathbf{A}$ belongs to W, then $c \odot \mathbf{A} = c\mathbf{A}$ can have entries $c \cdot a_{ij}$ that need not be real numbers. It follows that $c \odot \mathbf{A}$ need not belong to W, so W is not a complex vector subspace.

Linear Independence and Basis in Complex Vector Spaces

The notions of linear combinations, spanning sets, linear dependence, linear independence, and basis are unchanged for complex vector spaces, except that we use complex scalars (see Sections 3.5 and 3.6).

Example 7 Let V be the complex vector space C^3. Let

$$\mathbf{X}_1 = (1, i, 0), \qquad \mathbf{X}_2 = (i, 0, 1 + i), \qquad \text{and} \qquad \mathbf{X}_3 = (1, 1, 1)$$

(a) Determine whether $\mathbf{X} = (-1, -3 + 3i, -4 + i)$ is a linear combination of $\{\mathbf{X}_1, \mathbf{X}_2, \mathbf{X}_3\}$.
(b) Determine whether $\{\mathbf{X}_1, \mathbf{X}_2, \mathbf{X}_3\}$ spans C^3.
(c) Determine whether $\{\mathbf{X}_1, \mathbf{X}_2, \mathbf{X}_3\}$ is a linearly independent subset of C^3.
(d) Is $\{\mathbf{X}_1, \mathbf{X}_2, \mathbf{X}_3\}$ a basis for C^3?

Solution
(a) We proceed as in Example 1 of Section 3.5. We form a linear combination of $\mathbf{X}_1$, $\mathbf{X}_2$, and $\mathbf{X}_3$ with unknown coefficients c_1, c_2, and c_3, respectively, and set it equal to $\mathbf{X}$:

$$c_1\mathbf{X}_1 + c_2\mathbf{X}_2 + c_3\mathbf{X}_3 = \mathbf{X}.$$

If we substitute the vectors $\mathbf{X}_1$, $\mathbf{X}_2$, $\mathbf{X}_3$, and $\mathbf{X}$ into this expression, we obtain the linear system

$$
\begin{aligned}
c_1 + \qquad ic_2 + c_3 &= -1 \\
ic_1 \qquad\qquad + c_3 &= -3 + 3i \\
(1 + i)c_2 + c_3 &= -4 + i.
\end{aligned}
$$

We next investigate the consistency of this system by using elementary row operations to transform its augmented matrix to either row echelon or reduced row echelon form. The row echelon form is (verify)

$$
\begin{bmatrix}
1 & i & 1 & \vdots & -1 \\
0 & 1 & 1 - i & \vdots & -3 + 4i \\
0 & 0 & 1 & \vdots & -3
\end{bmatrix},
$$

which implies that the system is consistent, hence $\mathbf{X}$ is a linear combination of $\mathbf{X}_1$, $\mathbf{X}_2$, $\mathbf{X}_3$. In fact, back substitution gives $a_1 = 3$, $a_2 = i$, and $a_3 = -3$.
(b) Let $\mathbf{X} = (a, b, c)$ be an arbitrary vector of C^3. We form the linear combination

$$c_1\mathbf{X}_1 + c_2\mathbf{X}_2 + c_3\mathbf{X}_3 = \mathbf{X}$$

and solve for c_1, c_2 and c_3. The resulting linear system is

$$
\begin{aligned}
c_1 + \quad\quad ic_2 + c_3 &= a \\
ic_1 \quad\quad\quad + c_3 &= b \\
(1 + i)c_2 + c_3 &= c
\end{aligned}
$$

Transforming the augmented matrix to row echelon form, we obtain (verify)

$$
\begin{bmatrix}
1 & i & 1 & \vdots & a \\
0 & 1 & 1 - i & \vdots & b - ia \\
0 & 0 & 1 & \vdots & -c + (1 + i)(b - ia)
\end{bmatrix}.
$$

Hence we can solve for c_1, c_2, c_3 for any choice of complex numbers a, b, c, which implies that $\{X_1, X_2, X_3\}$ spans C^3.

(c) Proceeding as in Example 8 of Section 3.5, we form the equation

$$
c_1 X_1 + c_2 X_2 + c_3 X_3 = 0
$$

and solve for c_1, c_2 and c_3. The resulting homogeneous system is

$$
\begin{aligned}
c_1 + \quad\quad ic_2 + c_3 &= 0 \\
ic_1 \quad\quad\quad + c_3 &= 0 \\
(1 + i)c_2 + c_3 &= 0.
\end{aligned}
$$

Transforming the augmented matrix to row echelon form, we obtain (verify)

$$
\begin{bmatrix}
1 & i & 1 & \vdots & 0 \\
0 & 1 & 1 - i & \vdots & 0 \\
0 & 0 & 1 & \vdots & 0
\end{bmatrix},
$$

and hence the only solution is $c_1 = c_2 = c_3 = 0$, showing that $\{X_1, X_2, X_3\}$ is linearly independent.

(d) Yes, because X_1, X_2, and X_3 span C^3 [part (b)] and they are linearly independent [part (c)].

Just as in real vector spaces, the questions of spanning sets, linearly independent or linearly dependent sets, and basis in a complex vector space are resolved by using an appropriate linear system. The definition of the dimension of a complex vector space is the same as that given in Section 3.6. In discussing the dimension of a complex vector space like C^n, we must adjust our intuitive picture. For example, C^1 consists of all complex multiples of a single nonzero vector. This collection can be put into a one-one correspondence with the complex numbers themselves, that is, all the points in the complex plane (see Figure A.1). Since the elements of a two-dimensional real vector space can be put into a one-one correspondence with the points of R^2 (see Section 3.1), we see that a complex vector space of dimension one has a geometric model that is in one-one correspondence with a geometric model of a two-dimensional real vector space. Similarly, a complex vector space of dimension two is the same, geometrically, as a four-dimensional real vector space.

Complex Eigenvalues and Eigenvectors

Let V be an n-dimensional complex vector space and let $\mathbf{L}: V \rightarrow V$ be a linear transformation. If $S = \{\mathbf{X}_1, \mathbf{X}_2, \ldots, \mathbf{X}_n\}$ is a basis for V such that $\mathbf{L}$ is represented by the $n \times n$ complex matrix $\mathbf{A}$, then for $\mathbf{X}$ in V.

$$\mathbf{L}(\mathbf{X}) = \mathbf{A}\mathbf{X}. \tag{1}$$

Imitating the definition in Section 5.1, we say that the complex number λ is an **eigenvalue** of $\mathbf{L}$ if there exists a nonzero vector $\mathbf{X}$ in V such that

$$\mathbf{L}(\mathbf{X}) = \lambda\mathbf{X}. \tag{2}$$

Every nonzero complex vector $\mathbf{X}$ satisfying this equation is called an **eigenvector** of $\mathbf{L}$ **associated with the eigenvalue** λ. Substituting Equation (1) into Equation (2), we have the relationship

$$\mathbf{A}\mathbf{X} = \lambda\mathbf{X}. \tag{3}$$

If λ is a complex scalar and $\mathbf{X} \neq \mathbf{0}$ is a vector in the complex vector space V such that Equation (3) is satisfied, then we say that λ is an **eigenvalue** of the complex matrix $\mathbf{A}$ and $\mathbf{X}$ is an **eigenvector** of $\mathbf{A}$ **associated with** λ. Equation (3) can be rewritten as the homogeneous system

$$(\lambda\mathbf{I}_n - \mathbf{A})\mathbf{X} = \mathbf{0}. \tag{4}$$

This homogeneous system has a nontrivial solution $\mathbf{X}$ if and only if

$$|\lambda\mathbf{I}_n - \mathbf{A}| = 0$$

has a solution. Following the development in Section 5.1, we will call $|\lambda\mathbf{I}_n - \mathbf{A}|$ the **characteristic polynomial** of the matrix $\mathbf{A}$, which is again a complex polynomial of degree n in λ. As in Section 5.1, the eigenvalues of the complex matrix $\mathbf{A}$ are the complex roots of the characteristic polynomial. According to the Fundamental Theorem of Algebra, an nth-degree polynomial has exactly n roots, if multiple roots are counted. Even if $|\lambda\mathbf{I}_n - \mathbf{A}|$ has only one root of multiplicity n, we then know that the characteristic polynomial of any complex matrix $\mathbf{A}$, or equivalently of any linear transformation $\mathbf{L}$ mapping a complex vector space into itself always has at least one root (possibly complex). This is quite different from the case of real vector spaces discussed in Section 5.1, because there are real matrices that do not have real eigenvalues (see Example 7 in Section 5.1).

Example 8 Let

$$\mathbf{A} = \begin{bmatrix} 0 & 1 \\ -1 & 0 \end{bmatrix}.$$

Then the characteristic polynomial of $\mathbf{A}$ is

$$|\lambda\mathbf{I}_2 - \mathbf{A}| = \lambda^2 + 1.$$

If we interpret **A** as representing a linear transformation between real vector spaces, then **A** has no (real) eigenvalues because the roots of $f(\lambda) = \lambda^2 + 1$ are i and $-i$. However, if we interpret **A** as representing a linear transformation between complex vector spaces, then we know that there is always at least one (complex) eigenvalue, and in this case the eigenvalues are i and $-i$. The corresponding (complex) eigenvectors are obtained by finding a nontrivial solution of the homogeneous systems

$$(i\mathbf{I}_2 - \mathbf{A})\mathbf{X} = \mathbf{0} \qquad \text{and} \qquad (-i\mathbf{I}_2 - \mathbf{A})\mathbf{X} = \mathbf{0},$$

respectively. An eigenvector associated with eigenvalue i is $\begin{bmatrix} 1 \\ i \end{bmatrix}$ and an eigenvector associated with eigenvalue $-i$ is $\begin{bmatrix} -1 \\ i \end{bmatrix}$. (Verify.)

Every real $n \times n$ matrix can be considered as an $n \times n$ complex matrix. Hence the eigenvalue and eigenvector computations for real matrices carried out in Chapter 5 can "be fixed up" by permitting complex roots of the characteristic polynomial to be called eigenvalues. Of course, this means that we would need to use complex arithmetic to solve for the associated eigenvector. However, this "fix" does not apply when we consider a linear transformation **L:** $V \rightarrow V$ where V is a real vector space, because a complex number with nonzero imaginary part does not belong to the set of scalars (the real numbers) in this case. Thus, if $\lambda = a + bi$, $b \neq 0$, is a root of the characteristic polynomial of **L**, the equation

$$\mathbf{L}(\mathbf{X}) = \lambda\mathbf{X}$$

cannot hold. However, the information derived from such a procedure can reveal fundamental properties of **L**.

So if, as in Chapter 5, we restrict ourselves to real vector spaces, we must permit only real numbers to be eigenvalues. The eigen-problem for symmetric matrices, Section 5.2, needs no "fix" because the eigenvalues of any real symmetric matrix are all real numbers (see Theorem 5.4).

A.2 EXERCISES

1. Solve using Gauss–Jordan reduction.

    (a) $(1 + 2i)x_1 + (-2 + i)x_2 = 1 - 3i$
    $(2 + i)x_1 + (-1 + 2i)x_2 = -1 - i.$

    (b) $2ix_1 - (1 - i)x_2 = 1 + i$
    $(1 - i)x_1 + x_2 = 1 - i.$

    (c) $(1 + i)x_1 - x_2 = -2 + i$
    $2ix_1 + (1 - i)x_2 = i.$

2. Transform the given augmented matrix of a linear system to row echelon form and solve by back substitution.

    (a) $\left[\begin{array}{ccc:c} 2 & i & 0 & 1 - i \\ 0 & 3i & -2 + i & 4 \\ 0 & 0 & 2 + i & 2 - i \end{array} \right].$

    (b) $\left[\begin{array}{ccc:c} i & 2 & 1 + i & 3i \\ 0 & 1 - i & 0 & 2 + i \\ 0 & 0 & 3 & 6 - 3i \end{array} \right].$

3. Solve by Gaussian elimination with back substitution.

(a) $ix_1 + (1 + i)x_2 \qquad = i$
$(1 - i)x_1 + \qquad x_2 - ix_3 = 1$
$\qquad ix_2 + \quad x_3 = 1.$

(b) $x_1 + \quad ix_2 + (1 - i)x_3 = 2 + i$
$ix_1 \qquad + (1 + i)x_3 = -1 + i$
$\qquad 2ix_2 - \qquad x_3 = 2 - i.$

4. Compute the determinant and simplify as much as possible.

(a) $\begin{vmatrix} 1 + i & -1 \\ 2i & 1 + i \end{vmatrix}$.   (b) $\begin{vmatrix} 2 - i & 1 + i \\ 1 + 2i & -(1 - i) \end{vmatrix}$.

(c) $\begin{vmatrix} 1 + i & 2 & 2 - i \\ i & 0 & 3 + i \\ -2 & 1 & 1 + 2i \end{vmatrix}$.   (d) $\begin{vmatrix} 2 & 1 - i & 0 \\ 1 + i & -1 & i \\ 0 & -i & 2 \end{vmatrix}$.

5. Find the inverse of each of the following matrices if possible.

(a) $\begin{bmatrix} i & 2 \\ 1 + i & -i \end{bmatrix}$.   (b) $\begin{bmatrix} 2 & i & 3 \\ 1 + i & 0 & 1 - i \\ 2 & 1 & 2 + i \end{bmatrix}$.

6. Determine whether or not the following subsets W of the vector space of all 2×2 matrices with complex entries are complex vector subspaces.

(a) W is the set of all 2×2 complex matrices with zeros on the main diagonal.

(b) W is the set of all 2×2 complex matrices that have diagonal entries with real part equal to zero.

(c) W is the set of all symmetric 2×2 complex matrices.

7. Let $W = \text{span } \{X_1, X_2, X_3\}$, where

$$X_1 = (-1 + i, 2, 1), \quad X_2 = (1, 1 + i, i)$$
$$X_3 = (-5 + 2i, -1 - 3i, 2 - 3i).$$

(a) Does $X = (i, 0, 0)$ belong to W?

(b) Is the set $\{X_1, X_2, X_3\}$ linearly independent or linearly dependent?

8. Let X_1, X_2, X_3 be a basis for a complex vector space V. Determine whether Y is in span $\{Y_1, Y_2\}$.

(a) $Y_1 = iX_1 + (1 - i)X_2 + 2X_3$
$Y_2 = (2 + i)X_1 + 2iX_2 + (3 - i)X_3$
$Y = (-2 - 3i)X_1 + (3 - i)X_2 + (-2 - 2i)X_3.$

(b) $Y_1 = 2iX_1 + X_2 + (1 - i)X_3$
$Y_2 = 3iX_1 + (1 + i)X_2 + 3X_3$
$Y = (2 + 3i)X_1 + (2 + i)X_2 + (4 - 2i)X_3.$

9. Find the eigenvalues and associated eigenvectors of the following complex matrices.

(a) $A = \begin{bmatrix} 1 & 1 \\ -1 & 1 \end{bmatrix}$.   (b) $A = \begin{bmatrix} 1 & i \\ -i & 1 \end{bmatrix}$.

(c) $A = \begin{bmatrix} 2 & 0 & 0 \\ 0 & 2 & i \\ 0 & -i & 2 \end{bmatrix}$.

10. For each of the parts in Exercise 9, find a matrix P such that $P^{-1}AP = D$, a diagonal matrix. For part (c), find three different matrices P that diagonalize A.

THEORETICAL EXERCISES

T.1. An $n \times n$ complex matrix A is called **Hermitian** if $\overline{A^T} = A$. This is equivalent to saying that $a_{ij} = \overline{a_{ji}}$ for all i and j.

(a) Prove that the diagonal entries of a Hermitian matrix must be real.

(b) Prove that every Hermitian matrix A can be written as $A = B + iC$, where B is real and symmetric and C is real and skew symmetric (see Exercise T.14 in Section 1.3). [*Hint:* Consider $B = (A + \overline{A})/2$ and $C = (A - \overline{A})/2i$.]

(c) Prove or disprove: The set W of all $n \times n$ Hermitian matrices is a complex vector subspace of the vector space of all $n \times n$ matrices with complex entries.

(d) Prove or disprove: The set W of all $n \times n$ Hermitian matrices is a real vector subspace of the real vector space of all $n \times n$ matrices with complex entries.

(e) Prove that every real symmetric matrix is Hermitian.

T.2. An $n \times n$ complex matrix $\mathbf{A}$ is called **unitary** if $(\overline{\mathbf{A}^T})\mathbf{A} = \mathbf{A}(\overline{\mathbf{A}^T}) = \mathbf{I}_n$. This is equivalent to saying that $\overline{\mathbf{A}^T} = A^{-1}$.

   (a) If $\mathbf{A}$ is unitary, show that $\mathbf{A}^T$ is unitary.

   (b) If $\mathbf{A}$ is unitary, show that $\mathbf{A}^{-1}$ is unitary.

   (c) Prove or disprove: The set W of all $n \times n$ unitary matrices is a complex vector subspace of the vector space of all $n \times n$ matrices with complex entries.

T.3. (a) Prove that if $\mathbf{A}$ is Hermitian (see Exercise T.1), then the eigenvalues of $\mathbf{A}$ are real.

   (b) Verify that $\mathbf{A}$ in part (c) of Exercise 9 is Hermitian.

   (c) Are the eigenvectors associated with an eigenvalue of a Hermitian matrix guaranteed to be real vectors? Explain.

T.4. Prove that an $n \times n$ complex matrix $\mathbf{A}$ is unitary (see Exercise T.2) if and only if the columns (and rows) of $\mathbf{A}$ form an orthonormal set with respect to the complex dot product in C^n. (*Hint:* See Theorem 5.6.)

T.5. Let $\mathbf{A}$ be an $n \times n$ complex matrix.

   (a) Show that $\mathbf{A}$ can be written as $\mathbf{B} + i\mathbf{C}$, where $\mathbf{B}$ and $\mathbf{C}$ are Hermitian.

   (b) $\mathbf{A}$ is called **normal** if

$$\overline{\mathbf{A}^T}\mathbf{A} = \mathbf{A}\overline{\mathbf{A}^T}$$

   Show that $\mathbf{A}$ is normal if and only if

$$\mathbf{BC} = \mathbf{CB}$$

   (*Hint:* Consider $\mathbf{B} = (\mathbf{A} + \overline{\mathbf{A}^T})/2$ and $\mathbf{C} = (\mathbf{A} - \overline{\mathbf{A}^T})/2i$.)

T.6. (a) Prove that any Hermitian matrix is normal.

   (b) Prove that any unitary matrix is normal.

   (c) Find a 2×2 normal matrix that is neither symmetric, Hermitian, nor unitary.

Appendix B:
The Computer in
Linear Algebra

It would be difficult at this time to ignore the use of the computer in linear algebra, since most computations involving linear algebra in real problems are carried out on computers. This relationship is briefly explored in this appendix. First, we provide a list of 63 *computer projects* grouped by sections corresponding to the text. These projects are of widely differing degrees of difficulty, and the student is expected to have adequate programming skills to handle these tasks. Many of them can be carried out on microcomputers and on hand-held programmable calculators. Next, we discuss the availability of *canned software* to implement linear algebra techniques. The use of many of these programs requires some programming skill, although some of them only require that the user know how to use a terminal for entering the data. Finally, we outline some of the features of APL, a relatively new programming language, which is especially suitable for matrix manipulations and is easy to learn, and we make several remarks about BASIC, a general-purpose computer language, which is being widely used on microcomputers and on time-sharing systems.

B.1

COMPUTER PROJECTS

Write programs to carry out the following projects.

CHAPTER 1

Section 1.2
1. Add two matrices.
2. Multiply a matrix by a scalar.

3. Multiply two matrices.
4. Form the transpose of a matrix.

Section 1.4

5. Transform an $m \times n$ matrix to reduced row echelon form.
6. Solve a linear system by Gauss–Jordan reduction.

Section 1.5

7. Find the inverse of a given $n \times n$ matrix.

CHAPTER 2

Section 2.1

8. Given n, list all the permutations of $S = \{1, 2, \ldots, n\}$ and determine which are even and which are odd.

Section 2.2

9. Evaluate the determinant of a given $n \times n$ matrix.
10. Find the inverse of a matrix by cofactors (inefficient for $n > 4$).
11. Solve $\mathbf{AX} = \mathbf{B}$, where $\mathbf{A}$ is $n \times n$, by Cramer's rule (inefficient for $n > 4$).

CHAPTER 3

Section 3.2

12. Find the inner product of two vectors in R^n.
13. Find the cosine of the angle and the angle between two nonzero vectors in R^n.
14. Find the length of a given vector in R^n.
15. Determine whether a triangle with three given vertices in R^3 is a right triangle.
16. Determine whether a triangle with three given vertices in R^3 is isosceles.
17. Determine whether a quadrilateral with four given vertices in R^3 is a parallelogram.
18. Determine whether a quadrilateral with four given vertices in R^3 is a rectangle.
19. Determine whether a quadrilateral with four given vertices in R^3 is a square.

Section 3.3

20. Find the cross product of two vectors in R^3.

Section 3.5

21. Test a given set of vectors in R^n for linear independence.
22. Determine whether a given vector $\mathbf{X}$ in R^n is a linear combination of a given set of vectors in R^n.
23. If a given vector $\mathbf{X}$ in R^n is a linear combination of a given set of vectors in R^n, find the coefficients of the linear combination.

Section 3.6

24. Extend a given set of linearly independent vectors in R^n to a basis for R^n.
25. If a given set of r vectors in R^n spans a subspace V of R^n of dimension m, find a basis for V.
26. Find a basis for the solution space of a given homogeneous system $\mathbf{AX} = \mathbf{0}$.

Section 3.7

27. Find the rank of a given $m \times n$ matrix.

Section 3.8

28. Transform a given basis for a subspace V of R^n to an orthonormal basis for V by the Gram–Schmidt process.

CHAPTER 4

Section 4.1

29. Given an $n \times n$ matrix, code a given message and decode a given message.

Section 4.2

30. Find a basis for the kernel of a linear transformation $\mathbf{L}: R^n \to R^m$ given by $\mathbf{L(X)} = \mathbf{AX}$.
31. Find a basis for the range of a linear transformation $\mathbf{L}: R^n \to R^m$ given by $\mathbf{L(X)} = \mathbf{AX}$.

Section 4.3

32. Find the coordinate vector of a given vector $\mathbf{X}$ with respect to a given basis S for R^n.
33. Let $\mathbf{L}: R^3 \to R^4$ be the linear transformation defined by.

$$
\mathbf{L}\left(\begin{bmatrix} a_1 \\ a_2 \\ a_3 \end{bmatrix}\right) = \begin{bmatrix} a_1 + 2a_2 - a_3 \\ 2a_1 + a_3 \\ 3a_1 - a_2 \\ 2a_1 + a_2 + a_3 \end{bmatrix}.
$$

Let

$$
S = \left\{ \begin{bmatrix} 1 \\ 1 \\ 0 \end{bmatrix}, \begin{bmatrix} 0 \\ 1 \\ 2 \end{bmatrix}, \begin{bmatrix} 1 \\ 1 \\ 1 \end{bmatrix} \right\} \quad \text{and} \quad T = \left\{ \begin{bmatrix} 1 \\ 2 \\ -1 \\ 0 \end{bmatrix}, \begin{bmatrix} 0 \\ 0 \\ 1 \\ 0 \end{bmatrix}, \begin{bmatrix} 1 \\ 1 \\ 0 \\ 0 \end{bmatrix}, \begin{bmatrix} 0 \\ 1 \\ 1 \\ 1 \end{bmatrix} \right\}
$$

be bases for R^3 and R^4, respectively. Find the matrix representing $\mathbf{L}$ with respect to S and T.

Section 4.4

34. Given a 2×2 matrix $\mathbf{A}$ and a sample of points in the plane, compute the image of the set of points in the sample.

CHAPTER 5

Section 5.2

35. Determine if a given matrix is orthogonal.

CHAPTER 6

Section 6.2

36. Solve a standard linear programming problem by the simplex method.

Section 6.3

37. Set up the dual of a given standard linear programming problem.
38. Set up and solve the dual of a standard linear programming problem.

CHAPTER 7

Section 7.1

39. Find an equation of the line in R^2 passing through the given points $P_1(a_1, b_1)$, $P_2(a_2, b_2)$.
40. Find parametric equations of the line in R^3 through two given points.
41. Find equations in symmetric form of a line in R^3 through two given points.
42. Find an equation of the plane passing through three given points.

Section 7.2

43. Diagonalize a given quadratic form. (*Hint:* Involves finding eigenvalues numerically.)

Section 7.3

44. Given the adjacency matrix of a digraph with n vertices, find the number of ways in which a specified vertex P_i influences another specified vertex P_j in a specified number, r, of stages.
45. Given the adjacency matrix of a digraph G, find all cliques in G.
46. Given the adjacency matrix of a digraph G, determine whether G is strongly connected.
47. Given the adjacency matrix of a digraph G with n vertices, determine if there is a path from a specified vertex P_i to another specified vertex P_j.
48. Given the adjacency matrix of a digraph G with n vertices, determine the number of paths from a specified vertex P_i to another specified vertex P_j.

Section 7.4

49. Determine whether a given matrix game is strictly determined. If it is, find all saddle points.
50. Solve a matrix game with given 2×2 payoff matrix.

Section 7.5

51. Find the line of best fit for a given set of points.

Section 7.6

52. Given an exchange matrix $\mathbf{A}$, find a vector $\mathbf{P} \geq \mathbf{0}$ with at least one positive component such that $(\mathbf{I}_n - \mathbf{A})\mathbf{P} = \mathbf{0}$.

Section 7.7

53. Determine if a given transition matrix $\mathbf{T}$ is regular by computing powers of $\mathbf{T}$.
54. Given a regular transition matrix $\mathbf{T}$, compute the matrix $\mathbf{A}$ approached by $\mathbf{T}^n$ as $n \to \infty$.
55. Find the steady-state vector of a given regular matrix.

Section 7.8

56. Compute the Fibonacci sequence using the recursion relation $u_n = u_{n-1} + u_{n-2}$.
57. Compute the Fibonacci number u_n directly.

Section 7.9

58. Compute the solution to the homogeneous linear system $\mathbf{X}' = \mathbf{A}\mathbf{X}$, where $\mathbf{A}$ is a given $n \times n$ matrix. (*Hint:* Involves finding eigenvalues and eigenvectors.)

CHAPTER 8

Section 8.2

59. Reduce an $m \times n$ matrix to row echelon form.
60. Implement Gaussian elimination with partial pivoting.
61. Implement the **LU**-factorization method discussed in Section 8.3.
62. Give a linear system $\mathbf{A}\mathbf{X} = \mathbf{B}$, where $\mathbf{A}$ is nonsingular, implement the solution technique where $\mathbf{A}$ is first factored into its **LU**-decomposition, then the system is solved by a forward substitution followed by a back substitution.
63. Implement the Jacobi iteration method.
64. Implement the Gauss-Seidel iteration method.

Section 8.3

65. Implement the power method for finding the eigenvalue of largest absolute value and a corresponding eigenvector of a given matrix.
66. Implement the Jacobi method for finding the eigenvalues and eigenvectors of a given symmetric matrix.

B.2

LINEAR ALGEBRA SOFTWARE

There is an enormous amount of software implementing linear algebra techniques. Some of the codes available are well written and others are not. Care must be exercised in selecting software for use in the classroom as well as for scientific or engineering calculations. While the goals of classroom and scientific software vary, the underlying algorithms should exhibit reliability in the environment for which they are intended. Some software can be used for both classroom and professional calculations.

As the power and sophistication of personal computers increase, software originally developed for mainframe or minicomputers has been transported to the microcomputer. The increased availability of microcomputers has also stimulated the development of instructional software in various areas of mathematics. There are microcomputer software packages specifically designed for linear algebra. We give a short description of several software packages and indicate in Table B.1 computing environments in which they are currently available. There are numerous other packages available.

EISPACK and LINPACK

A collection of subroutines for solving eigenvalue and eigenvector problems, and problems involving linear systems, respectively. Both codes reflect fundamental theoretical developments and much practical experience. The codes are efficient, reliable, and are machine independent. Originally developed for mainframe computers, EISPACK and LINPACK represented a major step in the development of software for numerical analysis. FORTRAN source code is available on magnetic tape for a nominal charge from IMSL: International Mathematical and Statistical Libraries, 2500 ParkWest Tower One, 2500 City West Boulevard, Houston, TX 77042-3020.

IMSL

International Mathematical and Statistical Libraries has several products that contain linear algebra codes. **IMSL subroutine libraries** is a collection of over 500 mathematical and statistical subroutines applicable to a wide variety of applications. A substantial segment of this library deals with linear algebra computations. Many computer centers lease this library for their users. **MATH/PC-LIBRARY** is a similar collection of subroutines for use on an IBM microcomputer or compatible machine. **MATH/PROTRAN** also has linear system and eigenvalue and eigenvector codes incorporated into a problem-solving interface to library routines. IMSL, 2500 ParkWest Tower One, 2500 City West Boulevard, Houston, TX 77042-3020.

NAG

The Numerical Algorithms Group has several products that contain linear algebra codes. Collections of subroutines for a wide variety of applications are available in **NAG Fortran Library**, **NAG Algol Library**, and **NAG Pascal Library**. Libraries are available for mainframes, minicomputers, workstations, and supercomputers. The **NAG PC50 Library** is available for IBM and compatible microcomputers. National Algorithms Group Inc., 1101 31st Street, Suite 100, Downers Grove, IL 60515-1263.

Math Advantage

A library of over 170 scientific subroutines in C or FORTRAN. Available for IBM PCs, DEC VAX, and supercomputers. Primarily for professional applications. Quantitative Technology Corporation, 8700 SW Creekside Place, Beaverton, OR 97005.

MATLAB

Designed as a ''matrix laboratory'' for both classroom and professional applications involving linear algebra. The software is based on both the LINPACK and EISPACK projects, and evolved under the leadership of Cleve Moler. MATLAB features easy access to routines with no programming required. Complex arithmetic is handled automatically. Commands are short, mnemonic, and similar to the way linear algebra is written. Your own routines or commands can be added to MATLAB; hence it can be tailored to individual instructional or computational needs. MATLAB is available in mainframe, minicomputer, and microcomputer versions. Currently, the microcomputer version runs on IBM and compatible machines, and requires a mathematics co-processor chip and a minimum of 320K of memory. An attractive feature is the availability of both two- and three-dimensional graphics. A Macintosh version is under development. An exciting product with potential for use in many areas of mathematics instruction. The MathWorks, Inc., Suite 250, 20 North Main St., Sherborn, MA 01770.

Table B.1

	Mainframe/ Minicomputer	Micro- computer	Super- computer
EISPACK	×		
LINPACK	×		
IMSL	×	×	
NAG	×	×	×
Math Advantage	×	×	×
MATLAB	×	×	

FURTHER READING

Dongarra, J. J., J. R. Bunch, C. B. Moler, and **G. W. Stewart.** *LINPACK User's Guide.* Philadelphia: Society of Industrial and Applied Mathematics, 1979.

Garbow, B. S., J. M. Boyle, J. J. Dongarra, and **C. B. Moler.** *Matrix Eigensystem Routines— EISPACK Guide Extension,* Lecture Notes in Computer Science, Vol. 51. New York: Springer-Verlag, New York, Inc., 1977.

Hill, David R. *Experiments in Computational Matrix Algebra.* New York: Random House/Birkhäuser, 1987.

Rice, J. R. *Matrix Computations and Mathematical Software.* New York: McGraw-Hill Book Company, 1981.

B.3

APL

APL (**A** **P**rogramming **L**anguage) is a relatively new programming language, developed in 1962 by Kenneth Iverson. APL is an **interactive language** in the sense that the user communicates with the computer by means of a special typewriter **terminal** or cathode-ray-tube **terminal** linked to the computer by telephone lines. The terminal is used to enter the instructions to the computer as well as to receive the answers, which are usually received within seconds after the last instruction has been entered.

An easy language to learn, APL is especially suitable for matrix manipulations. We give several examples to illustrate the capability of the language in linear algebra.

Assign

The symbol ← means that we assign what follows it to what precedes it. This symbol is used to enter data into the computer: Thus

$$X \leftarrow 4$$

means that X is assigned the value 4.

Vectors

In APL a vector is denoted by a string of numbers. Thus if we type in

$$X \leftarrow 2\ 0\ 1\ 3$$

we enter the vector **X** = (2, 0, 1, 3) into the computer.

Vector Operations

To add vectors **X** and **Y**, we write $X + Y$

To form the product of a vector **X** by the scalar K, we write $K \times X$ or $X \times K$

Thus, if **X** = (2, 1, 0, 3) and **Y** = (3, 2, 1, 4), and we type in

$$X \leftarrow 2\ 1\ 0\ 3$$
$$Y \leftarrow 3\ 2\ 1\ 4$$
$$X + Y$$

we have formed **X** + **Y**. If we type in

$$X \leftarrow 2\ 1\ 0\ 3$$
$$K \leftarrow 3$$
$$K \times X$$

we form $K\mathbf{X}$.

Matrices

To enter the 3 × 2 matrix

$$\mathbf{A} = \begin{bmatrix} 2 & 1 \\ 3 & 2 \\ 0 & 1 \end{bmatrix}$$

into the computer we type in

$$A \leftarrow 3\ 2\rho\ 2\ 1\ 3\ 2\ 0\ 1$$

This instruction tells the computer to assign to **A** the 3 × 2 matrix with entries listed row by row after the special symbol ρ.

Matrix Operations

To form the sum of the matrices **A** and **B**, we write $A + B$

To form the product of a matrix **A** by a scalar K, we write $K \times A$

To form the product of the matrices **A** and **B**, we write $A + \cdot \times B$

To form the inverse of a square matrix **A**, we write $\boxdot A$

Thus let

$$\mathbf{A} = \begin{bmatrix} 2 & 1 & 3 \\ 4 & 6 & 4 \end{bmatrix}, \qquad \mathbf{B} = \begin{bmatrix} 2 & 6 & 8 \\ 3 & 5 & 2 \end{bmatrix},$$

$$\mathbf{C} = \begin{bmatrix} 2 & 3 \\ 4 & 1 \\ 2 & 5 \end{bmatrix}, \qquad \mathbf{D} = \begin{bmatrix} 1 & 2 \\ 3 & 4 \end{bmatrix},$$

and $K = 2$.

To form $\mathbf{A} + \mathbf{B}$, we type in

$$
\begin{array}{l}
\text{A} \leftarrow 2\ 3\rho\ 2\ 1\ 3\ 4\ 6\ 4 \\
\text{B} \leftarrow 2\ 3\rho\ 2\ 6\ 8\ 3\ 5\ 2 \\
\text{A} + \text{B}
\end{array}
$$

To form $K\mathbf{A}$, we type in

$$
\begin{array}{l}
\text{A} \leftarrow 2\ 3\rho\ 2\ 1\ 3\ 4\ 6\ 4 \\
\text{K} \leftarrow 2 \\
\text{K} \times \text{A}
\end{array}
$$

To form $\mathbf{AC}$, we type in

$$
\begin{array}{l}
\text{A} \leftarrow 2\ 3\rho\ 2\ 1\ 3\ 4\ 6\ 4 \\
\text{C} \leftarrow 3\ 2\rho\ 2\ 3\ 4\ 1\ 2\ 5 \\
\text{A} + \cdot \times \text{C}
\end{array}
$$

To form $\mathbf{D}^{-1}$, we type in

$$
\begin{array}{l}
\text{D} \leftarrow 2\ 2\rho\ 1\ 2\ 3\ 4 \\
\boxdot \ \text{D}
\end{array}
$$

In APL it is quite simple to solve a linear system of n equations in n unknowns. Thus, if

$$
\mathbf{E} = \begin{bmatrix} 1 & 2 \\ 3 & 4 \end{bmatrix} \quad \text{and} \quad \mathbf{F} = \begin{bmatrix} 8 \\ 9 \end{bmatrix},
$$

then to solve $\mathbf{EX} = \mathbf{F}$ we type in

$$
\begin{array}{l}
\text{E} \leftarrow 2\ 2\rho\ 1\ 2\ 3\ 4 \\
\text{F} \leftarrow 8\ 9 \\
\text{F} \boxdot \text{E}
\end{array}
$$

There are many other instructions in APL that enable the user to readily manipulate vectors and matrices.

FURTHER READING

Gilman, Leonard, and **Allen J. Rose.** *APL: An Interactive Approach,* 3rd ed. New York: John Wiley & Sons, Inc., 1980.

B.4

BASIC

BASIC is a general-purpose computer language that is being widely used on micro-computers and on time-sharing systems. Typical matrix operations are

MAT C = A + B	Form the sum **C** of two given matrices **A** and **B**.
MAT C = (K) ∗ A	Form the scalar multiple **C** of a given matrix **A** by a given scalar K.
MAT C = A ∗ B	Form the product **C** of two given matrices **A** and **B**.
MAT C = TRN(A)	Form the transpose **C** of a given matrix **A**.
MAT C = INV(A)	Form the inverse **C** of a given matrix **A**.

FURTHER READING

Kemeny, John G., and **Thomas E. Kurtz.** *BASIC Programming,* 3rd ed. New York: John Wiley & Sons, Inc., 1980.

Answers to Odd-Numbered Exercises and Chapter Tests

CHAPTER 1

Section 1.1, page 9

1. $x = 4$, $y = 2$.

3. $x = -4$, $y = 2$, $z = 10$.

5. $x = 2$, $y = -1$, $z = -2$.

7. $x = -20$, $y = \frac{1}{4}z + 8$, $z =$ any real number.

9. No solution.

11. $x = 5$, $y = 1$.

13. No solution.

15. 20 tons of each type of fuel.

17. 3.2 ounces of A, 4.2 ounces of B, 2.0 ounces of C.

Section 1.2, page 21

1. (a) $-3, -5, 4$.   (b) $4, 5$.   (c) $2, 6, -1$.

3. (a) $\begin{bmatrix} 58 & 4 \\ 66 & 4 \end{bmatrix}$.   (b) $\begin{bmatrix} 28 & 8 & 38 \\ 34 & 4 & 41 \end{bmatrix}$.

(c) $\begin{bmatrix} 5 & 10 & 15 \\ 10 & 5 & 20 \end{bmatrix}$.   (d) $\begin{bmatrix} -2 & 17 \\ 16 & 31 \end{bmatrix}$.

(e) Impossible.   (f) $\begin{bmatrix} -36 & 12 & -36 \\ -48 & -12 & -60 \\ -24 & -12 & -36 \end{bmatrix}$.

5. (a) $\begin{bmatrix} 1 & 2 \\ 2 & 1 \\ 3 & 4 \end{bmatrix}$, $\begin{bmatrix} 1 & 2 & 3 \\ 2 & 1 & 4 \end{bmatrix}$.

(b) $\begin{bmatrix} 5 & 4 & 5 \\ -5 & 2 & 3 \\ 8 & 9 & 4 \end{bmatrix}$.   (c) $\begin{bmatrix} 14 & 16 \\ 8 & 9 \end{bmatrix}$.

(d) $\begin{bmatrix} -6 & 10 \\ 11 & 17 \end{bmatrix}$.   (e) Impossible.

(f) $\begin{bmatrix} 18 & 6 & 25 \\ 10 & 4 & 15 \end{bmatrix}$.

7. $\mathbf{DI} = \mathbf{ID} = \mathbf{D}$.

9. $a = 0$, $b = 2$, $c = 1$, $d = 2$.

11. $\mathbf{A0} = \mathbf{0}$.

13. $\begin{aligned} -2x - y + \quad\quad 4w &= 5 \\ -3x + 2y + 7z + 8w &= 3 \\ x + \quad\quad\quad 2w &= 4 \\ 3x + \quad\quad z + 3w &= 6. \end{aligned}$

15. (a) $\begin{bmatrix} 3 & -1 & 2 \\ 2 & 1 & 0 \\ 0 & 1 & 3 \\ 4 & 0 & -1 \end{bmatrix}$.

(b) $\begin{bmatrix} 3 & -1 & 2 \\ 2 & 1 & 0 \\ 0 & 1 & 3 \\ 4 & 0 & -1 \end{bmatrix} \begin{bmatrix} x \\ y \\ z \end{bmatrix} = \begin{bmatrix} 4 \\ 2 \\ 7 \\ 4 \end{bmatrix}$.

(c) $\left[\begin{array}{ccc|c} 3 & -1 & 2 & 4 \\ 2 & 1 & 0 & 2 \\ 0 & 1 & 3 & 7 \\ 4 & 0 & -1 & 4 \end{array}\right]$.

17. (a) $\begin{bmatrix} 1 \\ 14 \\ 0 \\ 13 \end{bmatrix}$. (b) $\begin{bmatrix} 0 \\ 18 \\ 3 \\ 13 \end{bmatrix}$.

19. For each product P or Q, the daily cost of pollution control at plant X or at plant Y.

21. (a) \$103,400. (b) \$16,050.

Section 1.3, page 32

1. $\mathbf{A} + \mathbf{B} = \begin{bmatrix} 3 & 2 & -1 \\ 6 & 2 & 10 \end{bmatrix}$,

$\mathbf{A} + \mathbf{B} + \mathbf{C} = \begin{bmatrix} -1 & -4 & 0 \\ 8 & 5 & 10 \end{bmatrix}$.

3. $\mathbf{A}(\mathbf{B} + \mathbf{C}) = \begin{bmatrix} -10 & -8 & 16 \\ 10 & 14 & -28 \end{bmatrix}$.

5. $\mathbf{A}(r\mathbf{B}) = \begin{bmatrix} -6 & 18 & -42 \\ 9 & -27 & 0 \end{bmatrix}$.

7. $(\mathbf{AB})^T = \begin{bmatrix} 11 & 5 \\ 15 & -4 \end{bmatrix}$.

11. (a) $\begin{bmatrix} 30 & 20 \\ 10 & 20 \end{bmatrix}$. (b) $\begin{bmatrix} 247 & 206 \\ 103 & 144 \end{bmatrix}$.

13. (a) $\begin{bmatrix} \frac{16}{45} \\ \frac{29}{45} \end{bmatrix}$. (b) $\begin{bmatrix} \frac{3}{8} \\ \frac{5}{8} \end{bmatrix}$.

Section 1.4, page 47

1. A, E, G.

3. (a) $\begin{bmatrix} 2 & 0 & 4 & 2 \\ -1 & 3 & 1 & 1 \\ 3 & -2 & 5 & 6 \end{bmatrix}$.

(b) $\begin{bmatrix} 2 & 0 & 4 & 2 \\ -12 & 8 & -20 & -24 \\ -1 & 3 & 1 & 1 \end{bmatrix}$.

(c) $\begin{bmatrix} 0 & 6 & 6 & 4 \\ 3 & -2 & 5 & 6 \\ -1 & 3 & 1 & 1 \end{bmatrix}$.

5. Possible answers:

(a) $\begin{bmatrix} 4 & 3 & 7 & 5 \\ 2 & 0 & 1 & 4 \\ -2 & 4 & -2 & 6 \end{bmatrix}$.

(b) $\begin{bmatrix} 3 & 5 & 6 & 8 \\ -4 & 8 & -4 & 12 \\ 2 & 0 & 1 & 4 \end{bmatrix}$.

(c) $\begin{bmatrix} 4 & 3 & 7 & 5 \\ -1 & 2 & -1 & 3 \\ 0 & 4 & -1 & 10 \end{bmatrix}$.

7. $\begin{bmatrix} 1 & 0 & 0 & 0 \\ 0 & 1 & 0 & 0 \\ 0 & 0 & 1 & 0 \\ 0 & 0 & 0 & 1 \end{bmatrix}$.

9. (a) $x = -2 + r$, $y = -1$, $z = 8 - 2r$, $w = r$, $r = $ any real number.
(b) $x = 1$, $y = \frac{2}{3}$, $z = -\frac{2}{3}$.
(c) No solution.

11. (a) $a = -2$. (b) $a \neq \pm 2$. (c) $a = 2$.

13. (a) $a = \pm\sqrt{6}$. (b) $a \neq \pm\sqrt{6}$. (c) None.

15. (a) $x = -1$, $y = 4$, $z = -3$.
  (b) $x = 0$, $y = 0$, $z = 0$.

17. (a) $x = 1 - r$, $y = 2$, $z = 1$, $w = r$,
  r = any real number. (b) No solution.

19. $x = -r$, $y = 0$, $z = r$, r = any real number.

21. 30 chairs, 30 coffee tables, and 20 dining-room tables.

Section 1.5, page 62

1. $\mathbf{A}^{-1} = \begin{bmatrix} \frac{3}{8} & -\frac{1}{8} \\ \frac{1}{4} & \frac{1}{4} \end{bmatrix}$.

3. Nonsingular; $\mathbf{A}^{-1} = \begin{bmatrix} 4 & -1 \\ -3 & 1 \end{bmatrix}$.

5. (a) $\begin{bmatrix} \frac{1}{2} & -\frac{1}{4} \\ \frac{1}{6} & \frac{1}{12} \end{bmatrix}$. (b) $\begin{bmatrix} 0 & 1 & -1 \\ 2 & -2 & -1 \\ -1 & 1 & 1 \end{bmatrix}$.

(c) $\begin{bmatrix} \frac{7}{3} & -\frac{1}{3} & -\frac{1}{3} & -\frac{2}{3} \\ \frac{4}{9} & -\frac{1}{9} & -\frac{4}{9} & \frac{1}{9} \\ -\frac{1}{9} & -\frac{2}{9} & \frac{1}{9} & \frac{2}{9} \\ -\frac{5}{3} & \frac{2}{3} & \frac{2}{3} & \frac{1}{3} \end{bmatrix}$.

7. (a) $\begin{bmatrix} -2 & \frac{3}{2} \\ 1 & -\frac{1}{2} \end{bmatrix}$. (b) Singular.

(c) $\begin{bmatrix} \frac{3}{2} & -1 & \frac{1}{2} \\ \frac{1}{2} & 0 & -\frac{1}{2} \\ -\frac{3}{2} & 1 & \frac{1}{2} \end{bmatrix}$.

9. (a) Singular. (b) $\begin{bmatrix} 1 & -1 & 0 \\ 1 & -2 & 1 \\ -\frac{3}{2} & \frac{5}{2} & -\frac{1}{2} \end{bmatrix}$.

(c) $\begin{bmatrix} -1 & \frac{3}{2} & \frac{1}{2} \\ 1 & -\frac{3}{2} & \frac{1}{2} \\ 0 & \frac{1}{2} & -\frac{1}{2} \end{bmatrix}$.

11. (a) and (b).

13. $\begin{bmatrix} \frac{4}{5} & -\frac{3}{5} \\ -\frac{1}{5} & \frac{2}{5} \end{bmatrix}$.

17. (a) $\begin{bmatrix} -30 \\ 60 \\ 10 \end{bmatrix}$. (b) $\begin{bmatrix} 23 \\ -31 \\ -1 \end{bmatrix}$.

19. Yes.

21. $\lambda = -1$, $\lambda = 3$.

23. $\begin{bmatrix} \frac{1}{4} & 0 & 0 \\ 0 & -\frac{1}{2} & 0 \\ 0 & 0 & \frac{1}{3} \end{bmatrix}$.

25. $\mathbf{X} = \begin{bmatrix} 19 \\ 23 \end{bmatrix}$.

SUPPLEMENTARY EXERCISES, PAGE 64

1. $\begin{bmatrix} -1 & -3 \\ 26 & 6 \end{bmatrix}$.

3. $\begin{bmatrix} 19 & 10 \\ -6 & 1 \end{bmatrix}$.

5. (a) $\begin{bmatrix} 1 & 2 & -1 & 1 & \vdots & 7 \\ 2 & -1 & 0 & 2 & \vdots & -8 \end{bmatrix}$.

(b) $3x + 2y = -4$
  $5x + y = 2$
  $3x + 2y = 6$.

7. $x = 1$, $y = 2$, $z = -2$.

9. (a) $a = -3$. (b) $a \neq \pm 3$. (c) $a = 3$.

11. $x = -3r$, $y = r$, $z = 0$, r = any real number.

13. $\begin{bmatrix} -40 & 16 & 9 \\ 13 & -5 & -3 \\ 5 & -2 & -1 \end{bmatrix}$.

15. Yes.

17. $\mathbf{X} = \begin{bmatrix} 4 \\ 1 \\ 4 \end{bmatrix}$.

21. (a) $k \neq 15$. (b) None. (c) $k = 15$.

23. $a = 1, -1$.

25. (a) $k = 1$ $\mathbf{B} = \begin{bmatrix} b_1 \\ 0 \end{bmatrix}$

$k = 2$ $\mathbf{B} = \begin{bmatrix} b_{11} & b_{12} \\ 0 & 0 \end{bmatrix}$

$k = 3$ $\mathbf{B} = \begin{bmatrix} b_{11} & b_{12} & b_{13} \\ 0 & 0 & 0 \end{bmatrix}$

$k = 4$ $\mathbf{B} = \begin{bmatrix} b_{11} & b_{12} & b_{13} & b_{14} \\ 0 & 0 & 0 & 0 \end{bmatrix}$

(b) The answers are not unique. The only requirement is that row 2 of **B** have all zero entries.

27. (a) $\begin{bmatrix} 1 & \frac{1}{2} \\ 0 & 1 \end{bmatrix}$ (b) $\begin{bmatrix} 1 & 0 & 0 \\ 0 & 0 & 0 \\ 0 & 0 & 0 \end{bmatrix}$ (c) $\mathbf{I_4}$

CHAPTER TEST, PAGE 67

1. No solution.

2. (a) $a = 2, 3$. (b) $a \neq 2, 3$. (c) None.

3. $\begin{bmatrix} -\frac{1}{2} & 1 & \frac{3}{2} \\ \frac{1}{2} & 0 & -\frac{1}{2} \\ -\frac{1}{2} & 1 & \frac{1}{2} \end{bmatrix}$.

4. $-2, 3$

5. (a) $\begin{bmatrix} 3 & 6 & 5 \\ -2 & 2 & -8 \\ 0 & 5 & -3 \end{bmatrix}$. (b) $\mathbf{X} = \begin{bmatrix} -4 \\ 14 \\ 25 \end{bmatrix}$.

6. (a) F. (b) T. (c) F. (d) T. (e) F.

CHAPTER 2

Section 2.1, page 78

1. (a) 5. (b) 7. (c) 4. (d) 4. (e) 7. (f) 0.

3. (a) $-$. (b) $+$. (c) $-$. (d) $-$. (e) $+$. (f) $+$.

5. (a) 7. (b) 30. (c) -24. (d) 4.

7. There are 24 terms.

9. $|\mathbf{B}| = 3$, $|\mathbf{C}| = 9$, $|\mathbf{D}| = -3$.

11. (a) $\lambda^2 - 3\lambda - 4$. (b) $\lambda^2 - 5\lambda + 6$.

13. (a) $-1, 4$. (b) $2, 3$.

15. (a) 72. (b) 0. (c) -24.

17. (a) -30. (b) 0. (c) 6.

21. $-\frac{3}{2}$.

Section 2.2, page 91

1. $A_{11} = -11$, $A_{12} = 29$, $A_{13} = 1$,
$A_{21} = -4$, $A_{22} = 7$, $A_{23} = -2$,
$A_{31} = 2$, $A_{32} = -10$, $A_{33} = 1$.

3. (a) -43. (b) 75. (c) 0.

5. (a) 0. (b) -6. (c) -36.

9. (a) $\begin{bmatrix} 24 & -42 & -30 \\ 19 & -2 & -30 \\ -4 & 32 & 30 \end{bmatrix}$. (b) 150.

11. (a) Singular. (b) $\begin{bmatrix} \frac{2}{7} & -\frac{3}{7} \\ \frac{1}{7} & \frac{2}{7} \end{bmatrix}$.

(c) $\begin{bmatrix} \frac{1}{4} & -\frac{1}{20} & \frac{3}{20} \\ 0 & \frac{1}{5} & \frac{2}{5} \\ 0 & \frac{1}{10} & -\frac{3}{10} \end{bmatrix}$.

13. (a) $\begin{bmatrix} 0 & \frac{1}{2} \\ 1 & \frac{3}{2} \end{bmatrix}$. (b) $\begin{bmatrix} \frac{1}{4} & 0 & 0 \\ 0 & -\frac{1}{3} & 0 \\ 0 & 0 & \frac{1}{2} \end{bmatrix}$.

(c) $\begin{bmatrix} \frac{15}{14} & \frac{5}{28} & -\frac{9}{28} & -\frac{23}{14} \\ \frac{8}{7} & -\frac{1}{7} & -\frac{1}{7} & -\frac{9}{7} \\ \frac{3}{7} & \frac{1}{14} & \frac{1}{14} & -\frac{6}{7} \\ -\frac{4}{7} & \frac{1}{14} & \frac{1}{14} & \frac{8}{7} \end{bmatrix}$.

15. (d) is nonsingular.

17. (a) $1, 4$. (b) $-5, 0, 3$.

19. (a) Has only the trivial solution.
    (b) Has nontrivial solutions.

21. $x = 1$, $y = -1$, $z = 0$, $w = 2$.

23. No solution.

SUPPLEMENTARY EXERCISES, PAGE 96

1. (a) -24. (b) 24.

3. (a) $\frac{1}{5}$. (b) 80. (c) $\frac{16}{5}$. (d) $\frac{1}{80}$.

5. 0, -1, -4.

7. 172.

9. -218.

11. $\frac{1}{5}\begin{bmatrix} 0 & 5 & -5 \\ -2 & 7 & -4 \\ 1 & -1 & 2 \end{bmatrix}$.

13. $\lambda \neq -1, 0, 1$.

17. $a \neq 0$, $a \neq 2$.

CHAPTER TEST, PAGE 97

1. 17.

2. (a) 54. (b) $\frac{27}{2}$. (c) $\frac{1}{54}$.

3. $\frac{14}{3}$.

4. -3, 0, 3.

5. $x = 1$, $y = 0$, $z = -2$.

6. (a) T. (b) F. (c) F. (d) F. (e) T. (f) T.
    (g) T. (h) T. (i) F. (j) T.

CHAPTER 3

Section 3.1, page 112

1.

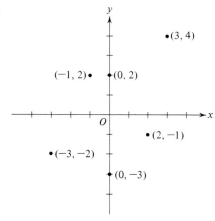

3. (a) $(1, 7)$.

5. (a) $\mathbf{X} + \mathbf{Y} = (0, 8)$; $\mathbf{X} - \mathbf{Y} = (4, -2)$;
    $2\mathbf{X} = (4, 6)$; $3\mathbf{X} - 2\mathbf{Y} = (10, -1)$.
    (b) $\mathbf{X} + \mathbf{Y} = (3, 5)$; $\mathbf{X} - \mathbf{Y} = (-3, 1)$;
    $2\mathbf{X} = (0, 6)$; $3\mathbf{X} - 2\mathbf{Y} = (-6, 5)$.
    (c) $\mathbf{X} + \mathbf{Y} = (5, 8)$; $\mathbf{X} - \mathbf{Y} = (-1, 4)$;
    $2\mathbf{X} = (4, 12)$; $3\mathbf{X} - 2\mathbf{Y} = (0, 14)$.

7. (a) $x = 2$. (b) $y = \frac{8}{3}$. (c) $x = 3$, $y = -2$.

9. (a) $\sqrt{5}$. (b) 5. (c) 2. (d) 5.

11. (a) $\sqrt{2}$ (b) 5. (c) $\sqrt{10}$ (d) $\sqrt{13}$.

13. (a) $(\frac{3}{5}, \frac{4}{5})$. (b) $(-2/\sqrt{13}, -3/\sqrt{13})$. (c) $(1, 0)$.

15. (a) -4. (b) 0. (c) 0. (d) -5.

17. (a) $\dfrac{-4}{\sqrt{5} \cdot \sqrt{13}}$. (b) 0. (c) 0. (d) -1.

21. (a) $\mathbf{i} + 3\mathbf{j}$. (b) $-2\mathbf{i} - 3\mathbf{j}$. (c) $-2\mathbf{i}$. (d) $3\mathbf{j}$.

23.

400 pounds

Resultant
500 pounds   300 pounds

Section 3.2, page 126

1. (a) $\mathbf{X} + \mathbf{Y} = (1, 3, -5)$, $\mathbf{X} - \mathbf{Y} = (1, 1, -1)$,
   $2\mathbf{X} = (2, 4, -6)$, $3\mathbf{X} - 2\mathbf{Y} = (3, 4, -5)$.
   (b) $\mathbf{X} + \mathbf{Y} = (3, 0, 6, -1)$,
   $\mathbf{X} - \mathbf{Y} = (5, -4, -4, 7)$,
   $2\mathbf{X} = (8, -4, 2, 6)$,
   $3\mathbf{X} - 2\mathbf{Y} = (14, -10, -7, 17)$.

3. (a) $x = \frac{1}{2}$, $y = \frac{3}{2}$. (b) $x = 4$, $y = 0$.
   (c) $x = -6$, $y = 1$, $u = 0$.

7.

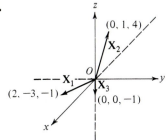

9. $(4, 2, 2)$.

11. (a) $\sqrt{29}$. (b) $\sqrt{14}$. (c) $\sqrt{5}$. (d) $\sqrt{30}$.

13. (a) $\sqrt{18}$. (b) $\sqrt{6}$. (c) $\sqrt{50}$. (d) $\sqrt{10}$.

15. (a) 0. (b) -5. (c) 3. (d) 0.

19. (a) 0. (b) $-\dfrac{1}{\sqrt{3}}$. (c) $\dfrac{1}{\sqrt{5}}$. (d) 0.

21. (a) $\mathbf{X}_1$ and $\mathbf{X}_2$, $\mathbf{X}_1$ and $\mathbf{X}_6$, $\mathbf{X}_2$ and $\mathbf{X}_3$,
   $\mathbf{X}_3$ and $\mathbf{X}_6$, $\mathbf{X}_4$ and $\mathbf{X}_6$.
   (b) $\mathbf{X}_1$ and $\mathbf{X}_3$.
   (c) None.

23. (a) $(2/\sqrt{14}, -1/\sqrt{14}, 3/\sqrt{14})$.
   (b) $(1/\sqrt{30}, 2/\sqrt{30}, 3/\sqrt{30}, 4/\sqrt{30})$.
   (c) $(0, 1/\sqrt{2}, -1/\sqrt{2})$.
   (d) $(0, -1/\sqrt{6}, 2/\sqrt{6}, -1/\sqrt{6})$.

25. (a) $\mathbf{i} + 2\mathbf{j} - 3\mathbf{k}$. (b) $2\mathbf{i} + 3\mathbf{j} - \mathbf{k}$.
   (c) $\mathbf{j} + 2\mathbf{k}$. (d) $-2\mathbf{k}$.

29. $1.08S$.

31. $\frac{1}{2}(\mathbf{H} + \mathbf{L})$.

Section 3.3, page 133

1. (a) $-15\mathbf{i} - 2\mathbf{j} + 9\mathbf{k}$.   (b) $-3\mathbf{i} + 3\mathbf{j} + 3\mathbf{k}$.
   (c) $7\mathbf{i} + 5\mathbf{j} - \mathbf{k}$. (d) $0\mathbf{i} + 0\mathbf{j} + 0\mathbf{k}$.

9. $\frac{1}{2}\sqrt{478}$.

11. $\sqrt{150}$.

13. 39.

Section 3.4, page 142

7. Not a vector space; (a), (c), (d), (e), (f), and (g) do
   not hold.

9. Vector space.

11. Not a vector space; (β), (d), (e), (f), and (g) do not
   hold.

13. Vector space.

15. (b).

17. (b) and (c).

23. (a) and (c).

25. (b), (c), and (e).

Section 3.5, page 152

1. (b) and (c).

3. (a) and (c).

5. (a) No. (b) No. (c) No. (d) No.

7. (a), (c), and (d).

9. (a) and (d).

11. No.

13. (a) $X_3 = 3X_1 + X_2 + X_4$.
(d) $X_2 = 2X_1 - X_3$.

15. $\lambda \neq \pm 2$.

Section 3.6, page 162

1. (a) and (d).

3. (a) and (d).

5. (c).

7. (a); $X = \frac{3}{2}X_1 + \frac{1}{2}X_2 - \frac{3}{2}X_3$.

9. Possible answer: $\{X_1, X_2\}$; dim $W = 2$.

11. Possible answer: $\{t^3 + t^2 - 2t + 1, t^2 + 1\}$;
dim $W = 2$.

13. (a) Possible answer: $\{(1, 1, 0), (0, 1, 1)\}$.
(b) Possible answer: $\{(1, 1, 0), (0, 0, 1)\}$.
(c) Possible answer: $\{(0, 1, 0), (0, 0, 1)\}$.
(d) Possible answer: $\{(1, 0, 2), (-1, 1, -1)\}$.

15. (a) 2. (b) 1. (c) 2. (d) 2.

17. (a) 4. (b) 3. (c) 3. (d) 4.

19. Possible answer: $\{(1, 0, 1, 0), (0, 1, -1, 0),$
$(1, 0, 0, 0), (0, 0, 0, 1)\}$.

21. Zero.

23. Possible answer: $\left\{ \begin{bmatrix} 1 \\ -2 \\ 1 \end{bmatrix} \right\}$; dim $W = 1$.

Section 3.7, page 172

1. Possible answer: $\{(1, 0, 0), (0, 1, 0), (0, 0, 1)\}$.

3. Possible answer: $\left\{ \begin{bmatrix} 1 \\ 0 \\ 1 \\ 0 \end{bmatrix}, \begin{bmatrix} 0 \\ 1 \\ 0 \\ 1 \end{bmatrix} \right\}$.

5. 3.

7. 2.

9. 3.

11. 2.

15. Singular.

17. Nonsingular.

19. Linearly dependent.

21. Nontrivial solution.

23. Has a solution.

25. Has no solution.

Section 3.8, page 183

1. (b).

3. $\{(1/\sqrt{2}, -1/\sqrt{2}, 0), (1/\sqrt{3}, 1/\sqrt{3}, 1/\sqrt{3})\}$.

5. $\{(1/\sqrt{3}, -1/\sqrt{3}, 0, 1/\sqrt{3}),$
$(5/\sqrt{42}, 1/\sqrt{42}, 0, -4/\sqrt{42}), (0, 0, 1, 0)\}$.

7. (a) $\{(1, 2), (-4, 2)\}$.
(b) $\{(1/\sqrt{5}, 2/\sqrt{5}), (-2/\sqrt{5}, 1/\sqrt{5})\}$.

9. $\{(\frac{2}{3}, -\frac{2}{3}, \frac{1}{3}), (\frac{2}{3}, \frac{1}{3}, -\frac{2}{3}), (\frac{1}{3}, \frac{2}{3}, \frac{2}{3})\}$.

11. Possible answer:
$\{(1/\sqrt{2}, 1/\sqrt{2}, 0, 0),$
$(3/\sqrt{22}, -3/\sqrt{22}, 0, 2/\sqrt{22}),$
$(1/\sqrt{11}, -1/\sqrt{11}, 0, -3/\sqrt{11})\}$.

13. $\dfrac{4}{\sqrt{5}}\left(\dfrac{1}{\sqrt{5}}, 0, \dfrac{2}{\sqrt{5}}\right) -$

$\qquad \dfrac{3}{\sqrt{5}}\left(-\dfrac{2}{\sqrt{5}}, 0, \dfrac{1}{\sqrt{5}}\right) -$

$\qquad\qquad 3(0, 1, 0) = (2, -3, 1).$

15. $\mathbf{Z} = (1, 0, 2, 3), \mathbf{Y} = (0, 0, 0, 0).$

17. 2.

SUPPLEMENTARY EXERCISES, PAGE 185

1. (a) $\sqrt{15}.$　(b) $3\sqrt{2}.$　(c) $\sqrt{43}.$　(d) $-5.$
(e) $-\frac{1}{3}\sqrt{\frac{5}{6}}.$

3. $c = \pm 3$

5. No.

7. Linearly dependent; possible answer:

$\qquad -t - 3 = (2t^2 + 3t + 1) - 2(t^2 + 2t + 2).$

9. Possible answer: $\{(1, 0, 1, 0), (0, 1, -2, 1)\};$
dimension is 2.

11. Possible answer: $\left\{\begin{bmatrix} 0 \\ -1 \\ 0 \\ 0 \\ 1 \end{bmatrix}, \begin{bmatrix} 7 \\ -4 \\ 0 \\ 1 \\ 0 \end{bmatrix}, \begin{bmatrix} -5 \\ 3 \\ 1 \\ 0 \\ 0 \end{bmatrix}\right\};$

$\qquad$ dimension is 3.

13. $\lambda \neq \pm 2.$

15. $a = 1.$

19. (a) m arbitrary and $b = 0.$　(b) $r = 0.$

23. (b) $k = 0$

25. 3

27. $\{(2, 1, 0), (-1, 0, 1)\}$

29. $(1, 2, 3) = -\sqrt{2}\left(\dfrac{1}{\sqrt{2}}, 0, -\dfrac{1}{\sqrt{2}}\right) +$

$\qquad 2(0, 1, 0) + 2\sqrt{2}\left(\dfrac{1}{\sqrt{2}}, 0, \dfrac{1}{\sqrt{2}}\right).$

31. (a) Possible answer: $\{(-1, 0, 1)\}$
(c) (i) $\mathbf{Z} = (\frac{1}{2}, 0, \frac{1}{2}),\ \mathbf{Y} = (\frac{1}{2}, 0, -\frac{1}{2})$
$\qquad$ (ii) $\mathbf{Z} = (2, 2, 2),\ \mathbf{Y} = (-1, 0, 1)$

CHAPTER TEST, PAGE 188

1. Yes.

2. Possible answer: $\left\{\begin{bmatrix} 0 \\ -1 \\ 1 \\ 0 \\ 0 \end{bmatrix}, \begin{bmatrix} 4 \\ -1 \\ 0 \\ 1 \\ 0 \end{bmatrix}, \begin{bmatrix} -2 \\ 0 \\ 0 \\ 0 \\ 1 \end{bmatrix}\right\}.$

3. Yes.

4. $\lambda = \pm 3.$

5. Possible answer:

$\left\{\left(\dfrac{1}{\sqrt{2}}, 0, -\dfrac{1}{\sqrt{2}}, 0\right),\right.$

$\qquad\left(\dfrac{1}{\sqrt{6}}, -\dfrac{2}{\sqrt{6}}, \dfrac{1}{\sqrt{6}}, 0\right),$

$\qquad\qquad\left.\left(\dfrac{1}{\sqrt{3}}, \dfrac{1}{\sqrt{3}}, \dfrac{1}{\sqrt{3}}, 0\right)\right\}.$

6. (a) T.　(b) F.　(c) F.　(d) F.　(e) T.　(f) F.
(g) T.　(h) F.　(i) F.　(j) T.

CHAPTER 4

Section 4.1, page 197

1. (b).

3. (a).

5. (a) $\begin{bmatrix} 1 \\ -19 \end{bmatrix}.$　(b) $\begin{bmatrix} a + b \\ -5a + 2b \end{bmatrix}.$

11. (a) 71, 52, 33, 47, 30, 26, 84, 56, 43, 99, 69, 55.
(b) Message: CERTAINLY NOT.

Section 4.2, page 207

1. (a) Yes. (b) No. (c) Yes. (d) No.
   (e) $\{(0, r)\}$, r = any real number.
   (f) $\{(r, 0)\}$, r = any real number.

3. (a) $\{(0, 0)\}$. (b) Yes. (c) No.

5. (a) Possible answer: $\left\{ \begin{bmatrix} -2 \\ 0 \\ 1 \\ 1 \\ 0 \end{bmatrix}, \begin{bmatrix} 0 \\ 1 \\ 0 \\ 0 \\ 0 \end{bmatrix} \right\}$.

   (b) Possible answer: $\left\{ \begin{bmatrix} 1 \\ 1 \\ 2 \\ 0 \end{bmatrix}, \begin{bmatrix} -1 \\ 0 \\ -1 \\ -1 \end{bmatrix}, \begin{bmatrix} -1 \\ -1 \\ -1 \\ 0 \end{bmatrix} \right\}$.

7. (a) Yes. (b) 1.

11. (a) No. (b) Yes. (c) Yes. (d) No.
    (e) Possible answer: $\{-t^2 - t + 1\}$.
    (f) Possible answer: $\{t^2, t\}$.

13. (a) Possible answer: $\{1\}$.
    (b) Possible answer: $\{t, 1\}$.

15. (a) 2. (b) 1.

17. 0.

19. 1.

Section 4.3, page 218

1. (a) $\begin{bmatrix} 1 \\ -2 \end{bmatrix}$. (b) $\begin{bmatrix} -1 \\ 2 \end{bmatrix}$. (c) $\begin{bmatrix} 2 \\ 5 \end{bmatrix}$. (d) $\begin{bmatrix} 0 \\ 1 \end{bmatrix}$.

3. (a) $\begin{bmatrix} 1 \\ 0 \\ 1 \end{bmatrix}$. (b) $\begin{bmatrix} -1 \\ 2 \\ 1 \end{bmatrix}$. (c) $\begin{bmatrix} 0 \\ 1 \\ 0 \end{bmatrix}$. (d) $\begin{bmatrix} -2 \\ 1 \\ 3 \end{bmatrix}$.

5. (a) $\begin{bmatrix} 1 \\ -1 \\ 1 \end{bmatrix}$. (b) $\begin{bmatrix} 0 \\ 1 \\ 1 \end{bmatrix}$. (c) $\begin{bmatrix} 1 \\ -2 \\ 1 \end{bmatrix}$. (d) $\begin{bmatrix} 0 \\ -1 \\ 0 \end{bmatrix}$.

7. (a) $t^2 - 3t + 2$. (b) $t^2 + 1$. (c) $t^2 - t + 2$.

9. (a) $\begin{bmatrix} 1 & -2 \\ 2 & 1 \\ 1 & 1 \end{bmatrix}$. (b) $\begin{bmatrix} \frac{7}{3} & -\frac{4}{3} \\ -\frac{2}{3} & \frac{5}{3} \\ \frac{2}{3} & -\frac{2}{3} \end{bmatrix}$. (c) $\begin{bmatrix} -3 \\ 4 \\ 3 \end{bmatrix}$.

11. (a) $[\mathbf{L(X)}]_T = \begin{bmatrix} 1 & 1 & 0 \\ 0 & 1 & -1 \end{bmatrix} [\mathbf{X}]_S$.

    (b) $[\mathbf{L(X)}]_{T'} = \begin{bmatrix} -1 & -\frac{1}{3} & 0 \\ 1 & \frac{2}{3} & 0 \end{bmatrix} [\mathbf{X}]_{S'}$.

    (c) $\begin{bmatrix} 3 \\ -1 \end{bmatrix}$.

13. (a) $\begin{bmatrix} 1 & 0 \\ 0 & 1 \\ 0 & 0 \\ 0 & 0 \end{bmatrix}$. (b) $\begin{bmatrix} 1 & 1 \\ 0 & 1 \\ 0 & -1 \\ 0 & 1 \end{bmatrix}$.

15. (a) $[\mathbf{L(X_1)}]_S = \begin{bmatrix} 2 \\ -1 \end{bmatrix}$, $[\mathbf{L(X_2)}]_S = \begin{bmatrix} -3 \\ 4 \end{bmatrix}$.

    (b) $\mathbf{L(X_1)} = \begin{bmatrix} 1 \\ 5 \end{bmatrix}$, $\mathbf{L(X_2)} = \begin{bmatrix} 1 \\ -10 \end{bmatrix}$.

    (c) $\begin{bmatrix} -2 \\ 25 \end{bmatrix}$.

17. (a) $\begin{bmatrix} 1 & 2 & 1 \\ 1 & 0 & 0 \\ 0 & 1 & 1 \end{bmatrix}$. (b) $\begin{bmatrix} 8 \\ 1 \\ 5 \end{bmatrix}$.

19. (a) $\begin{bmatrix} 0 & -1 \\ 1 & 0 \end{bmatrix}$. (b) $\begin{bmatrix} 0 & -1 \\ 1 & 0 \end{bmatrix}$.

    (c) $\begin{bmatrix} -\frac{1}{2} & -\frac{1}{2} \\ \frac{1}{2} & -\frac{1}{2} \end{bmatrix}$. (d) $\begin{bmatrix} 1 & -1 \\ 1 & 1 \end{bmatrix}$.

Section 4.4, page 227

1. (a) $\begin{bmatrix} -1 & 0 \\ 0 & 1 \end{bmatrix}$. (b) $\begin{bmatrix} 0 & 1 \\ 1 & 0 \end{bmatrix}$. (c) $\begin{bmatrix} 0 & -1 \\ 1 & 0 \end{bmatrix}$.

   (d) $\begin{bmatrix} \frac{\sqrt{3}}{2} & \frac{1}{2} \\ -\frac{1}{2} & \frac{\sqrt{3}}{2} \end{bmatrix}$.

3. $\begin{bmatrix} 1 & 0 \\ 0 & r \end{bmatrix}$.

5. (a) 6. (b) $(1, -2), (-3, 6), (11, -10)$. (c) 24.

SUPPLEMENTARY EXERCISES, PAGE 228

1. No.

3. $8t + 7$.

5. (a) Possible answer: $\{(1, 1, 1), (1, -1, 2)\}$.
(b) No.

7. (a) $\begin{bmatrix} 2 \\ 2 \\ -4 \end{bmatrix}$. (b) $2t^2 + 2t - 7$.

9. (a) $\begin{bmatrix} 2 \\ 1 \end{bmatrix}, \begin{bmatrix} -3 \\ 2 \end{bmatrix}$.
(b) $3t - 3, -t + 8$. (c) $-\frac{7}{3}t + \frac{35}{3}$.

11. $\begin{bmatrix} 1 & 0 & 0 \\ 0 & 1 & 0 \\ 0 & 0 & 1 \end{bmatrix}$

CHAPTER TEST, PAGE 230

1. $\begin{bmatrix} 2 \\ 7 \\ 4 \end{bmatrix}$.

2. (a) ker $\mathbf{L} = \{(0, 0, 0)\}$, so there is no basis.
(b) Yes.

3. (a) Possible answer: $\{(1, 1, 2), (1, -1, 1)\}$.
(b) No.

4. 2.

5. $\begin{bmatrix} 0 & \frac{3}{2} \\ 1 & -\frac{5}{2} \end{bmatrix}$.

6. (a) F. (b) T. (c) T. (d) F. (e) F.

CHAPTER 5

Section 5.1, page 249

1. $\lambda^3 - 4\lambda^2 + 7$.

3. $(\lambda - 4)(\lambda - 2)(\lambda - 3) = \lambda^3 - 9\lambda^2 + 26\lambda - 24$.

5. $f(\lambda) = (\lambda - 1)(\lambda - 3)(\lambda + 2); \lambda_1 = 1,$
$\lambda_2 = 3, \lambda_3 = -2;$

$$\mathbf{X}_1 = \begin{bmatrix} 6 \\ 3 \\ 8 \end{bmatrix}, \quad \mathbf{X}_2 = \begin{bmatrix} 0 \\ 5 \\ 2 \end{bmatrix}, \quad \mathbf{X}_3 = \begin{bmatrix} 0 \\ 0 \\ 1 \end{bmatrix}.$$

7. $f(\lambda) = \lambda^2 - 5\lambda + 6; \lambda_1 = 2, \lambda_2 = 3;$

$$\mathbf{X}_1 = \begin{bmatrix} 1 \\ -1 \end{bmatrix}, \quad \mathbf{X}_2 = \begin{bmatrix} 1 \\ -2 \end{bmatrix}.$$

9. $f(\lambda) = \lambda^3 - 5\lambda^2 + 2\lambda + 8; \lambda_1 = -1, \lambda_2 = 2,$
$\lambda_3 = 4;$

$$\mathbf{X}_1 = \begin{bmatrix} 1 \\ 0 \\ -1 \end{bmatrix}, \quad \mathbf{X}_2 = \begin{bmatrix} -2 \\ -3 \\ 2 \end{bmatrix}, \quad \mathbf{X}_3 = \begin{bmatrix} 8 \\ 5 \\ 2 \end{bmatrix}.$$

11. $f(\lambda) = (\lambda - 1)(\lambda + 1)(\lambda - 3)(\lambda - 2); \lambda_1 = 1,$
$\lambda_2 = -1, \lambda_3 = 3, \lambda_4 = -2;$

$$\mathbf{X}_1 = \begin{bmatrix} 1 \\ 0 \\ 0 \\ 0 \end{bmatrix}, \quad \mathbf{X}_2 = \begin{bmatrix} 1 \\ -1 \\ 0 \\ 0 \end{bmatrix},$$

$$\mathbf{X}_3 = \begin{bmatrix} 9 \\ 3 \\ 4 \\ 0 \end{bmatrix}, \quad \mathbf{X}_4 = \begin{bmatrix} 29 \\ 7 \\ 9 \\ -3 \end{bmatrix}.$$

13. Not diagonalizable; $\lambda_1 = \lambda_2 = 1$.

15. Diagonalizable; $\lambda_1 = 1, \lambda_2 = -1, \lambda_3 = 2$.

17. Not diagonalizable; $\lambda_1 = 1, \lambda_2 = 1, \lambda_3 = 3$.

19. $\mathbf{P} = \begin{bmatrix} 1 & -3 & 1 \\ 0 & 0 & -6 \\ 1 & 2 & 4 \end{bmatrix}; \lambda_1 = 4, \lambda_2 = -1, \lambda_3 = 1.$

21. $\mathbf{P} = \begin{bmatrix} 1 & 2 & 1 \\ 0 & 1 & 0 \\ 0 & 0 & -3 \end{bmatrix}$; $\lambda_1 = 3, \lambda_2 = 2, \lambda_3 = 0$.

23. Basis for eigenspace associated with $\lambda_1 = \lambda_2 = 2$

is $\left\{ \begin{bmatrix} 1 \\ 0 \\ 0 \\ 0 \end{bmatrix} \right\}$.

Basis for eigenspace associated with $\lambda_3 = \lambda_4 = 1$

is $\left\{ \begin{bmatrix} 3 \\ -3 \\ 1 \\ 0 \end{bmatrix} \right\}$.

25. $256 \begin{bmatrix} 3 & -5 \\ 1 & -3 \end{bmatrix}$.

Section 5.2, page 260

7. $\begin{bmatrix} 0 & 0 \\ 0 & 4 \end{bmatrix}$; $\mathbf{P} = \begin{bmatrix} 1/\sqrt{2} & 1/\sqrt{2} \\ -1/\sqrt{2} & 1/\sqrt{2} \end{bmatrix}$.

9. $\begin{bmatrix} 0 & 0 & 0 \\ 0 & 0 & 0 \\ 0 & 0 & 4 \end{bmatrix}$;

$\mathbf{P} = \begin{bmatrix} 1 & 0 & 0 \\ 0 & -1/\sqrt{2} & 1/\sqrt{2} \\ 0 & 1/\sqrt{2} & 1/\sqrt{2} \end{bmatrix}$.

11. $\begin{bmatrix} -2 & 0 & 0 \\ 0 & 1 & 0 \\ 0 & 0 & 1 \end{bmatrix}$;

$\mathbf{P} = \begin{bmatrix} 1/\sqrt{3} & -1/\sqrt{2} & -1/\sqrt{6} \\ 1/\sqrt{3} & 1/\sqrt{2} & -1/\sqrt{6} \\ 1/\sqrt{3} & 0 & 2/\sqrt{6} \end{bmatrix}$.

13. $\begin{bmatrix} 3 & 0 \\ 0 & 1 \end{bmatrix}$.

15. $\begin{bmatrix} 1 & 0 & 0 \\ 0 & 2 & 0 \\ 0 & 0 & 0 \end{bmatrix}$.

17. $\begin{bmatrix} 1 & 0 & 0 \\ 0 & 0 & 0 \\ 0 & 0 & 2 \end{bmatrix}$.

19. $\begin{bmatrix} 1 & 0 & 0 \\ 0 & 1 & 0 \\ 0 & 0 & 4 \end{bmatrix}$.

SUPPLEMENTARY EXERCISES, PAGE 261

1. $f(\lambda) = (\lambda + 2)(\lambda^2 - 8\lambda + 15)$; $\lambda_1 = -2$, $\lambda_2 = 3, \lambda_3 = 5$;

$$\mathbf{X}_1 = \begin{bmatrix} -35 \\ 12 \\ 19 \end{bmatrix}, \quad \mathbf{X}_2 = \begin{bmatrix} 0 \\ 3 \\ 1 \end{bmatrix}, \quad \mathbf{X}_3 = \begin{bmatrix} 0 \\ 1 \\ 1 \end{bmatrix}.$$

3. Yes.

5. Not diagonalizable; the roots of the characteristic polynomial are not all real.

7. For $\lambda = 0$, possible answer: $\left\{ \begin{bmatrix} 1 \\ 0 \\ 0 \end{bmatrix} \right\}$;

for $\lambda = 2$, possible answer: $\left\{ \begin{bmatrix} 1 \\ 0 \\ 2 \end{bmatrix}, \begin{bmatrix} 0 \\ 1 \\ 0 \end{bmatrix} \right\}$.

9. $\mathbf{P} = \begin{bmatrix} -1/\sqrt{2} & 1/\sqrt{6} & 1/\sqrt{3} \\ 0 & -2/\sqrt{6} & 1/\sqrt{3} \\ 1/\sqrt{2} & 1/\sqrt{6} & 1/\sqrt{3} \end{bmatrix}$,

$\mathbf{D} = \begin{bmatrix} 0 & 0 & 0 \\ 0 & 0 & 0 \\ 0 & 0 & 3 \end{bmatrix}$.

11. $\lambda = 0, \lambda = 1$

CHAPTER TEST, PAGE 262

1. Not diagonalizable; $\lambda_1 = 1, \lambda_2 = \lambda_3 = 2$.

2. Check that $\mathbf{A}\mathbf{A}^T = \mathbf{I}_3$.

3. P = $\begin{bmatrix} -2 & 1 & 1 \\ 1 & 2 & 0 \\ 2 & 0 & 1 \end{bmatrix}$,

D = $\begin{bmatrix} 9 & 0 & 0 \\ 0 & -9 & 0 \\ 0 & 0 & -9 \end{bmatrix}$.

4. (a) F. (b) F. (c) T. (d) T. (e) F.

CUMULATIVE REVIEW OF PART I, PAGE 264

1. F	**2.** T	**3.** F	**4.** F	**5.** F
6. T	**7.** T	**8.** T	**9.** F	**10.** T
11. T	**12.** F	**13.** F	**14.** F	**15.** T
16. T	**17.** F	**18.** F	**19.** F	**20.** T
21. T	**22.** F	**23.** T	**24.** T	**25.** F
26. T	**27.** F	**28.** F	**29.** F	**30.** T
31. T	**32.** T	**33.** T	**34.** T	**35.** T
36. F	**37.** F	**38.** F	**39.** F	**40.** T
41. F	**42.** F	**43.** F	**44.** T	**45.** T
46. F	**47.** T	**48.** T	**49.** T	**50.** T

CHAPTER 6

Section 6.1, page 285

1. Maximize $z = 120x + 100y$
subject to
$$2x + 2y \leq 8$$
$$5x + 3y \leq 15$$
$$x \geq 0, \ y \geq 0.$$

3. Maximize $z = 0.08x + 0.10y$
subject to
$$x + y \leq 6000$$
$$x \geq 1500$$
$$y \leq 4000$$
$$y \leq \tfrac{1}{2}x$$
$$x \geq 0, \ y \geq 0.$$

5. Maximize $z = 40{,}000x + 45{,}000y$
subject to
$$x + y \leq 30$$
$$y \geq 24$$
$$x \geq 2$$
$$x \leq 4$$
$$x \geq 0, \ y \geq 0.$$

7. Maximize $z = 4x + 6y$
subject to
$$x + 2y \leq 10$$
$$x + y \leq 7$$
$$x \geq 0, \ y \geq 0.$$

9. Minimize $z = 10x + 12y$
subject to
$$2x + 3y \geq 18$$
$$x + 3y \geq 12$$
$$80x + 60y \geq 480$$
$$x \geq 0, \ y \geq 0.$$

11.

13.

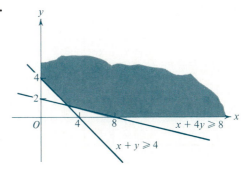

15. $x = \frac{8}{11}$, $y = \frac{45}{11}$, optimal value of $z = -\frac{21}{11}$.

17. Invest $2000 in bond A and $4000 in bond B; maximum return is $560.

19. Carry no containers from the Smith Corporation and 1500 containers from the Johnson Corporation, or

120 containers from the Smith Corporation and 1440 containers from the Johnson Corporation. In either case the maximum revenue is $900.

21. Use $\frac{5}{2}$ gallons of L and $\frac{3}{2}$ gallons of H; minimum cost is $2.25.

23. No solution.

25. (a).

27. Maximize $z = 3x_1 - x_2 + 6x_3$
subject to

$$2x_1 + 4x_2 + x_3 \leq 4$$
$$3x_1 - 2x_2 + 3x_3 \leq 4$$
$$2x_1 + x_2 - x_3 \leq 8$$
$$x_1 \geq 0, \, x_2 \geq 0, \, x_3 \geq 0.$$

29. Maximize $z = 2x_1 + 3x_2 + 7x_3$
subject to

$$3x_1 + x_2 - 4x_3 + x_4 \qquad\qquad = 3$$
$$x_1 - 2x_2 + 6x_3 \qquad + x_5 \qquad = 21$$
$$x_1 - x_2 - x_3 \qquad\qquad + x_6 = 9$$
$$x_1 \geq 0, \, x_2 \geq 0, \, x_3 \geq 0, \, x_4 \geq 0, \, x_5 \geq 0, \, x_6 \geq 0.$$

Section 6.2, page 303

1.

	x	y	u	v	w	z	
u	3	−2	1	0	0	0	7
v	2	5	0	1	0	0	6
w	2	3	0	0	1	0	8
	−3	−7	0	0	0	1	0

3.

	x_1	x_2	x_3	x_4	x_5	x_6	x_7	z	
x_5	3	−2	1	1	1	0	0	0	6
x_6	1	1	1	1	0	1	0	0	8
x_7	2	−3	−1	2	0	0	1	0	10
	−2	−2	−3	−1	0	0	0	1	0

5. $x = 2$, $y = 0$, optimal $z = 4$.

7. No finite optimal solution.

9. $x_1 = 0$, $x_2 = \frac{33}{20}$, $x_3 = \frac{27}{10}$, optimal $z = \frac{69}{10}$.

11. $x_1 = 0$, $x_2 = 0$, $x_3 = 49$, $x_4 = 41$,
optimal $z = 156$.

13. Carry no containers from the Smith Corporation and 1500 containers from the Johnson Corporation, or 120 containers from the Smith Corporation and 1440 containers from the Johnson Corporation. In either case the maximum revenue is $900.

15. Use 4 tons of gas, no coal and no oil; maximum energy generated is 2000 kilowatt hours.

Section 6.3, page 311

1. Minimize $z' = 7w_1 + 6w_2 + 9w_3$
subject to

$$4w_1 + 5w_2 + 6w_3 \geq 3$$
$$3w_1 - 2w_2 + 8w_3 \geq 2$$
$$w_1 \geq 0, \quad w_2 \geq 0, \quad w_3 \geq 0.$$

3. Maximize $z' = 7w_1 + 12w_2 + 18w_3$
subject to

$$2w_1 + 8w_2 + 10w_3 \leq 3$$
$$3w_1 - 9w_2 + 15w_3 \leq 5$$
$$w_1 \geq 0, \quad w_2 \geq 0, \quad w_3 \geq 0.$$

7. $w_1 = \frac{5}{7}$, $w_2 = 0$, $w_3 = 0$, optimal $z' = \frac{30}{7}$.

9. $w_1 = 2$, $w_2 = 0$, $w_3 = 0$, optimal $z' = 10$.

SUPPLEMENTARY EXERCISES, PAGE 312

1. $x = \frac{6}{5}$, $y = \frac{12}{5}$, optimal $z = \frac{48}{5}$.

3. $x = 0$, $y = 8$, or $x = 2$, $y = 7$, optimal $z = 800$.

5. $x_1 = \frac{18}{11}$, $x_2 = \frac{10}{11}$, optimal $z = \frac{158}{11}$.

CHAPTER TEST, PAGE 313

1. Plant no corn, plant 20 acres of wheat.

2. $x_1 = 1$, $x_2 = \frac{1}{3}$, $x_3 = 0$, optimal $z = 11$.

3. Maximize $z' = 8y_1 + 12y_2 + 6y_3$

subject to
$$y_1 + 2y_2 + 2y_3 \leqslant 3$$
$$4y_1 + 3y_2 + y_3 \leqslant 4$$
$$y_1 \geqslant 0, \quad y_2 \geqslant 0, \quad y_3 \geqslant 0.$$

CHAPTER 7

Section 7.1, page 321

1. (a) $-7x + 5y + 1 = 0$.   (b) $9x + 5y + 7 = 0$.
   (c) $-5x - 3y = 0$.   (d) $-7x + 3y - 6 = 0$.

3. (d).

5. (a) $x = 3 + 4t$, $y = 4 - 5t$, $z = -2 + 2t$,
$-\infty < t < \infty$.
   (b) $x = 3 - 2t$, $y = 2 + 5t$, $z = 4 + t$,
$-\infty < t < \infty$.
   (c) $x = t$, $y = t$, $z = t$, $-\infty < t < \infty$.
   (d) $x = -2 + 2t$, $y = -3 + 3t$, $z = 1 + 4t$,
$-\infty < t < \infty$.

7. (a) $\dfrac{x - 2}{2} = \dfrac{y + 3}{5} = \dfrac{z - 1}{4}$.

   (b) $\dfrac{x + 3}{8} = \dfrac{y + 2}{7} = \dfrac{z + 2}{6}$.

   (c) $\dfrac{x + 2}{4} = \dfrac{y - 3}{-6} = \dfrac{z - 4}{1}$.

   (d) $\dfrac{x}{4} = \dfrac{y}{5} = \dfrac{z}{2}$.

9. (a) $3x - 2y + 4z + 16 = 0$.
   (b) $y - 3z + 3 = 0$.
   (c) $-z + 4 = 0$.
   (d) $-x - 2y + 4z - 3 = 0$.

11. (a) $x = \frac{8}{13} + 23t$, $y = -\frac{27}{13} + 2t$, $z = 13t$,
$-\infty < t < \infty$.
   (b) $x = -\frac{28}{13} + 7t$, $y = -\frac{16}{13} - 22t$, $z = 13t$,
$-\infty < t < \infty$.
   (c) $x = -\frac{16}{3} + 5t$, $y = -\frac{8}{3} + 4t$, $z = -3t$,
$-\infty < t < \infty$.

13. Yes.

15. $(5, 1, 2)$.

19. $4x - 4y + z + 16 = 0$.

21. $7x + 2y - 2z - 19 = 0$.

23. $x = -2 + 2t$, $y = 5 - 3t$, $z = -3 + 4t$.

Section 7.2, page 331

1. (a) $\begin{bmatrix} x & y \end{bmatrix} \begin{bmatrix} -3 & \frac{5}{2} \\ \frac{5}{2} & -2 \end{bmatrix} \begin{bmatrix} x \\ y \end{bmatrix}$.

   (b) $\begin{bmatrix} x_1 & x_2 & x_3 \end{bmatrix} \begin{bmatrix} 2 & \frac{3}{2} & -\frac{5}{2} \\ \frac{3}{2} & 0 & \frac{7}{2} \\ -\frac{5}{2} & \frac{7}{2} & 0 \end{bmatrix} \begin{bmatrix} x_1 \\ x_2 \\ x_3 \end{bmatrix}$.

   (c) $\begin{bmatrix} x_1 & x_2 & x_3 \end{bmatrix} \begin{bmatrix} 3 & \frac{1}{2} & -1 \\ \frac{1}{2} & 1 & -2 \\ -1 & -2 & -2 \end{bmatrix} \begin{bmatrix} x_1 \\ x_2 \\ x_3 \end{bmatrix}$.

3. (a) $\begin{bmatrix} -1 & 0 & 0 \\ 0 & 2 & 0 \\ 0 & 0 & 0 \end{bmatrix}$.   (b) $\begin{bmatrix} 3 & 0 & 0 \\ 0 & 0 & 0 \\ 0 & 0 & 0 \end{bmatrix}$.

5. $x_1'^2 + 2x_2'^2$.

7. $4x_3'^2$.

9. y'^2.

11. $y_1^2 + y_2^2 - y_3^2$.

13. $y_1^2 - y_2^2$.

15. $y_1^2 + y_2^2 + y_3^2$, an ellipsoid.

17. Q_1, Q_2, and Q_4 are equivalent. The eigenvalues of
the matrices associated with the quadratic forms are
for Q_1: 0, 1, 2; for Q_2: 0, 1, 3; for Q_3: 0, 5, -5; for
Q_4: 0, 1, 5. The rank r and signature s of Q_1, Q_2,
and Q_4 are $r = 2$ and $s = 2p - r = 2$.

Section 7.3, page 347

1.

3. (a) and (c).

5. (a)

	P_1	P_2	P_3	P_4	P_5
P_1	0	1	0	0	0
P_2	1	0	1	0	1
P_3	1	0	0	1	0
P_4	0	1	0	0	0
P_5	0	0	0	1	0

(b)

	P_1	P_2	P_3	P_4	P_5	P_6
P_1	0	1	1	0	0	0
P_2	1	0	0	1	0	0
P_3	0	1	0	0	0	0
P_4	0	0	1	0	1	1
P_5	0	0	1	1	0	1
P_6	1	0	0	0	0	0

7. (b).

9. (a) One way: $P_2 \to P_5 \to P_1$.
(b) Two ways: $P_2 \to P_5 \to P_1 \to P_2$
$P_2 \to P_5 \to P_3 \to P_2$.

11. P_2, P_3, and P_4.

13. There is no clique.

15. (a) Strongly connected.
(b) Not strongly connected.

17. P_1, P_2, or P_3.

Section 7.4, page 364

1.

		C	
		2 fingers shown	3 fingers shown
R	2 fingers shown	-4	5
	3 fingers shown	5	-6

3.

		Firm B	
		Abington	Wyncote
Firm A	Abington	50	60
	Wyncote	25	50

5. (a) $\begin{bmatrix} 5 & 4 \\ 3 & -2 \end{bmatrix}$. (b) $\begin{bmatrix} 2 & 1 & 0 \\ 3 & 1 & -2 \\ 4 & 2 & -4 \end{bmatrix}$.

(c) $\begin{bmatrix} 3 & 4 & 5 \\ -2 & 5 & 1 \\ -1 & 0 & 1 \end{bmatrix}$. (d) $\begin{bmatrix} 5 & 2 & 4 & 2 \\ 0 & -1 & 2 & 0 \\ 3 & 2 & 3 & 2 \\ 1 & 0 & -1 & -1 \end{bmatrix}$.

7. (a) $\mathbf{P} = [1 \quad 0]$, $\mathbf{Q} = \begin{bmatrix} 0 \\ 1 \\ 0 \end{bmatrix}$, $v = 1$.

(b) $\mathbf{P} = [0 \quad 0 \quad 1]$,

$\mathbf{Q} = \begin{bmatrix} 1 \\ 0 \\ 0 \\ 0 \end{bmatrix}$, $v = 0$.

(c) $\mathbf{P} = [1 \quad 0]$ or $[0 \quad 1]$, $\mathbf{Q} = \begin{bmatrix} 0 \\ 1 \end{bmatrix}$, $v = 4$.

9. (a) $\frac{19}{36}$. (b) $\frac{1}{7}$.

11. $p_1 = \frac{9}{14}$, $p_2 = \frac{5}{14}$, $q_1 = \frac{1}{2}$, $q_2 = \frac{1}{2}$, $v = -\frac{1}{2}$.

13. $p_1 = 0$, $p_2 = \frac{4}{5}$, $p_3 = \frac{1}{5}$, $q_1 = 0$,
$q_2 = \frac{3}{5}$, $q_3 = \frac{2}{5}$, $v = \frac{22}{5}$.

15. $\mathbf{P} = [0 \quad \frac{3}{8} \quad \frac{5}{8} \quad 0]$, $\mathbf{Q} = \begin{bmatrix} \frac{5}{8} \\ \frac{3}{8} \\ 0 \\ 0 \end{bmatrix}$, $v = \frac{1}{8}$.

17. $\mathbf{P} = [\frac{1}{3} \quad \frac{1}{3} \quad \frac{1}{3}]$, $\mathbf{Q} = \begin{bmatrix} \frac{1}{3} \\ \frac{1}{3} \\ \frac{1}{3} \end{bmatrix}$, $v = 0$.

19. $\mathbf{P} = [\frac{2}{3} \quad \frac{1}{3}]$, $\mathbf{Q} = \begin{bmatrix} \frac{1}{2} \\ \frac{1}{2} \end{bmatrix}$, $v = 0$.

Section 7.5, page 373

1. $y = 0.4x + 0.6$.

3. $y = 0.086x + 3.114$.

5. (a) $y = 0.426x + 0.827$. (b) 5.087.

7. (a) $y = 0.974x - 2.657$. (b) 10.979.

Section 7.6, page 381

1. (b) and (d).

3. $\begin{bmatrix} 4 \\ 0 \\ 3 \end{bmatrix}$.

5.

	Farmer	Carpenter	Tailor
Farmer	$\frac{2}{5}$	$\frac{1}{3}$	$\frac{1}{2}$
Carpenter	$\frac{2}{5}$	$\frac{1}{3}$	$\frac{1}{2}$
Tailor	$\frac{1}{5}$	$\frac{1}{3}$	0

$\mathbf{P} = \begin{bmatrix} 75 \\ 75 \\ 40 \end{bmatrix}$.

7. Not productive.

9. Productive.

11. (a) $\begin{bmatrix} 18 \\ 16 \end{bmatrix}$. (b) $\begin{bmatrix} 12 \\ 8 \end{bmatrix}$.

Section 7.7, page 391

1. (b) and (c).

3. (a) $\mathbf{P}^{(1)} = \begin{bmatrix} 0.7 \\ 0.3 \end{bmatrix}$, $\mathbf{P}^{(2)} = \begin{bmatrix} 0.61 \\ 0.39 \end{bmatrix}$,

$\mathbf{P}^{(3)} = \begin{bmatrix} 0.583 \\ 0.417 \end{bmatrix}$.

(b) $\mathbf{T} > \mathbf{0}$, hence it is regular; $\mathbf{U} = \begin{bmatrix} 0.571 \\ 0.429 \end{bmatrix}$.

5. (a) and (d).

9. (a) $\mathbf{T} = \begin{bmatrix} & A & B \\ 0.3 & 0.4 \\ 0.7 & 0.6 \end{bmatrix} \begin{matrix} A \\ B \end{matrix}$.

(b) 0.364.

(c) $\mathbf{U} = \begin{bmatrix} \frac{4}{11} \\ \frac{7}{11} \end{bmatrix} = \begin{bmatrix} 0.364 \\ 0.636 \end{bmatrix}$.

11. (a) 0.69.
(b) 20.7 percent of the population will be farmers.

13. (a) 35 percent, 37.5 percent.
(b) 40 percent.

Section 7.8, page 397

3. (a) $u_8 = 34$. (b) $u_{12} = 233$. (c) $u_{20} = 10{,}946$.

Section 7.9, page 407

1. (a) $\mathbf{X}(t) = \begin{bmatrix} x_1(t) \\ x_2(t) \\ x_3(t) \end{bmatrix} = \begin{bmatrix} b_1 e^{-3t} \\ b_2 e^{4t} \\ b_3 e^{2t} \end{bmatrix} =$

$b_1 \begin{bmatrix} 1 \\ 0 \\ 0 \end{bmatrix} e^{-3t} + b_2 \begin{bmatrix} 0 \\ 1 \\ 0 \end{bmatrix} e^{4t} + b_3 \begin{bmatrix} 0 \\ 0 \\ 1 \end{bmatrix} e^{2t}$.

(b) $\begin{bmatrix} 3e^{-3t} \\ 4e^{4t} \\ 5e^{2t} \end{bmatrix} = 3 \begin{bmatrix} 1 \\ 0 \\ 0 \end{bmatrix} e^{-3t} + 4 \begin{bmatrix} 0 \\ 1 \\ 0 \end{bmatrix} e^{4t} + 5 \begin{bmatrix} 0 \\ 0 \\ 1 \end{bmatrix} e^{2t}$.

3. $\mathbf{X}(t) = b_1 \begin{bmatrix} 6 \\ 2 \\ 7 \end{bmatrix} e^{4t} + b_2 \begin{bmatrix} 0 \\ 7 \\ -1 \end{bmatrix} e^{-5t} + b_3 \begin{bmatrix} 0 \\ 0 \\ 1 \end{bmatrix} e^{2t}$.

5. $\mathbf{X}(t) = b_1 \begin{bmatrix} 1 \\ 0 \\ 0 \end{bmatrix} e^{5t} + b_2 \begin{bmatrix} 0 \\ 1 \\ 3 \end{bmatrix} e^{5t} + b_3 \begin{bmatrix} 0 \\ -3 \\ 1 \end{bmatrix} e^{-5t}$.

7. $\mathbf{X}(t) = b_1 \begin{bmatrix} 1 \\ 0 \\ 1 \end{bmatrix} e^{4t} + b_2 \begin{bmatrix} -3 \\ 0 \\ 2 \end{bmatrix} e^{-t} + b_3 \begin{bmatrix} 1 \\ -6 \\ 4 \end{bmatrix} e^{t}$.

9. $\mathbf{X}(t) = 220 \begin{bmatrix} 2 \\ 1 \end{bmatrix} + 20 \begin{bmatrix} 3 \\ -1 \end{bmatrix} e^{-5t}$

$= \begin{bmatrix} 440 + 60e^{-5t} \\ 220 - 20e^{-5t} \end{bmatrix}$.

SUPPLEMENTARY EXERCISES, PAGE 410

1. (b) and (c).

3. $y_1^2 + y_2^2$.

5. $\mathbf{P} = \begin{bmatrix} \frac{2}{3} & \frac{1}{3} & 0 \end{bmatrix}$, $\mathbf{Q} = \begin{bmatrix} 0 \\ \frac{1}{3} \\ \frac{2}{3} \end{bmatrix}$, $v = \frac{2}{3}$.

7. $p_1 = 12r$, $p_2 = 8r$, $p_3 = 9r$, $r =$ any real number.

9. 121,393.

CHAPTER TEST, PAGE 411

1. $x = 5 + 3t$, $y = -2 - 2t$, $z = 1 + 5t$,
$-\infty < t < \infty$.

2. $x - y + 1 = 0$.

3. $y_1^2 - y_2^2$, a hyperbola.

4. There are two cliques: P_1, P_3, P_5, and P_1, P_3, P_6.

5. $\mathbf{P} = \begin{bmatrix} \frac{3}{8} & \frac{5}{8} \end{bmatrix}$, $\mathbf{Q} = \begin{bmatrix} \frac{1}{8} \\ \frac{7}{8} \\ 0 \end{bmatrix}$, $v = \frac{11}{8}$.

6. $a_{11} = 2$ is a saddle point for any value of a.

7. (a) $y = \frac{31}{58} + \frac{173}{290}x$.
(b) Approximately 1247 calories.

8. \$5.65 million of steel, \$5.41 million of coal, and \$2.83 million of transportation.

9. (a) $\begin{bmatrix} 0.4 & 0.5 \\ 0.6 & 0.5 \end{bmatrix}$.   (b) 0.4546.   (c) $\frac{6}{11}$.

10. 1,346,269.

11. $\mathbf{X} = b_1 \begin{bmatrix} 1 \\ 1 \end{bmatrix} e^{5t} + b_2 \begin{bmatrix} -1 \\ 3 \end{bmatrix} e^{-3t}$.

CHAPTER 8

Section 8.1, page 417

1. 0.3472×10^2.

3. -0.2840×10^3.

5. 0.1230×10^1, 0.1230×10^1.

7. 0.5666×10^1, 0.5667×10^1.

9. $\varepsilon_a = 0.21 \times 10^{-1}$, $\varepsilon_r = 0.17 \times 10^{-2}$.

11. $\varepsilon_a = 0.11 \times 10^1$, $\varepsilon_r = 0.17 \times 10^{-3}$.

Section 8.2, page 427

1. $\begin{bmatrix} 1 & -2 & 0 \\ 0 & 1 & -1 \\ 0 & 0 & 1 \end{bmatrix}$.

3. $x_1 = 2$, $x_2 = -3$, $x_3 = 4$.

5. $x_1 \cong 0.004$, $x_2 \cong 1.006$, $x_3 \cong -2.000$.

7. $x_1 \cong 2.393$, $x_2 \cong 4.565$, $x_3 \cong -3.607$.

9. Solution to the first linear system is $x \cong 3.2305$, $y \cong 1.8571$. Solution to the second linear system is $x \cong -3.6821$, $y \cong 6.1429$.

11.

	Jacobi	Gauss–Seidel	Exact
x	-1.993	-2.004	-2
y	4.998	5.002	5

13.

	Jacobi	Gauss–Seidel	Exact
x_1	1.559	1.658	$\frac{5}{3}$
x_2	1.685	1.513	$\frac{3}{2}$
x_3	-0.313	-0.492	$-\frac{1}{2}$

15. (a) Exact solution is $x = -2$, $y = 3$.
(b) Exact solution is $x = -3$, $y = 2$.

Section 8.3, page 436

1. $\mathbf{X} = \begin{bmatrix} 1 \\ 2 \\ 1 \end{bmatrix}$.

3. $\mathbf{X} = \begin{bmatrix} 1 \\ 0 \\ 2 \\ -4 \end{bmatrix}$.

5. $\mathbf{L} = \begin{bmatrix} 1 & 0 & 0 \\ 2 & 1 & 0 \\ 2 & -2 & 1 \end{bmatrix}$, $\mathbf{U} = \begin{bmatrix} 2 & 3 & 4 \\ 0 & -1 & 2 \\ 0 & 0 & -2 \end{bmatrix}$, $\mathbf{X} = \begin{bmatrix} 4 \\ -2 \\ 1 \end{bmatrix}$.

7. $\mathbf{L} = \begin{bmatrix} 1 & 0 & 0 \\ 0.5 & 1 & 0 \\ 0.25 & -1.5 & 1 \end{bmatrix}$, $\mathbf{U} = \begin{bmatrix} 4 & 2 & 3 \\ 0 & -1 & 3.5 \\ 0 & 0 & 5.5 \end{bmatrix}$, $\mathbf{X} = \begin{bmatrix} 2 \\ -2 \\ -1 \end{bmatrix}$.

9. $\mathbf{L} = \begin{bmatrix} 1 & 0 & 0 & 0 \\ 0.5 & 1 & 0 & 0 \\ -1 & 0.2 & 1 & 0 \\ 2 & -0.4 & 2 & 1 \end{bmatrix}$, $\mathbf{U} = \begin{bmatrix} 2 & 1 & 0 & -4 \\ 0 & -0.5 & 0.25 & 1 \\ 0 & 0 & 0.2 & 2 \\ 0 & 0 & 0 & 2 \end{bmatrix}$, $\mathbf{X} = \begin{bmatrix} 0.5 \\ 2 \\ -2 \\ 1.5 \end{bmatrix}$.

Section 8.4, page 444

	Approximate		Exact	
	Eigenvalue	Eigenvector	Eigenvalue	Eigenvector
1.	-4.974	$[-1.000 \quad 0.527]^T$	-5	$[-1 \quad \frac{1}{2}]^T$
3.	6.000	$[1.000 \quad 0.667]^T$	6	$[1 \quad \frac{2}{3}]^T$
5.	9.974	$[1.000 \quad 0.244]^T$	10	$[1 \quad \frac{1}{4}]^T$
7.	3.998	$[0.707 \quad 0.707]^T$	4	$[1/\sqrt{2} \quad 1/\sqrt{2}]^T$
	0.000	$[-0.707 \quad 0.707]^T$	0	$[-1/\sqrt{2} \quad 1/\sqrt{2}]^T$
9.	6.001	$[0.666 \quad 0.666 \quad 0.334]^T$	6	$[\frac{2}{3} \quad \frac{2}{3} \quad \frac{1}{3}]^T$
	11.989	$[-0.333 \quad 0.666 \quad -0.666]^T$	12	$[-\frac{1}{3} \quad \frac{2}{3} \quad -\frac{2}{3}]^T$
	8.998	$[-0.666 \quad 0.331 \quad 0.667]^T$	9	$[-\frac{2}{3} \quad \frac{1}{3} \quad \frac{2}{3}]^T$

SUPPLEMENTARY EXERCISES, PAGE 445

1.

	Jacobi	Gauss–Seidel	Exact
x	−2.9999	−3.000	−3
y	1.9999	2.001	2

3. $L = \begin{bmatrix} 1 & 0 & 0 \\ -3 & 1 & 0 \\ 2 & 4 & 1 \end{bmatrix}$, $U = \begin{bmatrix} -2 & 1 & -2 \\ 0 & 4 & 3 \\ 0 & 0 & -3 \end{bmatrix}$, $X = \begin{bmatrix} -1 \\ -2 \\ 3 \end{bmatrix}$.

5.

	Approximate		Exact	
Eigenvalue	Eigenvector		Eigenvalue	Eigenvector
2.000	$[0.707 \quad -0.707 \quad 0.000]^T$		2	$[1 \quad 1 \quad 1]^T$
−1.001	$[0.578 \quad 0.578 \quad 0.577]^T$		−1	$[1 \quad -1 \quad 0]^T$
−2.000	$[-0.408 \quad -0.408 \quad 0.817]^T$		−2	$[-1 \quad -1 \quad 2]^T$

CHAPTER TEST, PAGE 446

1.

	Jacobi	Gauss–Seidel	Exact
x	4.002	4.002	4
y	−3.000	−3.000	−3

2. $L = \begin{bmatrix} 1 & 0 & 0 \\ 3 & 1 & 0 \\ -3 & 2 & 1 \end{bmatrix}$, $U = \begin{bmatrix} 2 & 2 & 3 \\ 0 & -1 & -2 \\ 0 & 0 & 3 \end{bmatrix}$, $X = \begin{bmatrix} 1 \\ -1 \\ -2 \end{bmatrix}$.

	Approximate		Exact	
Eigenvalue	Eigenvector		Eigenvalue	Eigenvector
3. 9.001	$[0.600 \quad 1.000]^T$		9	$[3 \quad 5]^T$
4. 3.000	$[0.577 \quad 0.577 \quad 0.577]^T$		3	$[1 \quad 1 \quad 1]^T$
−3.000	$[-0.707 \quad 0.707 \quad 0.000]^T$		−3	$[-1 \quad 1 \quad 0]^T$
−3.000	$[-0.408 \quad -0.408 \quad -0.816]^T$		−3	$[-1 \quad -1 \quad 2]^T$

APPENDIX A, PAGE 453

Section A.1, page 453

1. (a) $4 + 2i.$ (b) $-4 - 3i.$ (c) $11 - 2i.$
(d) $-3 + i.$ (e) $-3 + 6i.$ (f) $-2 - i.$
(g) $7 - 11i,$ (h) $-9 + 13i.$

3. (a)

(b)

(c)

(d)

5.

7. $\mathbf{A}^2 = \begin{bmatrix} -1 & 0 \\ 0 & -1 \end{bmatrix}$, $\mathbf{A}^3 = \begin{bmatrix} 0 & -i \\ -i & 1 \end{bmatrix}$,

$\mathbf{A}^4 = \begin{bmatrix} 1 & 0 \\ 0 & 1 \end{bmatrix}$, $\mathbf{A}^{4n} = \mathbf{I}_n$, $\mathbf{A}^{4n-1} = \mathbf{A}$, $\mathbf{A}^{4n+2} = \mathbf{A}^2$,

$\mathbf{A}^{4n+3} = \mathbf{A}^3.$

9. (a) $\begin{bmatrix} 0 & 0 \\ 0 & 0 \end{bmatrix}.$ (b) $\begin{bmatrix} 4 & 18 \\ 0 & 4 \end{bmatrix}.$

(c) $\begin{bmatrix} -5 & 5i \\ 5i & -5 \end{bmatrix}$. (d) $\begin{bmatrix} 4 & 7i \\ 0 & -3 \end{bmatrix}$.

11. (a) Possible answers: $\begin{bmatrix} 2 & 0 \\ 0 & 2 \end{bmatrix}, \begin{bmatrix} -1 & 0 \\ 0 & -1 \end{bmatrix}$.

Section A.2, page 462

1. (a) No solution.   (b) No solution.

(c) $x_1 = \dfrac{3 + 5i}{4}, \ x_2 = \dfrac{3 + 2i}{2}$.

3. (a) $x_1 = i, \ x_2 = 1, \ x_3 = 1 - i$.

(b) $x_1 = 0, \ x_2 = -i, \ x_3 = i$.

5. (a) $\frac{1}{5} \begin{bmatrix} 2 + i & 2 - 4i \\ 3 - i & -2 - i \end{bmatrix}$,

(b) $\frac{1}{6} \begin{bmatrix} i & 1 - 3i & 1 \\ -2 - 3i & 2i & 3 + 2i \\ 1 & 2i & -i \end{bmatrix}$.

7. (a) Yes.   (b) Linearly independent.

9. (a) The eigenvalues are $\lambda_1 = 1 + i, \ \lambda_2 = 1 - i$. Associated eigenvectors are

$$\mathbf{X}_1 = \begin{bmatrix} 1 \\ i \end{bmatrix} \quad \text{and} \quad \mathbf{X}_2 = \begin{bmatrix} 1 \\ -i \end{bmatrix}.$$

(b) The eigenvalues are $\lambda_1 = 0, \ \lambda_2 = 2$. Associated eigenvectors are

$$\mathbf{X}_1 = \begin{bmatrix} 1 \\ i \end{bmatrix} \quad \text{and} \quad \mathbf{X}_2 = \begin{bmatrix} 1 \\ -i \end{bmatrix}.$$

(c) The eigenvalues are $\lambda_1 = 1, \ \lambda_2 = 2, \ \lambda_3 = 3$. Associated eigenvectors are

$$\mathbf{X}_1 = \begin{bmatrix} 0 \\ 1 \\ i \end{bmatrix}, \ \mathbf{X}_2 = \begin{bmatrix} 1 \\ 0 \\ 0 \end{bmatrix}, \quad \text{and} \quad \mathbf{X}_3 = \begin{bmatrix} 0 \\ 1 \\ -i \end{bmatrix}.$$

Index

LIST OF FREQUENTLY USED SYMBOLS

$\mathbf{A} = [a_{ij}]$	An $m \times n$ matrix, p. 11
$[\mathbf{A} \vdots \mathbf{B}]$	An augmented matrix, p. 18
$\mathbf{A}^T$	The transpose of the matrix $\mathbf{A}$, p. 19
$\displaystyle\sum_{i=1}^{n} a_i$	Summation notation, p. 20
$\mathbf{0}$	The zero $m \times n$ matrix, p. 24
$\mathbf{I}_n$	The $n \times n$ identity matrix, p. 27
$\mathbf{A}^{-1}$	The inverse of the matrix $\mathbf{A}$, p. 51
$j_1 j_2 \cdots j_n$	A permutation of $S = \{1, 2, \ldots, n\}$, p. 67
$|\mathbf{A}|$	The determinant of the matrix $\mathbf{A}$, p. 70
$|\mathbf{M}_{ij}|$	The minor of a_{ij}, pp. 80–81
A_{ij}	The cofactor of a_{ij}, p. 81
adj $\mathbf{A}$	The adjoint of the matrix $\mathbf{A}$, p. 84
$\mathbf{U}, \mathbf{V}, \mathbf{X}, \mathbf{Y}, \mathbf{Z}$	Vectors in a vector space, pp. 99, 134
$\mathbf{0}$	The zero vector, pp. 107, 115
$-\mathbf{X}$	The negative of the vector $\mathbf{X}$, pp. 107, 116
$\|\mathbf{X}\|$	Length of the vector $\mathbf{X}$, pp. 107, 116
$\mathbf{X} \cdot \mathbf{Y}$	Inner product, pp. 109, 121
R^n	n-space; the vector space of all n-vectors, p. 114
$\|\mathbf{X} - \mathbf{Y}\|$	Distance between the vectors $\mathbf{X}$ and $\mathbf{Y}$ in R^n, pp. 120–121
θ	Angle between two nonzero vectors in R^n, p. 123
$\times$	The cross product operator, p. 128
V, W	Vector spaces, pp. 134–135
$p(t)$	A polynomial in t, p. 136
P_n	The vector space of all polynomials of degree $\leq n$ and the zero polynomial, pp. 136–137
$C(-\infty, \infty)$	The vector space of all real-valued functions, p. 140
$C[a, b]$	The vector space of all continuous real-valued functions on the interval $[a, b]$, p. 140
S, T, S', T'	Sets of vectors in a vector space, pp. 141, 158, 219
span $\{\mathbf{X}_1, \mathbf{X}_2, \ldots, \mathbf{X}_k\}$	The set of all vectors of the form $a_1\mathbf{X}_1 + a_2\mathbf{X}_2 + \cdots + a_k\mathbf{X}_k$, p. 141
$\mathbf{E}_i$	An n-vector whose ith component is 1 and all other components are zero, p. 147
$\{\mathbf{E}_1, \mathbf{E}_2, \ldots, \mathbf{E}_n\}$	The natural basis for R^n, p. 154
$\{\mathbf{i}, \mathbf{j}, \mathbf{k}\}$	The natural basis for R^3, pp. 147, 154
dim V	The dimension of the vector space V, p. 159